CW00350851

Pile Design and Construction Practice

Also available from Taylor & Francis

Piling Engineering 2nd ed
M. Tomlinson et *al*.   Hb: ISBN 0–415–26646–7

Design of Caisson Foundations for Offshore
Wind Turbines
B. Byrne et *al*.   Hb: ISBN 0–415–37056–6

An introduction to Geotechnical Processes
J. Woodward   Hb: ISBN 0–415–28645–X
                                                Pb: ISBN 0–415–28646–8

Introductory Geotechnical Engineering
Y. Fang   Hb: ISBN 0–415–30401–6
                                                Pb: ISBN 0–415–30402–4

Soil Liquefaction
M. Jefferies et *al*.   Hb: ISBN 0–419–16170–8

Information and ordering details

For price availability and ordering visit our website **www.tandf.co.uk/builtenvironment**

Alternatively our books are available from all good bookshops.

Pile Design and Construction Practice

Fifth edition

Michael Tomlinson and
John Woodward

Taylor & Francis
Taylor & Francis Group

LONDON AND NEW YORK

First published 1977,
reprinted with amendments 1981,
third edition 1987 by Palladian,
reprinted 1991, fourth edition 1994,
reprinted 1994, 1995, 1998 by E & FN Spon

Fifth Edition published 2008
by Taylor & Francis
2 Park Square, Milton Park, Abingdon, Oxon OX14 4RN

Reprinted 2009

Simultaneously published in the USA and Canada
by Taylor & Francis
270 Madison Ave, New York, NY 10016

*Taylor & Francis is an imprint of the Taylor & Francis Group,
an informa business*

© 1977, 1987, 1994, 2008 Michael Tomlinson and John Woodward

Typeset in Times New Roman by
Newgen Imaging Systems (P) Ltd, Chennai, India
Printed and bound in Great Britain by
CPI Antony Rowe, Chippenham, Wiltshire

All rights reserved. No part of this book may be reprinted or reproduced
or utilised in any form or by any electronic, mechanical, or other means,
now known or hereafter invented, including photocopying and recording,
or in any information storage or retrieval system, without permission in
writing from the publishers.
The publisher makes no representation, express or implied, with regard
to the accuracy of the information contained in this book and cannot
accept any legal responsibility or liability for any efforts or
omissions that may be made.

British Library Cataloguing in Publication Data
A catalogue record for this book is available from the British Library

Library of Congress Cataloging in Publication Data
Tomlinson, M. J. (Michael John)
    Pile design and construction practice / Michael Tomlinson and
John Woodward . – 5th ed.
        p. cm.
    Includes bibliographical references and index.
    1. Piling (Civil engineering) I. Woodward, John, 1936–
II. Title.
    TA780.T65 2007
    624.1'54–dc22   2006019427

ISBN10: 0–415–38582–2 (hbk)
ISBN10: 0–203–96429–2 (ebk)

ISBN13: 978–0–415–38582–4 (hbk)
ISBN13: 978–0–203–96429–3 (ebk)

Contents

Appendix: properties of materials

Preface to fifth edition

Piling rigs are a commonplace feature on building sites in cities and towns today. The continuing introduction of new, more powerful, and self-erecting machines for installing piled foundations has transformed the economics of this form of construction in ground conditions where, in the past, first consideration would have been given to conventional spread or raft foundations, with piling being adopted only as a last resort in difficult ground.

The increased adoption of piling is not only due to the availability of more efficient mechanical equipment. Developments in analytical methods of calculating bearing capacity and dynamic methods for load and integrity testing have resulted in greater assurance of sound long-term performance. Further economies in foundation and superstructure design are now possible because of the increased ability to predict movements of piles under load, thus allowing engineers to adopt with confidence the concept of redistribution of load between piles with consequent savings in overall pile lengths and cross-sectional dimensions, as described in this new edition.

Since the publication of the fourth edition of this book, Eurocode 7, Geotechnical Design, has been issued. As the name implies this code does not deal with all aspects of foundation design; there are extensive cross-references to other Eurocodes dealing with such matters as the general basis of design and the properties of constructional materials. The Code does not cover foundation design and particularly construction as comprehensively as the present British Standard 8004 Foundations, and the British National Annex to Eurocode 7 is yet to be published. The authors have endeavoured to co-ordinate the principles of both codes in this new book.

The authors are grateful to Professor Richard Jardine and his colleagues at Imperial College and Thomas Telford Limited for permission to quote from their book on the ICP method of designing piles driven into clays and sands based on extensive laboratory research and practical field testing of instrumented piles. Their work represents a considerable advance on previous design methods. The authors also gratefully acknowledge the help of Mr Ian Higginbottom in checking the proofs and Mr Tony Bracegirdle of the Geotechnical Consulting Group for his helpful comments on the application of Eurocode 7 to the design of piles and pile groups.

Many specialist piling companies and manufacturers of piling equipment have kindly supplied technical information and illustrations of their processes and products. Where appropriate the source of this information is given in the text.

In addition, the authors wish to thank the following for the supply of and permission to use photographs and illustrations from technical publications and brochures.

Abbey Pynford Foundation Systems Limited	Figure 2.15
ABI GmbH	Figures 3.1 and 3.2
American Society of Civil Engineers	Figures 4.6, 4.11, 4.12, 4.29, 4.39, 5.17, 5.30 and 6.29
Austrian Member Society, SMGE	Figure 6.20
Bachy-Soletanche	Figure 2.29a and b
Ballast Nedam Groep N.V.	Figures 9.22 and 9.23
Bauer Maschinen GmbH	Figure 3.27
The British Petroleum Company Limited	Figure 8.15
BSP International Foundations Limited	Figures 3.12 and 3.14
Building Research Establishment	Figure 10.2a and b
Canadian Geotechnical Journal	Figures 4.34, 4.36, 4.37, 5.20, 5.38 and 6.9
A. Carter	Figure 9.24
Cement and Concrete Association	Figure 7.13
Cementation Foundations Skanska Limited	Figures 3.30, 3.33, 9.31 and 11.5
Central Electricity Generating Board	Figure 2.18
CIRIA/Butterworth	Figures 4.11 and 5.28
Comité Français SMGE	Figure 6.30
Construction Industry Research and Information Association (CIRIA)	Figure 4.8
Danish Geotechnical Institute	Figures 5.6–5.10, 6.21 and 6.22
Dar-al-Handasah Consultants	Figure 9.14
Dawson Construction Plant Limited	Figure 3.5
Department of the Environment	Figure 10.1
DFP Foundation Products	Figure 2.28
Fondedile Foundations Limited	Figure 9.5
Frank's Casing Crew and Rental Inc	Figure 2.17
Fugro Limited	Figure 5.24
GeoDelft	Figure 5.23
The Geological Society	Figure 8.9
U G de Gijt	Figure 4.20
International Construction Equipment	Figure 3.3
International Society for Soil Mechanics and Foundation Engineering	Figures 3.38, 5.24, 5.25, 6.18, 6.30 and 9.20
Institution of Civil Engineers / Thomas Telford Limited	Figures 4.28, 5.26, 5.27, 5.33, 5.34, 5.35, 5.41, 5.42, 9.21, 9.24, 9.26 and 9.27
Land and Water, Den Haag	Figure 4.20
Liebherr Great Britain Limited	Figure 3.4
Menck GmbH	Figure 3.12
National Coal Board	Figures 2.18, 4.26 and 8.2
Numa Hammers	Figure 3.32
Offshore Technology Conference	Figures 4.16, 5.29 and 8.17

Pearson Education	Figures 4.16, 4.22, 4.25 and 5.6–5.10
Pentech Press	Figures 4.35 and 5.19
Professor Poulos	Figure 5.24
Russian Society, SMGE	Figure 5.25
Seacore Limited	Figures 3.7, 3.13 and 3.36
Sezai-Turkes-Feyzi-Akkaya Construction Company	Figure 4.26
Sheet Piling Contractors Limited	Figure 3.19
Soil Mechanics Limited	Figures 2.10 and 2.11
Springer-Verlag	Figure 4.21
Stent Foundations Limited	Figure 2.32
Swedish Geotechnical Society	Figures 5.22 and 5.28
Test Consult Limited	Figures 11.10 and 11.21
Thomson Publishing Services	Figures 3.37 and 4.30
TRL	Figures 9.17, 9.18 and 9.19
Trans-Tech Publications	Figures 6.34 and 6.35
University of Austin in Texas	Figures 6.25, 6.26, 6.27 and 6.28
United States Department of Transportation	Figure 4.33
John Wiley and Sons Incorporated	Figure 4.10
George Wimpey Limited	Figures 2.16, 2.18, 3.8, 8.2, 8.8 and 8.16
Wirth Maschinen und Bohrgerate-Fabrik GmbH	Figure 3.35

Figure 9.25 and the front cover are published with the permission of the Deep Foundations Institute as they were originally published in the DFI 2005 Marine Foundations Speciality Seminar proceedings. Copies of the full proceedings are available through DFI – Deep Foundations Institute, 326 Lafayette Avenue Hawthorne, NJ 07506; tel: 973-423-4030; fax: 973-423-4031; email: dfihq@dfi.org

Permission to reproduce extracts of British Standards is granted by BSI. British Standards can be obtained from BSI Customer Services, 389 Chiswick High Road, London W4 4AL; tel: +44 (0) 20 8996 9001; email: cservices@bsiglobal.com

Every effort has been made to contact copyright holders. Please advise the publisher of any errors or omissions and these will be corrected in subsequent edition.

MJT, Richmond-upon-Thames
JCW, Princes Risborough, 2006

Preface to first edition

Piling is both an art and a science. The art lies in selecting the most suitable type of pile and method of installation for the ground conditions and the form of the loading. Science enables the engineer to predict the behaviour of the piles once they are in the ground and subject to loading. This behaviour is influenced profoundly by the method used to install the piles and it cannot be predicted solely from the physical properties of the pile and of the undisturbed soil. A knowledge of the available types of piling and methods of constructing piled foundations is essential for a thorough understanding of the science of their behaviour. For this reason the author has preceded the chapters dealing with the calculation of allowable loads on piles and deformation behaviour by descriptions of the many types of proprietary and non-proprietary piles and the equipment used to install them.

In recent years substantial progress has been made in developing methods of predicting the behaviour of piles under lateral loading. This is important in the design of foundations for deep-water terminals for oil tankers and oil carriers and for offshore platforms for gas and petroleum production. The problems concerning the lateral loading of piles have therefore been given detailed treatment in this book.

The author has been fortunate in being able to draw on the world-wide experience of George Wimpey and Company Limited, his employers for nearly 30 years, in the design and construction of piled foundations. He is grateful to the management of Wimpey Laboratories Ltd. and their parent company for permission to include many examples of their work. In particular, thanks are due to P. F. Winfield, FIStructE, for his assistance with the calculations and his help in checking the text and worked examples.

Burton-on-Stather, 1977
M. J. T

Chapter 1

General principles and practices

1.1   Function of piles

Piles are columnar elements in a foundation which have the function of transferring load from the superstructure through weak compressible strata or through water, onto stiffer or more compact and less compressible soils or onto rock. They may be required to carry uplift loads when used to support tall structures subjected to overturning forces from winds or waves. Piles used in marine structures are subjected to lateral loads from the impact of berthing ships and from waves. Combinations of vertical and horizontal loads are carried where piles are used to support retaining walls, bridge piers and abutments, and machinery foundations.

1.2   Historical

The driving of bearing piles to support structures is one of the earliest examples of the art and science of a civil engineer. In Britain, there are numerous examples of timber piling in bridge works and riverside settlements constructed by the Romans. In mediaeval times, piles of oak and alder were used in the foundations of the great monasteries constructed in the fenlands of East Anglia. In China, timber piling was used by the bridge builders of the Han Dynasty (200 BC to AD 200). The carrying capacity of timber piles is limited by the girth of the natural timbers and the ability of the material to withstand driving by hammer without suffering damage due to splitting or splintering. Thus primitive rules must have been established in the earliest days of piling by which the allowable load on a pile was determined from its resistance to driving by a hammer of known weight and with a known height of drop. Knowledge was also accumulated regarding the durability of piles of different species of wood, and measures taken to prevent decay by charring the timber or by building masonry rafts on pile heads cut off below water level.

Timber, because of its strength combined with lightness, durability and ease of cutting and handling, remained the only material used for piling until comparatively recent times. It was replaced by concrete and steel only because these newer materials could be fabricated into units that were capable of sustaining compressive, bending and tensile forces far beyond the capacity of a timber pile of like dimensions. Concrete, in particular, was adaptable to in-situ forms of construction which facilitated the installation of piled foundations in drilled holes in situations where noise, vibration and ground heave had to be avoided.

Reinforced concrete, which was developed as a structural medium in the late nineteenth and early twentieth centuries, largely replaced timber for high-capacity piling for works on land. It could be precast in various structural forms to suit the imposed loading and ground conditions, and its durability was satisfactory for most soil and immersion conditions. The partial replacement of driven precast concrete piles by numerous forms of cast in-situ piles has been due more to the development of highly efficient machines for drilling pile bore-holes of large diameter and great depth in a wide range of soil and rock conditions, than to any deficiency in the performance of the precast concrete element.

Steel has been used to an increasing extent for piling due to its ease of fabrication and handling and its ability to withstand hard driving. Problems of corrosion in marine struc-tures have been overcome by the introduction of durable coatings and cathodic protection.

1.3   Calculations of load-carrying capacity

While materials for piles can be precisely specified, and their fabrication and installation can be controlled to conform to strict specification and code of practice requirements, the calculation of their load-carrying capacity is a complex matter which at the present time is based partly on theoretical concepts derived from the sciences of soil and rock mechanics, but mainly on empirical methods based on experience. Practice in calculating the ultimate carrying capacity of piles based on the principles of soil mechanics differs greatly from the application of these principles to shallow spread foundations. In the latter case the entire area of soil supporting the foundation is exposed and can be inspected and sampled to ensure that its bearing characteristics conform to those deduced from the results of exploratory boreholes and soil tests. Provided that the correct constructional techniques are used the disturbance to the soil is limited to a depth of only a few centimetres below the excavation level for a spread foundation. Virtually the whole mass of soil influenced by the bearing pressure remains undisturbed and unaffected by the constructional operations (Figure 1.1a). Thus the safety factor against general shear failure of the spread foundation and its settlement under the design working load can be predicted from a knowledge of the physical characteristics of the *undisturbed* soil with a degree of certainty which depends only on the complexity of the soil stratification.

The conditions which govern the supporting capacity of the piled foundation are quite different. No matter how the pile is installed, whether by driving with a hammer, by jetting, by vibration, by jacking, screwing or drilling, the soil in contact with the pile face, from which the pile derives its support by shaft friction, and its resistance to lateral loads, is com-pletely disturbed by the method of installation. Similarly, the soil or rock beneath the toe of a pile is compressed (or sometimes loosened) to an extent which may affect significantly its end-bearing resistance (Figure 1.1b). Changes take place in the conditions at the pile–soil interface over periods of days, months or years which materially affect the skin-friction resistance of a pile. These changes may be due to the dissipation of excess pore pressure set up by installing the pile, to the relative effects of friction and cohesion which in turn depend on the relative pile-to-soil movement and to chemical or electro-chemical effects caused by the hardening of the concrete or the corrosion of the steel in contact with the soil. Where piles are installed in groups to carry heavy foundation loads, the operation of driving or drilling for adjacent piles can cause changes in the carrying capacity and load/settlement characteristics of the piles in the group that have already been driven.

In the present state of knowledge, the effects of the various methods of pile installation on the carrying capacity and deformation characteristics cannot be calculated by the strict application of soil or rock mechanics theory. The general procedure is to apply simple empirical factors to the strength, density and compressibility properties of the undisturbed

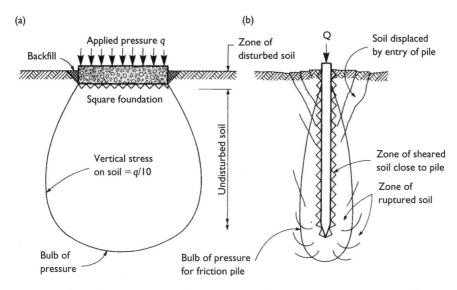

Figure 1.1 Comparison of pressure distribution and soil disturbance beneath spread and piled foundations (a) Spread foundation (b) Single pile.

soil or rock. The various factors which can be used depend on the particular method of installation and are based on experience and on the results of field loading tests.

The basis of the 'soil mechanics approach' to calculating the carrying capacity of piles is that the total resistance of the pile to compression loads is the sum of two components, namely shaft friction and base resistance. A pile in which the shaft-frictional component predominates is known as a friction pile (Figure 1.2a), while a pile bearing on rock or some other hard incompressible material is known as an end-bearing pile (Figure 1.2b). The need for adopting an adequate safety factor in conjunction with calculations to determine these components is emphasized by the statement by Randolph[1.1] 'that we may never be able to estimate axial pile capacity in many soil types more accurately than about ±30%'. However, even if it is possible to make a reliable estimate of total pile resistance, a further difficulty arises in predicting the problems involved in installing the piles to the depths indicated by the empirical or semi-empirical calculations. It is one problem to calculate that a precast concrete pile must be driven to a depth of, say, 20 m to carry safely a certain working load, but quite another problem to decide on the energy of the hammer required to drive the pile to this depth, and yet another problem to decide whether or not the pile will be irredeemably shattered while driving it to the required depth. In the case of driven and cast-in-place piles the ability to drive the piling tube to the required depth and then to extract it within the pulling capacity of the piling rig must be correctly predicted.

Time effects are important in calculating the resistance of a pile in clay; the effects include the rate of applying load to a pile and the time interval between installing and testing a pile. The shaft-frictional resistance of a pile in clay loaded very slowly may only be one-half of that which is measured under the rate at which load is normally applied during a pile loading test. The slow rate of loading may correspond to that of a building under

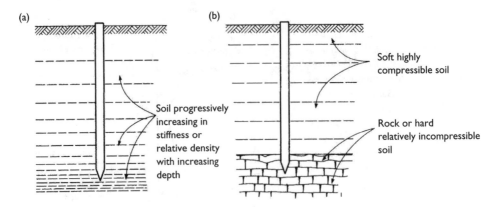

Figure 1.2 Types of bearing pile (a) Friction pile (b) End-bearing pile.

construction, yet the ability of a pile to carry its load is judged on its behaviour under a comparatively rapid loading test made only a few days after installation. Because of the importance of such time effects both in fine- and coarse-grained soils the only practicable way of determining the load-carrying capacity of a piled foundation is to confirm the design calculations by short-term tests on isolated single piles, and then to allow in the safety factor for any reduction in the carrying capacity with time. The effects of grouping piles can be taken into account by considering the pile group to act as a block foundation, as described in Chapter 5.

1.4 Dynamic piling formulae

The soil mechanics approach to calculating allowable working loads on piles is that of determining the resistance of static loads applied at the test-loading stage or during the working life of the structure. Methods of calculation based on the measurement of the resistance encountered when driving a pile were briefly mentioned in the context of history. Historically all piles were installed by driving them with a simple falling ram or drop hammer. Since there is a relationship between the downward movement of a pile under a blow of given energy and its ultimate resistance to static loading, when all piles were driven by a falling ram a considerable body of experience was built up and simple empirical formulae established from which the ultimate resistance of the pile could be calculated from the 'set' of the pile due to each hammer blow at the final stages of driving. However, there are many drawbacks to the use of these formulae with modern pile-driving equipment particularly when used in conjunction with diesel hammers. The energy of blow delivered to the pile by these types increases as the resistance of the ground increases. The energy can also vary with the mechanical condition of the hammer and its operating temperature. Simple dynamic formulae are now largely discredited as a means of predicting the resistance of piles to static loading unless the driving tests are performed on piles instrumented to measure the energy transferred to the pile head. If this is done the dynamic analyser (see Section 7.3) provides the actual rather than the assumed energy of blow enabling the dynamic formula to be used as a means of site control when driving the working piles. Dynamic pile formulae are allowed to be used by Eurocode EC7 provided that their validity has been demonstrated by experience in similar ground conditions or verified by static loading tests.

Steady progress has been made in developing analytical methods for calculating pile capacity. With increasing experience of their use backed by research, the soil mechanics approach can be applied to all forms of piling in all ground conditions, whereas even if a reliable dynamic formula could be established its use would be limited to driven piles only. However, dynamic formulae still have their uses in predicting the stresses within the material forming the pile during driving and hence in assessing the risk of pile breakage, and their relevance to this problem is discussed in Chapter 7.

1.5   Code of practice requirements

The uncertainties in the methods of predicting allowable or ultimate loads on piles are reflected in the numerous ways of defining these loads in the many codes of practice which cover piling. The British Standard Code of Practice BS 8004: 1986 (Foundations) defines the ultimate bearing capacity of a pile as 'The load at which the resistance of the soil becomes fully mobilized' and goes on to state that this is generally taken as the load causing the head of the pile to settle a depth of 10% of the pile width or diameter. BS 8004 does not define ultimate loads for uplift or lateral loading. Specific design information is limited to stating the working stresses on the pile material and the cover required to the reinforcement, the requirements for positional tolerance and verticality also being stated. No quantitative information is given on shaft friction or end-bearing values in soils or rocks, but many countries place limits on these values or on maximum pile loads in order to ensure that piles are not driven very heavily so as to achieve the maximum working load that can be permitted by the allowable stress on the cross-sectional area of the pile shaft.

A conflict can arise in British practice where structures, including foundation substructures, are designed to the requirements of BS 8110 and their foundations to those of BS 8004. In the former document partial safety factors are employed to increase the characteristic dead and imposed loads to amounts which are defined as the ultimate load. The ultimate resistance of the structure is calculated on the basis of the characteristic strength of the material used for its construction which again is multiplied by a partial safety factor to take into account the possibility of the strength of the material used being less than the designed characteristic strength. Then, if the ultimate load on the structure does not exceed its ultimate resistance to load, the *ultimate* or *collapse limit state* is not reached and the structure is safe. Deflections of the structure are also calculated to ensure that these do not exceed the maximum values that can be tolerated by the structure or user, and thus to ensure that the *serviceability limit state* is not reached.

When foundations are designed in accordance with BS 8004, the maximum working load is calculated. This is comparable to the characteristic loading specified in BS 8110, i.e. the most unfavourable combination of the dead and imposed loading. The resistance offered by the ground to this loading is calculated. This is based on *representative* shearing strength parameters of the soils or rocks concerned. These are not necessarily minimum or average values but are parameters selected by the engineer using his experience and judgement and taking into account the variability in the geological conditions, the number of test results available, the care used in taking samples and selecting them for test, and experience of other site investigations and of the behaviour of existing structures in the locality. The maximum load imposed by the sub-structure on the ground must not exceed the calculated resistance of the ground multiplied by the appropriate safety factor. The latter takes into account the risks of excessive total and differential settlements of the structure as well as allowing for uncertainties in the design method and in the values selected for the shearing strength parameters.

The settlements of the foundations are then calculated, the loading adopted for these calculations being not necessarily the same as that used to obtain the maximum working load. It is the usual practice to take the actual dead load and the whole or some proportion of the imposed load, depending on the type of loading, i.e. the full imposed load is taken for structures such as grain silos, but the imposed wind loading may not be taken into account when calculating long-term settlements.

There is no reason why this dual approach should not be adopted when designing structures and their foundations, but it is important that the designer of the structure should make an unambiguous statement of the loading conditions which are to be supported by the ground. If he provides the foundation engineer with a factored ultimate load, and the foundation engineer then uses this load with a safety factor of, say, 2.5 or 3 on the calculated shearing resistance of the ground, the resulting design may be over-conservative. Similarly, if the ultimate load is used to calculate settlements, the values obtained will be unrealistically large. The foundation engineer must know the actual dead load of the superstructure and sub-structure and he must have full details of the imposed loading, i.e. its type, distribution and duration.

Many of the conflicts between the design of structures and sub-structures to BS 8110 or similar structural codes, and the design of piled foundations to BS 8004 have been dealt with in BS EN 1997-1: 2004 Eurocode 7 (EC7), Geotechnical Design – Part 1 General rules[1.2] and BS EN 1992-1: 2004 Eurocode 2 (EC2), Design of Concrete Structures – Part 1-1 General rules and rules for buildings[1.3]. These two Eurocodes will partially supersede BS 8004 and BS 8110 (and other related geotechnical standards). However, until all the Eurocode packages for designing the various parts of a structure are available together with the National Annexes, the British Standards Institute advises that the current standards will remain valid for geotechnical investigation and design and concrete design 'until further notice'. This 'coexistence period' for EC7 is likely to last for several years before the current standards are modified or withdrawn.

EC7 has to be read in conjunction with BS EN 1990 Eurocode 0: Basis of Structural Design[1.4] and BS EN 1991-1 Eurocode 1: Part 1 Actions on Structures which ensure that partial factors are considered in a logical and uniform manner avoiding the application of global safety factors. In addition, reference has to be made to BSEN 1993-1: 2005 Eurocode 3 (EC3): Design of Steel Structures, Part 1-1 General Rules and Part 5: 2007 (EC3-5) Piling and BS EN 1995-1: 2004 Eurocode 5 (EC5): Design of Timber Structures, Part 1-1 General Rules. It should be noted that new European Standards have been published dealing with the 'execution of special geotechnical works' (bored piling, displacement piles, sheet piles, micropiles, etc.) which have the status of current British Standards (designated as 'BS EN').

Clause 7 of EC7 deals with piled foundations from the aspects of actions on piles from superimposed loading or ground movements, design methods for piles subjected to compression, tension and lateral loading, pile-loading tests, structural design and supervision of construction. In using Clause 7 of EC7 the designer is required to demonstrate that the sum of the ultimate limit state components of bearing capacity of the pile or pile group (*resistances 'R'*) exceeds the ultimate limit state design loading (*actions 'F'*) and that the serviceability limit-state is not reached.

The EC7 loading scenarios, defined as 'actions' in the Eurocodes, are designed to cover 'permanent unfavourable', 'permanent favourable' and 'variable' situations and require the application of different load factors depending on which of three 'design approaches' is being used. The National Annex documents (to be published separately from the Eurocodes) which are to be used in each country to conform to their individual practices will address within prescribed limits the design approach, partial factors, methods of calculating settlement and the procedures to be used where alternatives to EC7 are needed. Design Approach 1 with 'partial factor combinations 1 and 2' is to be adopted in the UK in which the factors are applied at source to actions and shear strengths, but the code notes that for *pile design* the partial factors must be applied to the bearing capacity or ground 'resistance'. When EC7 Design Approach 1 (DA1) is used, partial factor 'combination 1' usually governs the structural design values of actions for piles and 'combination 2' the geotechnical sizing.

References are made to EC7 in the chapters of this book dealing with pile design and to the BS EN standards for the execution of geotechnical works, but EC7 itself does not make specific recommendations on methods of pile design. Essentially it prescribes the succession of stages in the design process using conventional methods to determine end bearing, frictional resistance and displacement. References are therefore continued to BS 8004 and BS 8110 alongside the appropriate Eurocode. At the time of preparing this edition the application of EC7 is not mandatory in the UK, but in due course all geotechnical design will have to conform. If the reader wishes to apply the EC7 rules to current designs, a thorough study of the *Designer's Guide*[1.5] which sets out the step-by-step design process and takes account of the various qualifications to the application of the code rules is recommended. Whether or not the Eurocode is used for design in preference to present conventional methods it does provide a very useful design check, itemizing all the factors which can influence economic foundation design.

As the British National Annex stating the partial factors to be used in designing piles is not due to be published until 2008, the factors provided in the tables in Chapter 4 and used in worked examples are those given in Annex A to EC7. The reader should therefore check that the quoted values conform to the data in the National Annex when avaliable. Also, engineers designing foundations in EU countries other than the UK should consult the particular National Annexes for guidance on design procedures and partial factors. The selection of characteristic parameters and the application of EC7 partial factors to 'numerical methods' for complex foundation design are yet to be decided and are therefore not dealt with in this edition.

1.6 Responsibilities of engineer and contractor

In Britain and in many other countries piling is regarded as a specialist operation and the procedure for calling for tendered prices for this work may result in a division of responsibility which can lead to undesirable practices. When the engineer is wholly responsible for design or supervision of construction, the type, width and overall length of the piles will be specified based on the ground information. Detailed designs for concrete piles showing the reinforcement, concrete mix proportions, cover and cube crushing strengths will then be prepared. In the case of steel piles the standard sections, grade of steel and welding requirements will be specified. The engineer will decide on the depth of penetration of each pile from the results of preliminary calculations checked by field observations during driving. Responsibility will be accepted by the engineer for paying the contractor for any costs involved in shortening or lengthening piles, or of providing additional piles should the ground conditions differ from those envisaged or should the piles fail a loading test or fail to achieve the specified 'set' criterion when at the design length.

Quite a different procedure is adopted when the contractor is responsible for design. The engineer will provide the piling contractor with whatever ground information is available, and will state either the required working load on a single pile, or he may simply provide a building layout plan showing the column loads or the load per metre run from the load-bearing walls. In the latter case, the contractor will be responsible for deciding the required piling layout. In all cases the contractor will determine the type and required diameter and length of the piles, but will be careful to quote a price for lengthening the piles should the actual ground conditions differ from the information supplied at the time of tendering. The contractor's tender is usually accompanied by financial provisions to guarantee the performance and safety of the design.

The engineer may not always specify allowable working stresses on the pile shaft, minimum cube crushing strengths or minimum cement contents in concrete mixes. It may be considered the proper duty of the piling contractor to decide on these values since they may be governed by the particular piling process employed. The need to specify allowable working stresses and the crushing strength and minimum cement content of concrete piles is dealt with in Chapters 2 and 10. In all cases the engineer must specify the maximum permissible settlement at the working load and at some simple multiple, say 1.5 times or twice the working load, either on test piles or on working piles or both. This is essential as it is the only means that the engineer possesses of checking that the contractor's design assumptions and the piles as installed will fulfil their function in supporting the structure. Only the engineer can state the requirement for settlement at the working load from knowledge of the tolerance of the structure to total and differential settlement. It frequently happens that the maximum settlements specified are so unrealistically small that they will be exceeded by the inevitable elastic compression of the pile shaft, irrespective of any elastic compression or yielding of the soil or rock supporting the pile. However, the specified permissible settlement should not be so large that the safety factor is compromised (see Section 4.1.4) and it should be remembered that the settlement of a pile group is related to the settlement of a single pile within the group (Chapter 5). It is unrealistic to specify the maximum movement of a pile under lateral loading, since this can be determined only by field trials.

The above procedure for contractor-designed piling has been advantageous in that it has promoted the development of highly efficient piling systems. However, they have the drawback that they place the engineer in a difficult position when checking the contractor's designs and in deciding whether or not to approve a request for pile lengths that are greater than those on which the tendered price was based. If the engineer declines to authorize extra pile lengths the contractor will withdraw a guarantee of performance. Nevertheless, the engineer has a duty to the employer or client to check the specialist contractor's designs as far as practically possible (guidance regarding this is given in Chapter 4) to enquire as to whether or not the contractor has made proper provision for difficult ground conditions such as obstructions or groundwater flow, to check on site that the piles are being installed in a sound manner and that they comply with the requirements for test loading. In the interests of the client the engineer should not allow extra pile lengths if it is considered that the contractor is being over-cautious in his assessment of the conditions. However, a decision should not be made without test-pile observations or previous knowledge of the performance of piles in similar soil conditions.

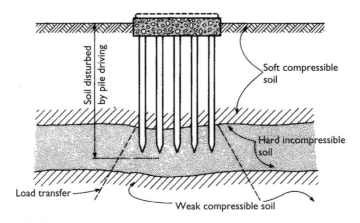

Figure 1.3 Pile group terminating in hard incompressible soil layer underlain by weak compressible soil.

The contractor's guarantee is usually limited to that of the load/settlement characteristics of a single pile and for soundness of workmanship, but his responsibilities regarding effects due to installation extend to the complete structure and to any nearby existing buildings or services. For example, if a building were to suffer damage due to the settlement of a group of piles and the settlement were due to the consolidation of a layer of weak compressible soil beneath the zone of disturbance caused by pile driving (Figure 1.3), the contractor could reasonably decline to accept responsibility. The engineer should have considered this in his overall design and specified a minimum pile length to take account of this compressible layer. On the other hand, a contractor is regarded as responsible for any damage to surrounding structures caused by vibrations or ground heave when driving a group of piles, or by any loss of ground when drilling for groups of bored and cast-in-place piles.

Because of the great importance of installation effects on pile behaviour, the various types of pile available and their methods of installation are first described in Chapters 2 and 3, before going on to discuss the various methods of calculating allowable loads on single piles and groups of piles in Chapters 4–6.

1.7 References

1.1 RANDOLPH, M. F. Science and empiricism in pile foundation design, *Geotechnique*, Vol. 53, No. 10, 2003, pp. 847–75.

1.2 BS EN 1997-1:2004 Eurocode 7: Geotechnical Design, Part 1 General Rules, British Standards Institution, London.

1.3 BS EN 1992-1:2004 Eurocode 2: Design of Concrete Structures, Part 1-1 General Rules and Rules for Buildings, British Standards Institution, London.

1.4 BS EN 1990 Eurocode 0: Basis of Structural Design, British Standards Institution, London.

1.5 FRANK, R., BAUDIN, C., DRISCOLL, R., KAVVADAS, M., KREBBS OVESEN, N., ORR, T., and SCHUPPENER, B. *Designers' Guide to EN 1997-1 Eurocode 7: Geotechnical Design* – General Rules, Thomas Telford, London, 2004.

Chapter 2

Types of pile

2.1 Classification of piles

The British Standard Code of Practice for Foundations (BS 8004: 1986) places piles in three categories. These are as follows:

Large displacement piles comprise solid-section piles or hollow-section piles with a closed end, which are driven or jacked into the ground and thus displace the soil. All types of driven and cast-in-place piles come into this category. Large diameter screw piles and rotary displacement auger piles are increasingly used for piling in contaminated land and soft soils.

Small displacement piles are also driven or jacked into the ground but have a relatively small cross-sectional area. They include rolled steel H- or I-sections and pipe or box sections driven with an open end such that the soil enters the hollow section. Where these pile types plug with soil during driving they become large displacement types.

Replacement piles are formed by first removing the soil by boring using a wide range of drilling techniques. Concrete may be placed into an unlined or lined hole, or the lining may be withdrawn as the concrete is placed. Preformed elements of timber, concrete or steel may be placed in drilled holes. Continuous flight auger (CFA) piles have become the dominant type of pile in the UK for structures on land.

Eurocode 7 (EC7)[1.2] does not categorize piles, but Clause 7 applies to the design of all types of load-bearing piles. When piles are used to reduce settlement of a raft or spread foundation (e.g. Love[2.1]), as opposed to supporting the full load from a structure, then the provisions of EC7 may not apply directly. A basic classification with examples of displacement piles is given in BS EN 12699: 2000 Execution of special geotechnical work – Displacement piles.

Types of piles in each of the BS 8004 categories can be listed as follows:

Large displacement piles (driven types)

(1) Timber (round or square section, jointed or continuous)
(2) Precast concrete (solid or tubular section in continuous or jointed units)
(3) Prestressed concrete (solid or tubular section)
(4) Steel tube (driven with closed end)
(5) Steel box (driven with closed end)
(6) Fluted and tapered steel tube

(7) Jacked-down steel tube with closed end
(8) Jacked-down solid concrete cylinder.

Large displacement piles (driven and cast-in-place types)

(1) Steel tube driven and withdrawn after placing concrete
(2) Steel tube driven with closed end, left in place and filled with reinforced concrete
(3) Precast concrete shell filled with concrete
(4) Thin-walled steel shell driven by withdrawable mandrel and then filled with concrete
(5) Rotary displacement auger and screw piles
(6) Expander body.

Small displacement piles

(1) Precast concrete (tubular section driven with open end)
(2) Prestressed concrete (tubular section driven with open end)
(3) Steel H-section
(4) Steel tube section (driven with open end and soil removed as required)
(5) Steel box section (driven with open end and soil removed as required).

Replacement piles

(1) Concrete placed in hole drilled by rotary auger, baling, grabbing, airlift or reverse-circulation methods (bored and cast-in-place)
(2) Tubes placed in hole drilled as above and filled with concrete as necessary
(3) Precast concrete units placed in drilled hole
(4) Cement mortar or concrete injected into drilled hole
(5) Steel sections placed in drilled hole
(6) Steel tube drilled down.

Composite piles

Numerous types of piles of composite construction may be formed by combining units in each of the above categories or by adopting combinations of piles in more than one category. Thus composite piles of a displacement type can be formed by jointing a timber section to a precast concrete section, or a precast concrete pile can have an H-section jointed to its lower extremity. Composite piles consisting of more than one type can be formed by driving a steel or precast concrete unit at the base of a drilled hole or by driving a tube and then drilling out the soil and extending the drill hole to form a bored and cast-in-place pile.

Selection of pile type

The selection of the appropriate type of pile from any of the above categories depends on the following three principal factors:

(1) The location and type of structure
(2) The ground conditions
(3) Durability.

Considering the first factor, some form of displacement pile is the first choice for a *marine structure*. A solid precast or prestressed concrete pile can be used in fairly shallow water, but in deep water a solid pile becomes too heavy to handle and either a steel tubular pile or a tubular precast concrete pile is used. Steel tubular piles are preferred to H-sections for exposed marine conditions because of the smaller drag forces from waves and currents. Large-diameter steel tubes are also an economical solution to the problem of dealing with impact forces from waves and berthing ships. Timber piles are used for temporary works in fairly shallow water. Bored and cast-in-place piles would not be considered for any marine or river structure unless used in a composite form of construction, say as a means of extending the penetration depth of a tubular pile driven through water and soft soil to a firm stratum.

Piling for a structure on *land* is open to a wide choice in any of the three categories. Bored and cast-in-place piles are the cheapest type where unlined or only partly lined holes can be drilled by rotary auger. These piles can be drilled in very large diameters and provided with enlarged or grout-injected bases, and thus are suitable to withstand high working loads. Augered piles are also suitable where it is desired to avoid ground heave, noise and vibration, i.e. for piling in urban areas, particularly where stringent noise regulations are enforced. Driven and cast-in-place piles are economical for land structures where light or moderate loads are to be carried, but the ground heave, noise and vibration associated with these types may make them unsuitable for some environments.

Timber piles are suitable for light to moderate loadings in countries where timber is easily obtainable. Steel or precast concrete-driven piles are not as economical as driven or bored and cast-in-place piles for land structures. Jacked-down steel tubes or concrete units are used for underpinning work.

For the design of foundations in *seismic situations* reference can be made to criteria in Eurocode 8 ENV 1998-5: 1994 Design of structures for earthquake resistance Part 5: Foundations, retaining walls and geotechnical aspects (EC8-5); these rules complement the information on soil–structure interaction given in EC7. The paper by Raison[2.2] refers to the checks required under EC8-1 for piles susceptible to liquefaction at a site in Barrow.

The second factor, *ground conditions*, influences both the material forming the pile and the method of installation. Firm to stiff fine-grained soils (silts and clays) favour the augered bored pile, but augering without support of the borehole by a bentonite slurry cannot be performed in very soft clays or in loose or water-bearing granular soils, for which driven or driven and cast-in-place piles would be suitable. Piles with enlarged bases formed by auger drilling can be installed only in firm to stiff or hard fine-grained soils or in weak rocks. Driven and driven and cast-in-place piles can neither be used in ground containing boulders or other massive obstructions, nor can they be used in soils subject to ground heave, in situations where this phenomenon must be prevented.

Driven and cast-in-place piles which employ a withdrawable tube cannot be used for very deep penetrations because of the limitations of jointing and pulling out of the driving tube. For such conditions a driven pile would be suitable. For hard driving conditions, for example, boulder clays or gravely soils, a thick-walled steel tubular pile or a steel H-section can withstand heavier driving than a precast concrete pile of solid or tubular section.

Some form of drilled pile, such as a drilled-in steel tube, would be used for piles taken down into a rock for the purpose of mobilizing resistance to uplift or lateral loads.

When piling in *contaminated land* using boring techniques, the disposal of arisings to licensed tips and measures to avoid the release of aerosols are factors limiting the type of pile which can be considered and can add significantly to the costs. Precautions may also be needed to avoid creating preferential flow paths while piling which could allow contaminated groundwater and leachates to be transported downwards. Hollow tubular steel piles can be expensive for piling in contaminated ground when compared with other displacement piles, but they are useful in overcoming obstructions which could cause problems when driving precast concrete or boring displacement piles. Large displacement piles are unlikely to form transfer conduits for contaminants, although untreated wooden piles may allow 'wicking' of volatile organics. End-bearing H-piles can form long-term flow conduits into aquifers (particularly when a driving shoe is needed) and it may be necessary for the piles to be hydraulically isolated from the contaminated zone.

The factor of *durability* affects the choice of material for a pile. Although timber piles are cheap in some countries they are liable to decay above groundwater level, and in marine structures they suffer damage by destructive mollusc-type organisms. Precast concrete piles do not suffer corrosion in saline water below the 'splash zone', and rich well-compacted concrete can withstand attack from quite high concentrations of sulphates in soils and groundwaters. Cast-in-place concrete piles are not so resistant to aggressive substances because of difficulties in ensuring complete compaction of the concrete, but protection can be provided against attack by placing the concrete in permanent linings of coated light-gauge metal or plastics. Check lists for durability of man-made materials in the ground are provided in Eurocode 2 (EC2) BS EN 1992-1: 2004 Design of Concrete Structures, Part 1-1: General rules and BS 8500 for concrete, and for steel in Eurocode 3 (EC3) BS EN 1993-1: 2005 Design of Steel Structures, Part 1-1: General rules and BS EN 1993 Part 5, Piling (EC3–5).

Steel piles can have a long life in ordinary soil conditions, if they are completely embedded in undisturbed soil, but the portions of a pile exposed to sea water or to disturbed soil must be protected against corrosion by cathodic means, if a long life is required. Corrosion rates can be derived from the corrosion tables published in EC3-5 Annex F. Recent work by *Corus Construction and Industrial*[2.3, 2.4] has refined guidelines for corrosion allowances for steel embedded in contaminated soil. 'Mariner grade' steel to ASTM standard can give performance improvement of 2 to 3 times that of conventional steels in marine splash zones.

Other factors influence the choice of one or another type of pile in each main classification, and these are discussed in the following pages, in which the various types of pile are described in detail. In UK practice specifications for pile materials, manufacturing requirements (including dimensional tolerances), workmanship and contract documentation are given in a publication of the *Institution of Civil Engineers*[2.5]. This specification is generally consistent with the requirements in EC7 and the associated standards for the 'Execution of special geotechnical works' – BS EN 1536: 1999 Bored piles, BS EN 12063: 1999 Sheet piling, BS EN 12699: 2000 Displacement piles and BS EN 14199: 2005 Micropiles.

Having selected a certain type or types of pile as being suitable for the location and type of structure for the ground conditions at the site and for the requirements of durability, the final choice is then made on the basis of *cost*. However, the total cost of a piled foundation is not simply the quoted price per metre run of piling or even the more accurate comparison of cost per pile per kN of working load carried. The most important consideration is the overall cost of the foundation work including the main contractor's costs and overheads.

It has been noted in Chapter 1 that a piling contractor is unlikely to quote a fixed price based on a predetermined length of pile. Extra payment will be sought if the piles are required to depths greater than those predicted at the tendering stage. Thus a contractor's previous experience of the ground conditions in a particular locality is important in assessing the likely pile length on which to base a tender. Experience is also an important factor in determining the extent and cost of a preliminary test piling programme. This preliminary work can be omitted if a piling contractor can give a warranty based on in-house knowledge of the site conditions that the engineer's requirements for load/settlement criteria can be met. The cost of test piling can then be limited to that of proof-loading selected working piles. In well-defined ground conditions and relatively light structural loads, the client may rely solely on the contractor's comprehensive warranty that the working piles meet the load-carrying requirement with an appropriate safety factor. It is a precept in EC7 that pile design should be related directly or indirectly to the results of static load tests and in certain cases such tests are mandatory. Where analytical calculations or interpretations of dynamic tests are used for design, the methods must have been validated against previous static load tests 'in comparable conditions'. EC7 introduces design by the 'observational method' in which the design is reviewed during construction and in response to monitoring during performance. This is not relevant to pile design, but a design method based on observed performance of comparable piled foundations is acceptable provided that it is 'supported by ground investigation and ground testing'.

In any case, preliminary test piling may be necessary to prove the feasibility of the contractor's installation method and to determine the load–settlement relationship for a given pile diameter and penetration depth. If a particular piling system is shown to be impracticable, or if the settlements are shown by the test loading to be excessive, then considerable time and money can be expended in changing to another piling system or adopting larger-diameter or longer piles. During the period of this preliminary work the main contractor continues to incur overhead costs and may well claim reimbursement of these costs if the test-piling work extends beyond the time allowed in the constructional programme. To avoid such claims it is essential to carry out a thorough ground investigation (as BS 5930 and EC7-Part 2 Ground investigation and testing), and it is desirable to conduct the preliminary test piling before the main contractor commences work on the site.

Finally, a piling contractor's resources for supplying additional rigs and skilled operatives to make up time lost due to unforeseen difficulties and technical ability in overcoming these difficulties are factors which may influence the choice of a particular piling system.

2.2   Driven displacement piles

2.2.1   Timber piles

In many ways, timber is an ideal material for piling. It has a high strength to weight ratio, it is easy to handle, it is readily cut to length and trimmed after driving, and in favourable conditions of exposure durable species have an almost indefinite life. Timber piles used in their most economical form consist of round untrimmed logs which are driven butt uppermost. The traditional British practice of using squared timber may have become established because of the purchase for piling work of imported timber which had been squared for general structural purposes in the sawmills of the country of origin. The practice of squaring the timber can be detrimental to its durability since it removes the outer sapwood which is absorptive to creosote or some other liquid preservative. The less absorptive heartwood is thus exposed and instead of a pile being encased by a thick layer of well-impregnated sapwood, there is only a thin layer of treated timber which can be penetrated by the hooks or slings used in handling the piles or stripped off by obstructions in the ground.

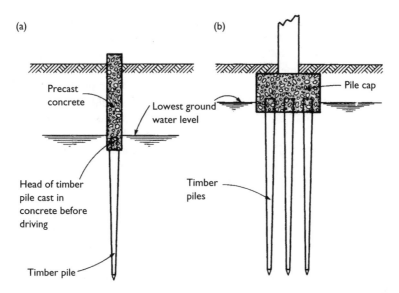

Figure 2.1 Protecting timber piles from decay (a) By precast concrete upper section above water level (b) By extending pile cap below water level.

Timber piles, when situated wholly below groundwater level, are resistant to fungal decay and have an almost indefinite life. However, the portion above groundwater level in a structure on land is liable to decay and BS EN 12699 prohibits the use of timber piles above free water level, unless creosote or other adequate protection is used. Therefore, it is the usual practice to cut off timber piles just below the lowest predicted groundwater level and to extend them above this level in concrete (Figure 2.1a). If the groundwater level is shallow the pile cap can be taken down below the water level (Figure 2.1b).

Timber piles in marine structures are liable to be severely damaged by the mollusc-type borers which infest the sea-water in many parts of the world, particularly in tropical seas. The severity of this form of attack can be reduced to some extent by using softwood impregnated with creosote or greatly minimized by the use of a hardwood of a species known to be resistant to borer attack. The various forms of these organisms, the form of their attack and the means of overcoming it are discussed in greater detail in Chapter 10.

Bark should be removed from round timbers where these are to be treated with preservative. If this is not done the bark reduces the depth of impregnation. Also the bark should be removed from piles carrying uplift loads by shaft friction in case it should become detached from the trunk, thus causing the latter to slip. Bark need not be removed from piles carrying compression loads or from fender piles of uncreosoted timber (hardwoods are not treated because they will not absorb creosote or other liquid preservatives).

Commercially available timbers which are suitable for piling include Douglas fir, pitch pine, larch and Western red cedar, in the softwood class, and greenheart, jarrah, opepe, teak and European oak in the hardwood class. The timber should be straight-grained and free from defects which could impair its strength and durability. BS 8004 states that a deviation in straightness from the centre-line of up to 25 mm on a 6 m chord is permitted for round logs but the centre-line of a sawn timber pile must not deviate by more than 25 mm from a straight line throughout its length. This is similar to the end to end centroid deviation allowed in BS EN 14081-2: 2005 for rectangular structural timber.

The requirements of BS 8004 for working stresses in timber piles merely state that these should not exceed the green permissible stresses in BS 5268-2 for compression parallel to the grain for the species and grade of timber being used. The Code suggests that suitable material will be obtained from stress grades SS and better. Grade stresses in accordance with BS 5268-2 are shown in Table 2.1, for various classes of softwood and hardwood suitable for piling work. The working stresses shown in Table 2.1 for the hardwoods are considerably higher than those of the comparable grades of softwood. It should be noted that the stresses in Table 2.1 are for dry timber. Timber piles are usually in a wet environment when the multiplying factors shown in Table 2.2 should be used to convert the dry stress properties to the wet conditions. When calculating the working stress on a pile, allowance must be made for bending stresses due to eccentric and lateral loading and to eccentricity caused by deviations in the straightness and inclination of a pile. Allowance must also be made for reductions in the cross-sectional area due to drilling or notching and to the taper on a round log. Eurocode 5 (EC5) BS EN 1995-1: 2004 Design of timber structures: Part 1-1 provides common rules on stresses which will be relevant to timber piling.

As a result of improved ability to predict and control driving stresses, BS EN 12699 allows the maximum compressive stress generated during driving to be increased to 0.8

Table 2.1 Grade stresses and moduli of elasticity of some softwoods and tropical hardwoods suitable for bearing piles

| Standard name | Grade | BS 5268: Part 2: 2002 (values in N/mm²) | | | | | | |
| | | Bending parallel to grain* | Tension parallel to grain* | Compression parallel to grain | Compression perpendicular to grain | Shear parallel to grain' | Modulus of elasticity | |
							Mean	Minimum
British larch	SS	7.5	4.5	7.9	2.1	0.82	10500	7000
Douglas fir	SS	6.2	3.7	6.6	2.4	0.88	11000	7000
Pitch pine	SS	10.5	6.3	11.0	3.2	1.16	13500	9000
Western red cedar	SS	5.7	3.4	6.1	1.7	0.63	8500	5500
(imported)	GS	4.1	2.5	5.2	1.6	0.63	7000	4500
Douglas fir-larch	SS	7.5	4.5	7.9	2.4	0.85	11000	7500
(Canada and USA)	GS	5.3	3.2	6.8	2.2	0.85	10000	6500
Greenheart	HS	26.1	15.6	23.7	5.9	2.6	21600	18000
Jarrah	HS	13.8	8.2	14.2	3.1	2.0	12400	8700
Opepe	HS	17.0	10.2	17.6	3.8	2.1	14500	11300
Teak	HS	13.7	8.2	13.4	3.1	1.7	10700	7400

Notes

* Stresses applicable to timber 300 mm deep (or wide).

' When the specifications specifically prohibit wane at bearing areas, the SS grade and HS grade comperssion perpendicular to the grain stress may be multiplied by 1.33 and used for all grades.

SS denotes special structural grade (visually stressed graded).

HS denotes special structural grade (machine stress graded).

All stresses apply to long-term loading.

Table 2.2 Modification factor K_2 by which dry stresses and moduli should be multiplied to obtain wet stresses and moduli applicable to wet exposure conditions

Property	Value of K_2
Bending parallel to grain	0.8
Tension parallel to grain	0.8
Compression parallel to grain	0.6
Compression perpendicular to grain	0.6
Shear parallel to grain	0.9
Mean and minimum modulus of elasticity	0.8

times the characteristic compressive strength measured parallel to the grain. While some increase in stress (up to 10%) may be permitted during driving if stress monitoring is carried out, it is advisable to limit the maximum load which can be carried by a pile of any diameter to reduce the need for excessively hard driving. This limitation is applied in order to avoid the risk of damage to a pile by driving it to some arbitrary 'set' as required by a dynamic pile-driving formula and to avoid a high concentration of stress at the toe of a pile end-bearing on a hard stratum. Damage to a pile during driving is most likely to occur at its head and toe.

The problems of splitting of the heads and unseen 'brooming' and splitting of the toes of timber piles occur when it is necessary to penetrate layers of compact or cemented soils to reach the desired founding level. This damage can also occur when attempts are made to drive deeply into dense sands and gravels or into soils containing boulders, in order to mobilize the required frictional resistance for a given uplift or compressive load. Judgement is required to assess the soil conditions at a site so as to decide whether or not it is feasible to drive a timber pile to the depth required for a given load without damage, or whether it is preferable to reduce the working load to a value which permits a shorter pile to be used. As an alternative, jetting or pre-boring may be adopted to reduce the amount of driving required. The temptation to continue hard driving in an attempt to achieve an arbitrary set for compliance with some dynamic formula must be resisted. Cases have occurred where the measured set achieved per blow has been due to the crushing and brooming of the pile toe and not to the deeper penetration required to reach the bearing stratum.

Damage to a pile can be minimized by reducing as far as possible the *number* of hammer blows necessary to achieve the desired penetration, and also by limiting the height of drop of the hammer to 1.5 m. This necessitates the use of a heavy hammer (but preferably less than 4 tonnes), which should at least be equal in weight to the weight of the pile for hard driving conditions and to one-half of the pile weight for easy driving. The lightness of a timber pile can be an embarrassment when driving groups of piles through soft clays or silts to a point bearing on rock. Frictional resistance in the soft materials can be very low for a few days after driving, and the effect of pore pressures caused by driving adjacent piles in the group may cause the piles already driven to rise out of the ground due to their own buoyancy relative to that of the soil. The only remedy is to apply loads to the pile heads until all the piles in the area have been driven.

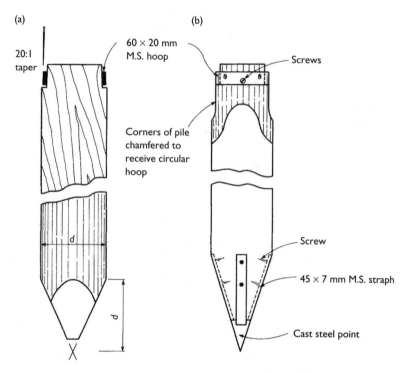

Figure 2.2 Protecting timber piles from splitting during driving (a) Protecting head by mild steel hoop (b) Protecting toe by cast steel point.

Heads of timber piles should be protected against splitting during driving by means of a mild steel hoop slipped over the pile head or screwed to it (Figure 2.2a and b). A squared pile toe can be provided where piles are terminated in soft to moderately stiff clays (Figure 2.2a). Where it is necessary to drive them into dense or hard materials a cast steel point should be provided (Figure 2.2b). As an alternative to a hoop, a cast steel helmet can be fitted to the pile head during driving. The helmet must be deeply recessed and tapered to permit it to fit well down over the pile head, allowing space for the insertion of hardwood packing.

Commercially available timbers are imported in lengths of up to 18 m. If longer piles are required they may be spliced as shown in Figure 2.3. A splice near the centre of the length of a pile should be avoided since this is the point of maximum bending moment when the pile is lifted from a horizontal position by attachments to one end or at the centre. Timber piles can be driven in very long lengths in soft to firm clays by splicing them in the leaders of the piling frame as shown in Figure 2.4. The abutting surfaces of the timber should be cut truly square at the splice positions in order to distribute the stresses caused by driving and loading evenly over the full cross-section.

2.2.2   Precast concrete piles

Precast concrete piles have their principal use in marine and river structures, i.e. in situations where the use of driven and cast-in-place piles is impracticable or uneconomical. For land

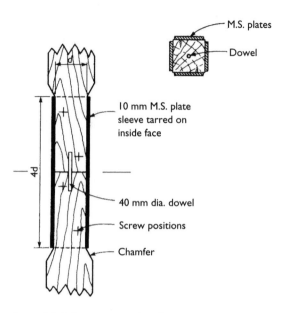

Figure 2.3 Splice in squared timber pile.

structures unjointed precast concrete piles are frequently more costly than driven and cast-in-place types for two main reasons:

(1) Reinforcement must be provided in the precast concrete pile to withstand the bending and tensile stresses which occur during handling and driving. Once the pile is in the ground, and if mainly compressive loads are carried, the majority of this steel is redundant.
(2) The precast concrete pile is not readily cut down or extended to suit variations in the level of the bearing stratum to which the piles are driven.

However, there are many situations for land structures where the precast concrete pile can be more economical. Where large numbers of piles are to be installed in easy driving conditions the savings in cost due to the rapidity of driving achieved may outweigh the cost of the heavier reinforcing steel necessary. Reinforcement may be needed in any case to resist bending stresses due to lateral loads or tensile stresses from uplift loads. Where high-capacity piles are to be driven to a hard stratum, savings in the overall quantity of concrete compared with cast-in-place piles can be achieved since higher working stresses can be used. Where piles are to be driven in sulphate-bearing ground or into aggressive industrial waste materials, the provision of sound high-quality dense concrete is ensured. The problem of varying the length of the pile can be overcome by adopting a jointed type.

From the above remarks it can be seen that there is still quite a wide range of employment for the precast concrete pile, particularly for projects where the costs of establishing a precasting yard can be spread over a large number of piles. The piles can be designed and man-ufactured in ordinary reinforced concrete, or in the form of pre-tensioned or post-tensioned prestressed concrete members. The ordinary reinforced concrete pile is likely to be preferred for a project requiring a fairly small number of piles, where the cost of establishing a production line

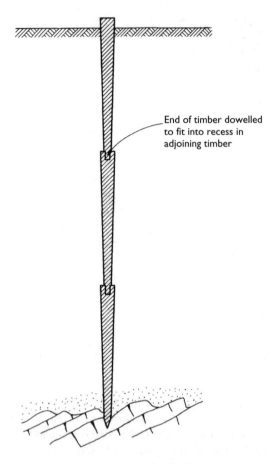

End of timber dowelled
to fit into recess in
adjoining timber

Figure 2.4 Splicing timber piles in multiple lengths.

for prestressing work on site is not justifiable and where the site is too far from an established factory to allow the economical transportation of prestressed units from the factory to the site.

Precast concrete piles in ordinary reinforced concrete are usually square or hexagonal and of solid cross-section for units of short or moderate length, but for saving weight long piles are usually manufactured with a hollow interior in hexagonal, octagonal, or circular sections. The interiors of the piles can be filled with concrete after driving. This is necessary to avoid bursting where piles are exposed to severe frost action. Alternatively, drainage holes can be provided to prevent water from accumulating in the hollow interior. To avoid excessive flexibility while handling and driving the usual maximum lengths of square section piles, and the range of working loads applicable to each size are shown in Table 2.3. Where piles are designed to carry the applied loads mainly in end-bearing, for example, piles driven through soft clays into medium-dense or dense sands, economies in concrete and reductions in weight for handling can be achieved by providing the piles with an enlarged toe. This is practised widely in the Netherlands where the standard enlargements are 1.5 to 2.5 times the shaft width with a length equal to or greater than the width of the enlargement.

Table 2.3 Working loads and maximum lengths for ordinary precast concrete piles of square section

Pile size (mm^2)	Range of working loads (kN)	Maximum length (m)
250	200–300	12
300	300–450	15
350	350–600	18
400	450–750	21
450	500–900	25

BS 8004 requires that piles should be designed to withstand the loads or stresses and to meet other serviceability requirements during handling, pitching, driving and in service in accordance with BS 8110 for the structural use of concrete (Table 2.4). EC7 requires the structural design of piles to conform to the serviceability requirements in the relevant material Eurocodes – EC2, EC3, and EC5 and the relevant National Annexes. EC2-1-1 provides common rules for concrete for building and civil engineering which are not very different from BS 8110 in terms of general design approach. Concrete performance, quality and production are subject to BS EN 206-1: 2000, which must be read in conjunction with the UK's complementary rules for strength classes, cover, etc. in BS 8500. If nominal BS 8110 mixes are adopted, a 40-grade concrete with a minimum 28-day cube strength of 40 N/mm^2 is suitable for hard to very hard driving and for all marine construction. For normal or easy driving, a 25-grade concrete is suitable (i.e. 28-day cube strength of 25 N/mm^2). Depending on exposure conditions defined in EC2-1-1 and appropriate cover to reinforcement, BS 8500 recommends strength classes of concrete from C20/25 (i.e. grading based on minimum characteristic strength of a *cylinder* at 28 days/minimum characteristic *cube* strength at 28 days in N/mm^2) in dry or permanently wet conditions to C45/55 in tidal splash zones. BS EN 12794: 2005 Precast concrete products – Foundation piles does not give specific requirements for the design strength of concrete for piles, but refers to BS EN 13369: 2004 Common rules for precast products, BS EN 206-1 and EC2. BS EN 12794 defines two classes of piles – 'Class 1' with distributed reinforcement or prestressed piles and 'Class 2' with single central reinforcing bar. The only difference this division makes is to the detailing of pile reinforcement. BS EN 13369 Clause 4.2.2 requires reinforced concrete products to have a minimum strength class of C20/25 and prestressed concrete a minimum of C30/37. Foundations in brownfield sites are not covered in BS 8500 and the recommendations in BRE Special Digest 1[2.6] should be followed for both in-situ foundation concrete and precast units. It should be noted that the design strengths in EC2 are based on the characteristic *cylinder* strengths. High stresses, which may exceed the handling stresses, can occur during driving and it is necessary to consider the serviceability limit of cracking. BS 8110 states that National Standards and Codes of Practice require cracks to be controlled to maximum widths close to the main reinforcement ranging from 0.3 mm down to 0.1 mm in an aggressive environment, or they require that crack widths shall at no point on the surface of the structure exceed a specified width, usually 0.3 mm. EC2-1-1 Clause 7.3 provides for maximum crack widths of 0.3–0.4 mm in reinforced concrete elements taking account of the proposed function of the structure and exposure of precast and prestressed elements. Application to precast piles is not considered, and neither BS EN 12699 nor BS EN 12794 comments on cracks due to driving.

Table 2.4 Basic code requirements for structural design of concrete and steel piles to BS 8004 and Eurocode 7

Pile type	BS 8004 Foundations	BS EN 1536 Bored piles	BS EN 12699 Displacement piles	BS EN 12794 Precast concrete products – piles
Precast concrete				
Concrete grade	As BS 8110 [25–40 N/mm² typically]	As BS EN 1992-1-1 and BS EN 206-1/BS 8500 for concrete insert elements C20/25–C45/55 (subject to exposure class)	As BS EN 12794	As BS EN 1992-1-1 BS EN 206-1/BS 8500 C35/45–C60/70 Also refer to BS EN13369
Cement content	300–400 kg/m³			
Reinforcing steel	As BS 8110	As ENV 1994-1-1 for load-bearing elements combining steel sections and concrete		As BS EN1008 and BS EN 10138
Calculated driving stress			$0.8\,f_{ck}^{d}$ [d] (10% increase if stresses monitored)	
Steel bearing				
Steel grade	As BS 4360 [43A, 50B typically]		As ENV 1993-5/ BS EN 10025 S275 and S355 typically	
Steel stress in compression	$0.3f_y$ [a] and $0.5f_y$ [b]			
Calculated driving stress			$0.9\,f_y$ (20% increase if stresses monitored)	
Driven cast-in-place				
Concrete grade	As BS 8110		As BS EN 1992-1-1 and BS EN 206-1/BS 8500 C20/25–C30/37 and >C25/30 for semi-dry mix	

Cement content	>300 kg/m^3	≥325 kg/m^3 in dry conditions ≥375 kg/m^3 in submerged conditions ≥350 kg/m^3 for semi-dry concrete
Concrete stress in compression	$0.25u_w$[c]	
Reinforcement	As BS 8110	As BS EN 10080 or ENV 1994-1-1 for steel lining and concrete core in compression

Bored cast-in-place

Concrete grade	As BS 8110	As BS EN 1992-1-1 and BS EN 206-1/BS 8500 C20/25–C30/37
Cement content	>300 kg/m^3 >400 kg/m^3 when under water or mud	≥325 kg/m^3 in dry conditions ≥375 kg/m^3 in submerged conditions
Concrete stress in compression	$0.25u_w$[c]	
Reinforcement	As BS EN 10080 or ENV 1994-1-1 for steel lining and concrete core in compression	

Notes
a Where f_{ly} is the characteristic yield strength of steel and where the safety factor on driving is less than 2.
b For jacked piles or where end-bearing piles are driven through relatively soft soils on to very dense granular soils or sound rock.
c u_w is concrete cube strength at 28 days.
d f_{ck} is the characteristic cylinder strength at 28 days.

To comply with the requirements of BS 8110 precast piles of either ordinary or prestressed concrete should have nominal cover to the reinforcement as shown in the following table:

Exposure conditions	Nominal cover (mm) for concrete grade of			
	25	30	40	50 and over
Buried concrete and concrete continuously under water	40	30	25	20
Alternate wetting and drying and freezing	50	40	30	25
Exposed to sea water and moorland water with abrasion			60	50

Using covers larger than required may lead to spalling of the concrete during driving.

In EC2-1-1 Clause 4.4 nominal cover to reinforcement is defined as $c_{nom} = c_{min} + \Delta c_{dev}$ where c_{min} is dependent on bond requirements or environmental conditions as summarized in the Code. Δc_{dev} allows for deviations, usually set at 10 mm but reduced where strict QA/QC procedures are in force. BS 8500 provides tabulated classifications for cover, characteristic concrete strength, cement content and type of cement combination depending on exposure conditions and type of steel corrosion; for example, for an intended life of 50 years and 20 mm maximum aggregate:

Corrosion due to	Exposure conditions[a]	Cement types[b]	Cover (mm) with concrete class and cement content (kg/m³)					
			$20+\Delta c$	$25+\Delta c$	$30+\Delta c$	$35+\Delta c$	$40+\Delta c$	$45+\Delta c$
Carbonation	Moderate humidity (XC3 and 4)	All except pfa >36%	C40/50 340	C32/40 300	C28/35 280	C25/30 260		
Chlorides from sea water	Tidal, splash and spray zones (XS3)	Group 5 and 6					C35/45 380	C32/40 360
		Group 4					C45/55 380	

Notes

a Degrees of exposure as defined in BS 8500 and EC2-1-1.

b Cement types 'CEM I to V' and combinations are defined in Table 1 of BS EN 197-1 and in Groups in Table A.17 of BS 8500. pfa (fly ash) as BS EN 450.

Neither BS EN 12794 nor BS EN 13369 comments on minimum cover for precast piles, other than advising that cover may be modified in accordance with BS EN 13369. For example, the cover may be reduced by 5 mm when using concrete class C40/50 or above, but requires an increase of 5 mm where achievement of dimensional tolerances may be a factor.

BS 8500 also specifies cement combinations and class of concrete to be used to resist attack on the concrete itself, again depending on exposure conditions. In addition, reference should be made to BRE Special Digest 1 and Section 10.3.

Table 2.5 BS 8004 requirements for longitudinal steel reinforcement, hoops, and links in precast piles

Longitudinal steel	Volume of steel at head and toe of pile	Volume of steel in body of pile	Other requirements
To provide for lifting, handling, and superstructure loads and for tensile forces caused by ground heave	0.6% gross volume over distance of 3 × pile width from each end	0.2% of gross volume spaced at not more than ½ × pile width	Lapping of short bars with main reinforcement to be arranged to avoid sudden discontinuity

BS EN 12794 states that longitudinal reinforcement shall be a minimum diameter of 8 mm with at least one bar placed in the corner of square piles; circular section piles shall have at least six bars placed around the periphery. Transverse reinforcement must be at least 4 mm in diameter depending on the pile diameter and the pile head must have a minimum of nine links in 500 mm. BS EN 12794 refers to BS EN 13369 for the quality of reinforcement and prestressing steel to be used, which in turn refers to other ENs, such as BS EN 10080 and to the national standards in the countries where the products are to be used. BS 4449 Steel for reinforcement of concrete, has been revised for use with BS EN 10080. Notwithstanding the new standards, users of reinforcing steel are advised to obtain third party certification such as the CARES scheme in the UK.

The proportion of main reinforcing steel (Table 2.5) in the form of longitudinal bars is determined by the bending moments induced when the pile is lifted from its casting bed to the stacking area. The magnitude of the bending moments depends on the number and positioning of the lifting points. Design data for various lifting conditions are dealt with in 7.2. In some cases the size of the externally applied lateral or uplift loads may necessitate more main steel than is required by lifting considerations. Lateral steel in the form of hoops and links is provided to prevent shattering or splitting of the pile during driving. In hard driving conditions it is advantageous to place additional lateral steel in the form of a helix at the head of the pile. The helix should be about two pile widths in length with a pitch equal to the spacing of the link steel at the head. It can have zero cover where the pile head is to be cut down for bonding to the cap. A design for a precast concrete pile to comply with BS 8004 for easy driving conditions is shown in Figure 2.5a. A design for a longer octagonal pile suitable for driving to end bearing on rock is shown in Figure 2.5b. The design of a prestressed concrete pile in accordance with the recommendations of BS 8110 is shown in Figure 2.6.

Prestressed concrete piles have certain advantages over those of ordinary reinforced concrete. Their principal advantage is in their higher strength to weight ratio, enabling long slender units to be lifted and driven. However, slenderness is not always advantageous since a large cross-sectional area may be needed to mobilize sufficient resistance in shaft friction and end bearing. The second main advantage is the effect of the prestressing in closing up cracks caused during handling and driving. This effect, combined with the high-quality concrete necessary for economic employment of prestressing, gives the prestressed pile increased durability which is advantageous in marine structures and corrosive soils.

Prestressed concrete piles should be made with designed concrete mixes of at least C35/45, but as noted above, account should be taken of the special exposure conditions quoted in EC2-1-1 and BS 8500 when deciding on the concrete class to be used. Minimum percentages of prestressing steel stipulated in BS EN 12794 are 0.1% of cross-sectional area

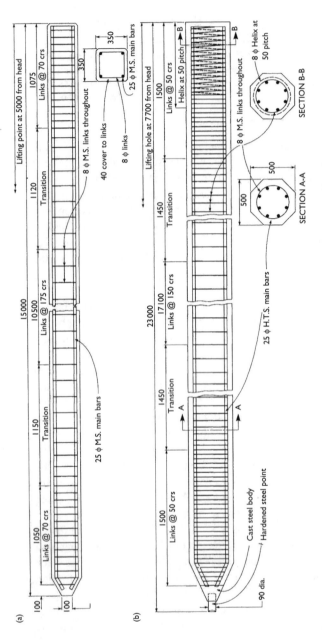

Figure 2.5 Design for precast concrete piles (a) 350 mm square pile 15 m long (b) 500 mm octagonal pile 23 m long.

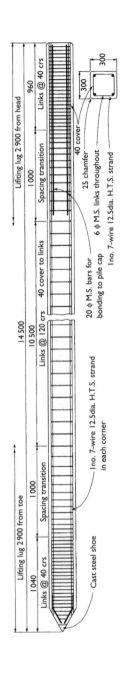

Figure 2.6 Design for prestressed concrete pile.

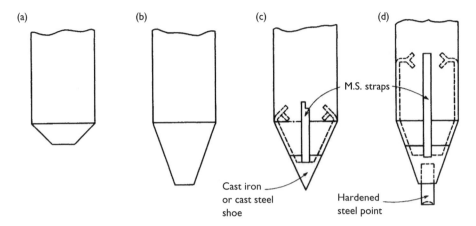

Figure 2.7 Shoes for precast (including prestressed) concrete piles (a) For driving through soft or loose soils to shallow penetration into dense granular or firm to stiff clays (b) Pointed end suitable for moderately deep penetration into medium-dense to dense sands and firm to stiff clays (c) Cast-iron or cast-steel shoe for seating pile into weak rock or breaking through cemented soil layer (d) 'Oslo point' for seating pile into hard rock.

in mm^2 for piles not exceeding 10 m in length, 0.01% cross-sectional area × pile length for piles between 10 and 20 m long and 0.2% for piles greater than 20 m long. It may be desirable to specify a maximum load which can be applied to a precast concrete pile of any dimensions. As in the case of timber piles this limitation is to prevent unseen damage to piles which may be over-driven to achieve an arbitrary set given by a dynamic pile-driving formula. BS EN 12699 limits the calculated stress (including any prestress) during driving of precast piles to 0.8 times characteristic concrete strength in compression at the time of driving; a 10% increase is permitted if the stresses are monitored during driving.

Concrete made with ordinary Portland cement (CEM I) is suitable for all normal exposure conditions but sulphate-resisting cement may be needed for aggressive ground conditions as discussed in Chapter 10.

Metal shoes are not required at the toes of precast concrete piles where they are driven through soft or loose soils into dense sands and gravels or firm to stiff clays. A blunt pointed end (Figure 2.7a) appears to be just as effective in achieving the desired penetration in these soils as a more sharply pointed end (Figure 2.7b) and the blunt point is better for maintaining alignment during driving. A cast-iron or cast-steel shoe fitted to a pointed toe may be used for penetrating rocks or for splitting cemented soil layers. The shoe (Figure 2.7c) serves to protect the pointed end of the pile.

Where piles are to be driven to refusal on a sloping hard rock surface, the 'Oslo point' (Figure 2.7d) is desirable. This is a hollow-ground hardened steel point. When the pile is judged to be nearing the rock surface, the hammer drop is reduced and the pile point is seated on to the rock by a number of blows with a small drop. As soon as there is an indication that a seating has been obtained the drop can be increased and the pile driven to refusal or some other predetermined set. The Oslo point was used by George Wimpey and Co. on the piles as illustrated in Figure 2.5b, which were driven on to hard rock at the site of the Whitegate Refinery, Cork. A hardened steel to BS 970 with a Brinell hardness of 400 to 600 was

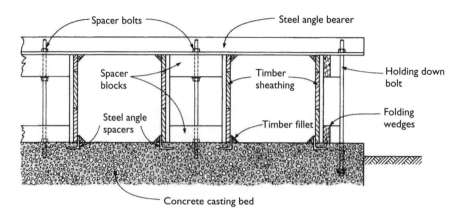

Figure 2.8 Timber formwork for precast concrete piles.

employed. The 89 mm point was machined concave to 12.7 mm depth and embedded in a chilled cast-iron shoe. Flame treatment of the point was needed after casting into the shoe to restore the hardness lost during this operation.

Piles may be cast on mass concrete beds using removable side forms of timber or steel (Figure 2.8). The reinforcing cage is suspended from bearers with spacing forks to maintain alignment. Spacer blocks to maintain cover are undesirable. The stop ends must be set truly square with the pile axis to ensure an even distribution of the hammer blow during driving. Vibrators are used to obtain thorough compaction of the concrete and the concrete between the steel and the forms should be worked with a slicing tool to eliminate honeycombed patches. The casting beds must be sited on firm ground in order to prevent bending of the piles during and soon after casting. After removing the side forms the piles already cast may be used as side forms for casting another set of piles in between them. If this is done the side forms should be set to give a trapezoidal cross-section in order to facilitate release. Piles may also be cast in tiers on top of each other, but a space between them should be maintained to allow air to circulate (Figure 2.9). Casting in tiers involves a risk of distortion of the piles due to settlement of the stacks. In addition, the piles which are first to be cast are the last to be lifted which is in the wrong order, since the most-mature piles should be the first to be lifted and driven.

Where piles are made in a factory, permanent casting beds can be formed in reinforced concrete with heating elements embedded in them to allow a 24-hour cycle of casting and lifting from the moulds. This method of construction was used by Soil Mechanics Ltd. to cast prestressed concrete piles at Drax Power Station in Yorkshire[2.7] where the large number of piles cast (18 500) justified the establishment on site of an elaborate casting yard such as would be used in a precast concrete factory. The reinforced concrete formwork is shown in Figure 2.10. This type, which does not have removable side forms, necessitates the embedment of lifting plugs or loops into the top of the piles.

The layout of the casting yard at Drax is shown in Figure 2.11. The strand reels were set on carriers at one end of the four rows of casting beds, with the winches for tensioning the strand at the opposite end. Each casting bed had five lines of forms. The provision of electric heating elements enabled the concrete to achieve its release strength of 27.6 N/mm^2 in 40 to 48 hours. An average of 300 piles per week, with a peak of 400 in a week, were manufactured. Two coats of whitewash were used as a release agent, as it was found that mould oil did not

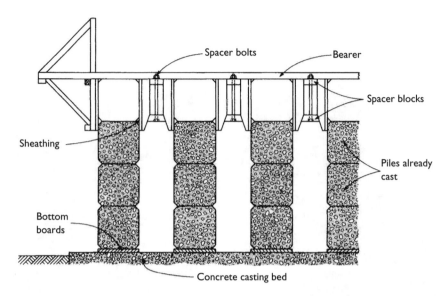

Figure 2.9 Casting precast concrete piles in tiers.

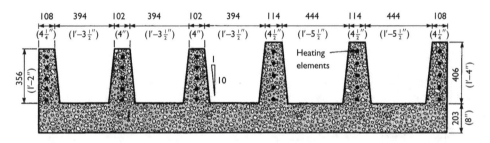

Figure 2.10 Heated concrete moulds for prestressed concrete piles.

give a sufficiently thick coating to prevent the piles from occasionally 'locking-in' to the moulds, in spite of a 1 in 10 taper on the sides. The oil also contaminated the prestressing strand.

When piles are cast within wooden side forms the latter should be removed as soon as possible, and wet curing by water spray and hessian maintained for a seven-day period. As soon as crushing tests on cubes indicate that the piles are strong enough to be lifted they should be slightly canted by careful levering with a bar and packing with wedges to release the suction between the pile and the bed. The lifting slings or bolt inserts may then be fixed and the pile lifted for transporting to the stacking area. This operation of first canting and lifting must be undertaken with great care since the piles have still only a comparatively immature strength and any cracks or incipient cracks formed at this stage will open under driving stresses.

The piles should be clearly marked with a reference number, length, and date of casting at or before the time of lifting, to ensure that they are driven in the correct sequence.

Figure 2.11 Casting yard for prestressed concrete piles at Drax Power Station.

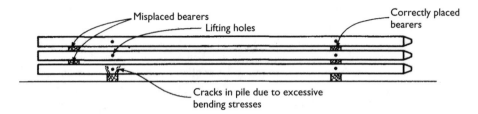

Figure 2.12 Misplaced packing in stacks of precast concrete piles.

Timber bearers should be placed between the piles in the stacks to allow air to circulate around them. They should be protected against too-rapid drying in hot weather by covering the stack with a tarpaulin or polyethylene sheeting. Care must be taken to place the bearers only at the lifting positions. If they are misplaced there could be a risk of excessive bending stresses developing and cracking occurring, as shown in Figure 2.12.

Prestressed concrete piles of hollow cylindrical section are manufactured by centrifugal spinning in diameters ranging from 400 to 1800 mm in lengths up to 40 m.

The precautions for driving precast concrete piles are described in Section 3.4.2, and the procedures for bonding piles to caps and ground beams and lengthening piles are described in Sections 7.6 and 7.7.

One of the principal problems associated with precast concrete piles is unseen breakage due to hard driving conditions. These conditions are experienced in Sweden where the widely used jointed or unjointed precast concrete piles are driven through soft or loose soils onto hard rock. On some sites the rock surface may slope steeply, causing the piles to deviate from a true line and break into short sections near the toe. Accumulations of boulders over bedrock can also cause the piles to be deflected with consequent breakage. Because of these experiences the Swedish piling code recommended the provision of a central inspection hole in test piles and sometimes in a proportion of the working piles. A check for deviation of the pile from line can be made by lowering a steel tube down the hole. If the tube can be lowered to the bottom of the hole under its own weight the pile should not be bent to a radius which would impair its structural integrity. If the tube jams in the hole, it is the usual practice to bring an inclinometer to the site to record the actual deviation, and hence to decide whether or not the pile should be rejected and replaced. The testing tube also detects deviations in the position or alignment of a jointed pile.

Breakages are due either to tensile forces caused by driving with too light a hammer in soft or loose soils, or to compressive forces caused by driving with too great a hammer drop on to a pile seated on a hard stratum; in both cases the damage occurs in the buried portion of the pile. In the case of compression failure it occurs by crushing or splitting near the pile toe. Such damage is not indicated by any form of cracking in the undriven portion of the pile above ground level. The provision of a central test hole will again enable crushing of the pile due to failure in compression to be detected.

2.2.3 Jointed precast concrete piles

The disadvantages of having to adjust the lengths of precast concrete piles either by cutting off the surplus or casting on additional lengths to accommodate variations in the depth to a hard bearing stratum will be evident. These drawbacks can be overcome by employing jointed piles in which the adjustments in length can be made by adding or taking away short lengths of pile which are jointed to each other by devices capable of developing the same bending and tensile resistance as the main body of the pile. BS EN 12794 defines pile joints in four classes, Class A to Class D, depending on whether the pile is used in compression, tension, or bending and the impact load test to be applied to verify the static design calcu-lations. If the pile joint satisfies the impact and bending tests then the ultimate capacity of the joint is 'identical' to the calculated static bearing capacity. Annex ZA to this standard deals with the CE marking of foundation pile units and the presumption of fitness for the intended use.

The 'Hercules' pile, originally developed in Sweden, is available in the UK from Stent Foundations Ltd in two square sizes with standard lengths of 6.1, 9.2, and 12.2 m, and properties as shown in Table 2.6. C45/55 concrete is normally used. The precast concrete units are locked together by a steel bayonet-type joint to obtain the required bending and tensile resistance and a rock shoe incorporating an Oslo point may be used (Figure 2.7d). A length is chosen for the initial driving which is judged to be suitable for the shallowest predicted penetration in a given area. Additional lengths are locked on if deeper penetra-tions are necessary, or if very deep penetrations requiring multiples of the standard lengths are necessary.

Other types of jointed precast concrete piles include the 'Centrum' pile manufactured and installed by Aarslef Piling in the UK using C40/50 concrete and rigid welded reinforcement cages in varying lengths from 4 to 18 m in square sections from 200 to 600 mm. Lengths greater than 4 m for the 200 and 250 mm sections can be jointed using a single locking pin driven horizontally into locking rings in the joint box (four locking pins for the larger sections), which are designed to provide a degree of pre-tensioning to the joint (Figure 2.13). Depending on the length, section, and joint used and the ground conditions, working loads up to 1200 kN in compression and 180 kN in tension are possible.

'RB' precast square concrete piles made and installed by Roger Bullivant Ltd are available in four sizes with working load capabilities (depending on ground conditions) from 200 kN for the nominal 150 mm square section to 1200 kN for the 355 mm square pile, in lengths of 1.5, 3, and 4 m. The standard joint for the limited tensile and bending capability is a

Table 2.6 Dimensions and properties of square section 'Hercules' piles as manufactured in the UK

Type of pile	S 550	S 730/750
Maximum safe working load[a] (kN)	600	1200
Side dimension (mm)	235	270/275
Cross-sectional area (mm^2)	55 225	72 900/75 625
Perimeter (mm)	940	1080/1100
Volume (m^3/m)	0.055	0.073/0.076
Mass (kg/m)	137	182/190
Surface area (m^2/m)	0.94	1.08/1.10

Note
a Safe working load is dependent on length of the pile and soil properties.

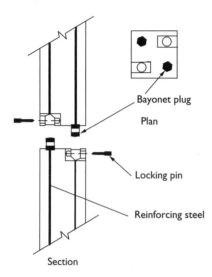

Bayonet plug

Plan

Locking pin

Reinforcing steel

Section

Figure 2.13 Typical locking pin joint for precast concrete pile.

simple spigot and socket type bonded with epoxy resin with each pile length bedded on a sand/cement mortar. Special joints and pile reinforcement can be provided as needed to resist bending moments and tension forces.

Precast concrete piles which consist of units joined together by simple steel end plates with welded butt joints are not always suitable for hard driving conditions, or for driving on to a sloping hard rock surface. Welds made in exposed site conditions with the units held in the leaders of a piling frame may not always be sound. If the welds break due to tension waves set up during driving or to bending caused by any deviation from alignment, the pile may break up into separate units with a complete loss of bearing capacity (Figure 2.14). This type of damage can occur with keyed or locked joints when the piles are driven heavily, for example, to break through thin layers of dense gravel. The design of the joint is, in fact, a critical factor in the successful employment of these piles, and tests to check bending, tension, and compression capabilities should be carried out for particular applications. However, even joints made from steel castings require accurate contact surfaces to ensure that stress concentrations are not transferred to the concrete.

The '*Presscore*' pile developed and installed by Abbey Pynford plc is a jointed precast concrete pile consisting of short units which are jacked into the soil. The concrete in the pile units and precast pile cap is 60 N/mm^2 and a reinforcing bar can be placed through the centre of the units (Figure 2.15). On reaching the required bearing depth the annulus around the pile is grouted through ports in the units. The use of jacked-in piles for underpinning work is described in Chapter 9.

A high strength cylindrical precast pile, 155 mm diameter and 1 m long, was developed in Canada for underpinning a 90-year-old building in Regina[2.8]. The segments were cast using steel fibre reinforced concrete with a 28-day compressive strength of 90 N/mm^2 and

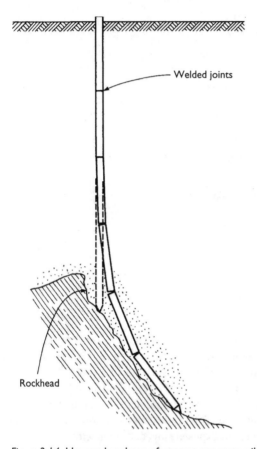

Welded joints

Rockhead

Figure 2.14 Unseen breakage of precast concrete piles with welded butt joints.

steel fibre content of 40 kg/m^3. Each segment was reinforced with four steel wires (9 mm) welded to a steel wire circumferential coil. Recesses were provided at each end of the segment and stainless steel rods connected each segment to form the joint. Hydraulic jacks with a capacity of 680 kN reacted against a new pile cap and as each segment was jacked down the next segment was screwed and tensioned onto the connecting rod. The required 600 kN pile capacity was achieved at depths ranging from 11 to 13 m.

2.2.4   Steel piles

Steel piles have the advantages of being robust, light to handle, capable of carrying high compressive loads when driven on to a hard stratum, and capable of being driven hard to a deep penetration to reach a bearing stratum or to develop a high frictional resistance, although their cost per metre run is high compared with precast concrete piles. They can be designed as small displacement piles, which is advantageous in situations where ground heave and lateral displacement must be avoided. They can be readily cut down and extended

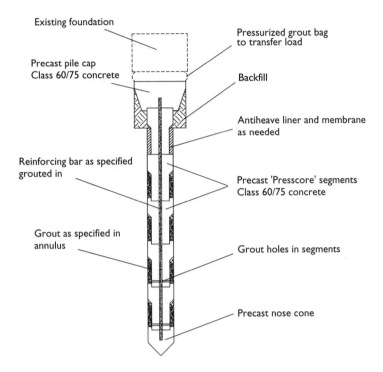

Existing foundation

Precast pile cap
Class 60/75 concrete

Pressurized grout bag
to transfer load

Backfill

Antiheave liner and membrane
as needed

Reinforcing bar as specified
grouted in

Precast 'Presscore' segments
Class 60/75 concrete

Grout as specified in
annulus

Grout holes in segments

Precast nose cone

Figure 2.15 Presscore pile (courtesy Abbey Pynford Foundation Systems Ltd).

where the level of the bearing stratum varies; also the head of a pile which buckles during driving can be cut down and re-trimmed for further driving. They have a good resilience and high resistance to buckling and bending forces.

Types of steel piles include plain tubes, box-sections, box piles built up from sheet piles, H-sections, and tapered and fluted tubes. Hollow-section piles can be driven with open ends. If the base resistance must be eliminated when driving hollow-section piles to a deep penetration, the soil within the pile can be cleaned out by grabbing, by augers, by reverse water-circulation drilling, or by airlift (see Section 3.4.3). It is not always necessary to fill hollow-section piles with concrete. In normal undisturbed soil conditions they should have an adequate resistance to corrosion during the working life of a structure, and the portion of the pile above the sea bed in marine structures or in disturbed ground can be protected by cathodic means, supplemented by bituminous or resin coatings (see Section 10.4). Concrete filling may be undesirable in marine structures where resilience, rather than rigidity, is required to deal with bending and impact forces.

Where hollow-section piles are required to carry high compressive loads they may be driven with a closed end to develop the necessary end-bearing resistance over the pile base area. Where deep penetrations are required they may be driven with open ends and with the interior of the pile closed by a stiffened steel plate bulkhead located at a predetermined height above the toe. An aperture should be provided in the bulkhead for the release of water,

Figure 2.16 Fabrication yard for steel tubular piles at Milford Haven.

silt or soft clay trapped in the interior during driving. In some circumstances the soil plug within the pile may itself develop the required base resistance (Section 4.3.3).

The facility of extending steel piles for driving to depths greater than predicted from soil investigation data has already been mentioned. The practice of welding-on additional lengths of pile in the leaders of the piling frame is satisfactory for land structures where the quality of welding may not be critical. A steel pile supported by the soil can continue to carry high compressive loads even though the weld is partly fractured by driving stresses. However, this practice is not desirable for marine structures where the weld joining the extended pile may be above sea-bed level in a zone subjected to high lateral forces and corrosive influences. Conditions are not conducive to first-class welding when the extension pile is held in leaders or guides on a floating vessel, or on staging supported by piles swaying under the influence of waves and currents. It is preferable to do all welding on a prepared fabrication bed with the pile in a horizontal position where it can be rotated in a covered welding station (Figure 2.16). The piles should be fabricated to cover the maximum predicted length and any surplus length cut off, rather than be initially of only medium length and then be extended. Cut-off portions of steel piles usually have some value as scrap, or they can be used in other fabrications. However, there are many situations where in-situ welding of extensions cannot be avoided. The use of a stable jack-up platform (Figure 3.7) from which to install the piles is then advantageous.

Where very long lengths of steel tubular piles are required to be driven, as in the case of offshore petroleum production platforms, they cannot be handled in a single length by cranes. They can be driven by underwater hammers, but for top-driven piles a pile connector is a useful device for joining such lengths of pile without the delays which occur when making welded joints. The Frank's Double Drive Shoulder Connector (Figure 2.17) was developed in the USA for joining lengths of oil well conductor pipe and can be adapted for making connections in piles upto 914 mm diameter. It is a pin and box joint which is flush with the OD and ID of the pile, with interlocking threads which pull the pin and box surfaces together. The joint is usually welded on to the steel pipe, not formed on the pipe ends. Long steel tubular piles driven within the tubular members of a jacket-type structure are redundant above their point of connection by annular grouting to the lower part of the tubular sleeve. This redundant part of the pile, which acts as a 'dolly' or follower for the final stages of driving, can be cut off for reuse when driving other piles.

Where large steel tubular piles are required to carry compressive loads only, the 'Advance' purpose-made splicing devices manufactured by the Associated Pile & Fitting Corporation of USA can be used. The splicer consists of an external collar which is slipped on to the upper end of the pile section already driven and is held in position by an internal lug. The next length of pile is then entered into the collar and driven down. The APF 'Champion' splicer is used for H-piles and consists of a pair of channel sections set on the head of the pile length already driven to act as a guide for placing and then welding-on the next length.

Steel tubular piles are the preferred shape when soil has to be cleaned out for subsequent placement of concrete, since there are no corners from which the soil may be difficult to dislodge by the cleaning-out. They are also preferred for marine structures where they can be fabricated and driven in large diameters to resist the lateral forces in deep-water structures. The circular shape is also advantageous in minimizing drag and oscillation from waves and currents (Sections 8.1.3 and 8.1.4). The hollow section of a tubular pile is also an advantage when inspecting a closed-end pile for buckling. A light can be lowered down the pile and if it remains visible when lowered to the bottom, no deviation has occurred. If a large deviation

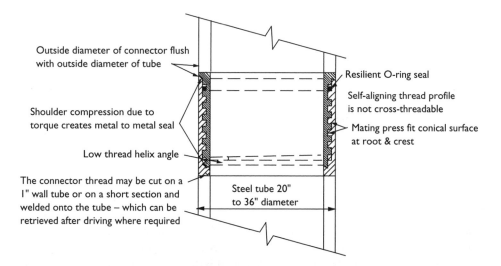

Figure 2.17 Schematic arrangement of Frank's 'Double Drive Shoulder Connector'.

is shown by complete or partial disappearance of the light, then measures can be taken to strengthen the buckled section by inserting a reinforcing cage and placing concrete.

Steel tubes are manufactured in Britain in standard outside diameters ranging up to 2 134 mm. The Japanese steel industry produces tubes in the standard range of 318.5 to 3 048 mm OD. Tubes for piles are manufactured as seamless, spirally welded, and longitudinally welded units. There is nothing to choose between the latter two types from the aspect of strength to resist driving stresses. In the spiral welding process the coiled steel strip is continuously unwound and spirally bent cold into the tubular. The joints are then welded from both sides. In the longitudinally welding process a steel plate is cut and bevelled to the required dimensions then pressed or rolled into tubular form and welded along the linear joints. The spiral method has the advantage that a number of different sizes can be formed on the same machine, but there is a limitation on the plate thickness that can be handled by particular machines. There is also some risk of weld 'unzipping' from the pile toe under hard driving conditions. This can be prevented by a circumferential shoe of a type described later. Piles driven in exposed deep water locations are fabricated from steel plate in thicknesses up to 62 mm by the longitudinal welding process. Special large-diameter piles can be manufactured by the process.

Economies in steel can be achieved by varying the wall thickness and quality of the steel. Thus in marine structures the upper part of the pile can be in mild steel which is desirable for welding on bracing and other attachments; the middle section can be in high-tensile steel with a thicker wall where bending moments are greatest, and the lower part, below sea bed, can be in a thinner mild steel or high-tensile steel depending on the severity of the driving conditions. The 1.3 m OD steel tubular piles used for breasting dolphins for the Abu Dhabi Marine Areas Ltd tanker berth at Das Island were designed by The British Petroleum Company to have an upper section of 24 mm in thickness, a middle section of 30 mm in thickness, and a lower section of 20 mm in thickness. The overall length was 36.6 m.

Light spirally welded mild steel tubular piles in the range of sizes and nominal working loads listed in Table 2.7 are widely used for lightly loaded structures, usually driven by a drop hammer acting on a plug of concrete in the bottom of the pile (see Section 3.2). These piles, known as 'cased piles', are designed to be filled with concrete after driving. Extension tubes can be welded to the driven length to increase penetration depth. Roger Bullivant Ltd provides thicker wall tubes for cased piles from 125 to 346 mm diameter with up to 10 mm wall section for top driving of the pile. If piles have to be spliced a special compression joint is needed for driving. Working loads claimed range from 350 to 1250 kN depending on ground conditions. In countries where heavy timbers are scarce, cased piles have to some extent replaced timber piling for temporary stagings in marine or river work. Here, the end of each pile is closed by a flat mild steel plate welded circumferentially to the pile wall.

Concrete-filled steel tubular piles need not be reinforced unless required to carry uplift or bending stresses which would overstress a plain concrete section cast in the lighter gauges of steel. Continuity steel is usually inserted at the top of the pile to connect with the ground beam or pile cap.

Steel box piles are fabricated by welding together trough-section sheet piles (Krupp, Hoesch, and Arcelor/Arbed combinations (CAZ, CAU)) or specially rolled trough plating (Arcelor/Arbed and Peine types). The Larssen and Frodingham sections are no longer produced by Corus. The types fabricated from sheet piles are useful for connection with sheet piling forming retaining walls, for example to form a wharf wall capable of carrying heavy compressive loads in addition to the normal earth pressure. However, if the piles rotate during driving there can be difficulty in making welded connections to the flats. Plain flat steel plates can also be welded together to form box piles of square or rectangular section.

The MV pile consists of either a steel box section (100 mm) or H-section fitted with an enlarged steel shoe to which a grout tube is attached. The H-pile is driven with a hammer or vibrator while grout is injected at the driving shoe. This forms a fluidized zone along the pile shaft and enables the pile to be driven to the deep penetration required for their principal use as anchors to retaining walls. The hardened grouted zone around the steel provides the necessary frictional resistance to enable them to perform as anchors.

H-section piles have a small volume displacement and are suitable for driving in groups at close centres in situations where it is desired to avoid substantial ground heave or lateral displacement. They can withstand hard driving and are useful for penetrating soils containing cemented layers and for punching into rock. Their small displacement makes them suitable for driving deeply into loose or medium dense sands without the 'tightening' of the ground that occurs with large displacement piles. They were used for this purpose for the Tay Road Bridge pier foundations, where it was desired to take the piles below a zone of deep scour on the bed of the Firth of Tay. Test piles 305 × 305 mm in section were driven to depths of up to 49 m entirely in loose becoming medium-dense to dense sands, gravels, cobbles and boulders, which is indicative of the penetrating ability of the H-pile.

The ability of these piles to be driven deeply into stiff to very stiff clays and dense sands and gravels on the site of the Hartlepools Nuclear Power Station is illustrated in Figure 2.18. On this site driving resistances of 355 × 368 mm H-piles were compared with those of precast concrete piles of similar overall dimensions. Both types of pile were driven by a Delmag D-22 diesel hammer (see Section 3.1.4). Although the driving resistances of both types were roughly the same to a depth of about 14 m (indicating that the ends of the H-piles were plugged solidly with clay), at this level the heads of the concrete piles commenced to spall and they could not be driven below 14.9 m, whereas the H-piles were driven on to 29 m without serious damage, even though a driving resistance of 0.5 mm/blow was encountered. Three of the H-piles were loaded to 3000 MN without failure but three of the precast concrete piles failed at test loads of between 1100 and 1500 MN.

Table 2.7 Dimensions and nominal working loads for typical concrete-filled cased piles using light-gauge tubes

Internal diameter (mm)	Area of concrete (mm)^2	Working load (kN) for ordinary soil[a]	Working load (kN) for rock, etc.[b]
254	50670	150	200
305	72960	300	350–450
356	99300	400	500–650
406	129700	500	600–850
457	164100	650	800–1000
508	202700	800	1000–1300
559	245200	1000	1250
610	291800	1200	1500

Notes
a Ordinary soil – sand, gravel, or very stiff clay.
b Rock, etc. – rock, very dense sand or gravel, very hard marl or hard shale.

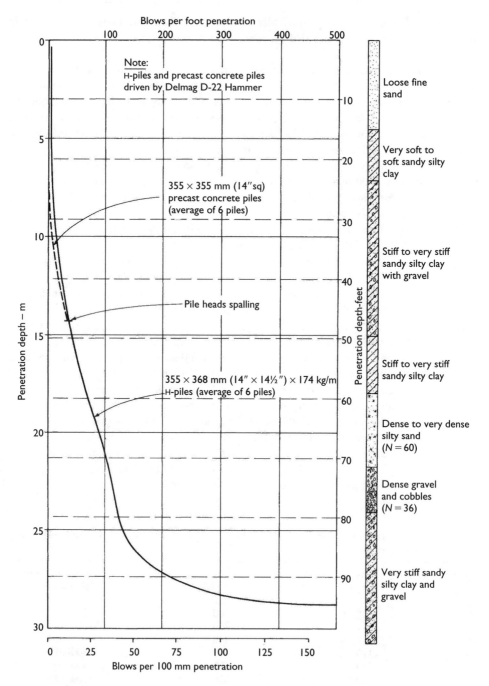

Figure 2.18 Comparison of driving resistance of 355 × 355 mm precast concrete piles and 355 × 368 mm H-section piles driven into glacial clays, sands, and gravels in Hartlepools Power Station.

Because of their relatively small cross-sectional area, H-piles cannot develop a high end-bearing resistance when terminated in soils or in weak or broken rocks. In Germany and Russia it is frequently the practice to weld short H-sections on to the flanges of the piles near their toes to form 'winged piles' (Figure 2.19a). These provide an increased cross-sectional area in end bearing without appreciably reducing their penetrating ability. The bearing capacity of tubular piles can be increased by welding T-sections on to their outer periphery when the increased capacity is provided by a combination of friction and end bearing on the T-sections (Figure 2.19b). This method was used to reduce the penetration depth of 1067 mm OD tubular steel piles used in the breasting dolphins of the Britoil Marine Terminal in Cromarty Firth. A trial pile was driven with an open end through 6.5 m of loose silty sand for a further 16 m into a dense silty sand with gravel and cobbles. The pile was driven by a Menck MRB 1000 single acting hammer with a 1.25 m drop of the 10 tonne ram. It will be seen from Figure 2.20 that there was only a gradual increase in driving resistance finishing with the low value of 39 blows/200 mm at 22.6 m penetration.

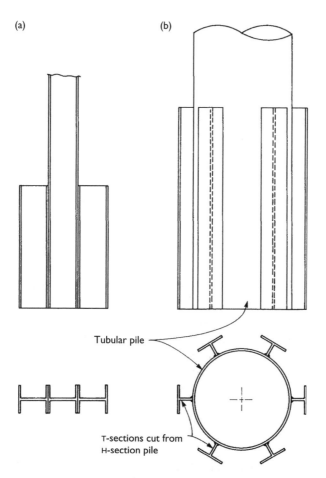

Figure 2.19 Increasing the bearing capacity of steel piles with welded-on wings (a) H-section wings welded to H-section pile (b) T-section wings welded to tubular pile.

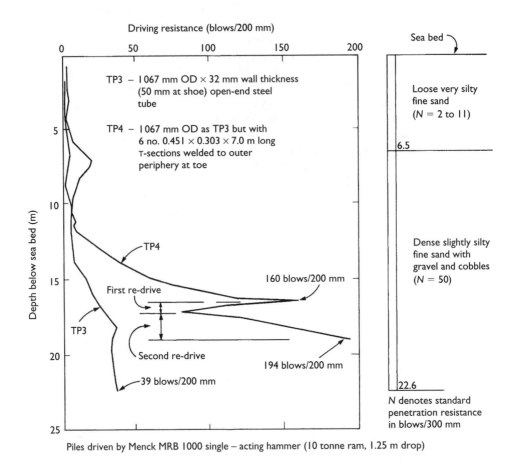

Figure 2.20 Comparison of driving resistance of open-end plain and winged tubular steel piles at Britoil Tanker Terminal, Cromarty Firth.

The pile was then cleaned out and plugged with concrete but failed under a test load of 6300 kN.

It was evident from the driving records that the plain piles showed little evidence of developing base resistance by plugging and would have had to be driven much deeper to obtain the required bearing capacity. In order to save the cost and time of welding-on additional lengths of pile it was decided to provide end enlargements in the form of six $0.451 \times 0.303 \times 7.0$ m long T-sections welded to the outer periphery in the pattern shown in Figure 2.19b. The marked increase in driving resistance of the trial pile is shown in Figure 2.20. The final resistance was approaching refusal at 194 blows/200 mm at 19 m below sea bed. The winged pile did not fail under the test load of 6300 kN.

A disadvantage of the H-pile is a tendency to bend about its weak axis during driving. The curvature may be sharp enough to cause failure of the pile in bending. Bjerrum[2.9] recommended that any H-pile having a radius of curvature of less than 366 m after driving should

(a)   (b)

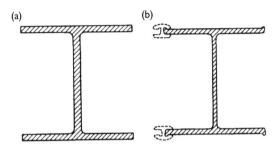

Figure 2.21 Types of H-section steel piles (a) Universal bearing pile (UK, European, and USA manufacture) (b) Peine pile (Hoesch).

be regarded as incapable of carrying load. A further complication arises when H-piles are driven in groups to an end bearing on a dense coarse-grained soil (sand and gravel) or weak rock. If the piles bend during driving so that they converge, there may be an excessive concentration of load at the toe and a failure in end bearing when the group is loaded. The authors observed a deviation of the toes of H-piles of about 500 mm after they had been driven only 13 m through sands and gravels to an end bearing on sandstone at Nigg Bay in Scotland.

The curvature of H-piles can be measured by welding a steel angle or channel to the web of the pile. After driving, an inclinometer is lowered down the square-shaped duct to measure the deviation from the axis of the pile. This method was used by Hanna[2.10] at Lambton Power Station, Ontario, where 305 and 355 mm H-piles that were driven through 46 m of clay into shale had deviated 1.8 to 2.1 m from the vertical with a minimum radius of curvature of 52 m. The piles failed under a test load, and the failure was attributed to plastic deformation of the pile shaft in the region of maximum curvature.

In the UK H-piles are rolled to BS 4 Part 1: 1993 and BS EN 10056 as universal bearing piles (Figure 2.21a). *Peine piles* are broad-flanged H-sections rolled by Hoesch. They are rolled with bulbs at the tips of the flanges (Figure 2.21b). Loose clutches ('locking bars') are used to interlock the piles into groups suitable for dolphins or fenders in marine structures. They can also be interlocked with the old Larssen sections to strengthen sheet-pile walls. The Arbed-HZ and PU (Arcelor) piles are of similar design.

The Monotube pile fabricated by the Monotube Pile Corporation of USA is a uniformly tapering hollow steel tube. It is formed from steel which is cold-worked to a fluted section having a tensile yield strength of 345 N/mm² or more. The strength of the fluted section is adequate for the piles to be driven from the top by hammer without an internal mandrel or concrete filling. The tubes have a standard tip diameter of 203 mm and the shaft diameter increases to 305, 356, 406 or 457 mm at rates of taper which can be varied to suit the required pile length. An upper section of uniform diameter can be fitted (Figure 2.22), which is advantageous for marine work where the fluted section has satisfactory strength and resilience for resisting wave forces and impact forces from small to medium-size ships. The tubes are fabricated in 3, 5, 7 and 9 gauge steel and taper lengths can be up to 23 m. The heavier gauges enable piles to be driven into soils containing obstructions without the tearing or buckling which can occur with thin steel shell piles.

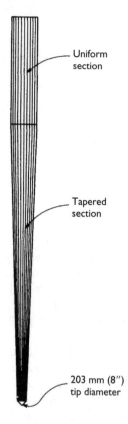

Uniform
section

Tapered
section

203 mm (8″)
tip diameter

Figure 2.22 Union Monotube pile (Union Metal Manufacturing Co.).

The '*Soilex*' system, developed in Sweden, uses the patented expander body to form an enlarged bulb to displace and compact the soil. The expander body consists of a thin folded sheet metal tube which, after insertion into the soil, is inflated by injecting concrete or grout under controlled pressure to form a bulb 5 to10 times the original diameter. Installation may be by conventional drilling, driving, jacking or vibration methods or placement in a pre-formed hole, the pile shaft geometry above the bulb being determined by the method of installation. The tube dimensions before expansion range from 70 to 110 mm square up to 3 m long which following inflation provides end-bearing areas of 0.12 to 0.5 m². In Borgasund, Sweden, fifty-seven 11 m long Soilex piles using a 110 mm expander body welded to 168 mm diameter thick wall tube were installed by a vibrator in a predrilled hole in medium dense sand below new railway bridge abutments. Approximately 0.5 m³ of concrete was used to inflate the expander body to form an 800 mm diameter bulb producing a pile which had an estimated ultimate load capacity of 1100 kN, limited by the strength of the concrete-infilled steel shaft. The system is also useful for underpinning where short piles are appropriate and as tension ground anchors; in all cases the spacing of piles is critical to avoid interference.

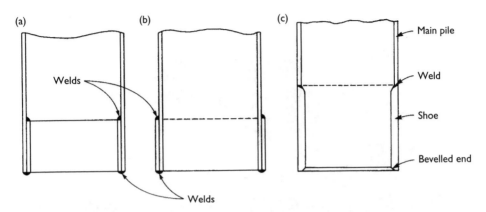

Figure 2.23 Strengthening toe of steel tubular piles (a) Internal stiffening ring (b) External stiffening ring (c) Thick plate shoe.

2.2.5 Shoes for steel piles

No shoes or other strengthening devices at the toe are needed for tubular piles driven with open ends in easy to moderately easy driving conditions. Where open-ended piles have to be driven through moderately resistant layers to obtain deeper penetrations, or where they have to be driven into weak rock, the toes should be strengthened by welding-on a steel ring. The internal ring (Figure 2.23a) may be used where it is necessary to develop the full external frictional resistance of the pile shaft. An external ring (Figure 2.23b) is useful for reducing the friction to enable end-bearing piles to be driven to a deep penetration, but the uplift resistance will be permanently reduced. Hard driving through strongly resistant layers or to seat a pile onto a rock may split or tear the ring shoe of the type shown in Figure 2.23a and b. For hard driving it is preferable to adopt a welded-on thick plate shoe designed so that the driving stresses are transferred to the parent pile over its full cross-sectional area (Figure 2.23c).

A shoe of this type can be stiffened further by cruciform steel plates (Figure 2.24a). Buckling and tearing of an external stiffening ring occurred when 610 mm OD steel tube piles were driven into the sloping surface of strong limestone bedrock (Figure 2.24b).

Steel box piles can be similarly stiffened by plating unless they have a heavy wall thickness such that no additional strengthening at the toe is necessary. Steel tubular or box piles designed to be driven with closed ends can have a flat mild steel plate welded to the toe (Figure 2.25a) when they are terminated in soils or weak rocks. The flat plate can be stiffened by vertical plates set in a cruciform pattern. Where they are driven on to a sloping hard rock surface, they can be provided with Oslo points as shown in Figure 2.25b.

Steel H-piles may have to be strengthened at the toe for situations where they are to be driven into strongly cemented soil layers, or soil containing cobbles and boulders. The strengthening may take the form of welding-on steel angles (Figure 2.26a), or purpose-made devices such as the 'Pruyn Point' manufactured in the USA by the Associated Pile and Fitting Corporation (Figure 2.26b) or the 'Strongshoe' and 'Jet shoe' manufactured in the UK by Dawson Construction Plant Ltd.

Figure 2.24 (a) Strengthening shoe of tubular steel pile by cruciform plates (b) Buckling and tearing of welded-on external stiffening ring to tubular steel pile driven on to sloping rock surface.

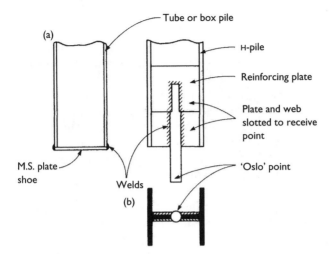

Figure 2.25 Shoes for steel piles (a) Flat plate shoe for tubular or box pile (b) 'Oslo' point for H-section pile.

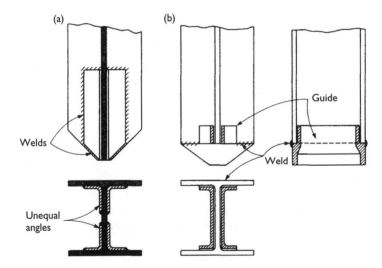

Figure 2.26 Strengthening toe of H-section pile (a) Welded-on steel angles (b) Pruyn Point (Associated Pile and Fitting Corporation).

2.2.6   *Working stresses for steel piles*

BS 8004 requires steel for piles to conform to BS 4360 Grades 43A, 50B, 'or other grades to the approval of the engineer'. The code limits the compression working stresses on tubular, box, and H-section piles, in steel conforming to BS 4360, to 30% of yield strength where the safety factor on driving resistance is not greater than 2. For jacked piles or where end-bearing piles are driven through relatively soft soils on to very dense granular soils or sound rock the limit is 50% of yield strength. Eurocodes are based on limit state design, hence BS EN 1993 Part 5: Piling (EC3–5) makes no reference to working stresses. BS EN 12699 allows for the peak calculated stress in steel piles during driving to be 0.9 times the characteristic yield strength of the steel; it is stated that this may be increased by 20% if the stresses are monitored during driving. BS 5950-1: 2000, the current code of practice for steelwork design in buildings simply states that foundation design should be in accordance with BS 8004 and the design strength of steel may be taken as the yield stress for steel thickness less than 16 mm. EC3 calls up a suite of EN codes for the production and composition of steel and manufacture of steel sections by hot rolling and cold forming which apply to bearing and sheet pile design. For example, unalloyed hot rolled structural steels, grades, designations and strengths are defined in BS EN 10025: 2004. The Steel Construction Institute's *H-Pile Design Guide*, 2005[2.11] is based on limit state design as provided in the Eurocodes, and therefore does not consider prescribed limits on steel working stresses. The *Guide* makes reference to the offshore industry's recommended practice for steel tubular piles based on North Sea experience as described in the ICP Design Methods for Driven Piles in Sands and Clays (see Chapter 4).

The American Petroleum Institute[3.5] states that the dynamic stresses during driving should not exceed 80–90% of yield strength depending on specific circumstances such as previous experience and confidence in the method of analysis.

The selection of a grade of steel for a particular task depends on the environmental conditions as well as on the design working stresses. For piles wholly embedded in the ground, or for piles in river and marine structures which are not subjected to severe impact forces, particularly in tropical or temperate waters, a mild steel conforming to Grade 43A (minimum yield strength 247 N/mm^2) or a high-tensile steel to Grade 50 (minimum yield strength 355 N/mm^2) should be satisfactory. The BS EN 10025-2: 2004 designated equivalent grades are S270GP (270 N/mm^2) and S355GP (355 N/mm^2) now normally used for bearing piles. Corus tubular sections suitable for general piling correspond to BS EN 10210 (hollow sections) grades S275J2H and S355J2H, both with Charpy impact values of 27J at $-20°$C, (i.e. T_{27J}), well above the BS 4360 requirement. Cold-formed, welded tubes in accordance with BS EN 10219 (hollow sections) for non-alloy steels have similar yield strength and impact designations. Steel sheet piles should conform to the yield strengths in BS EN 10248 for hot rolled sections or to BS EN 10249 for cold-formed.

Piles for deep-water platforms or berthing structures for large vessels are subjected to high dynamic stresses from berthing impact and wave forces. In water at zero or sub-zero temperatures, there is a risk of brittle fracture under dynamic loading, and the effects of fatigue damage under large numbers of load repetitions and also of salt water corrosion need to be considered. Steels must be selected to have a high impact value when tested at low temperatures. Corus produces a special steel tube for offshore applications, 335 NH 'Modified', with a yield strength of 355 N/mm^2 and mechanical and chemical properties superior to BS EN 10210 and BS 4360 grades 50E and 55E. The Charpy impact value is 60 J at $-50°$C. Piles or bracing members for deep-water structures may be required to be fabricated from plates 30 mm or more in thickness. The steel for such plates should have a brittle fracture resistance at low temperatures similar to the above impact values and note must be taken of the maximum thicknesses allowed in BS 5950 for each grade of steel at normal and lower temperatures. High-tensile alloy steel conforming to Grades 55C or E in BS 4360 or to grades above S460Q in BS EN 10137 High Yield Plates can meet these requirements.

Steel designations to BS EN 10025, 10210, and 10219 are defined in BS EN 10027-1 with the following nomenclature:

S	Structural steel	JR	Charpy impact value 27 J at +20°C
E	Engineering steel	J0	Charpy impact value 27 J at 0°C
275 or 355	Minimum yield strength in N/mm²	J2	Charpy impact value 27 J at −20°C
W	Improved atmospheric corrosion resistance	K2	Charpy impact value 40 J at −20°C
N	Normalized	N	Charpy impact value 40 J at >−20°C
		NL	Charpy impact value 27 J at >−50°C
Q	Quenched and tempered		
H	Hollow section		

Below are some steel grades for foundations as BS 4630 compared with nearest equivalent grades in BS EN 10025[2.12]:

For BS 8004 design BS 4360 grades	For BS EN 1993-5 (EC3-5) design Equivalent BS EN 10025-2: 2004 grades
40A	Equivalent not available
40B	Lowest grade available is S235JR in Part 2 of BS EN 10025 (non-alloy structural steels)
43A	Equivalent not available
43B	Lowest grade available is S270JR in Part 2
50	S355N (in Part 3 *normalized* fine grain structural steels)
50B	S355JR (in Part 2)
50EE	S355NL (in Part 3)
55C	S460N (in Part 3)
55EE	S460NL (in Part 3)

2.3   Driven and cast-in-place displacement piles

2.3.1   General

Driven and cast-in-place piles are installed by driving to the desired penetration a heavy-section steel tube with its end temporarily closed. A reinforcing cage is next placed in a tube which is filled with concrete. The tube is withdrawn while placing the concrete or after it has been placed. In other types of pile, thin steel shells or precast concrete shells are driven by means of an internal mandrel, and concrete, with or without reinforcement, is placed in the permanent shells after withdrawing the mandrel.

Driven and cast-in-place piles have the principal advantage of being readily adjustable in length to suit the desired depth of penetration. Thus in the withdrawable-tube types the tube is driven only to the depth required by the ground conditions. In the shell types, the length of the pile can be easily adjusted by adding or taking away the short units which make up the complete shell. In conditions favourable for their employment, where the required penetration depth is within the capability of the piling rig to pull out the tube, and there are no restrictions on ground heave or vibrations, withdrawable tube piles can be installed more cheaply than any other type of driven or bored pile for comparable capacities. They also have the advantage, which is not enjoyed by all types of shell pile, of allowing an enlarged base to be formed at the toe. However, some codes of practice, notably that of New York City, forbid the use of a wholly uncased shaft for all forms of driven and cast-in-place pile, where these are installed in soft to firm clays or in loose to medium-dense sands and materials such as uncompacted fill. These restrictions are the result of unfortunate experiences resulting from lifting of the concrete while pulling out the driving tube, and of squeezing ('waisting') the unset concrete in the pile shaft where this is formed in soft clays or peat. The firms that install these proprietary types of pile have adopted various techniques for avoiding these troubles, such as inserting permanent light-gauge steel shells before placing the concrete. However, such expedients increase the cost of the withdrawable-tube piles to the extent that their advantage in price over shell piles may be wholly or partially lost. The soundness of the uncased type of pile depends on the skill and integrity of the operatives manning the piling rig.

Piling rigs have not yet been developed to install driven and cast-in-place piles of the very large diameters which are possible with driven thick-walled steel tubes or bored and cast-in-place piles. Thus the working loads are limited to the light to medium range. Also the withdrawable-tube or thin-shell types are unsuitable for marine structures, but they can be employed in marine situations if they are extended above the sea bed as columns or piers in steel or precast concrete.

Problems associated with ground heave when installing driven and cast-in-place piles in groups are discussed in Section 5.7.

2.3.2   Withdrawable-tube types

Descriptions of the various types of driven and cast-in-place piles are given in CIRIA report *A Review of bearing pile types*[2.13]. The methods of installing these piles are essentially the same. A piling rig consisting of a mast, leaders and winch on a track or roller-mounted frame (see Section 3.1) is used to support and guide the withdrawable tube. The original *Franki* pile is now mainly available in Australia and Southeast Asia, for pile working loads

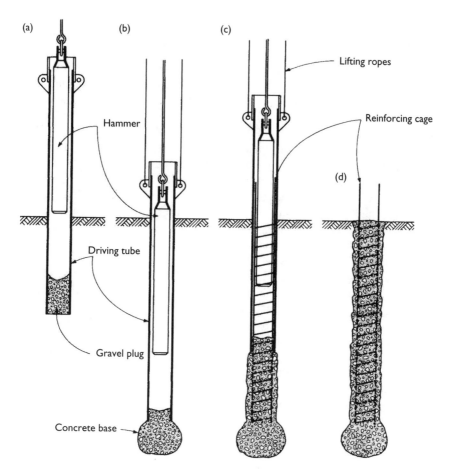

Figure 2.27 Stages in installing an open-ended Franki pile (a) Driving piling tube (b) Placing concrete in piling tube (c) Compacting concrete in shaft (d) Completed pile.

from 250 to 1500 kN (subject to ground conditions). This employs an internal drop hammer acting on a plug of gravel at the bottom of the drive tube. The drive tube (248 to 600 mm diameter with a wall thickness of 32 mm) is carried down with the plug until the required toe level is reached when the tube is restrained from further penetration by rope tackle. Then the gravel plug and batches of dry concrete are hammered out to form a bulb or enlarged base to the pile. The reinforcing cage is then inserted, followed by placing a semi-dry (no-slump) concrete in batches as the drive tube is pulled out in stages. After each stage of withdrawal the concrete is compacted by the internal hammer (Figure 2.27). The operations of driving by internal hammer and concreting in stages are slower than the top driving method described above. Hence, these techniques are used only when there are economic advantages, for example, when the enlarged base adds appreciably to the bearing capacity of the pile.

In the Cementation Foundations Skanska version of the withdrawal tube pile, the heavy wall section tube has its lower end closed by an expendable steel plate or shoe and is driven from the top by a five-tonne hydraulic hammer. On reaching the required toe level, as predetermined by calculation or as determined by measurements of driving resistance, the hammer is lifted off and a reinforcing cage is lowered down the full length of the tube. A highly workable self-compacting concrete is then placed in the tube through a hopper, followed by raising the tube by a hoist rope operated from the pile frame. The tube may be filled completely with concrete before it is lifted or it may be lifted in stages depending on the risks of the concrete jamming in the tube. The length of the pile is limited by the ability of the rig to pull out the drive tube. This restricts the length to about 20 to 30 m. Pile diameters range from 285 to 525 mm with working loads up to 1500 kN.

In a further variation of the Franki technique, the gravel plug (or dry concrete plug) can be hammered out at several intermediate stages of driving to form a shell of compact material around the pile shaft. This technique is used in very soft clays which are liable to squeeze inwards when withdrawing the tube. Composite Franki piles are formed by inserting a precast concrete pile or steel tube into the driving tube and anchoring it to the base concrete plug by light hammer blows. The drive tube is then withdrawn.

A full-length reinforcing cage is always advisable in the driven and cast-in-place pile. It acts as a useful tell-tale against possible breaks in the integrity of the pile shaft caused by arching and lifting of the concrete as the tube is withdrawn. BS EN 12699 requires minimum reinforcement of 0.5% of the pile cross-section or four 12 mm diameter bars over the top 4 m of all such piles; with minimum cover of 50 mm where the casing is withdrawn, 75 mm where reinforcement is installed after concreting (or where subject to ground contaminants), and 40 mm where there is permanent lining.

The problem of inward squeezing of soft clays and peats or of bulging of the shafts of piles from the pressure of fluid concrete in these soils is common to cast-in-place piles both of the driven and bored types. A method of overcoming this problem is to use a permanent light gauge steel lining tube to the pile shaft. However, great care is needed in withdrawing the drive tube to prevent the permanent liner being lifted with the tube. Even a small amount of lifting can cause transverse cracks in the pile shaft of sufficient width to result in excessive settlement of the pile head under the working load. The problem is particularly difficult in long piles when the flexible lining tube tends to snake and jam in the drive tube. Also where piles are driven in large groups, ground heave can lift the lining tubes off their seating on the unlined portion of the shaft. Snaking and jamming of the permanent liner can be avoided by using spacers such as rings of sponge rubber.

In most cases the annulus left outside the permanent liner after pulling the drive tube will not close up. Hence, there will be no frictional resistance available on the lined portion. This can be advantageous because dragdown forces in the zone of highly compressible soils and fill materials will be greatly reduced. However, the ability of the pile shaft to carry the working load as a column without lateral support below the pile cap should be checked. Problems concerned with the installation of driven and cast-in-place piles are discussed further in Section 3.4.5.

Allowable stresses on the shafts of these piles are influenced by the need to use easily workable self-compacting mixes with a slump in the range of 130 to 180 mm and to make allowances for possible imperfections in the concrete placed in unseen conditions. BS EN 12699 for driven displacement piles requires the rules on the concreting of bored piles (see Section 2.4.2) using self-compacting concrete as recommended in BS EN 1536 to apply to all cast-in-place displacement piles unless otherwise specified. BS 8004 limits the working stress to 25% of the 28-day cube strengths, but BS EN 12699 specifies concrete strength classes of C20/25 to C30/37 which are 25% stronger than the *cube* strengths usually adopted in the UK under BS 8004, that is, a range of 20–30 N/mm². EC2-1-1 Clause 3 refers to *characteristic cylinder strengths* for the determination of design compressive strengths, and if the 25% limit is applied the allowable stresses range from 5 to 7.5 N/mm² (i.e. similar to the BS 8004 limits, but for the stronger mixes). For these values, allowable loads for piles of various shaft diameters are as shown in the following table:

Nominal shaft diameter (mm)	Allowable working load (kN)
300	350–500
350	450–700
400	600–900
450	800–1000
500	1000–1400
600	1400–2000

The higher ranges in the above table should be adopted with caution, particularly in difficult ground conditions.

Maximum working loads are as shown in the following table:

Nominal shaft diameter (mm)	Nominal maximum working load (kN)
350	440
400	590
450	740
500	930
550	980
600	1500
715	2000

The spacing of bars in the reinforcing cage should give ample space for the flow of concrete through them. Bars of 5 mm diameter in the form of a spiral or flat steel hoops used for lateral reinforcement should not be spaced at centres closer than 100 mm (80 mm when using <20 mm aggregate).

The *Vibrex pile* installed by Fundex Verstraeten BV employs a diesel or hydraulic hammer to drive the tube which is closed at the end by a loose sacrificial plate. An external ring vibrator is then employed to extract the tube after the reinforcement cage and concrete has been placed. A variation of the technique allows an enlarged base to be formed by using the hammer to drive out a charge of concrete at the lower end of the pile. The Vibrex pile is formed in shaft diameters from 350 to 600 mm.

The *Fundex pile* also installed by Fundex group (versions of this and the other types are available from American Piledriving Inc in the USA) is a form of screwpile (see Section 2.3.5). A helically screwed drill point is held by a bayonet joint to the lower end of the piling tube. The latter is then rotated by a hydraulic motor on the piling frame and at the same time forced down by hydraulic rams. On reaching founding level, a reinforcing cage and concrete are placed in the tube which is then withdrawn leaving the sacrificial drill point in the soil. This limits the disturbance at the pile base. The *Tubex pile* also employs the screwed drill point, but the tubes are left in place for use in very soft clays when 'waisting' of the shaft must be avoided. The tube can be drilled down in short lengths, each length being welded to the one already in place. Thus the pile is suitable for installation in conditions of low headroom, for example, for underpinning work. This pile can also be installed with simultaneous grout injection which leaves a skin of grout around the tube and increases bearing capacity.

The speciality of the *Vibro pile* (from Terracon in the Netherlands) is the method used to compact the concrete in the shaft by alternate upward and downward blows of a hammer on the driving tube. The upward blow of the hammer operates on links attached to lugs on top of the tube. This raises the tube and allows concrete to flow out. On the downward blow the concrete is compacted against the soil. The blows are made in rapid succession which keeps the concrete 'alive' and prevents its jamming in the tube.

2.3.3 Shell types

Types employing a *metal shell* generally consist of a permanent light gauge steel tube in diameters from 150 to 500 mm with wall thickness up to 6 mm and are internally bottom driven by a drop hammer acting on a plug of dry concrete (care being taken not to burst the tube). The larger diameter tubes are usually fabricated to the estimated length and handled into a piling frame with a crane. Smaller diameter, spirally welded tube can be manually placed on the rig leader and welded in sections to produce the required depth during installation. On reaching the bearing layer the hammer is removed, any reinforcement inserted, and a high slump concrete placed to produce the pile. Working loads up to 1 200 kN are possible.

In France cased piles varying in diameter from 150 to 500 mm are installed by welding a steel plate to the base of the tubular section to project at least 40 mm beyond the outer face of the steel. As the pile is driven down, a cement/sand mortar with a minimum cement content of 500 kg/m^3 is injected into the annulus formed around the pile by the projecting plate through one or more pipes having their outlet a short distance above the end plate. The rate of injection of the mortar is adjusted by observing the flow of mortar from the annulus at the ground surface. The working load is designed to be carried by the steel section. The working stress permitted of 160 N/mm^2 is higher than the value normally accepted for steel piles using EN24-1 steel, because of the protection given to the steel by the surrounding mortar. Steel H or box sections can be given mortar protection in a similar manner.

The well-known Raymond 'Step-Taper' Piles which consisted of helically corrugated light-steel shells are no longer available. The *'TaperTube'* pile (Figure 2.28), a steel shell similar to the Monotube but without the flutes, has been developed by DFP Foundation Products and Underpinning & Foundation Constructors of the USA. It uses a heavier wall thickness of 9.5 mm in 247 N/mm^2 grade hot-rolled steel to form a 12-sided polygon tapering from 609 to 203 mm at the cast steel point over lengths of 3 to 10 m. Where tube extensions are needed the top of the polygon can be formed into a circle for butt welding; this provides

Figure 2.28 The *TaperTube* pile.

improved axial uplift resistance. After top driving is completed the tapered shell pile is filled with concrete. Ultimate bearing capacities up to 4000 kN and lateral resistance to 200 kN have been determined in pile tests.

The *West* shell pile is no longer available in the UK, and this type of short precast cylindrical concrete shell has generally gone out of favour with the development of improved precast pile joints and CFA piling techniques.

2.3.4   Working stresses on driven and cast-in-place piles

It can be seen from the above brief descriptions that driven and cast-in-place piles encompass a wide variety of shapes, combinations of materials and installation methods. A common feature of nearly all types is an interior filling of concrete placed in-situ, which forms the main load-carrying component of the pile. Whether or not any load is allowed to be carried by the steel shell depends on its thickness and on the possibilities of corrosion or tearing of the shell. As noted in Section 2.3.2, BS 8004 limits the working stress in the concrete to 25% of the characteristic cube strength at 28 days with a minimum cement content of 300 kg/m^3. While the required strength classes in BS EN 12699 are apparently higher than BS 8004, the working stresses are lower than for precast concrete piles to take account of possible deficiencies in workmanship during placing the concrete, or reductions in section of the pile shaft due to 'waisting' or buckling of the shells. When semi-dry concrete is tamped during installation the concrete class should be at least C25/30 with a minimum cement content of 350 kg/m^3.

Where steel tubes or sections are used as part of the load carrying capability or reinforcement of the pile, BS EN 12699 requires Eurocode 4 (EC4) BS EN 1994: 2004 Design of Composite Steel and Concrete Structures Part 1-1 General rules to be applied.

Steel shell piles ('pipe piles') are more widely used in the USA than elsewhere and most of the American codes require the shells to be at least 2.5 mm thick before they can be permitted to carry a proportion of the load. Frequently a wall thickness of 3 mm is required.

2.3.5   Rotary displacement auger piles

Displacement auger piles and *screw piles* are drilled piles, but the soil is displaced and compacted as the auger head is rotated into the ground to form the stable pile shaft, with little soil being removed from the hole. The methods were mainly developed in the 1960s in Belgium from continuous flight auger (CFA) techniques (see Section 2.4.2) and are now widely available. The original system is the cast-in-place '*Atlas*' pile in which the special single flight auger head is screwed and jacked into the ground on a thick-walled steel tube using a specially designed rotary high torque rig. The helical shape of the pile shaft produced by screwing in the auger flange is maintained as the auger is back-screwed to form a stable hole into which the reinforcement cage is placed prior to concreting. Other proprietary displacement piles such as the '*ScrewSol*' pile by Bachy Soletanche (Figure 2.29a and b) which produces a helical flanged pile shaft in weak soils and the '*Spire*' system by SEFI in France for a straight shaft also use specially shaped augers on the end of the drill tube to compact the soil and inject concrete. A helical 'threaded' shaft is also produced by Bachy Soletanche with a reduced pitch and shorter flange for use in London Clay. The amount of reinforcement which can be inserted is limited. The use of the thin-flanged hollow stem CFA augers in short lengths to form the shaft helix has not been successful.

(a)

(b)

Figure 2.29 (a) The *ScrewSol* tapered auger and tight-fit follower tube (b) Cleaned-off section of an excavated *ScrewSol* pile.

Rigs are similar to the high torque, instrumented CFA pile units, but the power required to install screw piles can be 20% greater than that required for equivalent CFA piles; additional pull-down is usually necessary. As only a small amount of material is removed as the auger is initially inserted, the screw pile is particularly useful for foundations in contaminated ground. The method of concreting is either by injection through the auger tip during rotation out of the hole, which can improve shaft friction, or by tremie depending on the system.

Design of displacement screw piles should be based on a detailed knowledge of the ground using pressuremeter tests, CPTs and SPTs, and pile test data in the particular soil. Care is required in selecting the effective diameter of the helical shaft for determination of shaft friction and end-bearing capacity. Bustamante and Gianeselli[2.14] have provided a useful simplified method of predetermining the carrying capacity of helical shaft piles based on a series of tests and recommend that a design diameter of 0.9 times the outside diameter of the auger flange should be used for calculating both base and shaft resistance for 'thin' flanges. For thick flanges (say 40 mm deep 75 mm wide), the outside diameter of the helix is appropriate. Depending on the ground conditions and the size of the helical flanges formed, savings of 30% in concrete volume compared with the equivalent bored pile are claimed. Typical pile dimensions are 500 mm outside auger diameter and 350 mm shaft diameter and lengths of 30 m are possible. The technique is best suited to silty sands and sandy gravels with SPT N-values between 10 and 30; above $N = 50$ there is likely to be refusal with currently available rigs, and unacceptable heave and shearing will occur in most clays.

An enlarged pile base can be formed with straight shafted displacement piles, such as the *Penpile* by Pennine Vibropile Ltd, in granular soils and weak chalk to improve end-bearing capacity.

Guidance on installation of displacement screw piles in BS EN 12699 is limited, but comprehensive trials of different types of pile at Limelette[2.15] in Belgium during 2000 and 2002 in stiff dense sand, together with earlier trials in stiff clay, have produced significant data on design, installation and performance of screw piles (including references to EC7 design procedures and CPT testing). Two main conclusions were that the bearing capacity is of similar magnitude as that for full displacement piles, and the prediction of bearing capacity was in good agreement with load tests, irrespective of the method used.

Solid steel shaft helical screw piles are used mainly in the USA for foundations in collapsible and expansive soils as described by Black and Pack[2.16]. The central shaft up to 57 mm^2 has one or more circular steel plates up to 350 mm diameter, shaped into a single helix, welded to it. As the shaft is rotated into the soil, the leading edge of the helix bites into the soil, transferring the rotational force into axial thrust; extension shafts with plates are added as needed. Reduced drag-down and ultimate load-carrying capacities of up to 890 kN in compression or tension are claimed in these soil conditions. In addition to selecting soil parameters, care is needed during design and installation to consider the effects of groundwater around the shaft, corrosion and buckling. (See also Section 9.2.2 for use in underpinning.)

2.4   Replacement piles

2.4.1   General

Replacement piles are installed by first removing the soil by a drilling process and then constructing the pile by placing concrete or some other structural element in the drilled hole.

The simplest form of construction consists of drilling an unlined hole and filling it with concrete. However, complications may arise such as difficult ground conditions, the presence of groundwater, or restricted access. Such complications have led to the development of specialist piling plant for drilling holes and handling lining tubes, but unlike the driven and cast-in-place piles, very few proprietary piling systems have been promoted. This is because the specialist drilling machines are available on sale or hire to any organization which may have occasion to use them. The resulting pile as formed in the ground is more or less the same no matter which machine, or method of using the machine, is employed. There have been proprietary systems such as the Prestcore pile, which incorporates precast units installed in the pile borehole, but these methods are largely obsolete.

There are two principal types of replacement pile. These are bored and cast-in-place piles, and drilled-in tubular (including caisson) piles. A general description of the two types now follows. Mechanical plant for installing the piles and methods of construction are described in Section 3.3.

2.4.2 Bored and cast-in-place piles

In stable ground an unlined hole can be drilled by hand or mechanical auger. If reinforcement is required, a light cage is then placed in the hole, followed by the concrete. In loose or water-bearing soils and in broken rocks casing is needed to support the sides of the bore-hole, this casing being withdrawn during or after placing the concrete. In stiff to hard clays and in weak rocks an enlarged base can be formed to increase the end-bearing resistance of the piles (Figure 2.30). The enlargement is formed by a rotating expanding tool. Hand excavation is now uneconomic because of stringent statutory health and safety regulations, even in piles with a large shaft diameter. A sufficient cover of stable fine-grained soil must be left over the top of the enlargement in order to avoid a 'run' of loose or weak soil into the unlined cavity, as shown in Figure 2.30.

Bored piles drilled by hand auger are limited in diameter to about 355 mm and in depth to about 5 m. They can be used for light buildings such as dwelling houses, but even for these light structures hand methods are used only in situations where mechanical augers, as described in Section 3.3.1, are not available.

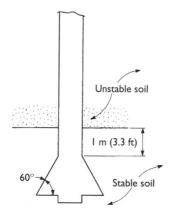

Figure 2.30 Under-reamed base enlargement to a bored and cast-in-place pile.

The versatile, light cable percussion tripod rigs can bore piles up to 600 mm diameter 10 m deep with working loads up to 1200 kN in suitable ground conditions. Temporary casing can be driven to cut off unstable ground and reinforcement inserted prior to concreting.

Bored piles drilled by mechanical spiral-plate or bucket augers or by grabbing rigs can drill piles with a shaft diameter up to 7.3 m. Standard plate auger boring tools for use with kelly bar rigs (see Section 3.3.4) range from 600 to 3650 mm. Rigs with telescopic kelly bars can reach 70 m depth and 102 m exceptionally. Under-reaming tools can form enlarged bases in stable soils up to 7.3 m in diameter. Rotary drilling equipment consisting of drill heads with multiple rock roller bits have been manufactured for drilling shafts up to 8 m in diameter.

In a stable dry bore, concreting is carried out from a hopper over the pile with a short length of pipe to direct flow into the centre of the reinforcement, ensuring that segregation does not occur. When concreting boreholes under flooded conditions or under stabilizing fluid a full length tremie pipe (6 times the maximum diameter of the aggregate or 150 mm diameter whichever is the greater) is essential. For reasons of economy and the need to develop shaft friction, it is the normal practice to withdraw the casing during or after placing the concrete. As in the case of driven and cast-in-place piles this procedure requires care and conscientious workmanship by the operatives in order to prevent the concrete being lifted by the casing, and resulting in voids in the shaft or inclusions of collapsed soil.

The shafts of bored and cast-in-place piles are liable to 'necking' or 'waisting' in soft clays or peats. Sometimes a permanent casing of light spirally welded metal is provided over the portion of the shaft within these soil types, but this measure can cause problems in installation (see Section 3.4.6). EC2-1-1 Clause 2.3.4 recommends tolerances in cross-sectional dimensions for cast-in-place piles (see Chapter 4). The design diameter for bored piles with helical flanged shafts should be determined as noted in Section 2.3.5.

Reinforcement is not always needed in bored and cast-in-place piles unless uplift loads are to be carried (uplift may occur due to the swelling and shrinkage of clays). Reinforcement may also be needed in the upper part of the shaft to withstand bending moments caused by any eccentricity in the application of the load, or by bending moments transmitted from the ground beams (Section 7.9). However, it is often a wise precaution to use a full-length reinforcing cage in piles where temporary support by casing is required over the whole pile depth. As noted in Section 2.3.2, the cage acts as a warning against the concrete lifting as the casing is extracted. The need to allow ample space between the bars for the flow of concrete is again emphasized. EC2-1-1 Clause 4 specifies minimum cover in respect of environmental conditions and BS EN 1536 requires 60 mm cover for piles greater than 600 mm diameter and 50 mm for piles less than 600 mm. Where reinforcement is required, BS EN 1536 follows EC2-1-1 Clause 9.8.5 rules for longitudinal reinforcement areas for bored piles depending on the cross-sectional area.

Pile cross-section: A_c	Minimum area of longitudinal reinforcement: A_s
$A_c \leq 0.5 \text{ m}^2$	$A_s \geq 0.005\, A_c$
$0.5 \text{ m}^2 < A_c \leq 1.0 \text{ m}^2$	$A_s \geq 25 \text{ cm}^2$
$A_c > 1.0 \text{ m}^2$	$A_s \geq 0.0025\, A_c$

Transverse reinforcement is also specified in BS EN 1536. Un-reinforced bored piles can be considered as Clause 12 of EC2-1-1, subject to serviceability and durability requirements, but BS EN 1536 requires minimum longitudinal reinforcement of four 12 mm diameter bars, unless the design demonstrates otherwise.

Over 1100 large diameter bored piles were installed at Canary Wharf by Bachy Soletanche in London Docklands ranging from 900 to 1500 mm and to depths of 30 m through terrace gravels, Lambeth clays, sands and gravels, and Thanet sands. It was possible to bore the piles without the aid of drilling fluids due to the low water table in the Thanet beds. Once the piles had reached the required depth using temporary casing, the shaft was filled with bentonite slurry to minimize the risk of pile collapse during concreting operations. The reinforcement cage was inserted to which were attached tubes-à-manchette for pile base grouting two days after concreting.

When using bentonite or other drilling fluids to support the sides of boreholes or diaphragm walls, the bond of the reinforcement to the concrete may be affected. Research by Jones and Holt[2.17] comparing the bond stresses in reinforcement placed under bentonite and polymer fluids indicated that it is acceptable to use the BS 8110 values of ultimate bond stress provided that the cover to the bar is at least twice its diameter when using deformed bars under bentonite. The results for the polymers investigated showed that the code bond stresses could be reduced by a divisor of 1.4. EC2-1-1 Clause 4 includes for a minimum cover factor dependent on bond requirements and Section 8 gives a reduction factor of 0.7 to apply to the ultimate bond stress where 'good' bond conditions do not exist – compatible with the Jones and Holt data for polymers. It also covers laps between bars using the reduced bond stress as appropriate. BS EN 1536 states that only ribbed bars shall be used for main reinforcement where a stabilizing fluid, bentonite or polymer, is used.

It is easier to remove drill cuttings from polymer stabilizing fluids for reuse compared with bentonite slurries. They are also better suited for drilling large diameter piles and shafts where the hole has to be stabilized for up to 36 hours of drilling time. The filter cake formation on the sides of the hole is limited and the sides do not soften to the same extent as with bentonite slurry support.

Barrettes can be an alternative to large diameter bored and cast-in-place piles where in addition to vertical loads, high lateral loads, or bending moments have to be resisted. They are constructed using diaphragm wall techniques to form short discrete lengths of rectangular wall, and interconnected Ell-, Tee-shapes and cruciforms to suit the loading conditions in a wide variety of soils and rock to considerable depths. The 'Hydrofraise' reverse circulation rig (see Section 3.3.6) is particularly well adapted to form barrettes, as verticality is accurately controlled and the time for construction is reduced compared with grab rigs thereby avoiding the potential for the excavation to collapse. Barrettes are usually only economical when the rig is mobilized for the construction of other basement walls.

Continuous flight auger or *auger-injected piles*, generally known as CFA piles, are installed by drilling with a rotary continuous-flight auger to the required depth. In stable ground above the water table the auger can be removed and a high slump concrete pumped through a flexible pressure hose that has been fed down to the bottom of the unlined hole. This type of pile is referred to as cast-in-place. In unstable or water-bearing soils a flight auger is used with a hollow stem temporarily closed at the bottom by a plug. After reaching the final level a high slump concrete is pumped down the hollow stem and, once sufficient pressure has built up, the auger is withdrawn at a controlled rate, removing the soil and forming a shaft of fluid concrete extending to ground level (Figure 2.31). Thus the walls of the borehole

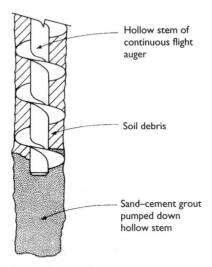

Figure 2.31 Pumping grout to form an auger-injected pile.

are continually supported by the spiral flights and the soil within them, and by the concrete. Reinforcing steel can be pushed into the fluid concrete to a depth of about 12 m. Exceptionally, reinforcing cages up to 17 m long were pushed down into the 30 m long piles for the foundations of the approach viaducts to the Dartford Bridge. Vibrators may be used to assist penetration. The shaft diameters range from the minipile sections (about 100 mm in which sand–cement grout may be injected in place of concrete) up to 1.5 m exceptionally. Concrete is usually mixed with a plasticizer to improve its 'pumpability', and an expanding agent is used in grout to counter the shrinkage while it is setting and hardening. Pile capacities up to 7 500 kN are possible depending on ground conditions and pile dimensions.

In granular soils a hollow-stem auger can be used in conjunction with wing drill bits to mix the soil in place with a cement grout pumped down the stem. This process is well developed for encapsulating contaminants in landfill.

The CFA pile has considerable advantages over the conventional bored pile in water-bearing and unstable soils. Temporary casing is not needed, and the problems of concreting under-water are avoided. The drilling operations are quiet and vibrations are very low making the method suitable for urban locations. However, in spite of these considerable advantages the CFA pile depends for its integrity and load-bearing capacity, as much as any other in-situ type of pile, on strict control of workmanship. This is particularly required where a high proportion of the load is carried in end-bearing. Because it is not possible to check the stratification and quality of the soil during installation as with conventional bored piles, considerable research and development has been undertaken by piling companies into the use of computerized instrumentation to monitor the process and ensure the quality and integrity of CFA piles. For example, a computer screen is positioned in the drilling rig cab in front of the operator which continuously displays the boring and concreting parameters. During the boring operation the depth of auger, torque applied, speed of rotation, and penetration rate are displayed. During concreting a continuous record of concrete pumping pressure and flow rate is shown, and on completion the results are provided on a printout of the pile log which records the construction parameters and under- or over-supply of concrete (Figure 2.32). Most specifications for CFA piles require the rig to be provided with such instrumentation, although, because of anomalies which inevitably exist in ground conditions, some authorities require all CFA piles to be tested by non-destructive integrity tests. Regular calibration of the instrumentation is essential[2.5]. In certain ground conditions doubts may exist as to whether or not the injected material has flowed-out to a sufficient extent to cover the whole drilled area at the pile toe. For this reason it may be advisable either to assume a base diameter smaller than that of the shaft or to adopt a conservative value for the allowable end-bearing pressure. In addition, 'polishing' of the shaft can occur in stiff clays due to over-rotation and 'flighting' (i.e. vertical movement of the soil on the auger relative to the soil on the wall of the borehole) in loose silty sands where over-rotation disturbs the surrounding soil and can reduce shaft resistance by 30%.

The instrumentation systems which have been successfully used to record the pile installation are now being applied to control the CFA process, taking some of the decision-making away from the operator in the cab, particularly to ensure that the target volume concrete is achieved throughout the pile length during withdrawal.

The CFA pile is best suited for ground conditions where the majority of the working load is carried by shaft friction, and the ground is free from large cobbles and boulders. The standard CFA system may have difficulty in penetrating stiff clayey soils and glacial till, with 'refusal' encountered before reaching the design depth and problems of flighting, shaft waisting and discontinuities occurring. Bustamante *et al.*[2.18] have shown that the *double rotary* CFA system can overcome such conditions by installing a temporary casing using a second rotary head on the rig while simultaneously drilling in the auger. The results indicated that stiff marl could be effectively penetrated, the verticality was better controlled, and the overall performance was similar to conventional bored and CFA piles. The shaft friction capacity of CFA piles in chalk has been assessed by Lord *et al.*[2.19] It is considered that there should be little difficulty in forming satisfactory CFA piles in better quality structured chalk, but in chalks with low penetration resistance there may be problems of softening and hole instability, particularly below water table.

Further information on installation and monitoring of CFA piles is given in a paper by Fleming[2.20].

Concrete materials and mix proportions for cast-in-place piles generally are specified in BS EN 1536 and strength grades should range between C20/25 and C30/37. Cement contents equal to or greater than 325 kg/m^3 are required for placement in dry conditions and equal to or greater than 375 kg/m^3 in submerged conditions; water/cement ratios are specified to be less than 0.6 and have good flow and self-compaction properties. As noted in Section 2.3.2, these mixes are stronger than BS 8004 requirements which limits the working stress in the concrete to 25% of the characteristic cube strength at 28 days. Structural design stresses in EC7 are specified to conform to EC2, EC3, and EC5 for the relevant material; for example, EC2-1-1 Clause 3 defines the *ultimate* design compressive stress of concrete in piles as the characteristic *cylinder* strength divided by a partial factor of 1.5 × 1.1.

2.4.3   Drilled-in tubular piles

The essential feature of the drilled-in tubular pile is the use of a tube with a medium to thick wall, which is capable of being rotated into the ground to the desired level and is left

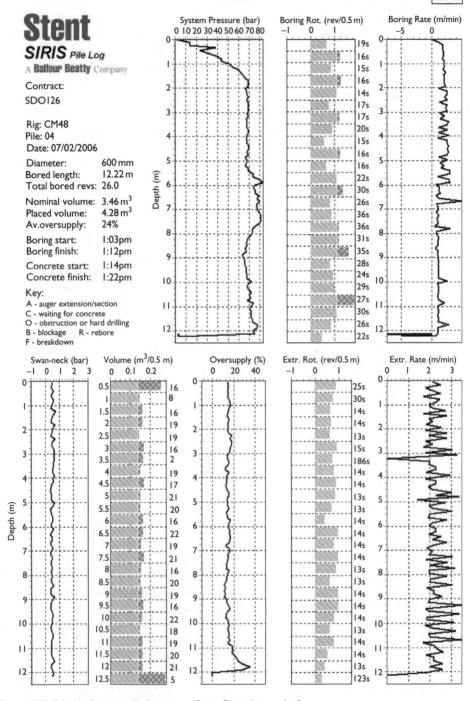

Figure 2.32 Pile log for CFA pile (courtesy Stent Foundations Ltd).

permanently in the ground with or without an in-filling of concrete. Soil is removed from within the tube as it is rotated down, by various methods including grabbing, augering and reverse circulation, as described in Section 3.3.5. The tube can be continuously rotated by a hydraulically powered rotary table or be given a semi-rotary motion by means of a casing oscillator.

The drilled-in tubular pile is a useful method for penetrating ground containing boulders or other obstructions, heavy chisels being used to aid drilling. It is also used for founding in hard formations, where a 'rock socket' capable of resisting uplift and lateral forces can be obtained by drilling and grouting the tubes into the rock, under-reaming as necessary. In this respect the drilled-in tubular pile is a good type for forming berthing structures for large ships. These structures have to withstand high lateral and uplift loads for which a thick-walled tube is advantageous. In rock formations the resistance to these loads is provided by injecting a cement grout to fill the annulus between the outside of the tube and the rock forming the socket. Code of practice requirements for these and other forms of drilled-in tubular piles are as given in Section 2.3.4 for cast-in-place piles.

Where tubular steel piles have to be driven and sealed into a pre-formed hole ready to drill a rock socket, care must be taken not to over-drive the pile. 'Curtain folds' and ovality can occur, potentially compromising the load-bearing capacity and are difficult to rectify to produce an acceptable pile. It is preferable to use an under-reamer or hole opener to match the outside diameter of the pile before finally driving to seal the tube.

In the USA steel H-sections are lowered inside the drilled-in tubes and concrete is placed within the tubes to develop full end bearing on the pile and to ensure full interaction between tube, H-section 'core' and concrete. Because of the area of steel provided by the combined steel and concrete sections, very high loads can be carried by these 'caisson' piles where they are end bearing on a hard rock formation.

2.5 Composite piles

Various combinations of materials in driven piles or combinations of bored piles with driven piles can be used to overcome problems resulting from particular site or ground conditions. The problem of the decay of timber piles above groundwater level has been mentioned in Section 2.2.1. This can be overcome by driving a composite pile consisting of a precast concrete upper section in the zone above the lowest predicted ground-water level, which is joined to a lower timber section by a sleeved joint of the type shown in Figure 2.3. The same method can be used to form piles of greater length than can be obtained using locally available timbers.

Alternatively, a cased borehole may be drilled to below water level, a timber pile pitched in the casing and driven to the required depth, and the borehole then filled with concrete. Another variation of the precast concrete–timber composite pile consists of driving a hollow cylindrical precast pile to below water level, followed by cleaning out the soil and driving a timber pile down the interior.

In marine structures a composite pile can be driven that consists of a precast concrete upper section in the zone subject to the corrosive influence of sea-water and a steel H-pile below the soil line. The H-section can be driven deeply to develop the required uplift resistance from shaft friction. EC4 requirements apply to the design of composite steel and concrete structures.

Generally, composite piles are not economical compared with those of uniform section, except as a means of increasing the use of timber piles in countries where this material is readily available. The joints between the different elements must be rigidly constructed to withstand bending and tensile stresses, and these joints add substantially to the cost of the pile.

Where timber or steel piles are pitched and driven at the bottom of drilled-in tubes, the operation of removing the soil and obtaining a clean interior in which to place concrete is tedious and is liable to provoke argument as to the standard of cleanliness required.

2.6   Minipiles and micropiles

Minipiles are defined as piles having a diameter of less than 300 mm. Generally, they range in shaft diameter from 50 to 300 mm, with working loads in the range of 50 to 500 kN. The term 'micropile' is given to those in the lower range of diameter. Neither micropiles nor minipiles are specifically mentioned in EC7, but if load bearing, it should be assumed that the EC7 design rules apply. BS EN 14199: 2005 Execution of special geotechnical works – Micropiles covers requirements for using steel bars or tubes for reinforcement, concrete and grout materials, and the use of additives. They can be installed by a variety of methods. Some of these are as follows:

(1) Driving small-diameter steel tubes followed by injection of grout with or without withdrawal of the tubes
(2) Driving thin wall shells in steel or reinforced concrete which are filled with concrete and left in place
(3) Drilling holes by rotary auger, continuous flight auger, or percussion equipment followed by placing a reinforcing cage and in-situ concrete in a manner similar to conventional bored pile construction (Section 2.4.2) and
(4) Jacking-down steel tubes, steel box-sections, or precast concrete sections. The sections may be jointed by sleeving or dowelling.

One of the principal uses of minipiles is for installation in conditions of low headroom, such as underpinning work (Section 9.2.2), or for replacement of floors of buildings damaged by subsidence. Where minipiles are used for underpinning in clays susceptible to heave and shrinkage, it is advisable to insert a sleeve into a pre-bored hole over the top 2 to 3 m of the shaft. In this case the pile must be considered as a column over the sleeved length and designed accordingly.

Pali radice or 'root piles' (Section 9.2.2) are a form of grouted minipile used mainly for underpinning where the piles are installed at appropriate angles through the structure or foundations to transfer load to competent strata.

2.7   Factors governing choice of type of pile

The advantages and disadvantages of the various forms of pile described in Sections 2.2 to 2.5 affect the choice of pile for any particular foundation project and these are summarized in the following subsections:

2.7.1   Driven displacement piles

Advantages

(1) Material forming pile can be inspected for quality and soundness before driving
(2) Not liable to 'squeezing' or 'necking'

(3) Construction operations not affected by groundwater
(4) Projection above ground level advantageous to marine structures
(5) Can be driven in very long lengths
(6) Can be designed to withstand high bending and tensile stresses
(7) Can be redriven if affected by ground heave
(8) Pile lengths in excess of 25 m are common and pile loads over 10 000 kN are feasible for large diameter piles.

Disadvantages

(1) Unjointed types cannot readily be varied in length to suit varying levels of bearing stratum
(2) May break during driving, necessitating replacement piles
(3) May suffer unseen damage which reduces carrying capacity
(4) Uneconomical if cross-section is governed by stresses due to handling and driving rather than by compressive, tensile or bending stresses caused by working conditions
(5) Noise and vibration due to driving may be unacceptable
(6) Displacement of soil during driving may lift adjacent piles or damage adjacent structures
(7) End enlargements, if provided, destroy or reduce shaft friction over shaft length
(8) Cannot be driven in conditions of low headroom.

2.7.2   Driven and cast-in-place displacement piles

Advantages

(1) Length can easily be adjusted to suit varying levels of bearing stratum
(2) Driving tube driven with closed end to exclude groundwater
(3) Enlarged base possible
(4) No spoil to remove; important on contaminated sites
(5) Formation of enlarged base does not destroy or reduce shaft friction
(6) Material in pile not governed by handling or driving stresses
(7) Noise and vibration can be reduced in some types by driving with internal drop-hammer
(8) Reinforcement determined by compressive, tensile or bending stresses caused by working conditions
(9) Concreting can be carried out independently of the pile driving
(10) Pile lengths up to 25 m and pile loads to around 1500 kN are common.

Disadvantages

(1) Concrete in shaft liable to be defective in soft squeezing soils or in conditions of artesian water flow where withdrawable-tube types are used
(2) Concrete cannot be inspected after installation
(3) Concrete may be weakened if artesian groundwater causes piping up shaft of pile as tube is withdrawn
(4) Length of some types limited by capacity of piling rig to pull out driving tube
(5) Displacement may damage fresh concrete in adjacent piles, or lift these piles or damage adjacent structures
(6) Noise and vibration due to driving may be unacceptable

(7) Cannot be used in river or marine structures without special adaptation

(8) Cannot be driven with very large diameters

(9) End enlargements are of limited size in dense or very stiff soils

(10) When light steel sleeves are used in conjunction with withdrawable driving tube, shaft friction on shaft will be destroyed or reduced.

2.7.3   Bored and cast-in-place replacement piles

Advantages

(1) Length can readily be varied to suit variation in levels of bearing stratum

(2) Soil or rock removed during boring can be inspected for comparison with site investigation data

(3) In-situ loading tests can be made in large-diameter pile boreholes, or penetration tests made in small boreholes

(4) Very large (up to 7.3 m diameter) bases can be formed in favourable ground

(5) Drilling tools can break up boulders or other obstructions which cannot be penetrated by any form of displacement pile

(6) Material forming pile is not governed by handling or driving stresses

(7) Can be installed in very long lengths

(8) Can be installed without appreciable noise or vibration

(9) No ground heave

(10) Can be installed in conditions of low headroom

(11) Pile lengths up to 50 m over 3 m in diameter with working loads over 30 000 kN are feasible.

Disadvantages

(1) Concrete in shaft liable to squeezing or necking in soft soils where conventional types are used

(2) Special techniques needed for concreting in water-bearing soils

(3) Concrete cannot be inspected after installation

(4) Enlarged bases cannot be formed in coarse-grained soils

(5) Cannot be extended above ground level without special adaptation

(6) Low end-bearing resistance in coarse-grained soils due to loosening by conventional drilling operations

(7) Drilling a number of piles in a group can cause loss of ground and settlement of adjacent structures.

2.7.4   Choice of pile materials

Timber is cheap relative to concrete or steel. It is light, easy to handle, and readily trimmed to the required length. It is very durable below groundwater level but is liable to decay above this level. In marine conditions softwoods and some hardwoods are attacked by wood-boring organisms, although some protection can be provided by pressure impregnation. Timber piles are unsuitable for heavy working loads, typical maximum being 600 kN.

  Concrete is adaptable for a wide range of pile types. It can be used in precast form in driven piles, or as insertion units in bored piles. Dense, well-compacted good-quality concrete can withstand fairly hard driving and it is resistant to attack by aggressive substances in the soil, or in sea water or groundwater. However, concrete in precast piles is liable to damage (possibly unseen) in hard driving conditions. Concrete with good workability, using plasticizers as appropriate, should be placed as soon as possible after boring cast-in-place piles. Weak, honeycombed concrete in cast-in-place piles is liable to disintegration when aggressive substances are present in soils or in groundwater.

Steel is more expensive than timber or concrete but this disadvantage may be outweighed by the ease of handling steel piles, by their ability to withstand hard driving, by their resilience and strength in bending, and their capability to carry heavy loads. Steel piles can be driven in very long lengths and cause little ground displacement. They are liable to corrosion above the soil line and in disturbed ground, and they require cathodic protection if a long life is desired in marine structures. Long steel piles of slender section may suffer damage by buckling if they deviate from their true alignment during driving.

2.8 Reuse of existing piled foundations

As the redevelopment of city sites continues, it is inevitable that many will be underlain with deep and complex foundations from the previous buildings. A foundation system that has already been tested and 'proved' by supporting the existing load could provide considerable economic advantage for a new structure on the same site. Clearly the foundations must be investigated thoroughly and shown to have an adequate factor of safety against failure and settlement for the new loads. Where an increase in load is to be applied or where new foundations have to be compatible with the old, the observational method can be adopted to ensure robustness of design and construction.

A comprehensive investigation into the problems that may be posed by the existence of old foundations and the potential solutions has been completed recently by a research consortium co-ordinated by the Building Research Establishment (the 'RuFUS project') and a *Handbook* published[2.21] giving guidance on the following:

- Why and when to consider reusing foundations
- Decision models to manage risk when reuse is considered
- Investigation, assessment, and performance of old foundations
- Upgrading performance and combining old foundations with new; and
- Measurement of performance.

2.9 References

2.1 LOVE, J. P. The use of settlement reducing piles to support a flexible raft structure in West London, *Proceedings of the Institution of Civil Engineers, Geotechnical Engineering*, Vol. 156, 2003, pp. 177–81.

2.2 RAISON, C. A. North Morcambe Terminal, Barrow: pile design for seismic conditions, *Proceedings of the Institution of Civil Engineers, Geotechnical Engineering*, Vol. 137, 1999, pp. 149–63.

2.3 MARSH, E. and CHAO, W. T. The durability of steel in fill soils and contaminated land, *Corus Research, Development & Technology, Swindon Technology Centre*, Report No. STC/CPR OCP/CKR/0964/2004/R, 2004.

2.4 A corrosion protection guide for steel bearing piles in temperate climates, *Corus Construction and Industrial*, Scunthorpe, 2005.

2.5 The Institution of Civil Engineers, *Specification for Piling and Embedded Retaining Walls*, Thomas Telford Ltd, London, 1996. (2nd edition in preparation 2006)

2.6 BRE Special Digest 1: 2005, Concrete in aggressive ground, *Building Research Establishment*, Watford, 2005.

2.7 GRABHAM, F. R., BRITT, G. B., and ROBERTS, W. K. Prestressed piles. Notes for Informal Discussion held at Institution of Civil Engineers, 25 January 1971.

2.8 HASSAM, T., RIZKALLA, S., RITAL, S., and PARMENTIER, D. Segmental precast concrete piles – a solution for underpinning, *Concrete International*, August 2000, p. 41.

2.9 BJERRUM, L. Norwegian experiences with steel piles to rock, *Geotechnique*, Vol. 7, No. 2, 1957, pp. 73–96.

2.10 HANNA, T. H. Behaviour of long H-section piles during driving and under load, *Ontario Hydro Research Quarterly*, Vol. 18, No. 1, 1966, pp. 17–25.

2.11 BIDDLE, A. R. *H-Pile Design Guide*. SCI Publication P355. Steel Construction Institute, Ascot, 2005.

2.12 European structural steel standard EN 10025: 2004. Explanation and comparison to previous standard, *Corus Construction and Industrial*, Scunthorpe, 2005.

2.13 WYNNE, C. P. A review of bearing pile types, *Construction Industry Research and Information Association*, CIRIA Report PGI, 2nd ed., 1988.

2.14 BUSTAMANTE, M. and GIANESELLI, L. Installation parameters and capacity of screwed piles, *Proceedings of Third International Seminar on Deep Foundations on Bored and Augered Piles*. Van Impe W. F. and Haegeman (eds). AA Balkema, Rotterdam, 1998, pp. 95–108.

2.15 HOLEYMAN, A. and CHARUE, N. International pile capacity prediction event at Limelette in *Belgian screw pile technology design and recent developments. Proceedings of Second Symposium on Screw Piles, Brussels*. Maertens J. and Huybrechts N. (eds) Swets and Zeitlinger, Lisse, 2003, pp. 215–34.

2.16 BLACK, D. R. and PACK, J. S. Design and performance of helical screw piles in collapsible and expansive soils in arid regions of the United States, *Proceedings of Ninth International Conference on Piling and Deep Foundations*. Presses du l'école natural des Ponts et Chaussées, 2002, pp. 469–76.

2.17 JONES, A. E. K. and HOLT, D. A. Design of laps for deformed bars in concrete under bentonite and polymer drilling fluids, *The Structural Engineer*, Vol. 80, No. 18, 2004, pp. 32–8.

2.18 BUSTAMANTE, M., GIANESELLI, L., and SALVADOR, H. Double rotary CFA piles: performance in cohesive soils, *Proceedings of Ninth International Conference on Piling and Deep Foundations*. Presses de l'école nationale des Ponts et Chaussées, 2002, pp. 375–81.

2.19 LORD, J. A., HAYWARD, T., and CLAYTON, C. R. I. Shaft friction of CFA piles in chalk. *CIRIA Project Report 86*. Construction Industry Research Information Association, London, 2003.

2.20 FLEMING, W. G. K. The understanding of continuous flight auger piling, its monitoring and control, *Proceedings of the Institution of Civil Engineers, Geotechnical Engineering*, Vol. 113, 1995, pp. 157–65.

2.21 Building Research Establishment. *The 'RuFUS Project' Handbook* and Conference Proceedings on the Re-use of foundations, CRC, London, 2006.

Chapter 3

Piling equipment and methods

There was a world-wide increase in the construction of heavy foundations in the period from 1950 to the 1970s as a result of developments in high office buildings, heavy industrial plants and shipyard facilities. The same period also brought the major developments of offshore oilfields. A high proportion of the heavy structures required for all such developments involved piled foundations, which brought about a great acceleration in the evolution of piling equipment. There were increases in the size and height of piling frames, in the weight and efficiency of hammers, and in the capacity of drilling machines to install piles of ever-increasing diameter and length. The development of higher-capacity machines of all types was accompanied by improvements in their mobility and speed of operation.

The development of piling equipment proceeded on different lines in various parts of the world, depending mainly on the influence of the local ground conditions. In Northern Europe the precast concrete pile continued to dominate the market and this led to the development of light and easily handled piling frames. These were used in conjunction with self-contained diesel hammers and winches, with the minimum of labour and without the need for auxiliary craneage, steam boilers, or air compressors. The stiff clays of the mid-western states of America and the Great Lakes area of Canada favoured large-diameter bored piles, and mobile rotary drilling machines were developed for their installation. In contrast, the presence of hard rock at no great depth in the New York area favoured the continuing development of the relatively slender shell piles driven by an internal mandrel. The growth of the offshore oil industry in many parts of the world necessitated the development of an entirely new range of very heavy single-acting steam and hydraulically powered hammers designed for driving large-diameter steel piles, guided by tubular-jacket structures. In the present day, the increasing attention which is being given to noise abatement is influencing the design of pile hammers and the trend towards forms of pile that are installed by drilling methods rather than by hammering them into the ground.

Great Britain has a wide variety of soil types and the tendency has been to adopt a range of piling equipment selected from the best types developed in other parts of the world for their suitability for the soil conditions in any particular region.

With the advances in the techniques of installing large-diameter bored piles, and the increasing acceptance of these types for the foundations of heavy structures, it did appear at one stage that the capability of the bored pile to carry very heavy loads would outstrip that of the driven pile. However, with the stimulus provided by the construction of marine facilities for large tankers and deep-water oil-production platforms, the driven type of pile can now be installed in very large diameters that approach and in some cases exceed those of the larger bored piles. Piles are also being used to support heavier loads and to deeper penetrations in difficult environmental conditions where ground investigations have been limited. This has resulted in on-the-job changes to techniques and equipment requiring a high degree of geotechnical expertise from the employer and contractor to complete such projects successfully.

The manufacturers of piling equipment and the range of types they produce are too numerous for all makes and sizes to be described in this chapter. The principal types of equipment in each category are described, but the reader should refer to manufacturers' handbooks for the full details of their dimensions and performance. The various items of equipment are usually capable of installing more than one of the many piling systems which are described in Chapter 2. Installation methods of general application are described in the latter part of this chapter.

All piling equipment should comply with the requirements in BS EN 996: 1996 Piling equipment – Safety requirements and BS EN 791: 1996 Drill rigs – Safety.

3.1   Equipment for driven piles

3.1.1   Piling frames

The piling frame has the function of guiding the pile at its correct alignment from the stage of first pitching in position to its final penetration. It also carries the hammer and maintains it in position co-axially with the pile. The essential parts of a piling frame are the *leaders* or *leads*, which are stiff members of solid, channel, box, or tubular section held by a lattice or tubular mast that is in turn supported at the base by a moveable carriage and at the upper level by backstays. The latter can be adjusted in length by a telescopic screw device, or by hydraulic rams, to permit the leaders to be adjusted to a truly vertical position or to be raked forwards, backwards, or sideways. Where piling frames are mounted on elevated stagings, *extension leaders* can be bolted to the bottom of the main leaders in order to permit piles to be driven below the level of the base frame.

The *piling winch* is mounted on the base frame or carriage. This may be a double-drum winch with one rope for handling the hammer and one for lifting the pile. A three-drum winch with three sheaves at the head of the piling frame can lift the pile at two points using the outer sheaves, and the hammer by the central sheave. Some piling frames have multiple-drum winches which, in addition to lifting the pile and hammer, also carry out the duties of operating the travelling, slewing and raking gear on the rig.

Except in special conditions, say for marine work, stand-alone piling frames have largely been replaced by the more mobile self-erecting hydraulic leaders on tracked carriages, or by the crane-mounted fixed or hanging leaders offered by the major piling hammer manufac-turers. In Europe the pile hammer usually rides on the front of the leader, whereas in the USA the practice is to guide the pile between the leaders. The pile head is guided by a cap or helmet which has jaws on each side that engage with U-type leaders. The hammer is similarly provided with jaws. The leaders are capable of adjustment in their relative positions to accommodate piles and hammers of various widths.

Self-erecting leaders on powerful hydraulic crawler carriages can be configured for a variety of foundation work. Initial erection and changing from drilling to driving tools can be rapidly accomplished and with the electronic controls now available the mast can be automatically aligned for accurate positioning. Some crawlers have expandable tracks to give added stability and can handle pile hammers with rams up to 12 tonne at 1:1 back rake.

A range of rigs (with leader extensions as available) are shown in the following table:

Maker	Type	Usable leader length (m)	Maximum capacity (pile plus hammer) (tonne)	Capacity pile winch (tonne)
ABI 'Mobilram'[a]	TM13/16	15.7	9	5
(Germany)	TM14/17V	16.7	10	5
	TM16/20	19.7	9	5
	TM18/22B	21.4	9	5
Banut	555	15.0	6	10
(Germany)	650	18.6	16	10
Junttan[b]	PM16	15.0	12.0	5
(Finland)	PM20	13.8	13.0	8
	PM20C	12.6	16.0	10
	PM23	12.6	14.0	8
	PM25	12.6	18.0	10
Liebherr	LRB 125	12.5	12.0	5
(Austria)	LRB155	24.0	15.0	8
	LRB 255	30.0	30.0	20
	LRB 400	42.0	40.0	30
	LRH 200[c]	40.0	20.0	25
	LRH 400[c]	48.0	35.0	30
	LRH 600[c]	60.0	65.0	35
Deiseko-PVE	3015	17.5	8.5	6
(Netherlands and	4017	21.8	13.0	6
USA)	5021	24.8	18.0	6
	6025	27.2	20.1	10
	8027	30.0	30.0	12.5

Notes
a Telescopic mast.
b Standard rigs, additional leader extensions available for pile lengths up to 30 m.
c Using swinging leader.

The ABI Mobilram TM series of telescopic leader masts (Figure 3.1) provide 'crowd' (pulldown) for driving piles up to 21 m long using vibratory drivers, but fold down for compact transport. The Banut 555 and 650 piling rigs (Figure 3.2) are not only specially designed for impact driving, primarily precast concrete piles but are also effective for most bearing and sheet piles. The hydraulic stays attached to the crawler enable forward rakes of up to 18° and 45° back rakes, together with lateral movement of up to 14° available on both units. The usable length given for the 650 unit relates to the Banut superRAM 6000 hydraulic hammer (see Table 3.2).

The Junttan PM hydraulic piling rigs with fixed leaders can drive piles ranging from 18 to 36 m long (for HHK hydraulic hammers with leader extensions), using hammer rams from 3000 to 12000 kg. Fore and aft rakes of 18° and 40° respectively and lateral raking up to 12° are available on the PM25 unit. Liebherr provide fixed leaders mounted on their own and others' crawler carriages. The LRB series can operate as pile driving rigs and rotary drills for CFA and kelly bored piles.

Figure 3.1 ABI Mobilram with telescopic leader fully extended driving tubular pile (courtesy ABI GmbH).

3.1.2 Crane-supported leaders

Although the hydraulic piling rig with its base frame and leaders supported by a stayed mast provides a reliable means of ensuring stability and control of the alignment of the pile, there are many conditions which favour the use of leaders suspended from a standard crawler crane. Rigs of this type have largely supplanted the frame-mounted leaders for driving long piles on land in the UK and the USA.

Fixed leaders are rigidly attached to the top of the crane jib by a swivel and to the lower part of the crane carriage by a 'spotter' or stay. Hydraulic spotters can extend and retract to control verticality and provide fore and aft raking; they can also move the leader from side to side. The ICE heavy-duty spotter provides 6 m of hydraulic movement fore and aft and

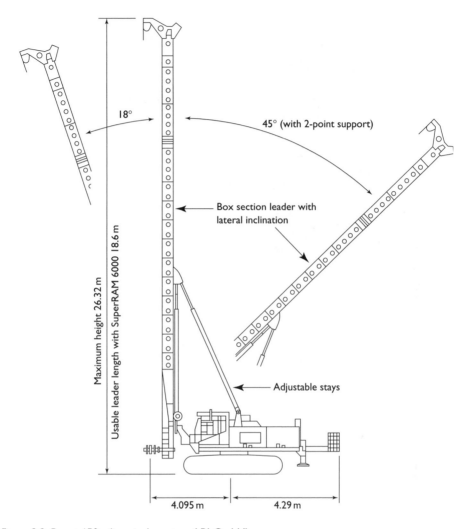

Maximum height 26.32 m

Usable leader length with SuperRAM 6000 18.6 m

18°

45° (with 2-point support)

Box section leader with
lateral inclination

Adjustable stays

4.095 m   4.29 m

Figure 3.2 Banut 650 piling rig (courtesy ABI GmbH).

an optional 35° leader rotation (Figure 3.3). In 'fixed extended' arrangements, the leaders extend above the top of the jib with a connector which allows freedom of movement. Leaders are usually provided in top and intermediate sections about 5 and 2.5 m long jointed together to provide the required leader height. As an alternative to spotters, simple hydraulic telescopic rams are used to enable raking piles to be driven, with a bottom stabbing point section bearing on the ground, levelled by hydraulic jack. BSP International Foundations Ltd fixed extended leaders have lattice panel lengths of 7 m for the 610 mm square section and 7.5 and 10 m for the 835 mm square section. The respective maximum lengths under the cat-head are 22.5 and 38 m, subject to crane jib length; maximum load for pile and hammer at a back rake of 1:12 with the 610 mm section is 12 tonne, and 21 tonne for the 835 mm section at a back rake of 1:3 using standard stays.

Figure 3.3 ICE 225 spotter with optional front lead rotation (courtesy International Construction Equipment).

Hanging and *swinging leaders* are suspended from the top of the crane jib with a head block which allows free movement to fit over a stabbed pile where guides are provided. Alternatively, there is a stabbing point at the bottom of the leader to fix the location for pile driving, either hanging vertically or swung to the required rake. The Liebherr LRH 600, 50 m long hanging leader has a maximum capacity of 65 tonne when used with the Liebherr HS 895 HD carriage, but account has to be taken of bending moments induced by the weight of the piling hammer when driving raked piles (Figure 3.4). Delmag 'European style' box leads constructed from tubular sections, lattice-braced, are light and stable and can be attached to any make of crane for driving piles at rakes up to 1:1 subject to bending moment considerations.

Backward and forward rakes in excess of 1:3 (Figure 3.5) are possible depending on the stability of the crawler crane. There is a practical limit to the length of pile which can be driven by a given type of rig and this can sometimes cause problems when operating the rig in the conventional manner without the assistance of a separate crane to lift and pitch the pile. The conventional method consists of first dragging the pile in a horizontal position close to the piling rig. The hammer is already attached to the leader and drawn up to the cat-head. The pile is then lifted into the leaders using a line from the cat-head and secured by toggle bolts. The helmet, dolly and packing are then placed on the pile head (Figure 3.20) and the assembly is drawn up to the underside of the hammer. The carriage of the piling rig is then slewed round to bring the pile over to the intended position and the stay and angle of the crane jib are adjusted to correct for verticality or to bring the pile to the intended rake.

In determining the size of leader whether rig-mounted, fixed or hanging it is always necessary to check the available height beneath the hammer when it is initially drawn up to the cat-head. Taking the example of leaders with a usable height of 20.5 m in conjunction with a hammer with an overall length of 6.4 m, after allowing a clearance of 1 m between the lifting lug on the hammer to the cathead and about 0.4 m for the pile helmet, the maximum length of pile which can be lifted into the leaders is about 12.7 m.

Figure 3.4 Liebherr LRH 400 48 m long swinging leader on HS 885 HD crane (courtesy Liebherr Great Britain Ltd).

Figure 3.5 US style 26 inch swinging leader supporting a Dawson HPH2400 hammer driving a 305 mm H-pile on 2:3 rake (courtesy Dawson Construction Plant Ltd).

Occasionally, it may be advantageous to use leaders independent of any base machine. Thus if only two or three piles are to be driven, say as test piles before the main contract, the leaders can be guyed to ground anchors and operated in conjunction with a separate petrol or diesel winch. Guyed leaders are slow to erect and move, and they are thus not used where many piles are to be driven, except perhaps in the confines of a narrow trench bottom where a normal rig could not operate.

3.1.3   Trestle guides

Another method of supporting a pile during driving is to use guides in the form of a moveable trestle. The pile is held at two points, known as 'gates', and the trestle is designed to be moved from one pile or pile-group position to the next by crane (Figure 3.6). The hammer is supported only by the pile and is held in alignment with it by leg guides on the hammer extending over the upper part of the pile shaft. Because of flexure of the pile during driving

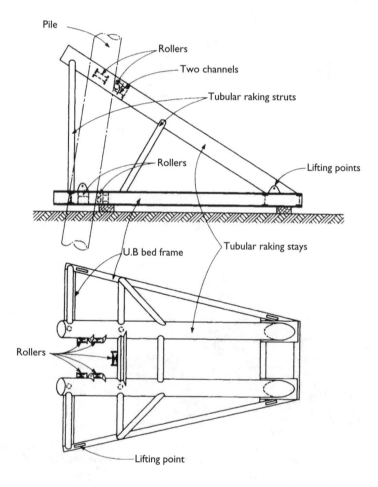

Figure 3.6 Trestle guides for tubular raking pile.

there is a greater risk, especially with raking piles, of the hammer losing its alignment with the pile during driving than in the case of piling frames which support and guide the hammer independently of the pile. For this reason the method of supporting the hammer on the pile in conjunction with trestle guides is usually confined to steel piles where there is less risk of damage to the pile head by eccentric blows. When driving long steel raking piles in guides it is necessary to check that the driving stresses combined with the bending stress caused by the weight of the hammer on the pile are within allowable limits.

Pile guides which are adjustable in position and direction to within very close limits are manufactured in Germany. Their principal use is for mounting on jack-up barges for marine piling operations. A travelling carriage or gantry is cantilevered from the side of the barge or spans between rail tracks on either side of the barge 'moon-pool'. The travelling gear is powered by electric motor and final positioning is by hydraulic rams. Hydraulically operated pile clamps or gates are mounted on the travelling carriage at two levels and are moved transversely by electric motor, again with final adjustment by hydraulic rams allowing the piles to be guided either vertically or to raking positions. Guides provided by hydraulic clamps on a guide frame fixed to the side of a piling barge are shown in Figure 3.7.

Trestle guides can be usefully employed for rows of piles that are driven at close centres simultaneously. The trestle shown in Figure 3.8 was designed by George Wimpey and Co. for the retaining wall foundations of Harland and Wolff's shipbuilding dock at Belfast[3.1].

Figure 3.7 Installing a 4 m dia monopile foundation for North Hoyle offshore wind farm with pile top rig and specially designed leader leg pile frame (courtesy Seacore Ltd).

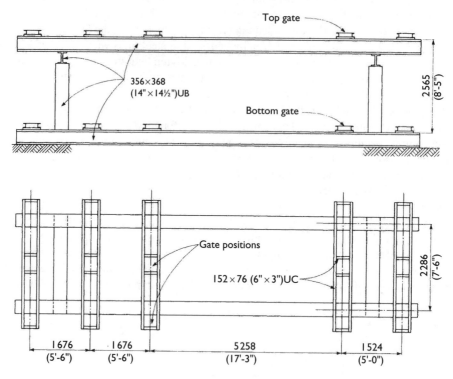

Figure 3.8 Trestle guides for multiple vertical piles.

Three rows of five 356 × 368 mm H-piles were pitched into the guides and were driven by a Delmag D22 hammer.

Guides can be used in conjunction with piling frames for a two-stage driving operation, which may be required if the piles are too long to be accommodated by the available height of frame. Guides are used for the first stage of driving, the piles carrying the hammer which is placed and held by a crane. At this stage the pile is driven to a penetration that brings the head to the level from which it can be driven by the hammer suspended in the piling frame. The latter completes the second stage of driving to the final penetration (Figure 3.9).

3.1.4 Piling hammers

The simplest form of piling hammer is the *drop hammer*, which is guided by lugs or jaws sliding in the leaders and actuated by the lifting rope. The drop hammer consists of a solid mass or assemblies of forged steel, the total mass ranging from 1 to 5 tonne. The striking speed is slower than in the case of single- or double-acting hammers, and when drop hammers are used to drive concrete piles there is a risk of damage to the pile if an excessively high drop of the hammer is adopted when the driving becomes difficult. There has been a revival of interest in the simple drop hammer because of its facility to be operated inside a sound-proofed box, so complying with noise abatement regulations (see Section 3.1.7).

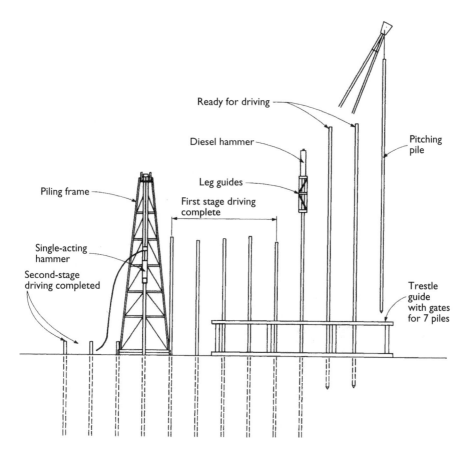

Figure 3.9 Driving piles in stages in conjunction with trestle guides.

Drop hammers are not used efficiently when operated from a pontoon-mounted piling frame working in open waters, since the height of the drop cannot be controlled when the pontoon is rising and falling on the waves. However, they can be used effectively in sheltered waters. The American Vulcan hammer, which has been designed to operate within the leaders, is shown in Figure 3.10.

The *single-acting hammer* is operated by steam or compressed air, which lifts the ram and then allows it to fall by gravity. BSP single-acting hammers of the type shown in Figure 3.11 range in mass from 2.5 to 6 tonne with a maximum height of fall of 1.37 m; a solenoid system can be used to control the drop of the hammer to avoid the operator fatigue of manual operation. The single-acting hammer is best suited to driving timber or precast concrete piles, since the drop of each blow of the hammer is limited in height and is individually controlled by the operator. The single-acting hammer is suitable for driving all types of pile in stiff to hard clays, where a heavy blow with a small drop is more efficient and less damaging to the pile than a large number of lighter blows. The steam or air supply for both single-acting and double-acting hammers should be at least 125% of the nominal consumption stated by the hammer manufacturer. The Menck MRBS offshore hammers (Figure 3.12)

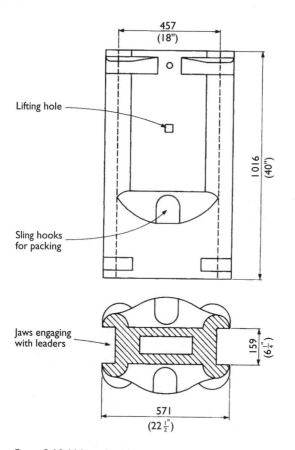

Figure 3.10 Vulcan drop hammer.

have masses ranging from 8.6 to 125 tonne with a maximum stroke of 1.75 m. They are fully automatic with infinitely variable stroke. By adding a belled out section beneath the hammer, Seacore have developed a rig capable of driving piles up to 4 m diameter into pre-drilled holes for the 'monopile' foundations for offshore wind turbine towers as in Figure 3.13.

The characteristics of the various types of single-acting hammer are shown in Table 3.1.

The ram of *hydraulic hammers* is raised by hydraulic fluid under high pressure to a predetermined height, and then allowed to fall under gravity or is forced down onto the pile head. In the BSP CX and CG hydraulic hammers (Figure 3.14), a hydraulic actuator is activated by a solenoid-operated control valve which raises the piston rod. At the required stroke height the flow of the hydraulic fluid is cut off. Pressures within the actuator then equalize allowing the ram to decelerate as it approaches the top of its stroke. The hammer then falls freely under gravity and repositions the piston rod for the next stroke. Using a remote control panel these hammers can deliver an infinitely variable stroke and blow rate within the limits stated. BSP also manufacture two double-acting hammers for driving steel sheet piles and bearing piles with blow rates of 84 blows per minute.

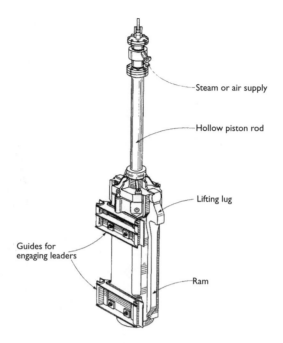

Figure 3.11 BSP International Foundations Ltd single-acting piling hammer.

Figure 3.12 MRBS air/steam single-acting hammer with stabilizing cage driving 54 inch diameter piles in legs of offshore jacket platform (courtesy Menck GmbH).

Figure 3.13 Driving a 4 m dia monopile foundation for North Hoyle offshore wind farm using a Menck MHU 500T hammer with large diameter pile sleeve and anvil adapter (courtesy Seacore Ltd).

Some free-fall hammers, for example, the Menck MHF series and the BSP CX series, have the option of additional acceleration by pressurizing the 'equalizing housing' above the piston, thereby increasing the energy by up to 20%.

The characteristics of various makes of hydraulic hammer are listed in Table 3.2. It is possible that the driving energy of hydraulic hammers may exceed the rated energy and this must be considered when analysing stresses during driving (see Section 2.2.6). Generally, these hammers have the advantage of being able to operate underwater, and because there is no exhaust they can be operated inside a sound-proofed box. Most modern hydraulic hammers can be fitted with electronic instrumentation giving a continuous display of depth, driving resistance and set. The largest APE hammers are equipped with digital radio GRL velocity sensing and energy monitoring control system.

Table 3.1 Characteristics of some single-acting piling hammers[a]

Maker	Type	Mass of ram (kg)	Maximum energy per blow (kJ)	Maximum striking rate (blows/min)
BSP International	3–2½ ton	2540	34	50
Foundations	2–3 ton	3050	41	50
Limited (UK)	4–4 ton	4060	54	50
	1–5 ton	5080	68	50
	1–6 ton	6110	82	50
MKT (USA)	MS-350	3500	42	40–50
	MS-500	5000	59	40–50
MENCK (Germany)	MRBS 850	8600	129	45
	MRBS 1100	11000	170	40
	MRBS 1800	17500	263	44
	MRBS 3000	30000	450	42
	MRBS 5000	50000	750	40
	MRBS 8800	88000	1320	36
	MRBS 12500	125000	2190	36
Vulcan (USA)	1	2270	20	60
	306	2950	26	60
	505	2270	34	46
	506	2950	44	46
	08	3630	35	50
	010	4540	44	50
	012	5443	53	50
	508	3630	54	41
	510	4540	68	41
	512	5440	81	41
	014	6350	57	59
	016	7370	66	58
	320	9070	81	55
	330	13610	122	54
	520	9070	135	42
	530	13610	203	42
	540	18550	277	48
	560	28350	424	47
	5100	45360	678	48
	5110	49900	746	39
	5150	68060	1012	46
	6300	136100	2440	42

Note

a Note that the information given in Tables 3.1 to 3.6 does not necessarily represent the full range of equipment by each maker. The makers listed in these tables should be contacted for full details and when making assessments of performance for particular applications. Because of market changes, some equipment will be obsolescent, but well-maintained used hammers not in current production may be available.

Underwater hydraulic hammers were developed specially for driving piles in deep water locations. The Menck MHU double-acting hammer range in Table 3.2 is designed specifically for underwater work; the S hammer series is for water depth up to 400 m, the T series for 2000 m and the U series for deep water to 3000 m. The MHU 3000 S with a ram weight of 280 tonne is the largest piling hammer ever constructed. A nitrogen shock absorber ring protects the hammer

Figure 3.14 BSP CX110 hydraulic piling hammer on Hitachi crane-mounted leader (courtesy BSP International Foundations Ltd).

from rebound forces and shock loads and will largely eliminate a tension wave in the pile (see Section 7.3). The MHU hammers are designed to operate either as free-riding units mounted on the pile with a slack lifting line, or to reduce weight on the guides they can be suspended from the floating crane with a heave compensator to maintain constant tension in the lifting line. Slender hammers can operate inside the pile or with a follower attached to the pile.

Double-acting (or *differential-acting*) hammers are steam- or air-operated both on the upstroke and downstroke, and are designed to impart a rapid succession (up to 300 blows per minute) of small-stroke blows to the pile. The double-acting hammer exhausts the steam or air on both the up- and down-strokes. In the case of the differential acting hammer, however, the cylinder is under equal pressure above and below the piston and is exhausted only on the upward stroke. The downward force is a combination of the weight of the ram and the difference in total force above and below the piston, the force being less below the piston because of the area occupied by the piston rod. These hammers are most effective in granular soils where they keep the ground 'live' and shake the pile into the ground, but they

Table 3.2 Characteristics of some hydraulic hammers

Maker	Type	Mass of ram (kg)	Maximum energy per blow (kJ)	Striking rate at maximum stroke height (blows/min)
American Piledriving Equipment (APE)[a] (USA)	13	3629	38	40–100
	14	4990	45	40–80
	H-5.4MT	5445	34	40–75
	H-7.2MT	7350	68	40–75
	225U	36287	305	30–60
	400U	36287	488	30–72
	750U	54431	847	20–40
BSP International Foundations Limited (UK)	CX40	3000	40	45–100
	CX50	4000	51	45–100
	CX85	7000	83	40–100
	CX110	9000	106	36–100
	CG180	12000	180	34–100
	CG210	14000	210	32–100
	CG240	16000	240	31–100
	CG270	18000	270	31–100
	CG300	20000	300	29–100
Banut SuperRAM (Germany)	3000	3000	35	100
	4000	4110	47	100
	5000	5060	58	100
	6000	6075	70	100
	6000XL	6110	82	100
	8000XL	8010	109	100
	10000XL	10020	117	100
	12000XL	12025	141	100
Dieseko PVE (Netherlands and USA)	3L	3050	36	45
	4L	4060	48	42
	5L	5080	60	40
	7L	7110	84	40
	10L	10160	120	40
	13L	13210	156	35
	16	16260	192	35
	20	20320	240	32
ICE (Netherlands and USA)	75	3401	41	48
	115	5215	62	45
	160	7256	87	45
	220	9977	119	45
	275	12471	149	45
IHC Hydrohammer (Netherlands)	S30	1625	30	65
	S35	3048	35	60
	S70	3556	70	50
	S90	4572	90	50
	S150	7620	150	44
	S200	10160	200	45
	S280	13818	280	45
	S500	25400	500	45
	S600	30480	600	36
	S900	45720	900	30

(*Table 3.2 continued*)

Table 3.2 Continued

Maker	Type	Mass of ram (kg)	Maximum energy per blow (kJ)	Striking rate at maximum stroke height (blows/min)
	S1200	60960	1200	30
	S1800	76200	1800	30
	S2300	116840	2300	30
	SC50[b]	3353	50	50
	SC75[b]	5791	75	50
	SC110[b]	8026	110	40
	SC150[b]	11176	150	40
	SC200[b]	13818	200	40
Junttan[c] (Finland)	HHK 4AL	4000	31	40–100
	HHK 5AL	5000	39	40–100
	HHK 5A	5000	59	40–100
	HHK 7A	7000	82	40–100
	HHK 9A	9000	106	40–100
	HHK 12A	12000	141	40–100
	HHK 14A	14000	164	40–100
	HHK 18A	18000	212	40–100
	HHK 5S	5000	74	30–100
	HHK 9S	9000	132	30–100
	HHK 14S	14000	206	30–100
	HHK 18S	18000	265	30–100
	HHK 25S	20000	294	30–100
MENCK (Germany)	MHF3-4	4000	40	50
	MHF3-5	5000	50	50
	MHF3-6	6000	60	45
	MHF3-7	7000	70	45
	MHF5-8	8000	80	45
	MHF5-10	10000	100	40
	MHF5-12	12000	120	40
	MHF10-15	15000	150	45
	MHF10-20	20000	200	40

Notes
a Free-fall hammer (many hammers now have assisted acceleration).
b SC series more suited to driving concrete piles.
c AL series hammers: 0.8 m stroke, A hammers: 1.2 m stroke, S hammers: 1.5 m stroke.

are not so effective in clays. Double-acting hammers have their main use in driving sheet piles and are not used for bearing piles in preference to diesel hammers. However, unlike the diesel hammer they can operate underwater, provided that the ram is fully enclosed. Leaders are usually necessary. The characteristics of various makes are shown in Table 3.3.

Diesel hammers are suitable for all types of ground except soft clays. They have the advantage of being self-contained without the need for separate power-packs, air compressors or steam-generators. They work most efficiently when driving into stiff to hard clays, and with their high striking rate and high energy per blow they are favoured for driving all types of bearing piles up to about 2.5 m in diameter. The principle of the diesel hammer is that as the falling ram compresses air in the cylinder, diesel fuel is injected into the cylinder and this is atomized by the impact of the ram on the concave base. The impact ignites the fuel

Table 3.3 Characteristics of some double-acting and differential-acting piling hammers

Maker	Type	Mass of ram (kg)	Maximum energy per blow (kJ)	Maximum striking rate (blows/min)
BSP International	600N[a]	227	4	250
Foundations Limited (UK)	700N[a]	385	7	225
	SL20da	1500	20	84
	SL30da	2500	30	84
Dawson Construction	HPH1200	1040	12	80–120
Plant (UK)	HPH1800	1500	19	80–120
	HPH2400	1900	24	80–120
	HPH4500	3500	45	80–120
	HPH6500	4650	65	80–120
	HPH10K	6500	100	80–120
MKT[a] (USA)	9B3	725	12	145
	10B3	1360	18	105
	11B3	2270	25	95
MENCK (Germany)	MHU220S[b]	11500	230	38
	MHU300S	16000	305	40
	MHU440S	24000	440	38
	MHU550S	30000	550	38
	MHU660S	36000	660	32
	MHU800S	45000	800	38
	MHU1000S	57600	1000	38
	MHU1700S	94000	1730	35
	MHU2100S	116500	2140	32
	MHU3000S	168000	3050	32
Vulcan[a, c] (USA)	30C	1360	10	133
	50C	2270	21	117
	65C	2950	27	117
	80C	3630	33	109
	85C	3670	35	111
	100C	4340	44	103
	140C	6350	49	101

Notes
a Air/steam operated.
b S denotes use in shallow water or onshore use; T version is for deep water.
c Differential acting.

and the resulting explosion imparts an additional 'kick' to the pile, which is already moving downwards under the blow of the ram. Thus the blow is sustained and imparts energy over a longer period than the simple blow of a drop or single-acting hammer. The ram rebounds after the explosion and scavenges the burnt gases from the cylinder. The well-known Delmag series of hammers (Figure 3.15) ranges from the D6 with a ram mass of 600 kg for driving piles up to 2000 kg to the 20 tonne ram of the D200 with a drop height of 3.4 m for piles up to 250 tonne. ICE and MKT manufacture double-acting diesel hammers with a striking rate of 70 to 90 blows per minute compared with the rates of 40 to 60 blows per minute attained by the comparable makes of single-acting diesel hammers. The characteristics of various makes of diesel hammer are shown in Table 3.4.

Figure 3.15 Delmag D30–20 diesel hammer on American style leaders with helmet for driving steel H-piles.

A difficulty arises in using the diesel hammer in soft clays or weak fills, since the pile yields to the blow of the ram and the impact is not always sufficient to atomize the fuel. Berminghammer have developed a high injection pressure, 'smokeless' diesel hammer which virtually eliminates the problem. The more resistant the ground, the higher the rebound of the ram, and hence the higher the energy of the blow. This can cause damage to precast concrete piles when driving through weak rocks containing strong bands. Although the height of drop can be controlled by adjusting, by a rope-operated lever, the amount of fuel injected, this control cannot cope with random hard layers met at varying depths, particularly when these are unexpected. The diesel hammer operates automatically and continuously at a given height of drop unless the lever is adjusted, whereas with the single-acting hammer every blow is controlled in height.

Table 3.4 Characteristics of some diesel piling hammers

Maker	Type	Mass of ram (kg)	Maximum energy per blow (kJ)	Maximum striking rate (blows/min)
Berminghammer	B2005	900	31	36–60
(Canada)	B23[a]	1270	31	82
	B3505	1800	62	36–60
	B4005	2300	78	36–60
	B4505	3000	103	36–60
	B5005	3400	118	36–60
	B5505	4182	146	36–60
	B6005	6000	212	35–60
	B6505	7990	273	36–60
	B6505C	10000	238	35–60
BSP International	DE30C	1360	36	42–54
Foundations	DE50C	2260	60	42–54
Limited (UK)				
Delmag (Germany)	D6-32	600	19	38–52
	D8–32	800	27	36–52
	D12-42	1280	46	35–52
	D16-32	1600	54	36–52
	D19-42	1820	66	35–42
	D21-42	2100	75	35–50
	D25-32	2500	90	35–52
	D30-32	3000	103	36–52
	D36-32	3600	123	36–53
	D46-32	4600	166	35–53
	D62-22	6200	224	35–50
	D100-13	10000	360	35–45
	D150-42	15000	512	36–52
	D200-42	20000	683	36–52
MKT (USA)	DE33/30/20C	1810	54	40–50
	DE-70/50C	3180	95	40–50
	DE-150/110	6800	203	40–50
	DA-35C[a]	1270	32	78–82
	DA-45[a]	1810	46	82
	DA-55C[a]	2270	57	78–82
ICE International	WICE 8	800	23.9	38–52
Construction Equipment	WICE 30	3000	94.8	37–52
(Netherlands and USA)	WICE 80	8000	266.8	36–45
	WICE 100	10000	333.5	36–45
	WICE 128	12500	426.5	36–45
	WICE 160	16000	550.0	36–45
	WICE 220	22000	733.0	36–45
	32S	1364	43.0	41–60
	60S	3175	98.9	41–59
	100S	4535	162.7	38–55
	120S	5440	202.0	38–55
	205S	9072	284.7	40–55
	I-12	1280	41.0	37–52
	I-30	3000	102.3	35–53
	I-46	4600	146.0	36–53

(Table 3.4 continued)

Table 3.4 Continued

Maker	Type	Mass of ram (kg)	Maximum energy per blow (kJ)	Maximum striking rate (blows/min)
	I-62	6623	223.7	35–50
	I-80	8030	288.0	35–45
	I-100	10000	360.0	35–45
	422[a]	1810	30.5	76–82
	520[a]	2300	40.6	80–84
	640[a]	2722	54.2	74–77

Note
a Double acting.

Because of difficulties in achieving a consistent energy of blow, due to temperature and ground resistance effects, the diesel hammer is being supplanted to a large extent by the hydraulic hammer, particularly when being used in conjunction with the pile-driving analyser (see Section 7.3) to determine driving stresses.

3.1.5   Piling vibrators

Vibrators consisting of pairs of exciters rotating in opposite directions can be mounted on piles when their combined weight and vibrating energy cause the pile to sink down into the soil (Figure 3.16). The two types of vibratory hammers, either mounted on leaders or as free hanging units, operate most effectively when driving small displacement piles (H-sections or open-ended steel tubes) into loose to medium-dense granular soils. Ideally a pile should be vibrated at or near to its natural frequency, which requires 100 Hz for a 25 m steel pile. Thus only the high-frequency vibrators are really effective for long piles,[3.2] and while resonant pile driving equipment is costly, high penetration rates are possible. Most types of vibrators operate in the low-frequency to medium-frequency range (i.e. 10 to 39 Hz). Vibrators mounted on the dipper arm of hydraulic excavators have high power to weight ratios and are useful for driving short lengths of small section tubular and H-piles, limited by the headroom under the bucket, say 6 m at best.

Rodger and Littlejohn[3.3] proposed vibration parameters ranging from 10 to 40 Hz at amplitudes of 1 to 10 mm for granular soil when using vibrators to drive piles with low point resistance, to 4 to 16 Hz at 9 to 20 mm amplitude for high point resistance piles. In fine soils frequencies in excess of 40 Hz and high amplitude will be needed but care must be exercised because of the potential changes in soil properties such as liquefaction and thixotropic transformation. Predicting the performance of vibratory pile driving is still not very reliable. Where specific test data are not available for the vibrator installing bearing piles or the pile is not bearing on a consistent rockhead, it may be advisable (as is common in the USA) to use the vibrator to install the pile to within 3 m of expected penetration and then use an impact hammer to drive to the bearing layer. Vibrators are not very effective in firm clays and cannot drive piles deeply into stiff clays. They are frequently used in bored pile construction for sealing the borehole casing into clay after pre-drilling through the granular overburden soils. After concreting the pile the vibrators are used to extract the casings and are quite efficient for this purpose in all soil types (see Section 3.4).

Vibrators have an advantage over impact hammers in that the noise and shock wave of the hammer striking the anvil is eliminated. They also cause less damage to the pile and have a

Figure 3.16 Driving a pile casing with a PVE 200 m free hanging vibrator.

very fast rate of penetration in favourable ground. It is claimed that a rate of driving averaging 18 m per minute may be achieved in loose to medium-dense granular soils. If the electric generator used to power the exciter motors is mounted in a well-designed acoustic chamber, the vibrators can be used in urban areas with far lower risk of complaints arising due to noise and shock-wave disturbance than when impact hammers are used. However, standard vibrators with constant eccentric moment have a critical frequency during starting and stopping as they change to and from the operating frequency, which may resonate with the natural frequency of nearby buildings. This can cause a short period of high amplitude vibrations which are quite alarming to the occupants. The development of high frequency (greater than 30 Hz), variable moment vibrators with automatic adjustment has virtually eliminated this start-up and shut-down 'shaking zone', reducing peak particle velocity to levels as low as 3 mm/s at 2 m (see Section 3.1.7). These are not as powerful as the standard units. Although there are limitations in respect of the soil types in which they can be used and notwithstanding the complexity of the machinery and its maintenance, the new range of resonant-free vibrators will generate greater driving force and displacement amplitude to overcome the toe resistance when driving longer and larger displacement piles[3.4].

Depth vibrators or 'poker' vibrators are used extensively for improving the bearing capacity and settlement characteristics of weak soils by vibro-compaction or vibro-replacement techniques. For vibro-compaction (or 'flotation') the vibrator is either flushed to the required depth using water jets or vibrated dry with air jets in partially saturated soil. As the poker is withdrawn the horizontal vibrations cause a compact cylinder of soil to be formed at depth as the soil particles are rearranged during densification, producing a depression at the surface which has to be filled with granular material during the process. In the vibro-displacement

Table 3.5 Characteristics of some pile-driving and extracting vibrators

Maker	Type	Frequency range (Hz)	Mass (kg)	Minimum power supply (KVA)
ABI (Germany)	MRZV 400	45	1800	160
	MRZV 500	46	1900	160
	MRZV 600	44	2300	190
	MRZV 800S	41	2750	290
	MRZV 925S	38	3800	330
	MRZV 600VS	44	1980	220
	MRZV 800VS	41	2550	290
	MRZV 925/16VS	38	2710	330
	MRZV 925/18VS	36	2730	360
	MRZV 1000/20VS	35	2750	400
American Piledriving Equipment (APE) (USA)	50	0–30	2630	194
	150	0–30	3084	261
	200	0–28	6167	470
	120V	6–33	2836	261
	170V	6–38	2900	339
	250V	6–33	4309	470
	400B	6–23	7260	745
	600B	6–23	10890	745
Bauer (Germany)	MR60V	0–24		179
	MR70V	0–24		209
	MR90V	0–24		336
	MR100V	0–24		370
Dawson Construction Plant Ltd[a] (UK)	EMV70	50	410	12
	EMV300	40	625	52
	EMV400	41	910	80
	EMV525	42	1150	120
Dieseko PVE (Netherlands and USA)	23M	27	2300	234
	38M	28	3400	392
	52M	28	4000	564
	110M	22	7000	784
	200M	23	19000	1130
	1420	33	1210	194
	2315	38	1880	234
	2520	33	3050	290
	2310VM	38	1450	290
	2332VM	38	4300	397
	2335VM	38	4400	784
	50VM	33	4750	564
	38VMR[b]	35	11500	
	20VMR[b]	35	6000	
ICE – International Construction Equipment (USA and Netherlands)	416	27	2350	193
	3220	33	3850	278
	815C	26	3950	346
	1412C	23	6400	470
	250NF	24	6415	657
	625	42	685	113
	1223	38	1400	184

Table 3.5 Continued

Maker	Type	Frequency range (Hz)	Mass (kg)	Minimum power supply (KVA)
	1423C	38	1700	209
	423	38	3750	320
	14RF	38	2420	213
	23RF	38	3900	287
	28RF	38	3800	417
	36RF	33	3900	431
	64RF	32	5000	663
	44-50	27	5487	429
	1412Btandem	22	13900	1193
P.T.C. (France)	7H5	33	650	50
	15HI	28	1250	108
	25HIA	29	2200	153
	30HIA	28	4120	193
	50HDI	25	3100	255
	60HD	28	4900	305
	75HD	25	11800	410
	100HD	23	8200	451
	175HD	23	13000	611
	265HD	23	19500	988
	7HF3	38	900	92
	15HF3	38	1300	210
	30HF3	38	2400	292
	46HF3	38	7400	447
	10HFV	38	1500	95
	15HFVS	38	2210	200
	23HFV	38	3610	222
	34HFV	38	5000	405
	60HFV	38	6360	425
	60HFVS	38	6360	662
Soilmec (Italy)	VS-4	30	1138	135
	VS-8	30	1901	277
	VS-16	30	3500	554

Notes
V Generally denotes variable moment vibrator.
a Mounted on excavator dipper arm.
b Ring vibrators with variable moment (total weight).

technique the vibrator forms a hole by compacting the soil to the required depth which is filled with graded stone and then compacted in lifts by the vibrator to form 'stone columns'. This process has been extended whereby concrete is injected into the hole at the tip of the poker and vibrated as it is withdrawn to provide a form of 'pile' (or 'vibro-concrete column') capable of carrying light loading when taken down to a competent stratum.

Types of vibrators suitable for driving bearing piles are shown in Table 3.5.

3.1.6 Selection of type of piling hammer

The selection of the most suitable type of hammer for a given task involves a consideration of the type and weight of the pile, and the characteristics of the ground into which the pile is to be driven. Single and double-acting hammers, hydraulic and diesel hammers are effective in all soil types and the selection of a particular hammer for the given duty is based on a consideration of the value of energy per blow, the striking rate and the fuel consumption. The noise of the pile-driving operation will also be an important consideration in the selection of a hammer. This aspect is discussed in Section 3.1.7.

A knowledge of the value of energy per blow is required to assess whether or not a hammer of a given weight can drive the pile to the required penetration or ultimate resistance without the need for sustained hard driving or risk of damage to the pile or hammer. The safety of operatives can be endangered if sustained hard driving causes pieces of spalled concrete or mechanical components to fall from a height. The employment of a dynamic pile-driving formula can, with experience, provide a rough assessment of the ability of a hammer with a known rated energy value to achieve a specific ultimate pile resistance to the time of driving (see Sections 1.4 and 7.3 for a further discussion of these formulae). However, the manufacturer's rated energy per blow is not always a reliable indication of the value to be used in a dynamic pile equation. The efficiency of a hammer can be very low if it is poorly maintained or improperly operated. Also the energy delivered by the hammer to the pile depends on the accuracy of alignment of the hammer, the type of packing inserted between the pile and the hammer, and on the condition of the packing material after a period of driving.

The increasing use of instruments to measure the stresses and acceleration at the head of a pile as it is being driven (see Section 7.3) has provided data on the efficiencies of a wide range of hammer types. Some typical values are as shown in the following table:

Hammer type	Efficiency of hammer/cushioning system (%)
Hydraulic	65–90
Drop (winch-operated)	40–55
Diesel	20–80

The wide range in values for the diesel hammer reflects the sensitivity to the type of soil or rock into which the pile is driven and the need for good maintenance. Present-day practice is to base the selection of the hammer on a driveability analysis using the Smith wave equation (see Section 7.3) to produce curves of the type shown in Figure 3.17. They show the results of an investigation into the feasibility of using a D100 diesel hammer to drive 2.0 m OD by 20 mm wall thickness steel tube piles through soft clay into a dense sandy gravel. The piles were to be driven with closed ends to overcome a calculated soil resistance of 17.5 MN at the final penetration depth. Figure 3.17 shows that a driving resistance (blow count) of 200 blows/250 mm penetration would be required at this stage. This represents a rather severe condition. A blow count of 120 to 150 blows/250 mm is regarded as a practical limit for sustained driving of diesel or hydraulic hammers. However, 200 blows/250 mm would be acceptable for fairly short periods of driving. Commercial computer programs based on wave equation models enable the piling engineer to predict driveability, optimize the selection of hammer, select energy level which will not damage the pile, and ensure that the correct dolly and adapters are used.

The American Petroleum Institute[3.5] states that if no other provisions are included in the construction contract, pile-driving refusal is defined as the point where the driving resistance exceeds either 300 blows per foot (248 blows/250 mm) for 1.5 consecutive metres or 800 blows per foot (662 blows/250 mm) for 0.3 m penetration. Figure 3.17 also shows the driving

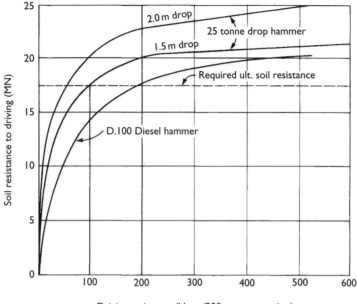

Figure 3.17 Pile driveability curves.

resistance curves for a 25 tonne drop hammer with drops of 1.5 or 2.0 m to be used as a standby to achieve the required soil resistance if this could not be obtained by the diesel hammer.

Vibratory hammers are very effective in loose to medium-dense granular soils and the high rate of penetration of low-displacement steel piles driven by vibratory hammers may favour their selection for these conditions.

3.1.7 Noise and vibration control in pile driving

The control of noise in construction sites is a matter of increasing importance in the present drive to improve environmental conditions, and the Noise at Work Regulations 2005 implement the latest European Directive for the protection of workers from the risks related to the exposure to noise. The requirements for employers to make an assessment of noise levels and take action to eliminate and control noise are triggered by three action levels: daily or weekly (5 days of 8 hours) personal noise exposures of 80 dBA as the lower level, 85 dBA as the upper level and a peak (single loud noise) of between 135 and 137 dBA. The exposure *limit* values are 87 and 140 dBA at peak; the method of calculating the various exposure levels is defined in the Regulations. Employers are required to reduce noise at source to a minimum by using appropriate working methods and equipment, but if noise levels cannot be controlled below the upper action level by taking reasonably practicable measures, hearing protectors which eliminate the risk must be provided. As these Regulations do not apply to control of noise to prevent annoyance or hazards to the health of the general public outside the place of work, the Environment Protection Act (EPA) and Control of Pollution Act provide the general statutory requirements to control noise and vibrations which are considered to be a legal nuisance.

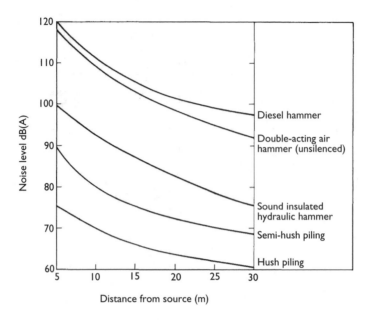

Figure 3.18 Typical noise levels for various pile-driving techniques.

Under the BS Code of Practice 5228 recommendations on noise exposure levels, no person (employee or the general public) should be exposed to a noise level of more than 85 dBA for eight hours a day in a five-day week (now superseded by the 2005 Regulations). It is recognized that the noise from many pile-driving methods will exceed 85 dBA but as the operations are not continuous through the working day, the observed noise level (or 'basic sound power level' as given in the Code) can be converted to an 'equivalent continuous sound pressure level' that takes into account the duration of the noise emission, distance from the source, screening and reflection.[3.6] Methods for predicting the impact of noise and the procedures for obtaining consent from the local authority under section 61 of the Control of Pollution Act for proposed noise control measures are detailed in BS 5228. EC3-5 does not now include recommendations on noise and vibration levels from piling.

Local authorities are empowered under the EPA and Control of Pollution Act to set their own standards of judging noise nuisance, and maximum day-time and night-time noise levels of 70 and 60 dBA respectively are frequently stipulated for urban areas (and as low as 40 dBA in sensitive areas). The higher of these values can be compared with field observations of pile-driving noise obtained from a number of sources and shown in Figure 3.18. Tables in Part 1 of BS 5228 also provide sound level data on various piling methods. Other information has shown that the attenuation of pile-driving impact noise to the 70 dBA level from the noisiest of the hammers requires a distance of more than 1000 m from the sound. Thus if a maximum sound level of 70 dBA is stipulated by a local authority, it is necessary to adopt some means of controlling noise emission in order to protect the general public whose dwellings or place of work are closer to the construction operations[3.7]. One method of doing this is to enclose the hammer and pile with a sound absorbent box. The Hoesch noise-abatement tower is formed of sandwiched steel plate/plastics construction and

Figure 3.19 Noise-abatement tower used for 'Hush piling' system.

consists of an outer 2 mm steel plate, a plastics layer 0.4 mm thick, and an inner 1.5 mm steel plate. The plates making up the box are jointed by a rubber insertion material, and the lid incorporates a sound-proofed air exhaust. A hinged door allows the pile and hammer to be pitched into the tower. The Hoesch tower reduced the noise from a Delmag D12 diesel hammer driving a sheet pile from 118 to 119 dBA at 7 m to 87 to 90 dBA at the same distance.

A tower of similar construction is shown in Figure 3.19. Shelbourne[3.8] described the use of the tower for driving 24 m steel H-piles by means of a 3-tonne drop hammer. Sound level measurements of 60 to 70 dBA were recorded 15 m from the tower, compared with values of 100 dBA before the noise-abatement system was adopted.

Surrounding only the lower part of the hammer by a shroud is not particularly effective. A reduction of only 3 to 4 dBA was obtained by shrouding a Delmag D22 hammer in this way. As noted in Section 3.1.4, the hydraulic hammer is a suitable type for enclosing in a sound-proof shroud.

Crane-mounted augers using kelly bars for bored piles (see Section 3.3.4) and large CFA rigs can produce sound power levels as high as 110 dBA, and are usually operated between 85% and 100% of the shift. This results in equivalent continuous sound pressure levels in excess of 80 dBA at 10 m. Acoustic enclosures are essential for the engines and power packs. The use of vibratory hammers for steel bearing piles has increased, but basic sound levels can still be around 120 dBA and even with conversions to equivalent sound, noise-abatement measures are necessary.

There is little evidence to show that ground-borne vibrations cause structural damage to buildings[3.9]. However, if there is concern then steps must be taken to survey buildings and measure vibrations induced by construction activity. BS 7385 describes methods of assessing vibrations in buildings and gives guidance on potential damage levels. The recommended thresholds in BS 5228 to avoid non-structural ('cosmetic') damage in residential property are peak particle velocity (ppv) of 10 mm/s for intermittent vibration and 5 mm/s for continuous vibrations at frequencies between 10 and 50 Hz. For heavy and stiff buildings the thresholds are 30 and 15 mm/s respectively. Protected buildings, buildings with existing defects and statutory services undertakings will be subject to specific lower limits. The human response to vibration should also be considered. Transmission of vibrations during piling depends on the strata, size, and depth of pile and hammer type, and predictions of the resulting ground frequency and ppv at distance from the source are difficult.

'Press-in' drivers such as the Dawson 'push–pull' unit with 2078 kN pressing force are becoming more common particularly for sheet piling, but many of the units can be adapted for installing box-type bearing piles and H-pile groups, particularly in clays. The advantages of these powerful, high pressure hydraulic drivers using 2 to 4 cylinders are the low noise levels (around 60 dBA) and the speed and vibration-less installation and extraction of piles. The drivers can be suspended from a crane or mounted on a hydraulic crawler rig with more than 20 tonnes of pulldown available on a rigid leader such as the Liebherr piling rig to assist the installation; hanging leaders are not suitable. The Giken press-in rig operates without a separate fixed leader relying on reaction from adjacent installed sheet piles; a service crane is needed to pitch the piles. In addition, in hard ground this unit can pre-drill a hole or apply water jets to assist in sheet piling.

3.1.8 Pile helmets and driving caps

When driving precast concrete piles, a helmet is placed over the pile head for the purpose of retaining in position a resilient 'dolly' or cap block that cushions the blow of the hammer and thus minimizes damage to the pile head. The dolly is placed in a recess in the top of the helmet (Figure 3.20). For easy driving conditions it can consist of an elm block, but for rather harder driving a block of hardwood such as oak, greenheart, pynkado or hickory is set in the helmet end-on to the grain. Plastic dollies are the most serviceable for hard-driving concrete or steel piles. The Micarta dolly consists of a phenolic resin reinforced with laminations of cross-grain cotton canvas. Layers of these laminates can be bonded to aluminium plates, or placed between a top steel plate and a bottom hardwood pad. The helmet should not fit tightly onto the pile head but should allow for some rotation of the pile, which may occur as it strikes obstructions in the ground.

Packing is placed between the helmet and the pile head to cushion further the blow on the concrete. This packing can consist of coiled rope, hessian packing, thin timber sheets, coconut matting, wallboards or asbestos fibre. The last-mentioned material has the advantage that it does not char when subjected to heat generated by prolonged driving. The packing must be inspected at intervals and renewed if it becomes heavily compressed and loses its resilience. Softwood packing should be renewed for every pile driven.

Williams[3.10] has described severe conditions for driving precast concrete piles at Uskmouth Power Station. He states that plastic dollies were used up to 40 times, compared with elm blocks which only lasted for a very few piles. The packing consisted of up to 125 mm of sawdust in jute bags, covered with two dry cement sacks placed at right-angles to each other

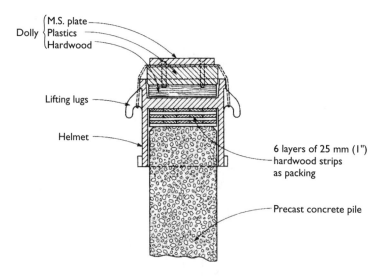

Figure 3.20 Dolly and helmet for precast concrete pile.

over the pile head. More recently, a thick cushion block of softwood, further softened by soaking, has been used for each pile to avoid damage when driving prestressed concrete piles.

Driving caps are used for the heads of steel piles but their function is more to protect the hammer from damage than to protect the pile. The undersides of the caps for driving box or H-section piles have projecting lugs to receive the head of the pile. Those for driving steel tubular piles (Figure 3.21) have multiple projections that are designed to fit piles over a range of diameters. They include jaws to engage the mating hammers.

Plastic dollies of the Micarta type have a long life when driving steel piles to a deep penetration into weak rocks or soils containing cemented layers. However, for economy contractors often cushion the pile heads with scrap wire rope in the form of coils or in short pieces laid cross-wise in two layers. These are replaced frequently as resilience is lost after a period of sustained driving. If dollies have to be changed while driving a pile, it should be noted that the blow count could change significantly.

3.1.9 Jetting piles

Water jets can be used to displace granular soils from beneath the toe of a pile. The pile then sinks down into the hole formed by the jetting, so achieving penetration without the use of a hammer. Jetting is a useful means of achieving deep penetration into a sandy soil in conditions where driving a pile over the full penetration depth could severely damage it. Jetting is ineffective in firm to stiff clays, however, and when used in granular soils containing large gravel and cobbles the large particles cannot be lifted by the wash water. Nevertheless, the sand and smaller gravel are washed out and penetration over a limited depth can be achieved by a combination of jetting and hammering. Air can be used for jetting instead of water, and bentonite slurry can be also used if the resulting reduced shaft friction is acceptable.

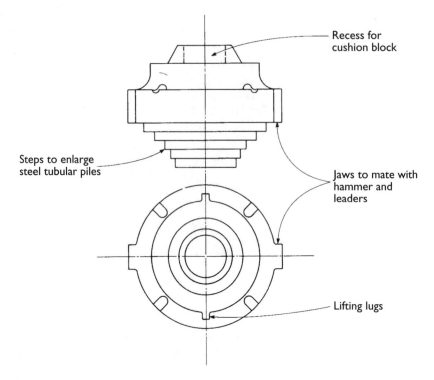

Recess for
cushion block

Steps to enlarge
steel tubular piles

Jaws to mate with
hammer and
leaders

Lifting lugs

Figure 3.21 Vulcan driving cap for steel tubular pile.

For jetting piles in clean granular soils a central jetting pipe is the most effective method, as this helps to prevent the pile from deviating off line. A 25 to 50 mm nozzle should be used with a 50 to 75 mm pipe (Figure 3.22). The quantity of water required for jetting a pile of 250 to 350 mm in size ranges from 15 to 60 l/s for fine sands through to sandy gravels. A pressure at the pump of at least 5 bars is required. The central jetting pipe is connected to the pump by carrying it through the side of the pile near its head. This allows the pile to be driven down to a 'set' on to rock or some other bearing stratum immediately after shutting down the jetting pump.

A central jetting pipe is liable to blockage when driving through sandy soils layered with clays, and the blockage cannot be cleared without pulling out the pile. A blockage can result in pipe bursting if high jetting pressures are used. An independent jetting pipe worked down outside the pile can be used instead of a central pipe, but the time spent in rigging the pipe and extracting it can cause such delays to pile driving as to be hardly worth the trouble involved. Open-ended steel tubular piles and box piles can be jetted by an independent pipe worked down the centre of the pile, and H-piles can be similarly jetted by a pipe operated between the flanges. Large-diameter tubular piles can have a ring of peripheral jetting pipes, but the resulting pile fabrication costs are high. Gerwick[3.11] has described the system for jetting 4 m diameter tubular steel piles for a marine terminal at Cook Inlet, Alaska. Sixteen 100 mm pipes were installed around the inner periphery of the pile. The nozzles were cut away at each side to direct the flow to the pile tip. Gerwick recommends that jetting nozzles

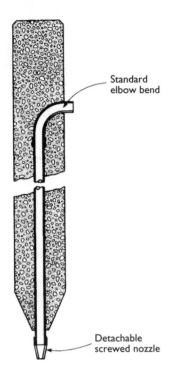

Standard
elbow bend

Detachable
screwed nozzle

Figure 3.22 Centrally placed jetting pipe.

should terminate about 150 mm above the pile tip. He gives the following typical requirements for jetting large diameter piles:

Jet pipe diameter   —   40 mm
Pressure   —   20 bar (at pump)
Volume   —   13 l/s per jet pipe

The large volume of water used in jetting can cause problems by undermining the piling rig or adjacent foundations as it escapes towards the surface. It can also cause a loss of shaft friction in adjacent piles in a group. Where shaft friction must be developed in a granular soil the jetting should be stopped when the pile has reached a level of about 1 m above the final penetration depth, the remaining penetration then being achieved by hammering the pile down. The jetting method is best suited to piles taken down through a granular overburden to end-bearing on rock or some other material resistant to erosion by wash water.

Water jetting is also used in conjunction with press-in and vibratory piling techniques to assist penetration of sheet piles in dense granular soil. A lance is fitted inside the pile pan and both are driven simultaneously into the ground. On reaching the required depth the lance is removed for reuse. Low injection rates are used at high pressure (5 l/s at 150 bar).

3.2   Equipment for installing driven and cast-in-place piles

The rigs used to install driven and cast-in-place piles are similar in most respects to the types described in Sections 3.1.1 to 3.1.3 but the firms who install proprietary types of pile usually make modifications to the rigs to suit their particular systems. The piling tubes are of heavy section, designed to be driven from the top by drop, single-acting, or diesel hammers, but the original Franki piles (Figure 3.23 and Section 2.3.2) are driven by an internal drop hammer. The leaders of the piling frames are often adapted to accommodate guides for a concreting skip (Figure 3.24).

Steel cased piles designed to be filled with concrete are driven more effectively by a hammer operating on the top, than by an internal drop hammer acting on a plug of concrete at the base.

Figure 3.23 Franki pile-driving rig.

Figure 3.24 Discharging concrete into the driving tube of a withdrawable tube pile. Concreting skip travelling on pile frame leaders.

This is because a hammer blow acting on top of the pile causes the tube to expand and push out the soil at the instant of striking, followed by a contraction of the tube. This frees the tube from some of the shaft friction as it moves downward under the momentum of the hammer. The flexure of the pile acting as a long strut also releases the friction at the moment of impact. However, when using an internal drop hammer, tension is induced in the upper part of the pile and the diameter contracts, followed by an expansion of the soil and an increase in friction as the pile moves downwards. Flexure along the piling tube does not occur when the hammer blow is at the base, and thus there is no reduction in friction from this cause. Tension caused by driving from the bottom can cause the circumferential cracking of reinforced concrete and thin-wall steel tubular piles.

Top driving has another advantage in allowing the pile to be driven with an open end, thus greatly reducing the end-bearing resistance during driving, but the soil plug will have to be drilled out if the concrete pile is to be cast in place as the tube is withdrawn. The bottom-driven pile demands a solid plug at the pile base at all stages, but produces a dry open shaft for concreting. In easy driving conditions bottom driving will give economy in the required thickness of the steel and considerable reduction in noise compared with top driving. For example, Cementation Foundations Skanska installed 508 mm diameter bottom-driven thin wall (6 mm) steel piles up to 15 m long in Cardiff Bay in preference to thicker-walled, top-driven, cased piles to reduce disturbance to residents. A 4 tonne drop hammer was used to drive the bottom plug to found in Mercia Mudstone; concreting was direct from the mixer truck or by skip.

Great care is necessary to avoid bursting of the tube by impact on the concrete when bottom driving through dense granular soil layers or into weak rocks containing bands of stronger rock. The concrete forming the plug should have a compacted height of not less than 2.5 times the pile diameter. In calculating the quantity of concrete required, allowance should be made for a volume reduction of 20% to 25% of the uncompacted height. The 1:2:4 concrete should be very dry with a water/cement ratio not exceeding 0.25 by weight. A hard aggregate with a maximum size of 25 mm should be used.

At least 10 initial blows should be given with hammer drops not exceeding 300 mm then increasing gradually. The maximum height of drop should never exceed the maximum specified for the final set which is usually between 1.2 and 1.8 m. Driving on a plug should not exceed a period of 1½ hours. After this time, fresh concrete should be added to a height of not less than the pile diameter and driving continued for a period of not more than 1½ hours before a further renewal. For prolonged hard driving it may be necessary to renew the plug every three-quarters of an hour.

3.3   Equipment for installing bored and cast-in-place piles

3.3.1   Power augers

Power-driven rotary auger drills are suitable for installing bored piles in clay soils. A wide range of machines is available using drilling buckets, plate and spiral augers, and continuous flight augers, mounted on trucks, cranes, and crawlers to bore open holes. The range of diameters and depths possible is considerable, from 300 to over 5000 mm and down to 100 m. Hydraulic power is generally used to drive either a rotary table, a rotating kelly drive on a mast or a top-drive rotary head; some tables are mechanically operated through gearing. Most units have additional pulldown or crowd capability. The soil is removed from spiral

Figure 3.25 Watson 5000 crane attachment power auger on elevated platform on a 40 tonne crane with 200 mm telescopic kelly for installing 2440 mm casings.

plate augers by spinning them after withdrawal from the hole and from buckets either by spinning or through a single or double bottom opening. It is an EU mandatory safety requirement that spoil from an auger should be removed at the lowest possible level during extraction to ensure that debris from the flights cannot fall onto personnel or damage machinery and to avoid rig instability. Hydraulically operated cleaners which can be rapidly adjusted to suit CFA diameters from 400 to 2000 mm are available.

The Calweld drilling machines comprise lorry-mounted bucket drills, with either hydraulic or mechanical drives, crane attachments and plate augers. The lorry-mounted hydraulic bucket drills range in size from the 42 LH with a 914 mm diameter bucket capable of drilling to about 29 m with a triple telescoping kelly or to 60 m with extensions, to the 5200 LH capable of operating a 3650 mm bucket and reaming gear to drill shafts to depths up to 52 m with a triple telescoping kelly. The mechanical drive models, 150C to 250C, have capabilities ranging from 914 mm diameter at 21 m depth to 3350 mm at 26 m. The ADL and ADM auger rigs are also lorry-mounted with rigid rotary tables producing 88 kNm torque for drilling 1800 mm diameter holes to 30 m using double telescoping kellys. Calweld crane attachment drills range in size from the 125 CH providing 169 kNm torque to the 400 CH 540 kNm torque all with large opening (686 mm) rotary tables. These rigs can operate with square or round quadruple kellys 21 m long and have standard bridge frames up to 5.18 m drilling radius. Watson produce drills for mounting on trucks, crawlers and as crane attachments (Figures 3.25 and 3.26) capable of producing shafts up to 3650 mm diameter to depths of 41 m also using quadruple kellys. The Casagrande RM21 mechanical rotary table crane attachment can drill a maximum diameter of 2500 mm to a depth of 82 m using maximum torque of 220 kNm. The Soilmec RT3 and R25 rigs are crane-mounted, mechanical and hydraulic

Figure 3.26 Watson 2100 truck-mounted auger drill.

drives respectively; maximum depth with mechanical drive is 102 m at 2500 mm diameter using maximum torque of 226 kNm. Soilmec truck-mounted rigs produce 100 kNm torque for drilling to 43 m at 1800 mm diameter.

Bauer has developed a powerful bucket auger unit (the 'Flydrill System') which integrates the hydraulic power packs and the rotary drive on one platform for mounting on top of a partially driven tubular pile. The rotary drive produces a torque of 462 kNm at 320 bar and two hydraulic crowd cylinders provide a pulldown of 40 tonne. The clamping device can exert a total force of 90 tonne to resist the torque and apply the pulldown. The system operates a triple telescope kelly with 3 and 4.4 m diameter buckets and was used for cleaning out and reaming below 4.75 m diameter tubular monopile foundations to allow driving to be completed to a depth of 61 m at the offshore wind farm in the Irish Sea off Barrow in Furness (Figure 3.27).

The range and capabilities of crawler-mounted hydraulic rotary piling rigs have increased significantly in recent years. The rigs in Table 3.6 are usually capable of installing CFA and rotary displacement piles as well as standard bored piles, but the height of the mast and stroke available may limit the depth achievable, hence the major manufacturers produce special long stroke rigs for CFA piles up to 34 m deep. For bored piles many rigs can accommodate casing oscillators and most have rams or winches to provide additional crowd and extraction forces, requiring robust masts and extendable tracks for stability. The major manufacturers produce double rotary heads (usually capable of rotating in opposite directions) as attachments for the more powerful piling rigs which enable casing up to 1000 mm diameter to be installed with the lower drive while augering with the top drive. The dual-rotary system from Foremost Industries of Canada operating on their DR 40 crawler

Figure 3.27 'Flydrill 5500' with bucket auger removing spoil from 4.75 m dia monopiles at the Barrow offshore wind farm site (courtesy Bauer Maschinen GmbH).

rig provides 30 kNm torque through the top drive for boring and 339 kNm torque on the lower rotary table for simultaneous casing up to 1000 mm diameter. The Liebherr pile driving rigs (see Section 3.1.1) have the option of running double rotary top drive or kelly tools for bored and CFA piles. In-cab electronic instrumentation and read-out to control positioning and drilling parameters is standard on most rigs.

Table 3.6 Some hydraulic self-erecting crawler boring rigs

Maker	Type	Standard stroke (m)	Main winch capacity (kN)[a]	Maximum diameter (mm)	Typical maximum depth (m)	Maximum torque (kNm)
Bauer	BG 15H	12	110	1500	40	151
(Germany)	BG 18H	14	140	1500	45	176
	BG 20H	15	170	1500	53	200
	BG 24	15.7	200	2000	57	233
	BG 24H	15.4	200	1700	54	233
	BG 28	19.3	250	2100	71	275
	BG 28H	18.4	250	1900	71	275
	BG 36	18.7	250	2500	68	367
	BG 40	19.7	300	3000	80	390
	BG 48	8.8	600	3000	100	482
Casagrande	B80	3.5	105	1300	38	100
(Italy)	B80 as CFA	11.15	105	600	16	100
	B125	3.8/11[b]	135	1500	50	112
	B125 as CFA	14.23	135	800	20	112
	B180HD	6/13.7[b]	200	1800	68	180
	B180HD as CFA	17.13	200	1000	22.7	180
	B250	6/12.5[b]	220	2000	68	217
	B250 as CFA	19.75	220	1200	23.7	217
	B300	6/12.5[b]	250	2000	68	250
	C600HD.H40 as CFA	21.25	280	1200	28.5	358
	C600HD HT H40	14[b]	280	2200	87	358
	C800 H50	14	280	2200	87	546
	C800 as CFA	28	280	1600	34	358
	C800DH	28/19.1	280	900	25/18.5	358/421
	C6[c]/M6A-1[c]	4				13.5
	C8[c]	8				17.9
	C14[c]	12				55
	CFA 425	19.4	135	900	25	112
	CFA 26	21.5	150	1200	27	137
Delmag	RH10	9.5	160	1450	15	100
(Germany)	RH12	12	200	1450	18	120
	RH14	12.5	200	1580	23	144
	RH20	14.2	300	1830	30	206
	RH26	15	420	1960	36	265
	RH32	15	420	1960	36	320
Foremost	DR 24	7.9	380	610		13/282 Dual
Industries	DR 40	8.4	380	1016		30/339 Dual
(Canada)	DR 610C	13.4	181	610		28/282 Dual
Soilmec	R 210	9	103	1200	40	100
(Italy)	R 312/200	10.5	133	1500	48	130
	R 416	11.4	150	1500	55	161
	R 516 HD	11.5	170	2000	61	180
	R 620	14.5	192	1800	66	201
	R 625	15.2	240	2500	77	240
	R 725	16.3	240	2000	77	240
	R 825		240	2500	77	240
	R 930		290	3000	77	305

Table 3.6 Continued

Maker	Type	Standard stroke (m)	Main winch capacity (kN)[a]	Maximum diameter (mm)	Typical maximum depth (m)	Maximum torque (kNm)
	R 940		320	3000	92	469
	CM 50 (CFA)	19.5	102	900	25	100
	CM 70 (CFA)	22.3	170	1000	28	154
	CM 700 (CFA)	22.5	80	1000	29	165
	CM 120 (CFA)	24.5	290	1400	30.5	305
	CM 1200 (CFA)	27.9	290	1400	33.5	305
Wirth	13-SV	13.5	120	1500	45	130
(Germany)	18	6.8	190	2500	45	176
	22-ZV	6.8[b]	180	1800	56	220
	22-SV	13.8	180	1800	56	220
	40	7.7	290	3000	36	400

Notes
a Pulling force.
b With pull-down winch.
c Various masts and rotary heads for these hydraulic crawlers.

Various types of equipment are available for use with rotary augers. The standard and rock augers (Figure 3.28a and b) have scoop-bladed openings fitted with projecting teeth. The coring bucket is used to raise a solid core of rock (Figure 3.28c) and the bentonite bucket (Figure 3.28d) is designed to avoid scouring the mud cake which forms on the wall of the borehole. The buckets on some Calweld machines can be lifted through the ring drive gear and swung clear to discharge the soil. Grabs can also be operated from the kelly bar.

Enlarged or under-reamed bases can be cut by rotating a belling-bucket within the previously drilled straight-sided shaft. The bottom-hinged bucket (Figure 3.29a) cuts to a hemispherical shape and because it is always cutting at the base it produces a clean and stable bottom. However, the shape is not so stable as the conical form produced by the top-hinged bucket (Figures 3.29b and 3.30), and the bottom-hinged arms have a tendency to jam when raising the bucket. The arms of the top-hinged type are forced back when raising the bucket, but this type requires a separate cleaning-up operation of the base of the hole after completing the under-reaming. Belling buckets normally form enlargements up to 3.7 m in diameter but can excavate to a diameter of 7.3 m with special attachments. Belling buckets require a shaft diameter of at least 0.76 m to accommodate them.

The essential condition for the successful operation of a rotary auger rig is a fine-grained soil which will stand without support until a temporary steel tubular liner is lowered down the completed hole, or a granular soil supported by a bentonite slurry or other stabilizing suspensions (also known as 'drilling muds' see Section 3.3.8). In these conditions fast drilling rates of up to 7 m per hour are possible for the smaller shaft sizes. Methods of installing piles with these rigs are described in Section 3.4.6.

3.3.2 Boring with casing oscillators

For drilling through sands, gravels, and loose rock formations, the pile boreholes may require continuous support by means of casing. For these conditions it is advantageous to use a casing oscillator which imparts a semi-rotating motion to the casing through clamps. Vertical

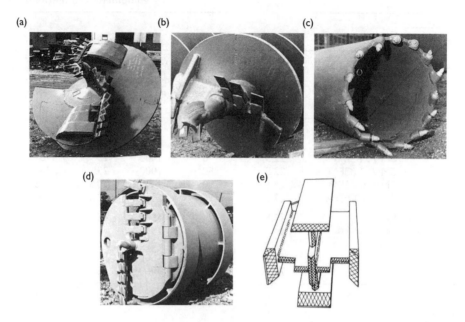

Figure 3.28 Types of drilling tools (a) Standard auger (b) Rock auger (c) Coring bucket (d) Bentonite bucket (e) Chisel.

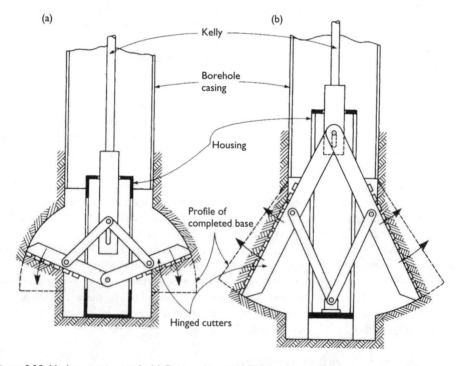

Figure 3.29 Under-reaming tools (a) Bottom hinge (b) Top hinge.

Figure 3.30 Top-hinged under-reaming bucket.

rams attached to the clamps enable the casing to be forced down, as the hole is deepened or raised as necessary. The semi-rotating motion is continuous, which prevents the casing from becoming 'frozen' to the soil, and it is continued while extracting the casing after placing the concrete. The essential feature of pile boring with a casing oscillator is that the special double-wall casing is always kept tight against the bottom of the hole. For this purpose the casing is jointed. The typical casings (e.g. the Bauer and Casagrande types) have male/female joints which are locked by inserting and tightening bolts manually (which can have safety implications) or by an automatic adapter lock to resist the high rotating or oscillating forces.

Hydraulic casing oscillators are available from most of the large rig manufacturers to attach to crane-mounted rigs or to rotary drills with diameters from 1000 to 3000 mm and torque capability up to, for example, 8350 kNm from the Soilmec VRM 3000, which has a clamping force of 4780 kN and lifting force of 7250 kN (Figure 3.31). Typically the rotation angle is 25°, but Soilmec and Leffer also manufacture 360° full rotation units with torque up to 7400 kN, designed to reduce friction losses and capable of depths to 70 m using carbide casing shoes. The material is broken up and excavated from the pile by hammer grabs hanging from a crane or attached to the drill rig, or removed by augers, grabs and down-the-hole hammers using crawler or crane attachment rigs. The German Hochstrasser-Weise oscillator system which operates on compressed air has been used to install casing up to 3000 mm vertically, and 2400 mm casing

Figure 3.31 Soilmec casing oscilator.

24 m deep at a 1 in 4 rake, but difficulties can occur when extracting the casing during concreting. The German Bade oscillators are used for piles up to 2500 mm in diameter.

Drilling and installing casing simultaneously ('duplex' drilling) through cobbles, boulders and rubble using special casing shoes and casing under-reamers attached to top drive, down-the-hole compressed air hammers has advanced significantly. For example, Numa Hammers of USA manufacture a range of drills capable of installing casing up to 1219 mm diameter to 15 m deep using a rotary percussive under-reamer which can be retracted to allow concreting of the pile as the casing is withdrawn (Figure 3.32).

3.3.3 Continuous flight auger drilling rigs

A typical continuous flight auger rig is shown in Figure 3.33. Drilling output with the rigs in Table 3.6 is greater than achievable for standard bored piles as the pile is installed in one continuous pass, hence the mast must have an adequate stroke for the auger under the rotary head. A kelly may be inserted through the rotary head to increase depth on some rigs. Most CFA rigs have crowd capability to assist in penetrating harder formations, and augers should be designed to suit the high torques available. Possible diameters range from 500 to 1400 mm to a maximum depth of 34 m.

Displacement auger piling is carried out with rigs similar to the high torque CFA equipment, but the diameter is limited to less than 600 mm by the shape of the displacement tool; maximum depth is around 28 m.

3.3.4 Drilling with a kelly

The kelly is a square or circular drill rod which is driven by keying into a rotary table either fixed to the rig near the ground surface, to the crane attachment or by a moveable drive head on the mast. The full range of drilling tools, plate and bucket augers, drag bits, compound rotary

Figure 3.32 Numa hammer with extending under-reaming drill bit for simultaneously drilling and inserting casing (courtesy Numa Hammers).

drill plate bits and tricone bits can be rotated by the kelly in most drillable ground conditions, subject to the available power. The kelly may be in sections, or more usefully, telescopic to make up the required length of drill string. Boreholes can be drilled as open holes, or supported by either excess hydrostatic head using drilling fluids (see Section 3.3.8) or by casing. The casing can be installed by oscillators or by the rotary drive with some of the larger rigs. Under-reamers and belling tools are expanded by an upward or downward force from the rotating kelly.

Figure 3.33 Installation of CFA piles in chalk with crane handling reinforcement cage (courtesy Cementation Foundations Skanska).

3.3.5   *Reverse-circulation drilling rigs*

Reverse-circulation drilling rigs operate on the principle of the airlift pump. Compressed air is injected near the base of the centrally placed discharge pipe. The rising column of air and water lifts the soil which has been loosened by rotating cutters, and the casing tubes are also rotated to keep them freely moving in the soil as they sink down while the boring advances. The reverse-circulation rig manufactured by Alfred Wirth and Co. of Germany is shown in Figure 3.34. The casing tubes and airlift riser pipe are rotated together or separately by means of a hydraulic rotary table or power swivel. Dual airlift drill pipes, maximum bore 300 mm, either flange jointed or flush, inject air through the annulus between the inner and outer tubes. The riser pipe is maintained centrally in the casing by one or more stabilizers, and the soil boring is effected by rock roller bits on a cutter head. The diameters of the latter range from 0.76 to 8.0 m. The injected air-flow and pressure and the point of injection all affect the efficiency of cuttings removal; air injection rate is up to 130 m^3 per minute at a pressure of 12 bar, requiring large air compressors. For offshore work the hole will be kept full of seawater, but on land drilling mud is used to convey the cuttings necessitating the use of mud tanks and cleaners (see Section 3.3.8). Also on land the reverse circulation system with mud may maintain a stable hole without the use of casing for cast-in-place piles. The more powerful self-erecting crawler rigs in Table 3.6 can be rigged for reverse circulation for holes up to 3300 mm diameter to 100 m depth. The Calweld reverse circulation rigs are manufactured to drill without reaming to diameters of up to 2.1 m and to depths up to 230 m.

Piletop rigs such as the Wirth PBA range (Figure 3.35) and the 'Teredo' units built and operated by Seacore Limited (Figure 3.36) using powerful top-drive swivels are more versatile than large rotary tables. The Wirth 1238 rig has a maximum power swivel torque of 380 kNm for

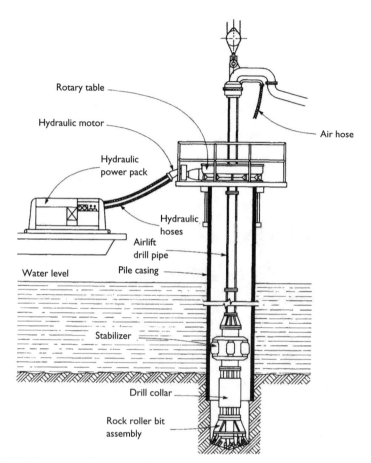

Figure 3.34 Wirth rotary table and rotating cutter.

drilling 4000 mm diameter holes with a drill string weighing 130 tonne. The largest Teredo rig, equipped with a 460 kNm power swivel, is capable of rock drilling up to 7000 mm diameter. The Bauer power auger in Figure 3.27 can be classified as a pile top rig, but has to be handled off the pile to discharge the bucket, requiring continuous service by a suitable crane.

Reverse-circulation rigs can drill at a fast rate in a wide range of ground conditions including weak rocks. They are most effective in granular soils and the large diameter of the airlift pipes enables them to lift large gravel, cobbles, and small boulders when drilling in glacial soils, or in jointed rocks which are broken up by the rock roller bits. Under-reamed bases can be provided in stiff clays or weak rocks by means of a hydraulically operated rotary enlarging tool mounted above the cutter head.

3.3.6   Large grab rigs

The use of diaphragm wall grabs to form barrettes in preference to large-diameter bored pile groups is well established. The grabs may be suspended from cranes or mounted on

Figure 3.35 Wirth PBA pile top rig (courtesy Wirth Maschinen und Bohrgerate-Fabrik GmbH).

purpose-built crawlers and excavate a square hole, Ell-, Tee-, or rectangular slots under bentonite or other support fluid. The 'Hydromill' or 'Hydrofraise' rig developed by Bachy-Soletanche is a reverse circulation down-the-hole milling machine with two contra-rotating cutter drums powered by hydraulic motors mounted on a heavy steel frame. The cuttings are removed from the slot in a bentonite or polymer slurry by a pump fitted above the drums to the de-sanding and cyclone plant at the surface for reconditioning and reuse. Over-break is minimal and the absence of vibration makes the system suitable for urban sites and close to existing buildings. Standard width is 600 mm but greater widths are possible for depths to 60 m. Walls have been constructed to 150 m deep, and a low head-room version is available.

3.3.7   Tripod rigs

Small-diameter piles with diameters of up to 600 mm installed in soils which require con-tinuous support by lining tubes can be drilled by tripod rigs. The drilling is performed in clays by a clay-cutter, which is a simple tube with a sharpened cutting edge, the tube being driven down under the impact of a heavy drill stem. The soil which jams inside the tube is

Figure 3.36 Pile top rig drilling 3.8 m dia piles for foundation strengthening to the Richmond-San Raphael bridge, California (courtesy Seacore Ltd).

prised out by spade when the cutter is raised to the surface. Drilling is effected in coarse-grained soils by means of a baler or 'shell', which is again a simple tube with a cutting edge and flap valve to retain the soil, the soil being drawn into the baler by a suction action when the tool is raised and lowered. If no groundwater is present in the pile borehole, water must be poured in, or a bentonite slurry may be used. This suction action inevitably causes loosening of the soil at the base of the pile borehole, thus reducing the base resistance. The loosening may be accompanied by settlement of the ground surface around the pile borehole. Rocks are drilled by chiselling and using a baler to raise the debris.

Tripod rigs are not as suitable as the spiral-plate or continuous auger types for drilling small-diameter piles in clays, except in situations where low headroom or difficult access would prevent the deployment of lorry-mounted or track-mounted augers. Methods of operating tripod rigs have been described by North-Lewis and Scott[3.12].

3.3.8   Drilling for piles with bentonite slurry

Lining tubes or casings to support the sides of pile boreholes are a requirement for most of the bored-pile installation methods using equipment described in Sections 3.3.1 to 3.3.7. Even in stiff fine-grained soils it is desirable to use casings for support since these soils are frequently fissured or may contain pockets of sand which can collapse into the boreholes, resulting in accumulations of loose soil at the pile toe, or discontinuities in the shaft.

Casings may be avoided by providing support to the pile borehole in the form of a slurry of bentonite clay. However, BS EN 1536 requires that an excavation under support fluid shall be protected by a lead-in tube or guide wall (for a barrette); a length of casing will also be required to carry a pile-top rig. Bentonite, or other montmorillonite clay with similar characteristics, has the property of remaining in suspension in water to form a stiff 'gel' when allowed to become static. When agitated by stirring or pumping, however, it has a mobile fluid consistency – i.e. it is 'thixotropic'. In a granular soil, the slurry penetrates the walls of the borehole and gels there to form a strong and stable 'filter-cake'. In a clay soil there is no penetration of the slurry but the hydrostatic pressure of the fluid, which has a density of around $1\,040$ kg/m^3, prevents collapse where the soil is weakened by fissures. Recommended rheological characteristics are provided in BS EN 1536.

When used in conjunction with auger or grab-type rigs the slurry is maintained in a state of agitation by the rotating or vertical motion of the drilling tools. When it becomes heavily contaminated with soil or diluted by groundwater it can be replaced by pumping-in fresh or reconditioned slurry. Toothed or bladed augers with double helix configurations and a flap in the carriage area help to retain spoil as the auger is withdrawn through the bentonite. Bentonite slurry is used most efficiently in conjunction with reverse-circulation rigs (see Section 3.3.5). The slurry is pumped into the outer annulus and the slurry–soil mixture that is discharged from the airlift riser pipe is allowed to settle in lagoons or tanks to remove soil particles. It is then further cleaned in a cyclone, and chemicals to aid gelling are added before the reconditioned slurry is pumped into a holding tank and then returned to the pile borehole.

Bentonite slurry is used in a simple and rather crude way in conjunction with rotary auger equipment when drilling pile boreholes through sands and gravels to obtain deeper penetration into stiff fine-grained soils. The hole must be augered through the sands and gravels without support, and then the casing is lowered down. It is uneconomical to provide screwed joints in large-diameter lining tubes and all joints are made by welding. To save time and cost in welding, the holes are drilled to the maximum possible depth before installing the first length of casing. In these conditions support may be provided while drilling by means of a bentonite slurry. Where the depth of coarse-grained soil is relatively small it is uneconomical to bring in high-speed mixers, slurry tanks, pumps and reconditioning plant for the normal employment of bentonite techniques. Instead, a few bags of the dry bentonite are dumped into the pile borehole and mixed with the groundwater or by adding water to form a crude slurry which is adequate to smear the wall of the borehole and give it the necessary short-term support. After drilling with this support through the granular overburden, the casing is lowered in one or more lengths and pushed down to seal it into the stiff fine-grained soil below. The thrust is provided either by the hydraulically operated crowd mechanism on the kelly-bar of the drilling machine or by means of a vibrator (see Section 3.1.5) mounted on the casing. This technique is known as 'mudding-in' the casing.

The use of a bentonite slurry to aid drilling with or without temporary lining tubes may cause some difficulties when placing concrete in the pile. The nature of these problems and the means of overcoming them are described in Section 3.4.8, and the effects of a bentonite slurry on shaft friction and end-bearing resistance of piles are discussed in Sections 4.2.3 and 4.3.6. For example, if the slurry becomes overloaded with solids from the excavation, a thick filter cake will be formed and may not be removed by scouring during concreting. In such cases it may be necessary to use a mechanical scraper to remove the excess filter cake prior to concreting. Reese et al.[3.13] recommend a minimum diameter of 600 mm for piles installed using slurry techniques, to avoid some of the problems associated with the method.

Polymer support fluids are more expensive than bentonite as an initial cost, but because they can be recycled without the frequent de-sanding required for bentonite, polymers can be economical for use on large projects and congested sites. Also the filter cake is much thinner and more easily scoured when placing concrete. When used for piling work on land or in river works, waste bentonite slurry has to be treated as 'hazardous' under pollution control regulations and disposed of accordingly, whereas polymers can be neutralized and, subject to desanding and approval from the water company, can be disposed of to existing drains.

3.3.9   Base and shaft grouting of bored and cast-in-place piles

When bored and cast-in-place piles are installed in granular soils, the drilling operation may loosen the soil surrounding the shaft and beneath the base of the pile borehole. Such loosening below the base can cause excessive working load settlements when the majority of the load is carried by end bearing. Base grouting is a means of restoring the original in-situ density and reducing settlements. Bolognesi and Moretto[3.14] described the use of stage grouting to compress the soil beneath the toes of 1.00 to 2.00 m bored piles supporting two bridges over the Parana River in Brazil, the piles being drilled with the aid of a bentonite slurry. The soil beneath the pile toes loosened by the drilling operations was subjected to a grouting pressure of up to 10 MN/m². The cement grout was introduced through a cylindrical metal basket pierced by a number of holes and filled with uniform gravel (Figure 3.37). The basket, with its upper surface covered by a rubber sheet, was lowered into the borehole suspended from the pile reinforcing cage. The pile was then concreted, followed by the injection of the grout into the basket through a 38 mm pipe set in the concrete of the shaft. The uplift caused by the grouting pressure was usually resisted by the shaft friction in the pile shaft, but in some cases the pile cap was constructed to provide additional dead-load resistance. Although Bolognesi and Moretto did not mention any weakening at the pile toe caused by the entrapment of bentonite slurry, as described by Reese et al.[3.13], the stage-grouting technique would be a useful method of expelling any slurry from beneath the toe of a pile.

The 'flat-jack' method of pressure grouting to compact soil beneath the base of a bored pile is similar. After completing the drilling, which can be performed underwater in favourable conditions, the reinforcing cage with a circular plate welded to the base is

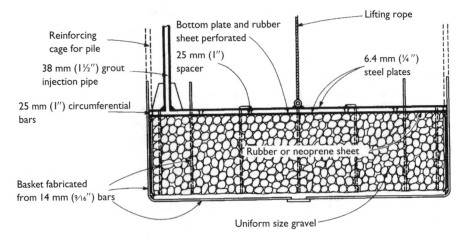

Figure 3.37 Preloading cell for compressing loosened soil beneath base of bored piles by grouting (after Bolognesi and Moretto[3.14]).

lowered to the bottom of the borehole. A flexible metal sheet covers the whole of the underside of this base plate. A grout injection pipe is connected to the space between the plate and the sheet, and a peripheral ring of grout pipes is attached to the reinforcing bars for a predetermined height above the pile base. All grout pipes are extended to a pump and metering unit at the ground surface. The pile is then concreted. After a waiting period to allow the concrete to harden a cement grout is injected into the peripheral injection pipes with the object of bonding the lower part of the pile shaft to the surrounding soil. A further period of a few days is allowed for the grout to harden, then the space between the metal sheet and steel plate is injected with grout under high pressure. The uplift on the steel plate is resisted by the peripheral grout–soil bond stress on the shaft and the soil beneath the flexible sheet is thus compressed. The height of the peripheral grouting above the pile base depends on the required base pressure and hence on the design base resistance of the pile. The important control criteria are the grout pressure (a function of the required ultimate pile capacity), an upward displacement limit to minimize degradation of the side friction, and a minimum grout injection volume to ensure lines are not clogged.

Direct injection of cement grout beneath the pile base was used to re-compress sand disturbed by drilling 1.2 m diameter bored piles supporting an office building at Blackwall Yard, London. Yeats and O'Riordan[3.15] described the installation of a 38.2 m deep test pile. The shaft was drilled by rotary auger under a bentonite slurry through the alluvium and stiff to hard clays of the London Clay and Woolwich and Reading formation (Lambeth group) into very dense Thanet Sands. The upper 31 m of the shaft were supported by casing. After completing the drilling four separate grout tube assemblies as shown in Figure 3.38 were lowered to the base of the borehole. The injection holes in the tubes were sleeved with rubber (tubes-à-manchette). The pile shafts were then concreted under bentonite, and 24 hours after this water was injected to crack the concrete surrounding the grout tubes. Base grouting commenced 15 days after concreting. The injections were undertaken in stages with pressures up to 60 bar and frequent checks to ensure the pile head did not lift by more than 1 mm.

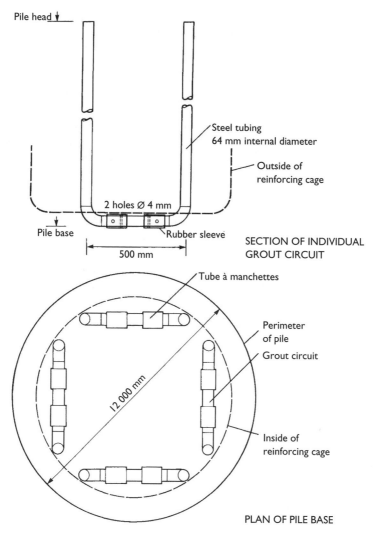

Figure 3.38 Arrangement of circuits for base-grouting of piles (after Yeats and O'Riordan[3.15]).

Similar base grouting techniques were used at six sites in the Docklands area of London beneath piles with diameters in the range of 0.75 to 1.5 m[3.16].

Part of the internal plugs to the 2.50 and 3.13 m OD driven tubular steel piles for the Jamuna River Bridge[3.17] were cleaned out by airlifting which loosened the soil at the base. In order to reconsolidate the remaining plug of sand a grid of tubes-à-manchette was placed in the hole above the plug and a layer of gravel placed by tremie to cover the grout tubes. A 7 m plug of concrete was placed over the gravel and 12 hours later, water was injected at a pressure of 20 bar to lift the sleeves. Cement grout (40 litres of water, 50 kg cement, 0.35 kg bentonite, and 0.5 kg plasticizer) was then injected into the gravel plug. Grouting was terminated when the pressure reached 50 bar, in order to ensure that uplift of the pile would not occur, or when 1000 litres of fluid had been injected to limit hydrofracture of the soil below the gravel.

Shaft grouting of cast-in-place piles and barrettes entails rupturing the outer skin of the pile and pushing it against the surrounding soil. This increase in lateral pressure is intended to cause local increases in the soil density which had become loosened or softened by the pile construction and thereby enhance the shaft resistance of the pile. When shaft grouting in granular soils, cementation of the soil particles may occur and voids and fissures become filled giving improved contact between pile and soil. The usual technique is to install 50 mm diameter steel tubes-à-manchette around the perimeter of the reinforcement cage for the depth to be treated, with non-return connections to the surface. The manchettes on the tubes at 1 m centres are cracked at pressures up to 80 bar and flushed with water after allowing the concrete to cure for 24 hours and each sleeve pressure grouted 10 to 15 days thereafter. Littlechild *et al.*[3.18] report on a series of tests on 20 shaft grouted, cast-in-place piles in soft marine clay underlain by alluvial deposits of stiff clay and dense to very dense sand in Bangkok. The measured shaft resistances for the shaft grouted piles, ranging from 150 to 320 kN/m^2, were approximately double those without shaft grouting. The test piles were reloaded more than one year after grouting and showed no loss of resistance in either the clay or sands. Core samples along the pile/grout interface showed grout infilling cracks and fissures in the concrete and a grout zone 20 to 30 mm around the pile with some cementation of the sands.

3.4   Procedure in pile installation

Each class of pile employs its own basic type of equipment and hence the installation methods for the various types of pile in each class are the same. Typical methods are described below to illustrate the use of the equipment described in the preceding sections of this chapter. Particular emphasis is given to the precautions necessary if piles are to be installed without unseen breakage, discontinuities or other defects. The installation methods described in this section are applicable mainly to vertical piles. The installation of raking piles whether driven or bored is a difficult operation and is described in Section 3.4.11.

BS EN 1536 and BS EN 12699 deal with the execution of bored and displacement piles respectively. However, in many respects the guidance on installation in these new codes is not as comprehensive as that contained in BS 8004. For example, BS EN 12699 does not comment on appropriate penetration, stroke, drop or weight of hammer, simply requiring that a suitable hammer or vibrator be used to achieve the required depth or resistance without damage to the pile.

3.4.1   Driving timber piles

Timber piles are driven by drop hammer or single-acting hammer after pitching them in a piling frame, in crane-suspended leaders or in trestle guides. The Swedish piling code requires the hammer to weigh at least 1.5 times the weight of the pile and helmet with a minimum of 1 tonne. Diesel hammers, unless they are of the light type used for driving trench sheeting, are too powerful and are liable to cause splitting at the toe of the pile. The heads of squared piles are protected by a helmet of the type shown in Figure 3.20. Round piles are driven with their heads protected by a steel hoop. A cap is used over the pile head and hoop, or packing can be placed directly on the head.

Care should be taken to prevent damage to the creosote protection by avoiding the use of slings or hooks which gouge the pile deeply. The damage caused by minor incisions is no more than the scratching caused by stones encountered while driving the piles.

3.4.2   Driving precast (including prestressed) concrete piles

The methods of handling the piles after casting and transporting them to the stacking area are described in Section 2.2.2. They must be lifted from the stacking positions only at the prescribed points. If designed to be lifted at the quarter or third points, they must not at any stage be allowed to rest on the ground on their end or head. Particular care should be taken to avoid over-stressing by impact if the piles are transported by road vehicles. Additional support points should be introduced if necessary.

A helmet of the type shown in Figure 3.20 and its packing are carefully centred on the pile, and the hammer position should be checked to ensure that it delivers a concentric blow. The hammer should preferably weigh not less than the pile. BS 8004 requires that the weight or power of the hammer should be sufficient to ensure a final penetration of about 5 mm per blow unless rock has been reached. Damage to the pile can be avoided by using the heaviest possible hammer and limiting the stroke. BS 8004 states that the stroke of a single-acting or drop hammer should be limited to 1.2 m and preferably to not more than 1 m. The Swedish piling code requires a drop hammer to weigh at least 3 tonne, except that 2-tonne hammers can be used for piles with a maximum length of 10 m and a maximum load of 450 kN, but a 4-tonne hammer should be used for long piles in compact materials. This code recommends that the drop of the hammer should be limited to 300 to 400 mm in soft or loose soils to avoid damage by tensile stresses. The drop should be limited to 300 mm when driving through compact granular soils.

The driving of the piles should be carefully watched, and binding by toggle bolts due to the pile rotating or moving off line should be eased. The drop of the hammer should be reduced if cracking occurs, and if necessary the hammer should be changed for a heavier one. After the completion of driving the pile heads should be prepared for bonding into the pile caps as described in Section 7.7. Hollow piles with a solid end may burst under the impact of the hammer if they become full of water, and holes should therefore be provided to drain off accumulated water. Where a soil plug is formed at the toe of an open-ended pile, water accumulation or arching of the soil within the pile may also result in bursting during driving. Further guidance is given in CIRIA Report PG8[3.19].

3.4.3   Driving steel piles

Because of their robustness steel piles can stand up to the high impact forces from a diesel hammer without damage other than the local distortion of the pile head and toe under hard driving. Open-ended tubular or box piles or H-piles can be driven to a limited penetration by a vibrator. By using rolled steel corner sections, plugged tube-bearing piles can be formed by driving a number of interlocking U-section sheet piles sequentially. As the resistance to driving is less than for welded box piles, vibrators or press-in pilers can be used to install high capacity piles to greater depths at sensitive sites where impact driving cannot be tolerated.

To achieve the required depth of penetration it is sometimes necessary to reduce the base resistance by removing the soil plug which forms at the bottom of an open-ended tubular or box pile. A sandy-soil plug can be removed by simple water jetting. A plug of clay or weak

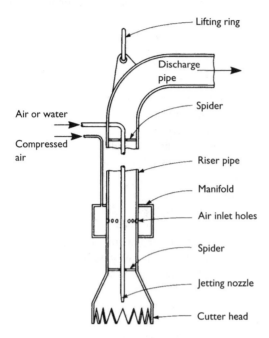

Figure 3.39 Airlift for cleaning-out soil from steel tubular piles.

broken rock can be removed by lowering the airlift device shown in Figure 3.39 down the tube, the soil or broken rock in the plug being loosened by dropping or rotating the riser pipe. A reverse-circulation rig with a rotating cutter (Figure 3.34) is an efficient means of removing soil if justified by the number and size of the piles. Crane-mounted power augers of the type shown in Figure 3.25 can only be used for cleaning after the pile has been driven down to its final level where there is space for the crane carrying the auger to be manoeuvred over the pile head. The self-erecting crawler rigs are more manoeuvrable and with the other methods described above can be used to drill below the pile toe and so ease the driving resistance. However, drilling below the toe also reduces the shaft friction and the method may have to be restricted to end-bearing piles. This aspect is discussed further in the section on piling for marine structures (Section 8.3). Because of the delays involved in alternate drilling and driving operations, it is desirable that any drilling to ease the driving resistance should be restricted to only one operation on each pile.

Difficulties arise when it is necessary to place a plug of concrete at the toe of the cleaned-out pile to develop high end-bearing resistance, or to transfer uplift loads from the superstructure to the interior wall of the hollow pile through a reinforcing cage. In such cases a good bond must be developed between the concrete filling and the interior of the steel pile. This requires any adherent soil which remains after removing the soil plug to be cleaned off the pile wall. A sandy soil can be effectively removed by water jetting or by airlifting, but an adherent clay may require high-pressure water jets to remove it. A rig can be used for this purpose that comprises a central airlift pipe and a base plate with jetting nozzles around the periphery. The assembly can be half-rotated as necessary and wire strand 'brushes' can be attached to the plate. However, the process is tediously slow since the jets tend to drill small-diameter holes in the clay. Equipment has yet to be devised which will quickly and effectively remove the clay adhering to the wall of a pile to a sufficient standard of cleanliness to achieve a good bond with a concrete plug. The procedure for placing the concrete plug in the cleaned-out pile or for completely filling a steel tubular or box pile is similar to that described below for shell piles.

3.4.4 Driving and concreting steel shell piles

Steel shell piles are driven by drop hammers or single-acting hammers acting on the head of an internal mandrel or core which is collapsed to allow it to be withdrawn before placing the concrete. Problems arise with heave when driving shell piles in groups, and distortion or collapse of the shells when driving past obstructions. Shell piles have the advantage that the interior of the shell can be inspected before concrete is placed. This can be done with the aid of light reflected down the pile by a mirror, or by a narrow beam lamp. Distortion of the shells can be detected by lowering a lamp down to the toe. If it disappears wholly or partially then distortion has occurred. This can be corrected by pulling up the shells and redriving them or, in the case of tapered shells, by inserting and redriving a new tapered shell assembly. The problem of heave is discussed in Sections 5.7 to 5.9.

Sometimes some leakage of groundwater occurs through shells in quantities which do not justify replacing the damaged units. The water can be removed from the shells before placing the concrete by pumping (if the depth to the pile toe is within the suction lift of the available pump), by an air lift or by baling. If, after removing the water, the depth of inflow is seen to be less than a few centimetres in 5 minutes the collected water can again be removed and concrete placed quickly to seal off the inflow. For higher rates of seepage the water should be allowed to fill the pile up to its rest level, and the concrete should then be placed by tremie-pipe as described in Section 3.4.8.

Concrete placed in 'dry' shell piles is merely dumped in by barrow or chute. It should be reasonably workable with a slump of 100 to 150 mm to avoid arching as it drops down a tapered shell or onto the reinforcing cage. The cement content should be such as to comply with the requirements in BS EN 1536 or with any special requirements for durability (see Section 10.3.1). The American Concrete Institute[6.13] states that vibration due to driving adjacent piles has no detrimental effect on fresh concrete in shell piles. Therefore concreting can proceed immediately after driving the shell even though adjacent shells are being driven, provided there are no detrimental effects due to ground heave or relaxation (see Section 5.7).

3.4.5 The installation of withdrawable-tube types of driven and cast-in-place piles

There are no standard procedures for installing driven and cast-in-place piles of the types which involve the driving and subsequent withdrawal of a casing tube. However, BS EN 12699 requires that cast-in-place displacement piles shall be concreted in the dry using high workability concrete or semi-dry concrete as appropriate to the methods for each type of pile as described in Section 2.3.2. Where the concrete is compacted by internal drop hammer a mix is required that is drier than that which is suitable for compaction by vibrating the piling tube. The workability and mix proportions of the concrete should be left to the piling contractor, subject to compliance with the requirements of BS EN 1536 and the needs regarding durability (see Section 10.3.1).

The procedures to be adopted for avoiding 'waisting' or 'necking' of the shaft, or the inclusion of silt pockets and laitance layers, are similar to those adopted for bored and cast-in-place piles and are described in the following section of this chapter. Precautions against the effects of ground heave are described in Section 5.8. Because the casing tube is, in all cases, driven down for the full length of the pile, it is essential to ensure that the interior of the tube is free of any encrustations of hardened concrete. Even small encrustations can cause the concrete to arch and jam as the tube is withdrawn. If the reinforcing steel is lifted with the tube the pile shaft is probably defective and should be rejected. Further guidance is given in CIRIA Report PG8[3.19].

3.4.6 The installation of bored and cast-in-place piles by power auger equipment

The employment of a power auger for the drilling work in bored and cast-in-place piles presupposes that the soil is sufficiently cohesive to stand unsupported, at least for a short time. Any upper soft or loose soil strata or water-bearing layers are 'cased-off' by drilling down a casing or pushing the tubes down into the pre-drilled hole by vibrator or the crowd mechanism on the kelly bar. If necessary, 'mudding-in' techniques are used at this stage (see Section 3.3.8). After the auger has reached the deeper and stiffer fine-grained soils, the bore-hole is taken down to its final depth without further support, until the stage is reached when a loosely fitting tube is lowered down the completed hole. This loose liner may be required for safety purposes when inspecting the pile base before placing the concrete; or if an enlarged base is required, the lining prevents the clay collapsing around the shaft over the period of several hours or more required to drill the under-ream. The loose liner may not be needed for straight-sided piles in weak rocks, or in stable unfissured clays, where there is no risk of collapse before or during the placing of the concrete. However, if the clays are in any degree fissured there is a risk of the walls collapsing during concreting, and thus leading to defects of the type shown in Figure 3.40. Lining tubes must be inserted in potentially unstable soils if a visual inspection is to be made of the pile base. Where it is required to lower inspection personnel into a shaft (which should be greater than 750 mm diameter) the requirements of BS 8008: 1996 (safety precautions and procedures for the construction and descent of machine-bored shafts for piling and other purposes) must be followed. Particular care is needed if an enlarged base has to be inspected or if concrete spreading has to be carried out by hand. The support can be in the form of a 'spider', consisting of a number of hinged arms mounted on a ring. The assembly is lowered down the shaft with the arms in a near-vertical closed position. They are then lifted upwards and outwards and locked in position to form a cone which is pulled up against the clay surface.

Favourable conditions for stability of the borehole are given by care in setting up the rig on a firm level base and attention to maintenance of verticality. Tilting of the rig or violent operation of the auger leads to misalignment and the need for corrective action by reaming the sides. 'Working Platforms for Tracked Plant'[3.20] provides guidance for the design and construction of ground-supported platforms for piling plant, although selection of design parameters to provide realistic mat thicknesses is still a problem. The requirements for the safety of operatives should be rigorously followed as detailed in the British Drilling Association Health and Safety Manual[3.21]. Casings protecting open pile boreholes should extend above ground level and should be provided with a strong cover.

The final cleaning-up operation before placing concrete in a bored pile consists of removing large crumbs of soil or trampled puddled clay from the pile base. Any lumps of clay adhering

Figure 3.40 Defective shaft of bored pile caused by collapse of clay after lifting casing.

to the walls of the borehole or to the lining tubes should be cleaned off. The reinforcing cage can then be placed and concreting commenced. The time interval between the final cleaning-up and placing concrete should not exceed 6 hours. If there is any appreciable delay the depth of the pile bottom should be checked against the measured drilled depth before placing the concrete to ensure that no soil has fallen into the hole. If the reinforcing cage extends only part-way down the hole it should be suspended from the top of the pile shaft before commencing to place the concrete.

The concrete used in the pile base and shaft should be easily workable with a slump of 100 to 180 mm as recommended in BS 1536. Such a mix is self-compacting and does not require ramming or vibrating. The mix proportions should be such as to ensure compliance with the requirements regarding strength and minimum cement content of BS EN 1536 or with any special requirements for durability (see Section 10.3.1). A dry mix should be used for the first few charges of concrete if the pile base is wet. The concrete in the shaft is fed through a hopper or chute placed centrally over the pile to direct it clear of the sides and the reinforcement. After completing concreting, the lining tubes are withdrawn. If a loose liner is used inside an upper casing, the former is lifted out as soon as the concrete extends above the base of the outer tube. A vibrator of the type described in Section 3.1.5 is a useful expedient for extracting the upper casings used to support soft clays or loose sand. The quantity of concrete placed in the shaft should allow for the outward slumping which takes place to fill the space occupied by the tube and any overbreak of the soil outside it. At this final stage there is inevitably some laitance which has risen to the top of the concrete. The laitance may be diluted and contaminated with water and silt expelled from around the casing as the concrete slumps outwards to fill the gap. Thus the level of the concrete should be set high so that this weak laitance layer can be broken away before bonding the pile head onto its cap. The terms of the contract should make it clear whether or not this removal should be performed by the piling contractor.

The concrete in a pile shaft may be required to be terminated at some depth below ground level, for example, when constructing from ground surface level, piles designed to support a basement floor. It is a matter of some experience to judge the level at which the concrete should be terminated, and it is difficult to distinguish between fluid concrete and thick laitance when plumbing the level with a float. Fleming and Lane[3.22] recommend the following tolerances for all conditions:

Concrete cast under water:	$+1.5$ to $+3$ m
Concrete cast in dry uncased holes:	$+75$ to $+300$ mm
Concrete cast in cased holes:	
the greater of	(a) $+75$ to $+300$ min $+$ cased length/15
or	(b) $+75$ to $+300$ mm $+$ [depth to casting level -900 mm/10]

The Institution of Civil Engineers *Specification for Piling*[2.5] specifies casting tolerances for three conditions of placing concrete in pile boreholes with and without temporary casing. The ground surface or piling platform level is defined as the 'commencing surface'. The three conditions refer to a situation where the cut-off level is at a depth Hm below the commencing surface such that H is from 0.15 to any depth for condition (a) and below or between 0.15 and 10 m for (b) and (c). The conditions are as follows:

(a) Concrete placed in dry boreholes using permanent casing or cut-off level in stable ground below base of casing: the casting tolerance in metres is specified to be $0.3 + H/10$

(b) Concrete placed in dry boreholes using temporary casing other than as (a) above: the casting tolerance in metres is specified to be $0.3 + H/12 + C/8$, where C is the length of temporary casing below the commencing surface

(c) Concrete placed under water or a drilling fluid: the casting tolerance in metres is specified to be $1.0 + H/12 + C/8$ where C is the length of temporary casing below the commencing surface.

The reader is referred to the ICE *Specification* for the various qualifications to the above tolerances. It will be noted that the casing length rather than the diameter is a factor which influences casting tolerances. This reflects the problems which occur when extracting the temporary casing.

The use of a permanent casing in the form of a light-gauge metal sleeve surrounding a pile shaft in soft clays or peats was described in Section 2.4.2. This sleeving cannot be used within a temporary lining tube where the latter has to be withdrawn in a long length by means of a vibrator or by jacking. This involves the risk of distortion or jamming of the sleeve, which is then lifted while raising the temporary tube with disastrous effects on the concrete in the pile shaft. The sleeve can be used within an outer temporary liner where the depth of soft clay is shallow, and it can be used in conjunction with a casing oscillator which keeps the outer tube free of any jamming by the sleeve. There are no problems of using the light-gauge sleeve where power auger drilling can be performed to produce a stable hole without employing a temporary outer lining tube.

Unfortunately, defects in a pile shaft of the type shown in Figure 3.41 are by no means uncommon, even when placing a workable concrete in the dry open hole of a large-diameter bored pile. Defects can take the form of large unfilled voids, or pockets of clay and silt in the concrete. Some causes of these defects are listed below:

(1) Encrustations of hardened concrete or soil on the inside of the lining tubes can cause the concrete to be lifted as the tubes are withdrawn, thus forming gaps in the concrete. *Remedy:* The tubes must be clean before they are lowered down the bore-hole.
(2) The falling concrete may arch and jam across the lining tube or between the tubes and the reinforcement. *Remedy:* Use a concrete of sufficient workability to slump easily down the hole and fill all voids.

Figure 3.41 Defective shaft of bored pile caused by cement being washed out of unset concrete.

(3) The falling concrete may jam between the reinforcing bars and not flow outwards to the walls of the borehole. *Remedy:* Ensure a generous space between the reinforcing bars. The cage should be stiff enough to prevent it twisting or buckling during handling and subsequent placing of concrete. Widely spaced stiff hoops are preferable to helical binding. Check that the bars have not moved together before the cage is lowered down the hole.

(4) Lumps of clay may fall from the walls of the borehole or lining tubes into the concrete as it is being placed. *Remedy:* Always use lining tubes if the soil around the borehole is potentially unstable and do not withdraw them prematurely. Ensure that adhering lumps of clay are cleaned off the tubes before they are inserted and after completing drilling.

(5) Soft or loose soils may squeeze into the pile shaft from beneath the base of the lining tubes as they are withdrawn, forming a 'waisted' or 'necked' shaft. *Remedy:* Do not withdraw the casing until the placing of the concrete is complete. Check the volume of concrete placed against the theoretical volume and take remedial action (removal and replacement of the concrete) if there is a significant discrepancy.

(6) If bentonite has been used for 'mudding-in', the hydrostatic pressure of the bentonite in the annulus, which is disturbed on lifting the casing, may be higher than that of the fluid concrete, thus causing the bentonite to flow into the concrete. This is a serious defect and is difficult to detect. It is particularly liable to happen if the concrete is terminated at some depth below the top of the 'mudded-in' casing. *Remedy:* Keep a careful watch on the level and density of the bentonite gel when the casing is lifted. Watch for any changes in level of the concrete surface and for the appearance of bentonite within the concrete. If inflow of the bentonite has occurred the defective concrete must be removed and replaced and the 'mudding-in' technique must be abandoned.

(7) Infiltration of groundwater may cause gaps, or honeycombing of the concrete. *Remedy:* Adopt the techniques for dealing with groundwater in pile boreholes described in Section 3.4.8.

Further guidance on the installation procedures is given in CIRIA Report PG2[3.23].

3.4.7   Installing continuous flight auger piles

CFA piles can be installed in a variety of soils, dry or waterlogged, loose or cohesive, and through weak rock. The soil is loosened on insertion of the auger and the borehole walls are supported by the auger flights filled with drill cuttings; bentonite support slurry is not used. The pile is concreted through a bottom or side exit at the tip of the hollow stem auger (100 or 127 mm bore) using a concrete pump connected by hose to a swivel on the rotary head as the auger is *slowly* rotated and withdrawn. Soil is brought to the surface on the auger blades. The concrete flow rate and feed pressure are continuously measured at the tip; reinforcement is pushed or vibrated into the fresh concrete. In order to avoid the problems of flighting and polishing (see Section 2.4.2) which reduce pile capacity, reliable instrumentation and experienced operators are essential.

For *rotary displacement auger piles* the displacement tool, which is mounted at the bottom of a drill tube, is rotated by the high torque top drive and forced into the ground by the rig crowd thereby compacting the wall of the hole. The pile is concreted through the auger tip as the tool is rotated out of the hole maintaining the profile. To form the various types of *screw piles*, discussed in Section 2.3.5, the thick-flanged continuous auger is screwed into the ground with limited crowd applied, although for less cohesive soil more thrust will be necessary to reach the required depth. The auger is rotated out of the hole as concrete is pumped through the tip to fill the helical profile of the pile, with only minimal soil being brought to the surface.

3.4.8   Concreting pile shafts under water

Groundwater in pile boreholes can cause serious difficulties when placing concrete in the shaft. A depth of inflow of only a few centimetres in, say, 5 minutes which has trickled down behind the lining tubes or has seeped into the pile base can be readily dealt with by baling or pumping it out and then placing dry concrete to seal the base against any further inflow. However, larger flows can cause progressive increases in the water content of the concrete, weakening it, and forming excess laitance.

A strong flow can even wash away the concrete completely. The defective piles shown in Figure 3.41 were caused by the flow of water under an artesian head from a fissured rock on which the bored piles were bearing after the boreholes had been drilled through a soft clay overburden. The lined boreholes were pumped dry of water before the concrete was placed, but the subsequent 'make' of water was sufficiently strong to wash away some of the cement before the concrete has set. The remedial action in this case was to place dry concrete in bags at the base of the pile borehole and then to drive precast concrete sections into the bags.

In all cases of strong inflow the water must be allowed to rise to its normal rest level and topped up to at least 1.0 m above this level to stabilize the pile base. BS EN 1536 requires that a tremie pipe be used for concreting in submerged conditions (water or slurry); the tremie bore must be 6 times the maximum size of the aggregate or 150 mm whichever is the greater. The maximum outside diameter of the pipe including joints should be less than 0.35 times the pile diameter or inner diameter of the casing. The tremie pipe must be clean and lowered to the bottom of the pile and lifted slightly to start concrete flow. A flap valve should be used on the end of the tremie pipe rather than a plug or polyethylene 'go-devil'. During concreting, the tremie tip must always be immersed in the concrete; 1.50 m below concrete surface for piles less than 1200 mm diameter and 2.50 m for piles greater than 1200 mm. If immersion is lost during concreting, special precautions are required before placement can continue; for example, steps must be taken to re-immerse the tremie so that any contamination will be above the final cut-off level. Other limits for the tremie are given for concreting barrettes.

Although a bottom-opening bucket is sometimes used instead of a tremie pipe for placing concrete in pile boreholes, the authors as a general rule condemn this practice. This is because the crane operator handling the bucket cannot tell, by the behaviour of the crane rope, whether or not he has lowered the bucket to the correct level into the fluid concrete before he releases the hinged flap. If he releases the bucket flap prematurely, the concrete will flow out through the water and the cement will be washed out. On the other hand, if he plunges the bucket too deeply it will disturb the concrete already placed when it is lifted out. The bottom-dumping bucket method has no advantage over the tremie pipe and the authors would use it only if a pile were large enough for the lowering and dumping to be controlled by a diver.

BS EN 1536 provides guidance for piles formed using the technique known as 'prepacked concrete' for underwater concreting, but it is not recommended here in preference to placing concrete by tremie pipe. This is because the water in a bored pile is rarely clean, and the silt stirred up by dumping the aggregate tends to get dispersed on to the surface of the stones. It is then displaced by the rising column of grout and tends to form layers or pockets of muddy laitance.

The procedure for drilling pile boreholes with support by a bentonite slurry is described in Section 3.3.8 and in CIRIA Report PG3[3.24]. Problems can be caused when placing concrete in a bentonite-filled hole. A tremie pipe must be used, and there must be a sufficient hydrostatic pressure of concrete in the pipe above bentonite level to overcome the external head of the slurry, to rupture the gel and to overcome friction in the tremie pipe. Sometimes a dispersing agent is added to the bentonite to break down the gel before placing the concrete. Where the mud becomes flocculated and heavily charged with sand (i.e. has a density greater than 1350 to 1400 kg/m^3) it should be replaced by a lighter mud before placing the concrete and the base of the pile cleaned of any debris. Circumferential steel should be kept to a minimum. The concrete in the piled foundations for the Wuya Bridge, Nigeria[3.25] was placed under bentonite. The piles were 18 to 21 m deep and a mud density of 1600 kg/m^3 was necessary to prevent the sides from collapsing. The concrete failed to displace the gel which was stiffened by the high ground temperatures and jamming occurred, especially when placing was suspended to remove each section of the tremie pipe. The problem was finally overcome by increasing the workability of the concrete by means of a plasticizer together with a retarder. The tremie pipe was lifted out as a single unit to avoid the delays in breaking the pipe joints.

3.4.9   The installation of bored and cast-in-place piles by grabbing, vibratory, and reverse-circulation rigs

The use of either grabbing, vibratory or reverse-circulation machines for drilling pile boreholes can involve continuous support by lining tubes which may or may not be withdrawn after placing the concrete. In all three methods the tubes may have to follow closely behind the drilling in order to prevent the collapse of the sides and the consequent weakening of shaft friction. The boreholes must be kept topped up with water in order to avoid 'blowing' of the pile bottom as a result of the upward flow of the groundwater. This is particularly necessary when drilling through water-bearing sand layers interbedded with impervious clays.

Grabbing in weak rocks can cause large accumulations of slurry in the boreholes which make it difficult to assess the required termination level of the pile in sound rock. The slurry should be removed from time to time by baling or by airlift pump with a final cleaning-up before placing the concrete.

The techniques of placing concrete in 'dry' holes or under water, are exactly the same as described in Sections 3.4.6 and 3.4.8.

3.4.10   The installation of bored and cast-in-place piles by tripod rigs

Pile boreholes in clays are drilled by a clay cutter operated from a tripod rig. Water should not be poured down the hole to soften a stiff clay, or used to aid removal of the clay from the cutter as this causes a reduction in shaft friction. When drilling in granular soils the lining tubes should follow closely behind the drilling to avoid overbreak, and the addition of water is needed to prevent 'blowing' and to facilitate the operation of the baler or shell.

Piles drilled by tripod rigs are relatively small in diameter, requiring extra care when placing the concrete as this is more likely to jam in the casing tubes when they are lifted. Curtis[3.26] suggests checking the concrete level by hanging a float on top of the concrete and comparing its measurement from the top of the tube with the amount of tube extracted. He also suggests that the position of the reinforcing cage should be checked by a 'tell-tale' wire and indicator. Problems can occur when placing concrete in raking piles. Internal ramming is impossible as the rammer catches on the reinforcing cage. A high slump concrete is necessary with special precautions being taken to prevent the reinforcement being lifted with the lining tubes.

3.4.11   The installation of raking piles

BS EN 1536 states that pile bores, whether drilled or driven, should be cased throughout their length if the rake is flatter than 1 horizontal to 15 vertical unless it can be shown that an uncased pile bore will be stable. Similarly, stabilizing fluids should not be used if the rake is flatter than 1 in 15 unless precautions are taken when inserting casing and concreting.

The advantages of raking piles in resisting lateral loads are noted in Chapters 6 and 8. However, the installation of such piles may result in considerable practical difficulties, and they should not be employed without first considering the method of installation and the ground conditions. If the soil strata are such that the piles can be *driven* to the full penetration depth without the need to drill out a soil plug or to use jetting to aid driving, then it should be feasible to adopt raking piles up to a maximum rake of 1 to 2. However, the efficiency of the hammer is reduced due to the friction of the ram in the guides. It may therefore be necessary to use a more powerful hammer than that required for driving vertical piles to the same penetration depth.

The vertical load caused by the pile and hammer on the leaders of the piling frame must be taken into consideration. Also when driving piles by guides without the use of leaders the bending stresses caused by the weight of the hammer on the upper end of the pile must be added to the driving stresses and a check should be made to ensure that the combined stresses are within allowable limits.

The principal difficulties arise when it is necessary to drill ahead of an open-ended pile to clear boulders or other obstructions, using the methods described in Section 3.3.5. When the drill penetrates below the shoe of the pile tube it tends to drop by gravity and it is then likely to foul the shoe as it is pulled out to resume further driving. Similarly, under-reaming tools are liable to be jammed as they are withdrawn. The risks of fouling the drilling tool are less if the angle of rake is small (say 1 in 10 or less) and the drill string is adequately centralized within the piling tube. However, the drill must not be allowed to penetrate deeply below the toe of the pile. This results in frequent alternations of drilling and driving with consequent delays as the hammer is taken off to enter the drill, followed by delays in entering and coupling up the drill string, and then removing it before replacing the hammer.

Difficulties also arise when installing driven and cast-in-place piles by means of an internal drop hammer, due to the friction of the hammer on the inside face of the driving tube. Installers of these piles state that a rake not flatter than 1 in 3.7 is possible.

Power augers can drill for pile boreholes at angles of rake of up to 1 in 3 but when casing is necessary to support the pile borehole the same difficulties arise with the jamming of the bucket or auger beneath the toe of the easing. As shown in Figure 3.2 the self-erecting leader rigs are capable of drilling open holes at rakes up to 1 in 1, but where casing has to be drilled, rakes flatter than 1 in 3 are difficult to manage.

The American Concrete Institute[6.13] recommends using an over-sanded mix for placing concrete in raking pile shells or tubes. A concrete mix containing 475 kg/m^3 of coarse aggregate with a corresponding increase in cement and sand to give a slump of 100 mm is recommended. This mix can be pumped down the raking tube.

3.4.12   Positional tolerances

It is impossible to install a pile, whether by driving, drilling or jacking, so that the head of the completed pile is always exactly in the intended position or that the axis of the pile is truly vertical or at the specified rake. Driven piles tend to move out of alignment during installation due to obstructions in the ground or the tilting of the piling frame leaders. Driving piles in groups can cause horizontal ground movements which deflect the piles. In the case of bored piles the auger can wander from the true position or the drilling rig may tilt due to the wheels or tracks sinking into a poorly prepared platform. However, controlling the positions of piles is necessary since misalignment affects the design of pile caps and ground beams (see Sections 7.8 and 7.9), and deviations from alignment may cause interference between adjacent piles in a group or dangerous concentrations of load at the toe. Accordingly, codes of practice specify tolerances in the position of pile heads or deviations from the vertical or intended rake. If these are exceeded, action is necessary either to redesign the pile caps as may be required or to install additional piles to keep the working loads within the allowable values.

Some codes of practice requirements are as follows:

BS 8004: Driven and cast-in-place, and bored and cast-in-place piles should not deviate by more than 1 in 75 from the vertical, or more than 75 mm from their designed position at the level of the piling rig. Larger tolerances can be considered for work over water or raking piles. A deviation of up to 1 in 25 is permitted for bored piles drilled at rakes of up to 1 in 4.

BS EN 1536: Plan location tolerances are given in Clause 7.2 for diameters of vertical and raking bored piles less than 1000 mm diameter: 100 mm, between 1000 and 1500 mm: 0.1 × diameter, and greater than 1500 mm: 150 mm. Deviation in inclination of vertical piles and piles designed for a rake less than 1 in 15 is limited to 20 mm/m run of pile. For piles designed with a rake of between 1 in 4 and 1 in 15 the deviation is limited to 40 mm/m.

BS EN 12699: The plan location tolerance (at working level) given in Clause 7.3 for vertical and raking displacement piles is 100 mm. Deviation for vertical and raking piles is 40 mm/m. The deviations in this code must be taken into account in the design. Both the new codes allow other tolerances to be specified.

BS 6349: Part 2 Code of Practice for Maritime Structures: A deviation of up to 1 in 100 is permitted for vertical piles driven in sheltered waters or up to 1 in 75 for exposed sites. The deviation for raking piles should not exceed 1 in 30 from the specified rake for sheltered waters or 1 in 25 for exposed sites. The centre of piles at the junction with the superstructure should be within 75 mm for piles driven on land or in sheltered waters. Where piles are driven through rubble slopes the code permits a positional tolerance of up to 100 mm, and for access trestles and jetty heads a tolerance of 75 to 150 mm is allowed depending on the exposure conditions.

Institution of Civil Engineers[2.5]: Positional – maximum deviation of centre point of pile to centre point on the setting out drawing not more than 75 mm, but additional tolerance for pile cut-off below ground level. Verticality – maximum deviation of finished pile from the vertical is 1 in 75. Maximum deviation of finished pile from the specified rake is 1 in 25 for piles raking up to 1:6 and 1 in 15 for piles raking more than 1:6. Relaxation permitted in exceptional circumstances subject to implications of this action. Other more stringent tolerances are specified for diaphragm walls and secant and contiguous piles.

American Concrete Institute Recommendations: The position of the pile head is to be within 75 to 150 mm for the normal usage of piles beneath a structural slab. The axis may deviate by up to 10% of the pile length for completely embedded vertical piles or for all raking piles, provided the pile axis is driven straight. For vertical piles extending above the ground surface the maximum deviation is 2% of the pile length, except that 4% can be permitted if the resulting horizontal load can be taken by the pile-cap structure. For bent piles the allowable deviation is 2% to 4% of the pile length depending on the soil conditions and the type of bend (e.g. sharp or gentle). Severely bent piles must be evaluated by soil mechanics' calculations or checked by loading tests.

The significance of positional tolerance to piling beneath deep basements is noted in Section 5.9.

3.5 Constructing piles in groups

So far only the installation of single piles has been discussed. The construction of groups of piles can have cumulative effects on the ground within and surrounding the pile group. These effects are occasionally beneficial but more frequently have deleterious effects on the load/settlement characteristics of the piles and can damage surrounding property. Precautions can be taken against these effects by the installation methods and sequence of construction adopted. Because the problems are more directly concerned with the bearing capacity and settlement of the group as a whole, rather than with the installation of the piles, they are discussed in Sections 5.7 to 5.9.

3.6 References

3.1 GEDDES, W. G. N., STURROCK, K. R., and KINDER, G. New shipbuilding dock at Belfast for Harland and Wolff Ltd, *Proceedings of the Institution of Civil Engineers*, Vol. 51, January 1972, pp. 17–47.

3.2 FAWCETT, A. The performance of the resonant pile driver, *Proceedings of the 8th International Conference, ISSMFE*, Moscow, Vol. 2.1, 1973, pp. 89–96.

3.3 RODGER, A. A. and LITTLEJOHN, G. A. Study of vibratory driving in granular soils, *Geotechnique*, Vol. 30, No. 3, 1980, pp. 269–93.

3.4 VIKING, K. The vibratory pile installation technique, *Pile Driver*, Pile Driving Contractors Association, USA, Spring 2005, pp. 27–35.

3.5 American Petroleum Institute, *Recommended Practice for Planning, Designing and Constructing Fixed Offshore Platforms*, API RP2A, 2000 edn.

3.6 WILLIS, A. J. and CHURCHER, D. W. How much noise do you make? A guide to assessing and managing noise on construction sites, *Construction Industry Research and Information Association (CIRIA)*, London, Report PR 70, 1999.

3.7 WELTMAN, A. J. (ed.), Noise and vibration from piling operations, *Construction Industry Research and Information Association (CIRIA)*, London, Report PSA/CIRIA PG7, 1980.

3.8 SHELBOURNE, H. Decibel rating-the important factor, *Construction News*, Piling and foundations supplement, 13 December 1973, p. 47.

3.9 Building Research Establishment. *Damage to Structures from Ground-borne Vibration*, BRE Digest 403, Watford, 1995.

3.10 WILLIAMS, N. S. Contribution to discussion on pile driving in difficult conditions, *Institution of Civil Engineers*, Works Construction Division, 1951.

3.11 GERWICK, B. C. *Construction of Offshore Structures*, Wiley-Interscience, New York, 1986.

3.12 NORTH-LEWIS, J. P. and SCOTT, I. D. Constructional control affecting the behaviour of piles with particular reference to small diameter bored cast in-situ piles, *Proceedings of the Conference on the Behaviour of Piles*, Institution of Civil Engineers, London, 1970, pp. 161–6.

3.13 REESE, L. C., O'NEILL, M. W., and TOUMA, F. T. Bored piles installed by slurry displacement, *Proceedings of the 8th International Conference, ISSMFE*, Moscow, Vol. 2.1, 1973, pp. 203–9.

3.14 BOLOGNESI, A. J. L. and MORETTO, O. Stage grouting preloading of large piles in sand, *Proceedings of the 8th International Conference, ISSMFE*, Moscow, Vol. 2.1, 1973, pp. 19–25.

3.15 YEATS, J. A. and O'RIORDAN, N. J. The design and construction of large diameter base-grouted piles in Thanet Sand, London, *Proceedings of the International Conference on Piling and Deep Foundations*, London, Vol. 1, Balkema, Rotterdam, 1989, pp. 455–61.

3.16 SHERWOOD, D. E. and MITCHELL, J. M. Base grouted piles in Thanet Sands, London, *Proceedings of the International Conference on Piling and Deep Foundations*, London, Vol. 1, – Balkema, Rotterdam, 1989, pp. 463–72.

3.17 TAPPIN, R. G. R., van DUIVENIJK, J., and HAQUE, M. The design and construction of Jamuna Bridge, Bangladesh, *Proceedings of the Institution of Civil Engineers*, Vol. 126, 1998, pp. 162–80.

3.18 LITTLECHILD, B. D., PLUMBRIDGE, G. D., and FREE, M. W. Shaft grouted piles in sand and clay in Bangkok, *Deep Foundations Institute 7th International Conference on piling and deep foundations*, DFI, Englewood Cliffs, NJ, 1998, pp. 1.7.1–1.7.8.

3.19 HEALY, P. R. and WELTMAN, A. J. Survey of problems associated with the installation of displacement piles, *Construction Industry Research and Information Association (CIRIA)*, London, Report PG8, 1980.

3.20 SKINNER, H. *Working Platforms for Tracked Plant*, BR470 Building Research Establishment, Watford, 2004.

3.21 British Drilling Association, *Health and Safety Manual for Land Drilling – Code of Safe Drilling Practice*, BDA 2002.

3.22 FLEMING, W. G. K. and LANE, P. F. Tolerance requirements and constructional problems in piling, *Proceedings of the Conference on the Behaviour of Piles*, Institution of Civil Engineers, London, 1970, pp. 175–8.

3.23 THORBURN, S. and THORBURN, J. Q. Review of problems associated with the construction of cast in-place concrete piles, *Construction Industry Research and Information Association (CIRIA)*, London, Report PG2, 1977.

3.24 FLEMING, W. K. and SLIWINSKI, Z. J. The use and influence of bentonite in bored pile construction, *Construction Industry Research and Information Association (CIRIA)*, London, Report PG3, 1977.

3.25 SAMUEL, R. H. The construction of Wuya Bridge, Nigeria, *Proceedings of the Institution of Civil Engineers*, Vol. 33, March 1966, pp. 353–80.

3.26 CURTIS, R. L. Constructional control affecting the behaviour of driven in-situ concrete piles, *Proceedings of the Conference on the Behaviour of Piles*, Institution of Civil Engineers, London, 1970, pp. 167–74.

Chapter 4

Calculating the resistance of piles to compressive loads

4.1 General considerations

4.1.1 The basic approach to the calculation of pile resistance

The numerous types of pile and the diversity in their methods of installation have been described in Chapters 2 and 3. Each different type and installation method disturbs the ground surrounding the pile in a different way. The influence of this disturbance on the shaft friction and end-bearing resistance of piles has been briefly mentioned (see Section 1.3). This influence can improve or reduce the bearing capacity of the piles, and thus a thorough understanding of how the piles are constructed is essential to the formulation of a practical method of calculating loading capacity.

The basic approach used in this chapter to calculate the resistance of piles to compressive loads is the 'static' or soil mechanics approach. Over the years much attention has been given by research workers to calculation methods based on 'pure' soil mechanics theory. They postulate that the interface friction on a pile shaft can be determined by a simple relationship between the coefficient of earth pressure 'at rest', the effective over-burden pressure and the drained angle of shearing resistance of the soil, but they recognize that the coefficient of earth pressure must be modified by a factor which takes into account the method of pile installation. Similarly, they believe that the end-bearing resistance of a pile can be calculated by classical soil mechanics theory based on the undisturbed shearing resistance of the soil surrounding the pile toe. The importance of the settlement of the pile or pile groups at the working load is recognized and methods have been evolved to calculate this settlement, based on elastic theory and taking into account the transfer of load in shaft friction from the pile to the soil.

The concepts of this research work commenced on quite simple lines, the two main groups, namely driven piles and bored piles, only being differentiated when considering pile behaviour. However, as the work progressed from the laboratory to the field, particularly in the study of the behaviour of instrumented full-scale piles, it was observed that there were very fundamental departures from classical soil mechanics theory, and the all-important effects of installation procedures on pile behaviour were realized. The installation of piles results in highly complex conditions developing at the pile–soil interface which are often quite unrelated to the original undisturbed state of the soil, or even to the fully remoulded state. The pore water pressures surrounding the pile can vary widely over periods of hours, days, months or years after installation, such that the simple relationships of shaft friction to effective overburden pressure are unrealistic. Similarly, when considering deformations of a pile group under its working load, any calculations of the transfer of load that are based on elastic theory which do not take account of soil disturbance for several diameters around the pile shaft and beneath the toe are unrealistic.

Therefore, while the authors base their approach to the calculation of pile carrying capacity on soil mechanics methods, this approach is simply an empirical one which relates known pile behaviour to simple soil properties such as relative density and undrained shearing strength. These can be regarded as properties to which empirical coefficients can be applied to arrive at unit values for the shaft friction and end-bearing resistances.

Observations made on full-scale instrumented piles have so far only served to reveal the extreme complexities of the problems, and have shown that there is no simple fundamental design method. The empirical or semi-empirical methods set out in this chapter have been proved by experience to be reliable for practical design of light to moderately heavy loadings on land-based or near-shore marine structures. Special consideration using more complex design methods are required for heavily loaded marine structures in deep water. The engineer is often presented with inadequate information on the soil properties. A decision then has to be made whether to base designs on conservative values with an appropriate safety factor without any check by load testing, or merely to use the design methods to give a preliminary guide to pile diameter and length and then to base the final designs on an extensive field testing programme with loading tests to failure. Such testing is always justified on a large-scale piling project. Proof-load testing as a means of checking workmanship is a separate consideration (see Section 11.4).

Where the effective overburden pressure is an important parameter for calculating the ultimate bearing capacity of piles (as is the case for coarse-grained soils) account must be taken of the unfavourable effects of a rise in groundwater levels. This may be local or may be a general rise, due for example to seasonal flooding of a major river, or a long-term effect such as the predicted large general rise in groundwater levels in Greater London.

4.1.2 The behaviour of a pile under load

For practical design purposes engineers must base their calculations of carrying capacity on the application of the load at a relatively short time after installation. The reliability of these calculations is assessed by a loading test which is again made at a relatively short time after installation. However, the effects of time on carrying capacity must be appreciated and these are discussed in Sections 4.2.4 and 4.3.8.

When a pile is subjected to a progressively increasing compressive load at a rapid or moderately rapid rate of application, the resulting load–settlement curve is as shown in Figure 4.1. Initially the pile–soil system behaves elastically. There is a straight-line relationship up to some point A on the curve and if the load is released at any stage up to this point the pile head will rebound to its original level. When the load is increased beyond point A there is yielding at, or close to, the pile–soil interface and slippage occurs until point B is reached, when the maximum shaft friction on the pile shaft will have been mobilized. If the load is released at this stage the pile head will rebound to point C, the amount of 'permanent set' being the distance OC. The movement required to mobilize the maximum shaft friction is quite small and is only of the order of 0.3% to 1% of the pile diameter. The base resistance of the pile requires a greater downward movement for its full mobilization, and the amount of movement depends on the diameter of the pile. It may

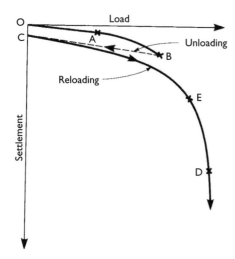

Figure 4.1 Load/settlement curve for compressive load to failure on pile.

be in the range of 10% to 20% of the base diameter. When the stage of full mobilization of the base resistance is reached (point D in Figure 4.1) the pile plunges downwards without any further increase of load, or small increases in load produce increasingly large settlements.

If strain gauges are installed at various points along the pile shaft from which the compressive load in the pile can be deduced at each level, the diagrams illustrated in Figure 4.2 are obtained, which show the transfer of load from the pile to the soil at each stage of loading shown in Figure 4.1. Thus when loaded to point A virtually the whole of the load is carried by friction on the pile shaft and there is little or no transfer of load to the toe of the pile (Figure 4.2). When the load reaches point B the pile shaft is carrying its maximum frictional resistance and the pile toe will be carrying some load. At Point D there is no further increase in the load transferred in friction, but the base load will have reached its ultimate value.

4.1.3   Determining allowable loads on piles using permissible stress methods

The loading corresponding to point D on the load/settlement curve in Figure 4.1 represents the ultimate resistance, or ultimate limit state, of the pile and is defined as the stage at which there is general shear failure of the soil or rock beneath the pile toe. However, this stage is of academic interest to the structural designer. A piled foundation has failed in its engineering function when the relative settlement between adjacent single piles or groups of piles causes intolerable distortion of the structural framework, or damage to claddings and finishes. This stage may be represented by some point such as E on the load/settlement curve (Figure 4.1). Thus structural failure will have occurred at a load lower than the ultimate resistance of the pile. Various criteria of assessing failure loads on piles from the results of loading tests are listed in Section 11.4.

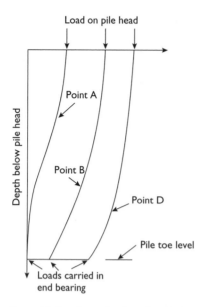

Figure 4.2 Load transfer from head of pile to shaft at points A, B and D on load/settlement curve in Figure 4.1.

The concept of the separate evaluation of shaft friction and base resistance forms the basis of all 'static' calculations of pile bearing capacity. The basic equation is

$$Q_p = Q_b + Q_s - W_p \qquad (4.1)$$

where Q_p is the ultimate resistance of the pile, Q_b is the ultimate resistance of the base, Q_s is the ultimate resistance of the shaft, and W_p is the weight of the pile. The components Q_s and Q_b of the failure load Q_p are shown at the final loading stage in Figure 4.2. Usually W_p is small in relation to Q_p and this term is generally ignored. However, it is necessary to provide for W_p in such situations as piles in marine structures in deep water where a considerable length of shaft extends above sea bed.

Permissible stress methods are adopted in BS 8004. The actual dead load of a structure and the most unfavourable combination of imposed loads are assumed to be applied to the ground. The foundation is assumed to be safe if the permissible stress on the soil or rock is not exceeded, taking into account the likely variable strength or stiffness properties of the ground and the effect of a varying groundwater level. In the case of piled foundations, uncertainty in the reliability of the calculation method is also taken into account. It is generally accepted that current methods cannot predict failure loads to a greater accuracy than plus or minus 60% of the value determined from a full-scale loading test taken to failure. Hence, the safety factors used to obtain the allowable load on a single pile from the calculated ultimate load are correspondingly high.

In BS 8004 a safety factor between 2 and 3 is generally adopted. Experience of a large number of loading tests on piles of diameter up to 600 mm taken to failure, both in sands and clays, has shown that if safety factor of 2.5 is taken on the ultimate resistance then the settlement of the pile head at the allowable load is unlikely to exceed 10 mm. For piles of diameters up to about 1000 mm, failure or ultimate loads as determined by loading tests are usually assumed to be the loads causing a pile head settlement of 10% of the diameter.

When using permissible stress methods for piles in groups it is accepted that a structure can suffer excessive distortion caused by group settlement long before an individual pile in the group has failed in bearing resistance. Hence a separate calculation is made of group settlement based on a realistic assessment of dead load and the most favourable or unfavourable combinations of imposed loading, using unfactored values of the compressibility of the ground in the zone influenced by the group loading (see Chapter 5).

Where piles are end bearing on a strong intact rock the concept of a safety factor against ultimate failure does not apply, since it is likely that the pile itself will fail as a structural unit before shearing failure of the rock beneath the pile toe occurs. The allowable loads are then governed by the safe working stress in compression and bending on the pile shaft (or the Eurocode regulations for the characteristic strength of the pile divided by the appropriate material factor) and the settlement of the pile due to elastic deformation and creep in the rock beneath the base of the pile, together with the elastic compression of the pile shaft.

4.1.4 Determining allowable loads in compression using the procedure in the Eurocode British Standard EN 1997-1:2004

This account of the procedure adopted in the above Eurocode (referred to in this and following chapters as Eurocode 7 or EC7)[1.2] is only a brief review of a lengthy document containing many provisos, exceptions, and cross-references to other Eurocodes. The background to the scope and purpose of EC7 is outlined in Chapter 1. If the engineer proposes to undertake the design of a foundation complying in all respects with the code requirements it is essential for the whole document to be studied together with the other relevant codes including BS EN 1990[1.3] and 1992[1.4]. The commentary by Frank *et al.*[1.5] is also helpful to a thorough understanding of the code. The main purpose of this section is to describe the use of partial factors and the associated 'design approach' for determining allowable pile loads (referred to in EC7). EC7 requires a structure, including the foundations, not to fail to satisfy its design performance criteria because of exceeding various limit states. The *ultimate limit state* can occur under the following conditions:

(a) Loss of equilibrium of the structure and the ground considered as a rigid body in which the strengths of the structural materials and the ground are insignificant in providing resistance (State EQU)
(b) Internal failure or excessive deformation of a structure and its foundation (State STR)
(c) Failure or excessive deformation of the ground in which the strengths of the soil or rock are significant in providing resistance (State GEO)
(d) Loss of equilibrium of a structure due to uplift by water pressure or other vertical actions (State UPL) and
(e) Hydraulic heave, internal erosion, and piping caused by hydraulic gradients (State HYD). State EQU could occur when a structure collapses due to a landslide or earthquake. This state is not considered further in this chapter. Design against occurrence of the other states listed above involves applying partial factors to the applied loads (actions) and to the ground resistance to ensure that reaching these states is highly improbable.

Serviceability limit states are concerned with ensuring that the deformations of a structure due to ground movements below the foundations do not damage the appearance or reduce the useful life of the structure, or cause damage to finishes, non-structural elements, machinery or other installations in the structure.

EC7 requires structures and their foundations to have sufficient *durability* to resist weakening from attack by substances in the ground or the environment.

The design methods using EC7 can be advantageous compared to permissible stress methods when designing structures using BS 8110 (Code of practice structural concrete) or BS 5950 (structural steelwork in buildings), or BS 5400 (bridges), which are all limit state codes, in that difficulty is avoided in achieving compatibility between them and the foundation code BS 8004 which is based on permissible stresses. When the use of BS 8004 for the foundations in conjunction with limit state codes for the superstructure cannot be avoided, the superstructure designer should specify clearly whether or not the loads applied to the foundations are factored or unfactored total loads, or are dead loads combined with factored imposed loading.

In addition to EC7, Eurocodes relevant to pile foundation design are the following:

- BS EN 1990 Basis of structural design
- BS EN 1991 Action on structures
- BS EN 1992 Design of concrete structures
- BS EN 1993 Design of steel structures
- BS EN 1994 Design of composite concrete and steel structures
- BS EN 1995 Design of timber structures
- BS EN 1996 Design of masonry structures
- BS EN 1998 Design of structures for earthquake resistance
- BS EN 1999 Design of aluminium structures

As a preliminary EC7 requires the structure to be considered in three categories of risk from the foundation aspect. Geotechnical Category 1 covers structures having negligible risk of failure or damage due to ground movements, or where enough is known about the ground conditions to adopt a routine method of design, provided that there are no risk problems associated with excavation below groundwater level.

Category 2 includes conventional structures and their foundations with no exceptional risk or difficult ground or loading conditions. Structures requiring piling come into this category provided that there is adequate geotechnical data based on routine methods of ground investigation.

Category 3 applies to all categories not coming within the scope of 1 and 2. It includes very large or unusual structures and those involving abnormal risks or exceptionally difficult ground or loading conditions, also structures in highly seismic areas and areas of site instability. EC7 (Clause 2.2) lists 15 geological and environmental features which need to be considered generally in foundation design. All of these are relevant to piled foundations for which the code prescribes three basic approaches to design. These are the following:

- Static load tests
- Empirical or analytical calculations
- Dynamic load tests.

Design by *prescription* and by the *observational method* are also referred to in the general part of the code. The prescriptive method applies to the tables of allowable bearing pressures for spread foundations in various classes of soils and rocks such as Table 1 of BS 8004. Similar tables are not generally available for piles except those giving allowable base pressures for piles bearing on rock.

The *observational method* is not usually relevant to piled foundation design. The method involves the observation during construction of the behaviour of the whole or part of the structure and its foundation. Typically, the method involves measuring the total and differential settlements as the loading increases and making any necessary modifications to the design if the movements are judged to become excessive. At this stage the piling would have been long completed and too late to make any changes to the design without demolishing the superstructure or introducing underpinning piles.

EC7 requires that geotechnical design by *calculation* should be in accordance with BS EN 1990:2002, Basis of structural design. It is emphasized that the quality of the information on the ground conditions is more significant than precision in calculation models and the partial factors employed. Also the interaction between the structure and the ground should be considered to ensure that the strains in the structure are compatible with the ground movements resulting from the applied loading.

EC7 defines *actions* on the foundations. They include earth and groundwater pressures as well as the dead and imposed loading from structure and ground movements such as soil swelling and shrinkage, frost action and drag-down. Duration of actions such as repetitive loading and time effects on soil drainage and compressibility are also required to be considered.

Ground properties are required to be obtained from field or laboratory tests, either directly or by correlation, theory or empiricism. The effects of time, stress level and deformation on the properties are to be taken into account.

Geometrical data to be considered include the slope of the ground surface, groundwater levels and structural dimensions.

Characteristic values of geotechnical parameters are selected from the available information, usually in the form of a site investigation report. A cautious appraisal of the data is made within the zone influenced by stresses transmitted to the ground (e.g. the zones beneath a pile group as shown in Figure 5.21). The selected values may be lower ones which are less than the most probable ones, or an upper range of values higher than the most probable ones. The latter selection would apply where high values have an unfavourable effect on foundation behaviour, for example, when considering drag-down on piles or differential settlement. Statistical evaluation of geotechnical data is permitted by EC7 provided that it conforms to local knowledge of comparable materials. A very cautious approach is necessary when selecting values from published tables of typical properties of soils and rocks.

Design values of actions are required to be determined in accordance with BS EN 1990. In the case of piled foundations the design value (F_d) can be assessed directly, or derived from representative values (F_{rep}) by the equation:

$$F_d = \gamma_F \, F_{rep} \qquad\qquad (4.2)$$

where γ_F is the partial factor for an action.

Actions can be either *structural*, i.e. the loads transmitted from a structure directly to the pile head or through a raft, or they may be *geotechnical*. The latter are caused by ground movements, for example axial compression loads on the embedded surface of a pile caused by negative skin friction (drag-down); or tension loads caused by swelling of the surrounding soil. Geotechnical actions can also occur from transversely applied loads such as those on piles supporting bridge abutments caused by surcharge from the adjacent approach embankments.

EC7 in Clause 7.3.2.1(3)P states that evaluation of geotechnical actions can be undertaken in two ways:

(a) by pile–soil interaction analyses when the degree of relative pile–soil movement is estimated and t–z curves are produced by computer to give the corresponding strains and axial forces in the pile shaft (Section 4.6). In the case of transversely applied actions a p–y analysis is performed (Section 6.3.5). Alternatively actions can be estimated from other forms of analysis, such as finite element analysis.
(b) The upper-bound force exerted on the pile by the ground movement is calculated and treated as an action.

Method (b) when applied to actions from negative skin friction can give over-conservative designs if due consideration is not given to variations in frictional forces over the depth of the pile shaft (Section 4.8).

Having determined the actions, treating structural and geotechnical actions separately, it is then necessary to show that the design value of resistance of the ground against the pile (R_{cd}) at the ultimate limit state is equal to or greater than the design value of the action (F_d). R_{cd} for example, the resistance to axial compression can be calculated by the 'model pile' method which assumes that a pile of the same penetration depth and cross-sectional dimensions as proposed for the project is installed at the location of each borehole or in-situ test. The two components of total pile resistance, that is, the shaft and base resistance, are calculated for the mean and minimum soil parameters for each borehole or test profile. The two components are then divided by a correlation factor, ξ, which depends on the number of ground test profiles on the project site or particular area of the site exhibiting homogeneous ground properties. Clause 7.6.2.3.5(P) does not make it clear whether the profiles represent mean or lower bound lines drawn through the plotted points of laboratory test results on samples from boreholes, or whether they refer only to profiles from in-situ tests such as the cone or pressuremeter test. The authors have assumed that profiles of laboratory test results can be used to obtain the correlation factors as shown in Table A10 of Annex A in EC7 (Table 4.6 in this book). The resulting characteristic pile resistances are given by the equations:

$$R_{ck\ mean} = (R_{b\ cal\ mean} + R_{s\ cal\ mean})/\xi_3 \qquad (4.3a)$$

and

$$R_{ck\ min} = (R_{b\ cal\ min} + R_{s\ cal\ min})/\xi_4 \qquad (4.3b)$$

where $R_{ck\ mean}$ and $R_{ck\ min}$ are the mean and minimum characteristic pile resistances respectively, and $R_{b\ cal}$ and $R_{s\ cal}$ are the calculated base and shaft resistances.

$R_{ck\ mean}$ is calculated from the arithmetic average of the total resistance $R_{c\ cal\ mean}$ obtained from each borehole or in-situ test profile on the project site or part of the site, while $R_{c\ cal\ min}$ is selected from each borehole or test profile showing minimum values. $R_{s\ cal}$ is calculated from the average of the ground properties over the depth of the pile shaft, and $R_{b\ cal}$ from the properties in the region of the pile base.

The correlation factors ξ_3 and ξ_4 in the above equations refer specifically to calculations of mean and minimum ground resistance based on the results of tests on borehole samples or parameters derived from in-situ tests. They are shown in Table 4.6. As described below when R_{ck} is obtained from static pile loading tests the factors are ξ_1 and ξ_2 (Table 4.7) and when R_{ck} is derived from dynamic impact tests the factors ξ_5 and ξ_6 are used.

The design action F_d in equation 4.2 and the design values of base and shaft resistance are obtained by dividing the characteristic values of R_{bk} and R_{sk} obtained from equations 4.3a and 4.3b by the partial factors γ_b and γ_s respectively. These factors are

for actions (A series): γ_G permanent and γ_Q variable from Table 4.1;
for ground properties (M series): γ_M from Table 4.2;
for ground resistances (R series): γ_R from Table 4.3 to 4.5; and
correlation factors from Table 4.6.

The above partial factors are selected by adopting one or all of three design approaches each with a different combination or set of the A, M, and R series. Thus

Design approach 1 (DA1), Combination 1: A1 + M1 + R1
Design approach 1 (DA1), Combination 2: A2 + (M1 or M2) + R4
Design approach 2 (DA2), A1 + M1 + R2
Design approach 3 (DA3), A1 (or A2) + M1 + R3.

The plus sign in the above list denotes 'combined with'. As design approach DA1 will be adopted for pile design by the British National Annex when published. DA2 and DA3 are not considered in this text.

Taking the case of a pile loaded axially in compression and considering the limit states STR or GEO, for DA1 Tables 4.2 and 4.3 show that the partial factors for ground properties and ground resistances are unity. Therefore to ensure that F_d is equal to or not greater than R_{cd} the partial factors are applied to the actions. Accordingly, it is essential that the field operations and laboratory testing techniques should be undertaken in a thorough manner with the appropriate standard of quality.

Approach DA1, Combination 2 provides for alternative material factors M1 or M2. The former is used for structural actions while the latter is applied to geotechnical actions caused by ground movements. Similarly, when approach DA3 is used the action factor A1 refers to

Table 4.1 Partial factors on actions (γ_F)

Action	Symbol	Set	
		A1	A2
Permanent			
Unfavourable	γ_G	1.35	1.0
Favourable		1.0	1.0
Variable			
Unfavourable	γ_G	1.5	1.3
Favourable		0	0

Notes
The partial factors shown in Tables 4.1 to 4.8 are those provided in Annex A of EC7. The mandatory factors to be applied under the British National Annex are due to be published in 2008, and should be consulted prior to undertaking design to EC7 rules.
Different factors will apply to bridges in Table 4.1

Table 4.2 Partial factors for soil parameters (γ_M)

Soil parameter	Symbol	Set	
		M1	M2
Angle of shearing resistance[a]	$\gamma_{\phi'}$	1.0	1.25
Effective cohesion	$\gamma_{c'}$	1.0	1.25
Undrained shear strength	γ_{cu}	1.0	1.4
Unconfined strength	γ_{qu}	1.0	1.4
Weight density	γ_γ	1.0	1.0

Note
a This factor is applied to $\tan\phi'$.

Table 4.3 Partial resistance factors (γ_R) for driven piles

Resistance	Symbol	Set			
		R1	R2	R3	R4
Base	γ_b	1.0	1.1	1.0	1.3
Shaft (compression)	γ_s	1.0	1.1	1.0	1.3
Total/combined (compression)	γ_t	1.0	1.1	1.0	1.3
Shaft in tension	$\gamma_{s;t}$	1.25	1.15	1.1	1.6

Table 4.4 Partial resistance factors (γ_R) for bored piles

Resistance	Symbol	Set			
		R1	R2	R3	R4
Base	γ_b	1.25	1.1	1.0	1.6
Shaft (compression)	γ_s	1.0	1.1	1.0	1.3
Total/combined (compression)	γ_t	1.15	1.1	1.0	1.6
Shaft in tension	$\gamma_{s;t}$	1.25	1.15	1.1	1.6

Table 4.5 Partial resistance factors (γ_R) for continuous flight auger piles

Resistance	Symbol	Set			
		R1	R2	R3	R4
Base	γ_b	1.1	1.1	1.0	1.45
Shaft (compression)	γ_s	1.0	1.1	1.0	1.3
Total/combined (compression)	γ_t	1.1	1.1	1.0	1.4
Shaft in tension	$\gamma_{s;t}$	1.25	1.15	1.1	1.6

Table 4.6 Correlation factors (ξ) to derive characteristic values from
ground test results

ξ	n						
	1	2	3	4	5	7	10
ξ_3	1.40	1.35	1.33	1.31	1.29	1.27	1.25
ξ_4	1.40	1.27	1.23	1.20	1.15	1.12	1.06

Notes
n denotes the number of test profiles.
ξ_3 denotes on the mean values of the calculated resistances from ground test results.
ξ_4 denotes on the minimum values of ground test results.

structural actions and A2 to geotechnical actions. The R3 factors in the DA3 approach are all unity requiring the ground resistances to be calculated directly from the characteristic ground properties in a single calculation to verify that R_{cd} is equal to or greater than F_d.

It will be noted from Tables 4.3 to 4.5 that higher factors are used for uplift loading. The application of these factors is discussed in Chapter 6.

Whichever of the design approaches or combinations of partial factors are used the factors for structural actions and geotechnical actions must be selected separately. EC7 gives no guidance on the factors to be used to obtain the design value of F_d where this is caused by geotechnical actions. Frank et al.[1.5] recommend that the material and resistance factors as shown in Tables 4.2 to 4.5 should be applied to the characteristic values of the geotechnical actions to obtain the design values. Because they are equal to unity or greater than unity they should be applied as *multipliers* to ensure that F_d (geotechnical) is equal to or less than the factored ground resistance.

An alternative to the model pile procedure is permitted. Characteristic values of the soil parameters over the penetration depth of the pile, as determined by field or laboratory testing, are used to obtain the components R_{bk} and R_{sk} characteristic of the whole site or homogeneous area of the site. Correlation factors are not applied to these components. It is likely that their sum, i.e. the characteristic total pile resistance, will be higher than that calculated using the model pile procedure. Clause 7.6.2.3(8) of EC7 states, 'If this alternative procedure is applied the values of the partial factors γ_b and γ_s recommended in Annex A may need to be corrected by a model factor larger than 1.0. The value of the model factor may be set by the National Annex.' The purpose of the model factor is to make the results of the alternative method compatible with the model pile method. The partial factor values shown in Tables 4.1 to 4.17 in this and other chapters of this book are as set out in Annex A of EC7 but correction factors to modify γ_b and γ_s are not shown in Annex A and the British National Annex has not been published at the time of preparing this edition.

One further step may need to be taken. If the superstructure or substructure supported by the piles is stiff enough to redistribute loads from the weaker to the stronger piles, Clause 7.6.2.3(7) allows the correlation factors ξ_3 and ξ_4 to be divided by 1.1 provided that ξ_3 is never less than 1.0.

Experimental models are not used in the day-to-day design of piled foundations. Scale models (including centrifugal models) have their uses as a general research tool, provided that they reproduce the pile installation method, and the findings are verified by full-scale tests and by experience.

Table 4.7 Correlation factors (ξ) to derive characteristic values from static pile load tests

ξ	n				
	1	2	3	4	≥ 5
ξ_1	1.40	1.30	1.20	1.10	1.00
ξ_2	1.40	1.20	1.05	1.00	1.00

Notes

n denotes the number of tested piles.

ξ_1 denotes on the mean values of measured resistances in static tests.

ξ_2 denotes on the minimum values of measured resistances in static load tests.

Pile loading tests using procedures described in Section 11.4.2 can be used directly to obtain design resistance values or to verify design resistances derived from empirical or analytical methods. If there is any doubt about the validity of such methods loading tests are required to be made on trial piles. The Geotechnical Risk Category, as previously described, should be considered when deciding on the programme for pile loading tests. At this stage the trial piles can also be used to check that the proposed installation method can achieve the design penetration depth without difficulty (particularly in the case of driven piles) and can produce a soundly constructed foundation. Loading tests are made on working piles at the project construction stage to confirm the experiences of pre-contract trials and as a routine check on the contractor's workmanship.

Whenever possible static pile loading tests should be taken to failure or to the stage where a failure can be reliably extrapolated from the load/settlement diagram. In cases where the failure load or ultimate limit state resistance R_{cm} cannot be interpreted from a continuously curving load/settlement diagram, Clause 7.6.1.1(3) of EC7 permits R_{cm} to be defined as the load applied to the pile head which causes a settlement of 10% of the pile diameter. Clause 7.5.2.1 recommends that tension tests should always be taken to failure because of doubts about the validity of extrapolation in uplift loading.

The correlation factors shown in Table 4.7 are applied to ultimate resistances (R_{cm}) obtained from loading tests on trial or working piles to obtain characteristic resistances R_{ck}. The partial factors in Tables 4.3 to 4.5 are then applied to arrive at the design resistances R_{cd}. When instrumented piles are used to measure the separate components of base and shaft resistances (R_{bk} and R_{sk}) the appropriate partial factors are used as shown in the tables. If the structure above the piles is stiff enough to redistribute loads from weaker to stronger piles the correlation factors may be reduced by the factor of 1.1 as noted above for calculations using analytical methods, provided that ξ_1 is never less than 1.0.

Clause 7.5.3. of EC7 states that *dynamic loading tests* may be used to estimate resistances to axial compression loads provided that there has been an adequate ground investigation and that the method has been calibrated against static loading tests on the same type of pile and of similar length and cross-section, and in comparable soil conditions. The equipment used for dynamic testing and the method of interpretation are described in Section 7.3.

At the stage of trial piling the correlation factors shown in Table 4.8 are applied to the test results to obtain characteristic and design resistances in the manner described above for static load tests. Table A.11 in Annex A of EC7 should be consulted for the various qualifications in the application of the correlation factors which depend on the instrumentation used in the tests.

Table 4.8 Correlation factors (ξ) to derive characteristic
values from dynamic impact tests

ξ	n				
	≥ 2	≥ 5	≥ 10	≥ 15	≥ 20
ξ_5	1.60	1.50	1.45	1.42	1.40
ξ_6	1.50	1.35	1.30	1.25	1.25

Notes

n denotes the number of tested piles.

ξ_5 denotes on the mean values of the measured resistances from
dynamic tests.

ξ_6 denotes on the minimum values of the measured resistances from
dynamic tests.

See Table A.11 of EN1997-1: 2004 for qualifications to validity of
above correlation factors.

Table 4.9 Design tolerances for diameters of uncased bored
piles

Nominal diameter (d_{nom})	Design diameter (d)
<400 mm	$d = d_{nom} - 20$ mm
$400 \leq d_{nom} \leq 1000$ m	$d = 0.95 d_{nom}$
$d_{nom} > 1000$ mm	$d = d_{nom} - 50$ mm

Geometrical data are concerned with the cross-sectional dimensions of piles. In the case
of precast concrete and manufactured steel sections, the dimensions are required to conform
to manufacturing tolerances as set out in BS EN 1990. These tolerances are insignificant in
relation to the uncertainties involved with soil properties and design methods and can be
ignored. Bored piles in which the concrete is placed in unlined boreholes or driven and cast-
in-place piles where the drive tube is extracted during or after placing the concrete may
undergo reductions in shaft diameter caused by 'waisting' or 'necking' as described in
Section 2.4.2.

BS EN 1992-1-1 specifies that the diameters to be used in design calculations should be
in accordance with the tolerances shown in Table 4.9.

Partial factors of unity are used when checking a foundation design for compliance with
serviceability limit-state criteria. The procedure for pile groups is discussed in Chapter 5.
The following sections of Chapter 4 will describe the use of partial factors in obtaining
values for the separate components of base and shaft resistance of driven and bored piles in
clays, sands, and rocks.

4.2   Calculations for piles in fine-grained soils

4.2.1   Driven displacement piles

When a pile is driven into a fine-grained soil (e.g. clays and clayey silts) the soil is displaced
laterally and in an upward direction, initially to an extent equal to the volume of the pile
entering the soil. The clay close to the pile surface is extensively remoulded and high pore-
water pressures are developed. In a soft clay the high pore pressures may take weeks or
months to dissipate. During this time the shaft friction and end-bearing resistance, in so far

as they are related to the effective overburden pressure (the total overburden pressure minus the pore water pressure), are only slowly developed. The soft clay displaced by the pile shaft slumps back into full contact with the pile. The water expelled from the soil is driven back into the surrounding clay, resulting in a drier and somewhat stiffer material in contact with the shaft. As the pore-water pressures dissipate and the re-consolidation takes place the heaved ground surface subsides to near its original level.

The effects in a stiff clay are somewhat different. Lateral and upward displacement again occurs, but extensive cracking of the soil takes place in a radial direction around the pile. The clay surrounding the upper part of the pile breaks away from the shaft and may never regain contact with it. If the clay has a fissured structure the radial cracks around the pile propagate along the fissures to a considerable depth. Beneath the pile toe, the clay is extensively remoulded and the fissured structure destroyed. The high pore pressures developed in the zone close to the pile surface are rapidly dissipated into the surrounding crack system and negative pore pressures are set up due to the expansion of the soil. The latter may result in an initially high ultimate resistance which may be reduced to some extent as the negative pore pressures are dissipated and relaxation occurs in the soil which has been compressed beneath and surrounding the lower part of the pile.

In permissible stress terminology the end-bearing resistance of the displacement pile (the term Q_b in equation 4.1) is calculated from the equation:

$$Q_b = N_c c_{ub} A_b \qquad (4.4)$$

where N_c is the bearing capacity factor, c_{ub} is the characteristic undisturbed undrained shear strength representative of the fissured strength at the pile toe, and A_b is the cross-sectional area of pile toe. The bearing capacity factor N_c is approximately equal to 9 provided that the pile has been driven at least to a depth of 5 diameters into the bearing stratum. It is not strictly correct to take the undisturbed strength for c_{ub} since remoulding has taken place beneath the toe. However, the greater part of the failure surface in end bearing shown in Figure 4.3 is in soil which has been only partly disturbed by the penetration of the pile. In a stiff fissured clay the gain in strength caused by remoulding is offset by the loss due to large displacement strains along a fissure plane. In the case of a soft and sensitive clay the full undisturbed cohesion should be taken only when the working load is applied to the pile after the clay has had time to regain its original shearing strength (i.e. after full dissipation of pore pressures); the rate of gain in the carrying capacity of piles in soft clays is shown in Figure 4.4. It may be noted that a period of a year is required for the full development of carrying capacity in the Scandinavian 'quick' clays. In any case the end-bearing resistance of a small-diameter pile in clay is only a small proportion of the total resistance and errors due to the incorrect assumption of c_{ub} on the failure surface are not of great significance.

In terms of 'pure' soil mechanics theory the *ultimate shaft friction* is related to the horizontal effective stress acting on the shaft and the effective remoulded angle of friction between the pile and the clay. Thus

$$\tau_s = \sigma'_h \tan \delta_r \qquad (4.5)$$

where τ_s is the unit shaft friction at any point, σ'_h is the horizontal effective stress, and δ_r is the effective remoulded angle of friction.

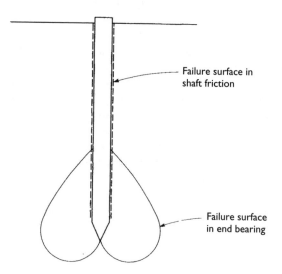

Figure 4.3 Failure surfaces for compressive loading on piles.

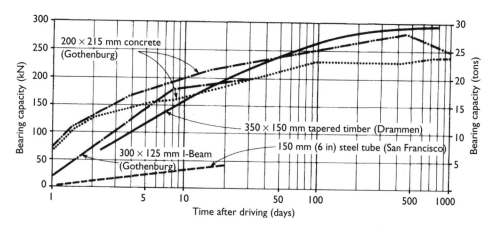

Figure 4.4 Gain in bearing capacity with increasing time after driving of piles into soft clays.

A further simplifying assumption is made that σ'_h is proportional to the vertical effective overburden pressure σ'_{vo}. Thus

$$\tau_s = K\sigma'_{vo} \tan \delta_r \tag{4.6}$$

The value of K is constantly changing throughout the period of installation of the pile and its subsequent loading history. In the case of a driven pile in a stiff clay K is initially very high, as a result of the energy transmitted by the hammer blows required to displace the clay around the pile. However, at this time σ'_{vo} is very low or even negative due to the high pore-water pressures induced by the pile driving. In the case of a bored pile, K is low as the soil swells at the time of drilling the hole, but it increases as concrete is placed in the shaft. Because of these constantly changing values of K, and the varying pore pressures (and hence values of σ'_{vo}), 'pure' soil mechanics methods cannot be applied to practical pile design for conventional structures without introducing empirical factors and simplified calculations to allow for these uncertainties.

A method has been developed at Imperial College, London, for determining the ultimate bearing capacity of piles driven into clays and sands. The method was developed primarily for piles carrying heavy compression and uplift loads in offshore platforms for petroleum exploration and production. The procedure for piles in clays is based on the use of rather complex and time-consuming laboratory tests, with the aim of eliminating many of the uncertainties inherent in the effective stress approach as noted above. The method for piles in clays and sands is described in Section 4.3.7.

In the case of piles which penetrate a relatively short distance into the bearing stratum of firm to stiff clay, that is piles carrying *light to moderate* loading, a sufficiently reliable method of calculating the ultimate shaft friction, Q_s, on the pile shaft is to use the equation:

$$Q_s = \alpha \bar{c}_u A_s \qquad\qquad (4.7)$$

where α is an adhesion factor, $\bar{c}_u$ is the characteristic or average undisturbed undrained shear strength of the soil surrounding the pile shaft, and A_s is the surface area of the pile shaft contributing to the support of the pile in shaft friction (usually measured from the ground surface to the toe).

The adhesion factor depends partly on the shear strength of the soil and partly on the nature of the soil above the bearing stratum of clay into which the piles are driven. Early studies[4.1] showed a general trend towards a reduction in the adhesion factor from unity or higher than unity for very soft clays, to values as low as 0.2 for clays having a very stiff consistency. There was a wide scatter in the values over the full range of soil consistency and these seemed to be unrelated to the material forming the pile.

Much further light on the behaviour of piles driven into stiff clays was obtained in the research project undertaken for the Construction Industry Research and Information Association (CIRIA) in 1969.[4.2] Steel tubular piles were driven into stiff to very stiff London clay and were subjected to loading tests at 1 month, 3 months and 1 year after driving. Some of the piles were then disinterred for a close examination of the soil surrounding the interface. This examination showed that the gap, which had formed around the pile as the soil was displaced by its entry, extended to a depth of 8 diameters and it had not closed up a year after driving. Between depths of 8 diameters and 14 to 16 diameters the clay was partly adhering to the pile surface, and below 16 diameters the clay was adhering tightly to the pile in the form of a dry skin 1 to 5 mm in thickness which had been carried down by the pile. Thus in the lower part of the pile the failure was not between the pile and the clay, but between the skin and surrounding clay which had been heavily sheared and distorted. Strain gauges mounted on the pile to record how the load was transferred from the pile to the soil showed the distribution of load in Figure 4.5. It may be noted that there was no transfer of load in the upper part of the pile, due to the presence of the gap. Most of the load was transferred to the lower part where the adhesion was as much as 20% greater than the undrained strength of the clay. For structures on land, the gap in the upper part of the pile shaft is of no great significance for calculating pile capacity because the greater part of the shaft friction is provided at lower levels. In any case much of the clay in the region of

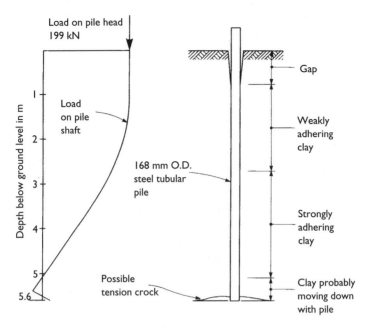

Figure 4.5 Load transfer from pile to stiff clay at Stanmore[4.2].

the gap is removed when excavating for the pile cap. The gap may be significant for relatively short piles with shallow capping beams for house foundations where these are required as a precaution against the effects of soil swelling and shrinkage caused by vegetation (Section 7.9). However, house foundation piles are usually bored and cast-in-place types where a gap is not formed during installation.

Recent research by Bond and Jardine[4.3] on extensively instrumented piles jacked into stiff London Clay confirmed the findings on the nature of the soil disturbance very close to the pile. Negative pore pressures were induced in the clay close to the pile wall and positive pressures further away from the pile. Equalization of pore pressures after installation was very rapid occurring in a period of about 48 hours. There was no change in shaft friction capacity after the equalization period as observed by periodic first-time loading tests over a 3½-month period.

Earlier research, mainly in the field of pile design for offshore structures, has shown that the mobilization of shaft friction is influenced principally by two factors. These are the over-consolidation ratio of the clay and the slenderness (or aspect) ratio of the pile. The over-consolidation ratio is defined as the ratio of the maximum previous vertical effective overburden pressure, σ'_{vc}, to the existing vertical effective overburn pressure, σ'_{vo}. For the purposes of pile design, Randolph and Wroth[4.4] have shown that it is convenient to represent the over-consolidation ratio by the simpler ratio of the undrained shear strength to the existing effective overburden pressure, c_u/σ'_{vo}. Randolph and Wroth showed that the c_u/σ'_{vo} ratio could be correlated with the adhesion factor, α. A relationship between these two has been established by Semple and Rigden[4.5] from a review of a very large number of pile loading tests, the majority of them being on open-end piles either plugged with soil or concrete. This is shown in Figure 4.6a for the case of a rigid pile, and where the shaft friction is calculated from the peak value of c_u. To allow for the flexibility and slenderness ratio of the pile it is necessary to reduce the values of α_p by a length factor, F, as shown in Figure 4.6b. Thus total shaft friction:

$$Q_s = F\alpha_p \bar{c}_u A_s \tag{4.8}$$

The slenderness ratio, L/B, influences the mobilization of shaft friction in two ways. First, a slender pile can 'whip' or flutter during driving causing a gap around the pile at a shallow depth. The second influence is the slip at the interface when the shear stress at transfer from the pile to the soil exceeds the peak value of shear strength and passes into the lower residual strength. This is illustrated by the shear/strain curve of the simple shear box test on a clay. The peak shear strength is reached at a relatively small strain followed by the much lower residual strength at long strain. It follows that when an axial load is applied to the head of a long flexible pile the relative movement between the pile and the clay at a shallow depth can be large enough to reach the stage of low post-peak strength at the interface. Near the pile toe the relative movement between the compressible pile and the compressible clay may not have reached the stage of mobilizing the peak shear strength. At some intermediate level the post-peak condition may have been reached but not the lowest residual condition. It is therefore evident that calculation of the skin friction resistance from the results of the peak undrained shear strength, as obtained from unconfined or triaxial compression tests in the laboratory, may overestimate the available friction resistance of long piles. The length factors shown in Figure 4.6b are stated by Semple and Rigden to allow both for the flutter effects and the residual or part-residual shear strength conditions at the interface. The effect of these conditions on the settlement of single piles is discussed in Section 4.6.

Where an overburden of soft clay is overlying a stiff clay adhesion, factors appropriate to each type should be selected and the shaft resistance calculated for the portion of the shaft embedded in each stratum. The length factor in Figure 4.6b is taken on the overall embedded length.

In marine structures where piles may be subjected to uplift and lateral forces caused by wave action or the impact of berthing ships, it is frequently necessary to drive the piles to much greater depths than those necessary to obtain the required resistance to axial compression loading only. To avoid premature refusal at depths which are insufficient to obtain the required uplift or lateral resistance, tubular piles are frequently driven with open ends. At the early stages of driving soil enters the pile when the pile is said to be 'coring'. As driving continues shaft friction will build up between the interior soil and the pile wall. This soil is acted on by inertial forces resulting from the blows of the hammer. At some stage the inertial forces on the core plus the internal shaft friction will exceed the bearing capacity of the soil at the pile toe calculated on the cross-sectional area of the open end. The plug is then carried down by the pile as shown in Figure 4.7a. However, on further driving and when subjected to the working load, the pile with its soil plug does not behave in the same way as one driven to its full penetration with the tip closed by a steel plate or concrete plug. This is because the soil around and beneath the open end is not displaced and consolidated to the same extent as that beneath a solid-end pile.

Comparative tests on open-end and closed-end piles were made by Rigden et al.[4.6] The two piles were 457 mm steel tubes driven to a penetration of 9 m into stiff glacial till in Yorkshire. A clay plug was formed in the open-end pile and carried down to occupy 40% of the final penetration depth. However, the failure loads of the clay-plugged and steel plate

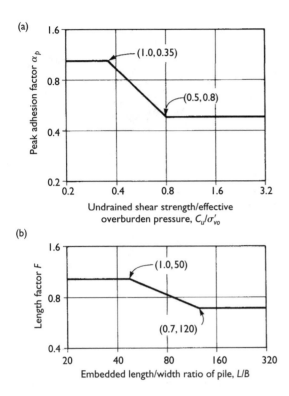

(a)

Peak adhesion factor α_p

(1.0, 0.35)

(0.5, 0.8)

0.2 0.4 0.8 1.6 3.2

Undrained shear strength/effective
overburden pressure, C_u/σ'_{vo}

(b)

Length factor F

(1.0, 50)

(0.7, 120)

20 40 80 160 320

Embedded length/width ratio of pile, L/B

Figure 4.6 Adhesion factors for piles driven to deep penetration into clays (after Semple and
Rigden[4.5]) (a) Peak adhesion factor versus shear strength/effective overburden pressure
(b) Length factor.

closed piles were 1160 and 1400 kN respectively. Evaluation of the ultimate shaft friction
and base resistances showed that the external shaft friction on the open-end piles was 20%
less than that on the closed-end piles.

Accordingly, it is recommended that where field measurements show that a clay plug
is carried down, the total ultimate bearing capacity should be calculated as the sum of the
external shaft friction (obtained from equation 4.8 and Figure 4.6) multiplied by a factor of
0.8, and the ultimate base resistance, Q_b, obtained from equation 4.4 multiplied by a factor
of 0.5 (see Section 4.3.9). Where an internal stiffening ring is provided at the toe of a steel
pile the base resistance should be calculated only on the net cross-sectional area of the steel.
Attempts to clean out the core of soil from within the pile and replace it by a plug of
concrete or cement–sand grout are often ineffective due to the difficulty of removing the
strongly adherent clay skin to provide an effective bond to the pile surface. Also on large
diameter piles the radial shrinkage of the concrete or grout plug can weaken the bond with
the pile. As already noted the majority of the pile tests used to derive the relationships in
Figure 4.6 were made on open-end piles plugged with soil or concrete. Hence, the shaft
friction derived from them already incorporates the effect of the open end.

Plug formation between the flanges and web of an H-section pile is problematical. The
possible plug formation at the toe of an H-pile is shown in Figure 4.7b. The mode of

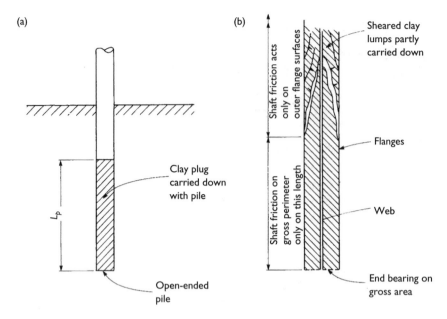

Figure 4.7 Formation of soil plug at toe of small displacement piles (a) Open-ended tube (b) H-section.

formation of a dragged-down soft clay or sand skin has not been studied. A gap has been observed around all flange and web surfaces of H-piles driven into stiff glacial till. An H-pile is not a good type to select if it is desired to develop shaft friction and end-bearing resistance in a stiff clay. The authors recommend calculating the shaft friction on the outer flange surfaces only, but plugging can be allowed for by calculating the end-bearing resistance on the gross cross-sectional area of the pile. Because of the conservative assumptions of shaft friction and the relatively low proportion of the load carried in end bearing the calculated resistance need not be reduced by the factor of 0.5 as recommended for tubular piles.

Applying the recommendations of EC7, for calculating the ultimate resistance of piles driven into clay, the permissible stress equation 4.1 becomes

$$R_{cd} = R_{bd} + R_{sd} \tag{4.9}$$

where R_{cd}, R_{bd}, and R_{sb} are the design values of the total, base and shaft resistances respectively.

The first step is to calculate the latter two components using the arithmetic average of the undrained shear strength c_u for each borehole or in-situ test profile. Thus:

$$R_{c\ cal\ mean} = 9 \times c_{ub\ mean} \times A_b + F \times \alpha_p \times \bar{c}_{us\ mean} \times A_s \tag{4.10}$$

and

$$R_{c\ cal\ min} = 9 \times c_{ub\ min} \times A_b + F \times \alpha_p \times \bar{c}_{us\ min} \times A_s \tag{4.11}$$

$R_{c\ cal\ mean}$ is calculated from equation 4.10 for each borehole or test profile using $\bar{c}_{ub\ mean}$ as the average value from tests in the region of the pile base, and $c_{us\ mean}$ as the average for each borehole or test profile over the penetration depth of the pile. $R_{c\ cal\ min}$ is calculated from equation 4.11 using the values of c_u at the base and over the penetration depth of the pile from the borehole or test pile showing the lowest results after reviewing all test data.

The characteristic total pile resistance is then obtained from the equation:

$$R_{ck} = (R_{b\ cal} + R_{s\ cal})/\xi \tag{4.12}$$

In equation 4.12 values of R_{ck} are obtained using the correlation factor ξ_3 for mean values of R_{ck} and ξ_4 for minimum values (see Table 4.6) to give

$$R_{ck} = R_{bk} + R_{sk} \tag{4.13}$$

where R_{ck}, R_{bk} and R_{sk} are characteristic values and R_{ck} is the lower of the factored components of the minimum or mean strength profiles.

R_{bk} and R_{sk} are divided by the partial factors γ_b and γ_s respectively after using DA1 design approach as described in Section 4.1.4 to give the design values:

$$R_{cd} = R_{bk}/\gamma_b + R_{sk}/\gamma_s = R_{bd} + R_{sd} \tag{4.14}$$

When the alternative procedure is used R_{cd} is given by

$$R_{cd} = (9 \times c_{ubk} \times A_b)/\gamma_b + (F \times \alpha_p \times \bar{c}_{usk})/\gamma_s \tag{4.15}$$

where c_{ubk} and $\bar{c}_{usk}$ are characteristic values of c_u for the base and shaft respectively.

The foregoing procedures are illustrated by Example 4.1 at the end of this chapter.

4.2.2   Driven and cast-in-place displacement piles

The end-bearing resistance of driven and cast-in-place piles terminated in clay can be calculated from equation 4.4. Where the piles have an enlarged base formed by hammering out a plug of gravel or dry concrete, the area A_b should be calculated from the estimated diameter of the base. It is difficult, if not impossible, for the engineer to make this estimate in advance of the site operations since the contractor installing these proprietary piles makes his own decision on whether to adopt a fairly shallow penetration and hammer out a large base in a moderately stiff clay, or whether to drive deeper to gain shaft friction, but at the expense of making a smaller base in the deeper and stiffer clay. In a hard clay it may be impracticable to obtain any worthwhile enlargement over the nominal shaft diameter. In any case, the base may have to be taken to a certain minimum depth to ensure that settlements of the pile group are not exceeded (see Section 5.2.2). The decision as to this minimum length must be taken or approved by the engineer.

The conditions for predicting shaft friction on the shaft are different from those with driven pre-formed piles in some important aspects. The effect on the soil of driving the piling tube with its end closed by a plug is exactly the same as with a steel tubular pile; the clay is remoulded, sheared, and distorted, giving the same conditions at the pile–soil interface as with the driven pre-formed pile. The clay has no chance to swell before the concrete is placed and the residual radial horizontal stress in the soil closes up any incipient gap caused by shrinkage of the concrete. Also the gap which may form around the upper part of the driving tube (or down the full

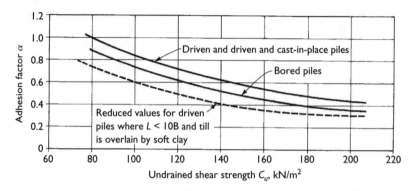

Figure 4.8 Adhesion factors for piles in glacial till (after Weltman and Healy[4.7]).

length of the driving tube if an enlarged detachable shoe is used to close its base) becomes filled with concrete. The tube, while being driven, drags down a skin of soft clay or sandy soil for a few diameters into the stiff clay and it is quite likely that this skin will remain interposed between the concrete and the soil, i.e. the skin is not entirely pulled out by adhering to the tube. However, in one important aspect there is a difference between the driven, and the driven and cast-in-place pile in that water migrates from the unset concrete into the clay and softens it for a limited radial distance. This aspect is discussed in greater detail in Section 4.2.3. Thus the adhesion factor for a driven and cast-in-place pile in a stiff clay may be slightly less than that for a driven pile in corresponding soil conditions. It will probably be greater over the length in a soft clay, however, since the concrete slumps outwards as the tube is withdrawn, producing an increase in effective shaft diameter.

The results of a number of loading tests on driven and driven and cast-in-place piles in glacial till have been reviewed by Weltman and Healy[4.7]. There appeared to be little difference in the $\alpha-c_u$ relationship for either type of pile. They produced the design curves shown in Figure 4.8 for the two types of driven pile including a curve for piles driven a short penetration into stiff glacial till overlain by soft clay. Their review also included a study of the shaft friction on bored piles in glacial till.

Trenter[4.8] recommended using the Weltman and Healy relationships and stated that it is essential to obtain 100 mm samples of the till suitable for strength tests.

When using the EC7 recommendations to determine the ultimate limit state resistance of driven and cast-in-place piles, the procedure described in Section 4.2.1 should be followed. Compliance with the shaft dimension tolerances in Table 4.9 should be observed.

4.2.3   Bored and cast-in-place non-displacement piles

The installation of bored piles using the equipment and methods described in Sections 3.3.1 to 3.3.6 and 3.4.6 causes changes in the properties of the soil on the walls of the pile borehole which have a significant effect on the frictional resistance of the piles. The effect of drilling is to cause a relief of lateral pressure on the walls of the hole. This results in swelling of the clay and there is a migration of pore water towards the exposed clay face. If the borehole intersects water-filled fissures or pockets of silt the water will trickle down the hole and form a slurry with the clay as the drilling tools are lowered down or raised from the hole. Water can also soften the clay if it trickles down from imperfectly sealed-off water-bearing strata above the clay, or if hose pipes are carelessly used at ground level to remove clay adhering to the drilling tools.

The effect of drilling is always to cause softening of the clay. If bentonite is used to support the sides of the borehole, softening of the clay due to relief of lateral pressure on the walls of the hole will still take place, but flow of water from any fissures will not occur. There is also a risk of entrapment of pockets of bentonite in places where overbreak has been caused by the rotary drilling operation. This would be particularly liable to occur in a stiff fissured clay.

After placing concrete in the pile borehole, water migrates from the unset concrete into the clay, causing further softening of the soil. The rise in moisture content due to the combined effects of drilling and placing concrete was observed by Meyerhof and Murdock[4.9], who measured an increase of 4% in the water content of London Clay close to the interface with the concrete. The increase extended for a distance of 76 mm from the interface.

This softening affects only the shaft. The soil within the zone of rupture beneath and surrounding the pile base (Figure 4.3) remains unaffected for all practical purposes and the end-bearing resistance Q_b can be calculated from equation 4.4, the value of the bearing capacity factor N_c again being 9. However, Whitaker and Cooke[4.10] showed that the fissured structure of London Clay had some significance on the end-bearing resistance of large bored piles, and they suggested that if a bearing capacity factor of 9 is adopted the characteristic shearing strength should be taken along the lower range of the graph of shearing strength against depth. If bentonite is used the effects of any entrapment of slurry beneath the pile base as described by Reese *et al.*[3.13] should be allowed for by an appropriate reduction in end-bearing resistance.

The effect of the softening on the shaft friction of bored piles in London Clay was studied by Skempton[4.11], who showed that the adhesion factor in equation 4.7 ranged from 0.3 to 0.6 for a number of loading test results. He recommended a value of 0.45 for normal conditions where drilling and placing concrete followed a reasonably rapid sequence. However, for short piles, where a large proportion of the shaft may be in heavily fissured clay, Skempton recommended the lower value of 0.3. Skempton observed that the actual unit shaft friction mobilized in London Clay did not exceed 100 kN/m^2, and this value should be taken as an upper limit when the unit resistance is calculated from 0.3 or 0.45 times the average undisturbed undrained shear strength. Alternatively, the curve for bored piles in Figure 4.8 can be used to obtain the adhesion factor for very stiff to hard clays.

The authors recommend that the same value of 0.3 should be used for small-diameter bored piles where there may be a long delay between drilling and placing the concrete, for example, where piles are drilled in the morning and the borehole is left unlined awaiting the arrival of the ready-mixed concrete truck at the end of the day. The factor of 0.3 should also be used for large bored piles with enlarged bases which may involve a long delay between first drilling and finally concreting the shaft, giving a long period for the swelling and softening of the clay on the sides of the shaft. It is believed that differences in the method of drilling, such as between the scoring or gouging of a plate auger and the smoothing of a bucket auger, can also cause differences in friction. However, the effects of soil swelling and water from the concrete are likely to be of much greater significance in controlling the adhesion factor.

Fleming and Sliwinski[4.12] reported no difference in the adhesion factor between bored piles drilled into clays in bentonite-filled holes and dry holes. In spite of this evidence it must be pointed out that if the use of a bentonite slurry to support an unlined hole in clay does not reduce the shaft friction this must mean that the rising column of concrete placed by tremie pipe beneath the slurry has the effect of sweeping the slurry completely off the wall of the borehole. It is difficult to conceive that this happens in all cases; therefore the adhesion factor α recommended for London Clay, or for other clays in Figure 4.8, should be reduced by 0.8 to allow for the use of bentonite unless a higher value can be demonstrated conclusively by loading tests.

In clays other than London Clay, where there is no information from loading tests or publications, the adhesion factors shown in the curve for bored piles in glacial till (Figure 4.8) can be used as a guide to pile design. The calculated pile capacity should be confirmed by field loading tests.

The procedure for checking the ULS resistance of bored piles in clay when using the EC7 rules is the same as described in Section 4.2.1 for driven piles. The γ_b and γ_s partial factors in equation 4.14 are used for conventional bored piles and continuous flight auger (CFA) piles as shown in Tables 4.4 and 4.5 respectively.

The greater part of the resistance of bored piles in clay is provided by shaft friction for which the component in equation 4.10 becomes $\alpha \bar{c}_{usk} A_s$. It will be noted that the value of γ_s is unity in the above tables. Hence, the engineer should give careful attention to the quality of the undisturbed sample and the laboratory testing techniques.

The higher value of 1.25 for γ_b in bored piles in equations 4.14 and 4.15 compared with unity for driven piles reflects the influence of the fissured structure of many stiff clays, and also takes into account possible inadequacies when cleaning out the base of the pile bore-hole before placing the concrete. The latter operation also involves the risk in soft clays of 'waisting' or necking when placing concrete in uncased boreholes or when extracting temporary casing. Allowances for possible reductions in pile diameter due to these causes are shown in Table 4.9.

When enlarged bases are provided on bored piles in a fissured clay there may be a loss of adhesion over part of the pile shaft in cases where appreciable settlements of the pile base are allowed to occur. The effect of such movements is to open a gap between the conical surface of the base and the overlying clay. The latter then slumps downwards to close the gap and this causes a 'drag-down' on the pile shaft. Arching prevents slumping of the full thickness of clay from the ground surface to the pile base. It is regarded as over-cautious to add the possible drag-down force to the working load on the pile, but nevertheless it may be prudent to disregard the supporting action on the pile of shaft friction over a height of two shaft diameters above the pile base, as shown in Figure 4.9.

Disregarding shaft friction over a height of two shaft diameters and taking an adhesion factor of 0.3 for the friction on the remaining length may make a pile with an enlarged base an unattractive proposition in many cases when compared with one with a straight shaft. However, the enlarged-base pile is economical if the presence of a very stiff or hard stratum permits the whole of the working load to be carried in end bearing. These piles can also be advantageous where the concept of yielding or 'ductile' piles is adopted for the purpose of achieving load distribution between piles as discussed in Sections 5.2.1 and 5.10. Enlarged bases may also be a necessity to avoid drilling down to or through a water-bearing layer in an otherwise impervious clay.

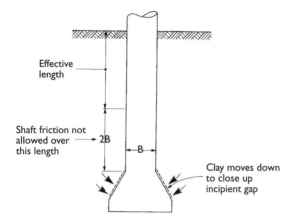

Figure 4.9 Effective shaft length for calculating friction on shaft of under-reamed pile.

Piles for marine structures are sometimes installed by driving a steel tube to a limited penetration below sea bed, followed by drilling-out the soil plug then continuing the drilled hole without further support by the pile tube. A bentonite slurry is sometimes used to support the borehole. On reaching the design penetration depth a smaller diameter steel tube insert pile is lowered to the bottom of the borehole and a cement–sand grout is pumped-in to fill the annulus around the insert pile. The grout is injected either through a small-diameter pipe or is pumped directly down the insert pile.

Kraft and Lyons[4.13] have shown that the adhesion factor used to calculate the shaft friction on the grout/clay interface is of the same order as that used for the design of conventional bored and cast-in-place concrete piles. Where bentonite is used as the drilling fluid a reduction factor should be adopted as discussed above. A considerable increase in the adhesion factor can be obtained if grout is injected under pressure at the pile–soil interface after a waiting period of 24 hours or more. Jones and Turner[4.14] report a two- to threefold increase in adhesion factor when post-grouting was undertaken around the shafts of 150 mm diameter micropiles in London clay. However, the feasibility of achieving such increases should be checked by loading tests before using them for design purposes particularly if there are doubts about the ability of the grouting process to achieve full coverage of the shaft area. The post-grouting technique is used as a first step around the shafts of bored piles where base grouting is used as described in Section 3.3.9.

4.2.4 The effects of time on pile resistance in clays

Because the methods of installing piles of all types have such an important effect on the shaft friction it must be expected that with time after installation there will be further changes in the state of the clay around the pile, leading to an increase or reduction in the friction. The considerable increase in resistance of piles driven into soft sensitive clays due to the effects of re-consolidation have already been noted in 4.2.1.

Bjerrum[4.15] has reported on the effects of time on the shaft friction of piles driven into soft clays. He observed that if a pile is subjected to a sustained load over a long period the shearing stress in the clay next to the pile is carried partly in effective friction and partly in effective cohesion. This results in a downward creep of the pile until such time as the frictional resistance of the clay is mobilized to a degree sufficient to carry the full shearing stress. If insufficient frictional resistance is available the pile will continue to creep downwards. However, the effect of long-period loading is to increase the effective shaft resistance as a result of the consolidation of the clay. It must therefore be expected that if a pile has an adequate safety factor as shown by a conventional short-term loading test, the effect of the permanent (i.e. long-term) working load will be to increase the safety factor with time. However, Bjerrum further noted that if the load was applied at a very slow rate there was a considerable reduction in the resistance that could be mobilized. He reported a reduction of 50% in the adhesion provided by a soft clay in Mexico City when the loading rate was reduced from 10 to 0.001 mm per minute, and a similar reduction in soft clay in Gothenburg resulting from a reduction in loading rate from 1 to 0.001 mm per minute. These effects must be taken into account in assessing the required safety factor if a pile is required to mobilize a substantial proportion of the working load in shaft friction in a soft clay.

No conclusive observations have been published on the effects of sustained loading on piles driven in stiff clays, but there may be a reduction in resistance with time. Surface water can enter the gap and radial cracks around the upper part of the pile caused by the entry of displacement piles, and this results in a general softening of the soil in the fissure system surrounding the pile. The migration of water from the setting and hardening concrete into the clay surrounding a bored pile is again a slow process but there is some evidence of a reverse movement from the soil into the hardened concrete[4.16]. Some collected data on reductions in resistance with time for loading tests made at a rapid rate of application on piles in stiff clays are as follows:

Type of pile	Type of clay	Change in resistance	Reference
Driven precast concrete	London	Decrease of 10% to 20% at 9 months over first test at 1 month	Meyerhof and Murdock[4.9]
Driven precast concrete	Aarhus (Septarian)	Decrease of 10% to 20% at 3 months over first test at 1 month	Ballisager[4.17]
Driven steel tube	London	Decrease of 4% to 25% at 1 year over first test at 1 month	Tomlinson[4.2]

It is important to note that the same pile was tested twice to give the reduction shown above. Loading tests on stiff clays often yield load/settlement curves of the shape shown in Figure 11.13b (Section 11.4.2). Thus the second test made after a time interval may merely reflect the lower 'long-strain' shaft friction which has not recovered to the original peak value at the time of the second test. From the above data it is concluded that the fairly small changes in pile resistance for periods of up to one year after equalization of pore pressure changes caused by installation are of little significance compared with other uncertain effects. An increase should be allowed only in the case of soft clays sensitive to remoulding.

4.3   Piles in coarse-grained soils

4.3.1   General

The classic formulae for calculating the resistance of piles in coarse soils follow the same form as equation 4.1. Expressed in the parameters of a coarse-grained soil ($c_u = 0$), the total pile resistance is given by the expression:

$$Q_p = N_q \sigma'_{vo} A_b + \frac{1}{2} K_s \sigma'_{vo} \tan \delta A_s \qquad (4.16)$$

where σ'_{vo} is the effective overburden pressure at pile base level, N_q is the bearing capacity factor, A_b is the area of the base of the pile, K_s is a coefficient of horizontal soil stress which depends on the relative density and state of consolidation of the soil, the volume displacement of the pile, the material of the pile and its shape, δ is the characteristic or average value of the angle of friction between pile and soil, and A_s is the area of shaft in contact with the soil. The factors N_q and K_s are empirical and based on correlations with static loading tests, δ is obtained from empirical correlations with field tests.

The factor N_q depends on the ratio of the depth of penetration of the pile to its diameter and on the angle of shearing resistance ϕ of the soil. The latter is normally obtained from the results on tests made in-situ (see Section 11.1.4). The relationship between the standard penetration resistance N and ϕ, as established by Peck et al.[4.18], and between the limiting static cone resistance, q_c and ϕ as established by Durgonoglu and Mitchell[4.19], are shown in Figures 4.10, and 4.11 respectively.

From tests made on instrumented full-scale piles, Vesic[4.20] showed that the increase of base resistance with increasing depth was not linear as might be implied from equation 4.16, but that *rate* of increase actually decreased with increasing depth. For practical design purposes it has been assumed that the increase is linear for pile penetrations of between 10 and 20 diameters, and that below these depths the unit base resistance has been assumed to be at a constant value. This simple design approach was adequate for ordinary foundation work where the penetration depths of closed-end piles were not usually much greater than 10 to 20 diameters. At these depths practical refusal was usually met when driving piles into medium dense to dense coarse soils.

However, the use of piled foundations for offshore petroleum production platforms has necessitated driving hollow tubular piles with open ends to very great depths below the sea bed to obtain resistance in shaft friction to uplift loading. The assumption of a constant unit base resistance below a penetration depth of 10 to 20 diameters has been shown to be over-conservative (see Section 4.3.7).

The value of N_q is obtained from the relationship between the drained angle of shearing resistance (ϕ') of the soil at the pile base and the penetration depth/breadth of the pile. The relationship developed by Berezantsev et al.[4.21] is shown in Figure 4.13. Vesic[4.20] stated that these N_q values gave results which most nearly conform to the practical criteria of pile failure. The alternative is to use the Brinch Hansen N_q factors shown in Figure 4.13. They should be multiplied by a shape factor of 1.3 to allow for the square or circular cross-section of the pile base. The Brinch Hansen factors may be over-conservative for some D/B ratios, as, for example, a D/B ratio of 20 and ϕ' values greater than 35°. It is important to note that the values of ϕ' obtained from SPT N-values should not be

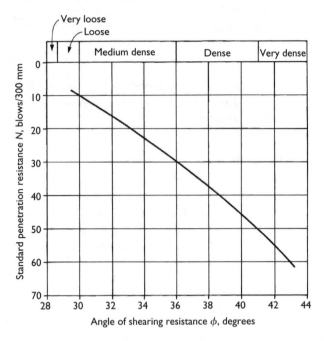

Figure 4.10 Relationship between standard penetration test N-values and angle of shearing resistance (after Peck *et al.*[(4.18)]).

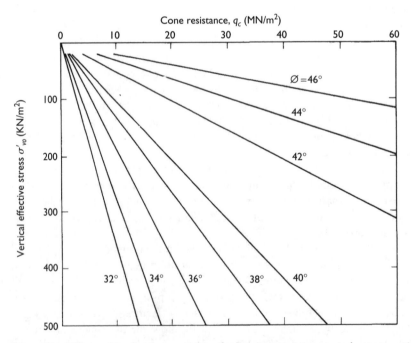

Figure 4.11 Relationship between angle of shearing resistance and cone resistance for an uncemented, normally consolidated quartz sand (after Durgonoglu and Mitchell[(4.19)]).

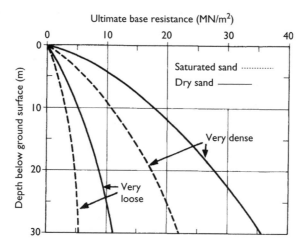

Figure 4.12 Approximate ultimate base resistance for foundations in sand (after Kulhawy[4.22]).

corrected for overburden pressure when relating them to the Berezantsev or Brinch Hansen N_q factors.

The base resistance of open-end piles driven into sands is low compared with closed-end piles, except when a plug of sand formed at the toe is carried down during driving. The mechanics and effects of plug formation are discussed in Section 4.3.3.

Kulhawy[4.22] calculated the ultimate base resistance for very loose and very dense sands in dry and saturated conditions (i.e. in the absence of groundwater and piles wholly below groundwater level) for a range of depths down to a penetration of 30 m. Unit weights of 18.1 and 19.7 kN/m³ were used for the dry loose and dense sands respectively. These values shown in Figure 4.12 may be used for preliminary design purposes in uniform sand deposits. For densities between very loose and very dense the base resistance values can be obtained by linear interpolation.

Reduction in the rate of increase in base resistance with increase in penetration depths is also shown by Berezantsev *et al.*[4.21] Their values of N_q related to ϕ and depth/width ratios are shown in Figure 4.13. Ultimate base resistance values using these factors have been calculated for a closed-end pile of 1220 mm diameter driven into loose sand having a uniform unit submerged weight of 7.85 kN/m³ in Figure 4.14. The angle of shearing resistance of the sand has been assumed to decrease from 30° at the soil surface to 28° at 30 m depth. It will be seen that the Berezantsev N_q values gave lower base resistance than those of Kulhawy. In Figure 4.13 the Berezantsev factors are compared with those of Brinch Hansen[5.4]. The latter have been adopted by the American Petroleum Institute[3.5].

A similar comparison was made for the 1 220 mm pile driven into a dense sand having a uniform unit submerged density of 10.8 kN/m³. The angle of shearing resistance was assumed to decrease from 40° at the soil surface to 37° at 30 m. Figure 4.14 shows that the Kulhawy base resistance values in this case were lower than those of Berezantsev.

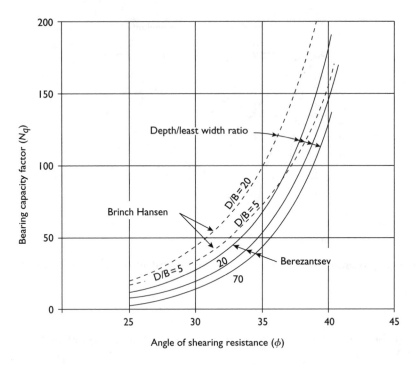

Figure 4.13 Bearing capacity factors of Berezantsev et al.[4.21] and Brinch Hansen[5.4].

The penetration depths in Figure 4.14 have been limited to 20 m for dense sands. This is because the pile capacity as determined by the base resistance alone exceeds the value to which the pile can be driven without causing excessive compression stress in the pile shaft. For example, taking a heavy section tubular pile with a wall thickness of 25 mm in high-yield steel and limiting the compression stress to twice the value given by the allowable working stress of 0.3 times the yield stress, the ultimate pile load is 9.7 MN. This is exceeded at 12 and 20 m penetration using the Berezantsev and Kulhawy factors respectively. The high base resistances which can be obtained in dense sands often make it impossible to drive piles for marine structures to a sufficient depth to obtain the required resistances to uplift and lateral loading. This necessitates using open-end piles, possibly with a diaphragm across the pile at a calculated height above the toe as described in Section 2.2.4.

The second term in equation 4.16 is used for calculating the friction on the pile shaft. The value of K_s is critical to the evaluation of the shaft friction and is the most difficult to determine reliably because it is dependent on the stress history of the soil and the changes which take place during installation of the pile. In the case of driven piles displacement of the soil increases the horizontal soil stress from the original K_o value. Drilling for bored piles can loosen a dense sand, and thereby reduce the horizontal stress.

When piles are driven into coarse-grained soils (gravels, sands and sandy silts) massive changes take place around the pile shaft and beneath its base. Loose soils are readily displaced in a radial direction away from the shaft. If the loose soils are water-bearing, vibrations from the pile hammer cause the soils to become 'quick' and the pile slips down easily. The

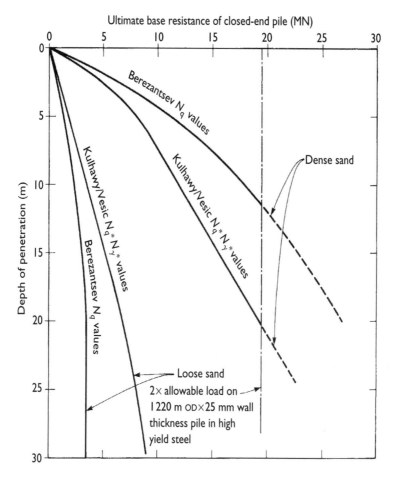

Figure 4.14 Base resistance versus penetration depth for 1220 mm diameter closed-end pile driven into sand (a) loose sand (b) dense sand.

behaviour is similar with bored piles, when the loosened sand (which may initially be in a dense state) slumps into the borehole. When piles are driven into medium-dense to dense sands, radial displacement is restricted by the passive resistance of the surrounding soil resulting in the development of a high interface friction between the pile and the sand. Continued hard driving to overcome the build-up of frictional resistance may cause degradation of angular soil particles with consequent reduction in their angle of shearing resistance. In friable sands, such as the detritus of coral reefs, crushing of the particles results in almost zero resistance to the penetration of open-end piles.

Driving a closed-end pile into sand displaces the soil surrounding the base radially. The expansion of the soil mass reduces its in-situ pore-pressure, even to a negative state, again increasing the shaft friction and greatly increasing the resistance to penetration of the pile. Tests on instrumented driven piles have shown that the interface friction increases exponentially with increasing depth as shown in Figure 4.15.

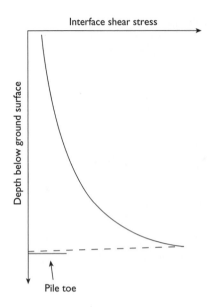

Figure 4.15 Distribution of interface friction on shaft of pile driven into sand.

Equation 4.6 applies to the shaft resistance of piles in coarse-grained soils. The factor K is governed by the following influences:

(1) The stress history of the soil deposit
(2) The ratio of the penetration depth to the diameter of the pile shaft
(3) The rigidity and shape of the pile and
(4) The nature of the material forming the pile shaft.

The stress history of the soil deposit is characterized by its coefficient of earth pressure at rest, K_o, in an undisturbed state. This is measured by field tests such as the standard penetration test (SPT), or cone penetration test (CPT) and by the pressuremeter (Section 11.1). In normally consolidated soils K_o is constant with depth and depends on the relative density of the deposit. Some typical values for a normally consolidated sand are

Relative density	K_o
Loose	0.5
Medium-dense	0.45
Dense	0.35

If the soil deposits are over-consolidated, that is, if they have been subjected to an overburden pressure at some time in their history, K_o can be much higher than the values shown above, say of the order of 1 to 2 or more. It is possible to determine whether or not the soil deposit is over-consolidated by reference to its geological history or by testing in the field using standard penetration tests or static cone tests. Normally consolidated soils show low penetration values at the surface increasing roughly linearly with depth. Over-consolidated soils show high values at shallow depths, sometimes decreasing at the lower levels.

When calculating τ_s in equation 4.6, the factor K in sands and other coarse-grained soils is denoted by K_s which is related to K_o, to the type of pile and to the installation method. Some typical values are shown in Table 4.10.

The angle of interface friction δ_r in equation 4.6 is obtained by applying a factor to the average effective angle of shearing resistance (ϕ') of the soil as determined from its relationship with SPT or CPT values as shown in Figure 4.10 and 4.11. The factor to obtain δ_r from the design ϕ' depends on the surface material of the pile. Factors established by Kulhawy[4.22] are shown in Table 4.11. They apply both to driven and bored piles. In the latter case ϕ' depends on the extent to which the soil has been loosened by the drilling process (Section 4.3.6). The CFA type of bored pile (Section 2.4.2) is advantageous in this respect.

Use of the K_s/K_o relationship in Table 4.10 to determine the shaft resistance of a pile driven into sand when using equation 4.16 does not reflect the exponential distribution of intergranular friction shown in Figure 4.15. A semi-empirical method based on cone resistance values has been developed at Imperial College, London. It is particularly suitable for piles driven to a deep penetration and is described in Section 4.3.6.

EC7 requires that the base resistance of tubular piles driven with open ends having an internal diameter greater than 500 mm should be the lesser of the shearing resistance between the soil plug and the pile interior, and the base resistance of the cross-sectional area of the pile at the toe.

4.3.2   Driven piles in coarse-grained soils

Driving piles into loose sands densifies the soil around the pile shaft and beneath the base. Increase in shaft friction can be allowed by using the higher values of K_s related to K_o from Table 4.10. However, it is not usual to allow any increase in the ϕ values and hence the bearing capacity factor N_q caused by soil compaction beneath the pile toe. The reduction in the rate of increase in end-bearing resistance with increasing depth has been noted above. A further reduction is given when piles are driven into soils consisting of weak friable particles such as calcareous soils consisting of carbonate particles derived from disintegrated corals and shells. The soil tends to degrade under the impact of hammer blows to a silt-sized material with a marked reduction in the angle of shearing resistance.

Because of these factors, published records for driven piles which have been observed from instrumented tests have not shown values of the ultimate base resistance much higher than 11 MN/m². The authors use this figure for closed-end piles as a practical peak value for ordinary design purposes but recognize that higher resistances up to a peak of 22 MN/m² may be possible when driving a pile into a dense soil consisting of hard angular particles. Such high values should not be adopted for design purposes unless proved by loading tests. Figure 4.14 shows that the base resistance of a closed-end pile driven into a dense sand can reach the maximum compressive stress to which the pile can be subjected during driving at a relatively short penetration. Therefore, if the peak base resistance of 11 MN/m² is used for design there is no advantage in attempting to drive piles deeply into medium-dense to dense soils with the risk of pile breakage in order to gain a small increase in shaft friction.

Table 4.10   Values of the coefficient of horizontal soil stress, K_s

Installation method	K_s/K_o
Driven piles, large displacement	1–2
Driven piles, small displacement	0.75–1.25
Bored and cast-in-place piles	0.70–1
Jetted piles	0.50–0.7

Table 4.11   Values of the angle of pile to soil friction for various interface conditions

Pile/soil interface condition	Angle of pile/soil friction, δ
Smooth (coated) steel/sand	$0.5\overline{\phi}$–$0.7\overline{\phi}$
Rough (corrugated) steel/sand	$0.7\overline{\phi}$–$0.9\overline{\phi}$
Precast concrete/sand	$0.8\overline{\phi}$–$1.0\overline{\phi}$
Cast-in-place concrete/sand	$1.0\overline{\phi}$
Timber/sand	$0.8\overline{\phi}$–$0.9\overline{\phi}$

H-section piles are not economical for carrying high compression loading when driven into sands. Plugging of the sand does not occur in the area between the web and flanges. The base resistance is low because of the small cross-sectional area. Accordingly the pile must be driven deeply to obtain worthwhile shaft friction. The latter is calculated on the total surface of the web and flanges in contact with the soil. At Nigg in Scotland soil displacements of only a few centimetres were observed on each side of the flanges of H-piles driven about 15 m into silty sand, indicating that no plugging had occurred over the full depth of the pile shaft. The base resistance of these piles can be increased by welding short stubs or wings (see Figure 2.19a) at the toe. Some shaft friction is lost on the portion of the shaft above these base enlargements.

The exponential distribution of interface friction shown in Figure 4.15 has been shown by the Imperial College research to be a function of the length to diameter ratio, or in the terms of the researchers to the ratio of the height above the toe to the pile radius (h/R). It follows that it is more advantageous to use a large-diameter pile with a relatively short embedment depth, rather than a small diameter with a deep penetration, but in some circumstances, however, it may be necessary to drive deeply to obtain the required resistance to uplift or lateral loading.

The maximum working stress on proprietary types of precast concrete jointed piles is in the range of 10 to 17 MN/m^2. Therefore, if the peak design ultimate resistance of 11 MN/m^2 is adopted the piles will have to develop substantial shaft friction to enable the maximum working load to be utilized. This is feasible in loose to medium-dense sands but impracticable in dense sands or medium-dense to dense sandy gravels. In the latter case peak base resistance values higher than 11 MN/m^2 may be feasible, particularly in flint gravels.

When using EC7 rules to determine the ULS resistance of piles driven into coarse-grained soils design approach DA1 in equation 4.16 becomes

$$R_{cd} = \frac{N_q \sigma'_{vok} \times A_b}{\xi \gamma_b} + \frac{0.5 K_s \sigma'_{vok} \tan \delta_r A_s}{\xi \gamma_s} \tag{4.17}$$

The partial factor γ_γ in Table 4.2 is applied to obtain σ'_{vok}, and N_q is derived from ϕ' using the relationship with SPT or CPT tests. The mean and minimum values of the standard penetration or cone resistance should be plotted against depth rather than basing the calculations on individual test results. The design ϕ' values obtained from the test profiles should be divided by the calibration factor of 1.05. The correlation factor ξ depends on the number of test profiles (Table 4.6).

N_q factors obtained from the Brinch Hansen relationship with ϕ' (Figure 4.13) should be multiplied by the shape factor of 1.3 as previously noted. The partial factors γ_b and γ_s for driven piles are shown in Table 4.3.

All set combinations using the DA1 approach should be analysed to obtain the design pile penetration depth until the engineer is sufficiently familiar with the EC7 procedures to be able to appreciate the combination critical to the particular problem. The M sets are not used because the ϕ' (and $\tan \delta_r$) are derived from in-situ tests.

4.3.3   Piles with open ends driven into coarse-grained soils

It was noted in Section 4.3.1 that it is frequently necessary to drive piles supporting off-shore petroleum production platforms to a very great depth below the sea bed in order to obtain the required resistance to uplift loading by shaft friction. Driving tubular piles with open ends is usually necessary to achieve the required penetration depth. Driving is relatively easy, even through dense soils, because with each blow of the hammer the overall pile diameter increases slightly thereby pushing the soil away from the shaft. When the hammer is operating with a rapid succession of blows the soil does not return to full contact with the pile. A partial gap is found around each side of the pile wall allowing the pile to slip down. Flexure of the pile in the stick-up length above sea bed also causes low resistance to penetration.

At some stage during driving a plug of soil tends to form at the pile toe after which the plug is carried down with the pile. At this stage the base resistance increases sharply from that provided by the net cross-sectional area of the pile shoe to some proportion (not 100%) of the gross cross-sectional area.

The stage when a soil plug forms is uncertain; it may form and then yield as denser soil layers are penetrated. It was noted in Section 2.2.4 that 1067 mm steel tube piles showed little indication of a plug moving down with the pile when they were driven to a depth of 22.6 m through loose becoming medium dense to dense silty sands and gravels in Cromarty Firth. No plugging, even at great penetration depths, may occur in uncemented or weakly cemented calcareous soils. Dutt et al.[4.23] described experiences when driving 1.55 m diameter steel piles with open ends into carbonate soils derived from coral detritus. The piles fell freely to a depth of 21 m below sea bed when tapped by a hammer with an 18-tonne ram. At 73 m the driving resistance was only 15 blows/0.3 m.

It should not be assumed that a solidly plugged pile will mobilize the same base resistance as one with a closed end. In order to mobilize the full resistance developed in friction on the inside face the relative pile–soil movement at the top of the plug must be of the order of ½% to 1% of the pile diameter. Thus with a large-diameter pile and a long plug a considerable settlement at the toe will be needed to mobilize a total pile resistance equivalent to that of a closed-end pile. Another uncertain factor is the ability of the soil plug to achieve sufficient resistance to yielding by arching of the plug across the pile interior. Research has shown that

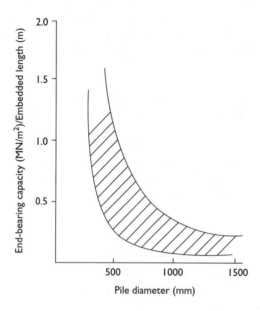

Figure 4.16 Reduction in end-bearing capacity of open-end piles driven into sand due to increase in diameter (after Hight et al.[4.24]).

the arching capacity is related principally to the pile diameter. Clearly it is not related to the soil density because the soil forming the plug is compacted by the pile driving. The estimated ultimate bearing resistances of sand-plugged piles obtained from published and unpublished sources have been plotted against the pile diameters by Hight *et al.*[4.24] Approximate upper and lower limits of the plotted points are shown in Figure 4.16. In most cases the piles were driven into dense or very dense soils and the test evidence pointed clearly to failure within the plug and not to yielding of the soil beneath the pile toe.

4.3.4   *Driven and cast-in-place piles in coarse-grained soils*

Both the base resistance and shaft friction of driven and cast in-situ piles can be calculated in the same way as described for driven piles in the preceding section. The installation of driven and cast in-situ types does not loosen the soil beneath the base in any way, and if there is some loosening of the soil around the shaft as the driving tube is pulled out the original state of density is restored, if not exceeded, as the concrete is rammed or vibrated into place while pulling out the tube. Loosening around the shaft must be allowed for if no positive means are provided for this operation. The provision of an enlarged base adds considerably to the end-bearing resistance of these piles in loose to medium-dense sands and gravels. The gain is not so marked where the base is formed in dense soils, since the enlargement will not greatly exceed the shaft diameter and, in any case, full utilization of the end-bearing resistance may not be possible because of the need to keep the compressive stress on the pile shaft within allowable limits (see Table 2.4).

4.3.5   Bored and cast-in-place piles in coarse-grained soils

If drilling for the piles is undertaken by baler (see Section 3.3.7) or by grabbing under water there is considerable loosening of the soil beneath the pile toe as the soil is drawn or slumps towards these tools. This causes a marked reduction in end-bearing resistance and shaft friction, since both these components must then be calculated on the basis of a low relative density ($\phi = 28°$ to $30°$). Only if the piles are drilled by power auger or reverse-circulation methods in conjunction with a bentonite slurry or by drilling under water using a base grouting technique as described in Section 3.3.9 can the end-bearing resistance be calculated on the angle of shearing resistance of the undisturbed soil. However, the effects of entrapping slurry beneath the pile toe[3.13] must be considered. Loading tests should be made to prove that the bentonite technique will give a satisfactory end-bearing resistance. If there are indications that the entrapment of slurry beneath the toe cannot be avoided, the appropriate reduction in resistance should be made. Fleming and Sliwinski[4.12] suggest that the shaft friction on bored piles, as calculated from a coefficient of friction and the effective lateral pressure, should be reduced by 10% to 30% if a bentonite slurry is used for drilling in a sand.

The effects of loosening of the soil by conventional drilling techniques on the interface shaft friction and base resistances of a bored pile in a dense sand is well illustrated by the comparative loading tests shown in Figure 4.17. Bored piles having a nominal shaft diameter of 483 mm and a driven precast concrete shell pile with a shaft diameter of 508 mm were installed through peat and loose fine sand into dense sand. The bored piles with toe levels at 4.6 and 9.1 m failed at 220 and 350 kN respectively, while the single precast concrete pile which was only 4 m long carried a 750 kN test load with negligible settlement.

When determining the ULS resistance of bored piles in coarse soils by EC7 rules, the direct use of in-situ tests is unpracticable because these tests measure the soil properties in a relatively undisturbed state compared with the gross disturbance which can occur when drilling the pile boreholes. Only when drilling is performed under a slurry can the undisturbed soil properties be used in the calculation, subject to applying a factor to take account of the process.

The most reliable method to obtain compliance with EC7 is to obtain the ULS resistance from static load tests. Dynamic tests are impractible because of the likely variations in the cross-sectional area of the shaft and different elastic properties between the concrete at the pile head and in the body of the shaft.

Design by calculation using the DA1 approach is similar to that described in Section 4.3.2 with N_q and $\tan \delta_r$ in equation 4.17 being obtained from ϕ' values based on SPT or CPT relationships and judgement used to estimate the reduction in ϕ' caused by the pile drilling. Values of K_s in equation 4.17 are obtained from Table 4.10 with the assumption that K_o represents the loosening of the sand.

4.3.6   The use of in-situ tests to predict the ultimate resistance of piles in coarse-grained soils

It has been noted that the major component of the ultimate resistance of piles in dense coarse soils is the base resistance. However, Figures 4.13 and 4.14 show that the values of N_q are

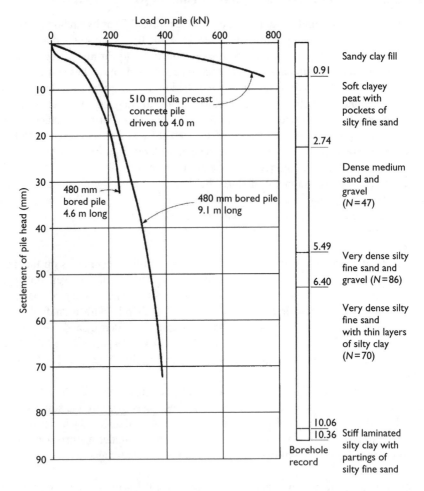

Figure 4.17 Comparison of compressive resistance of driven piles and bored and cast in-situ piles in dense to very dense coarse soils.

very sensitive to the values of the angle of shearing resistance of the soils. These values are obtained from in-situ tests made in boreholes, and if the boring method has loosened the soil, which can happen if incorrect techniques are used (see Section 11.1.4), then the base resistance of any form of driven pile is grossly underestimated. It is very unlikely that the boring method will compact the soil, and thus any over-estimation of the shearing resistance is unlikely.

A reliable method of predicting the shaft friction and base resistance of driven and driven and cast-in-place piles is to make static cone penetration tests at the site investigation stage (see Section 11.1.4). This equipment produces curves of cone penetration resistance with depth (Figure 4.18). Extensive experience with pile predictions based on the cone penetrometer in the Netherlands has produced a set of design rules which have been summarized by Meigh[4.25].

Table 4.12 Relationships between pile shaft friction and cone resistance
(after Meigh[4.25])

Pile type	Ultimate unit shaft friction
Timber	0.012 q_c
Precast concrete	0.012 q_c
Precast concrete enlarged base[a]	0.018 q_c
Steel displacement	0.012 q_c
Open-ended steel tube[b]	0.008 q_c
Open-ended steel tube driven into fine to medium sand	0.0033

Notes
a Applicable only to piles driven in dense groups otherwise use 0.003 where shaft
  size is less than enlarged base.
b Also applicable to H-section piles.

Although engineers in the Netherlands and others elsewhere base shaft friction values on the measured local sleeve friction (f_s), the authors prefer to use established empirical correlations between unit friction and cone resistance (q_c). This is because the cone resistance values are more sensitive to variations in soil density than the sleeve friction and identification of the soil type from the ratio of q_c to f_s is not always clear-cut. Empirical relationships of pile friction to cone resistance are shown in Table 4.12.

A limiting value of 0.12 MN/m^2 is used for the ultimate shaft friction. The values shown in Table 4.12 are applicable to piles under static compression loading and a safety factor of 2.5 is used for q_c values obtained from the electrical cone and 3.0 for the mechanical cone (see Chapter 11). A somewhat higher safety factor would be used for piles subjected to cyclic compression loading to allow for degradation of the assumed siliceous sand (see Section 6.2.2 for piles carrying uplift loading).

Cone-resistance values should not be used to determine the shaft friction of bored piles. This is because of the loosening of the soil caused by drilling as described in the preceding section.

The end-bearing resistance of piles is calculated from the relationship:

$$q_{ub} = \bar{q}_c \tag{4.18}$$

where $\bar{q}_c$ is the average cone resistance within the zone influenced by stresses imposed by the toe of the pile. This average value can be obtained by plotting the variation of q_c against depth for all tests made within a given area. An average curve is then drawn through the plots either visually or using a statistical method. The *allowable* base pressure is then determined from the value of the average curve at pile toe level divided by the appropriate safety factor (Figure 4.18). The value of the safety factor will depend on the scatter of results. It is normally 2.5 but it is a good practice to draw a lower bound line through the lower cone-resistance values, ignoring sharp peak depressions provided that these are not clay bands in a sand deposit. The allowable base pressure selected from the average curve should have a small safety factor when calculated from the lower bound q_c at the toe level (Figure 4.18a).

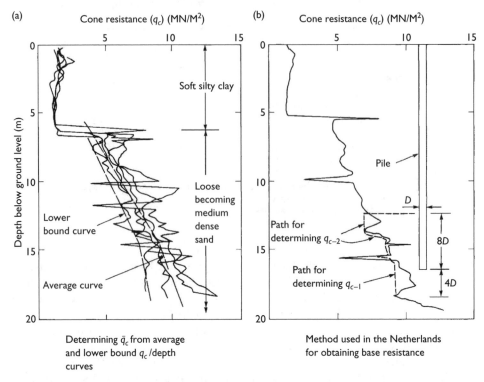

Figure 4.18 Use of static cone penetration tests (CPT) to obtain design values of average cone resistance ($\overline{q}_c$) in coarse soils.

The method generally used in the Netherlands is to take the average cone resistance $\overline{q}_{c-1}$ over a depth of up to four pile diameters below the pile toe, and the average $\overline{q}_{c-2}$ eight pile diameters above the toe as described by Meigh[4.25].

The ultimate base resistance is then

$$q_{ub} = \frac{\overline{q}_{c-1} + \overline{q}_{c-2}}{2} \tag{4.19}$$

The shape of the cone-resistance diagram is studied before selecting the range of depth below the pile to obtain $\overline{q}_{c-1}$. Where the q_c increases continuously to a depth of $4D$ below the toe, the average value of $\overline{q}_{c-1}$ is obtained only over a depth of $0.7D$. If there is a sudden decrease in resistance between $0.7D$ and $4D$ the lowest value in this range should be selected for $\overline{q}_{c-1}$ (Figure 4.18b). To obtain $\overline{q}_{c-2}$ the diagram is followed in an upward direction and the envelope is drawn only over those values which are decreasing or remain constant at the value at the pile toe. Minor peak depressions are again ignored provided that they do not represent clay bands; values of q_c higher than 30 MN/m² are disregarded over the $4D - 8D$ range.

Safety factors generally used in the Netherlands in conjunction with the '4D – 8D' method to obtain the allowable pile load are given by te Kamp[4.26] as:

Timber	1.7
Precast concrete, straight shaft	2.0
Precast concrete, enlarged shaft	2.5

An upper limit is placed on the value of the ultimate base resistance obtained by either of the methods shown in Figure 4.18. Upper limiting values depend on the particle-size distribution and over-consolidation ratio and are shown in Figure 4.19.

The relationship $q_b = q_c$ in equation 4.18 is valid for piles up to about 500 mm in diameter or breadth provided, when designing by permissible stress methods, that a pile head displacement of one-tenth of the diameter is taken as the criterion of failure and that a safety factor of 2.5 is adopted on the calculated total resistance. The reduction of the q_b/q_c ratio with increase in diameter is discussed in Section 4.3.7.

Cone-resistance values cannot be used to obtain the end-bearing resistance of bored and cast-in-place piles because of the loosening of the soil caused by drilling as described in the preceding chapter.

A further factor must be considered when calculating pile shaft friction and end-bearing resistance from CPT data. This is the effect of changes in overburden pressure on the q_c (and also local friction) values at any given level. Changes in overburden pressure can result from excavation, scour of a river or sea bed, or the loading of the ground surface by placing fill. The direct relationship between q_c and overburden pressure is evident from Figure 4.11. Taking the case of a normally consolidated sand when the vertical effective stress is reduced by excavation, the ratio of the horizontal to the vertical stress is also reduced, but not in the same proportion depending on the degree of unloading. The effects are most marked at shallow depths.

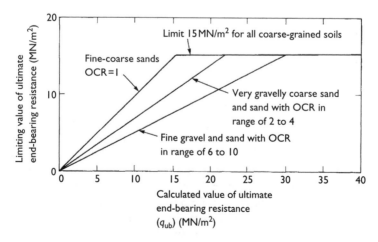

Figure 4.19 Limiting values of pile end-bearing resistance for solid end piles (after te Kamp[4.26]).

Small reductions in overburden pressure cause only elastic movements in the assembly of soil particles. Larger reductions cause plastic yielding of the assembly and a proportionate reduction of horizontal pressures. Broug[4.28] has shown that the threshold value for the change from elastic to elastoplastic behaviour of the soil assembly occurs when the degree of unloading becomes less than 0.4.

The effect of unloading on cone resistance values was shown by de Gijt and Brassinga[4.27]. Figure 4.20 shows q_c/depth plots before and after dredging to a depth of 30 m in the normally consolidated alluvial sands of the River Maas in connection with an extension to the Euroterminal in the Netherlands. Large reductions in overburden pressure within the zone 10 m below the new harbour bed caused the reduction in cone resistance. The difference between the observed new cone resistance and the mean line predicted by Broug did not exceed 5%.

The effects are most marked where the soil deposits contain weak particles such as micaceous or carbonate sands. Broug[4.28] described field tests and laboratory experiments on sands containing 2% to 5% of micaceous particles. These studies were made in connection with the design of piled foundations for the Jamuna River bridge in Bangladesh where scour depths of 30 to 35 m occur at times of major floods[3.17].

The static cone penetration test, which measures the resistance of the *undisturbed* soil, is used as a measure of the resistance to penetration of a pile into a soil which has been compacted by the pile driving. Heijnen[4.29] measured the cone resistance of a loose to medium-dense silty fine sand before and after installing driven and cast in-situ piles. The increase in resistance at various distances from the 1 m diameter enlarged base caused by the pile driving was as follows:

Distance from pile axis (m)	Increase in static cone resistance (%)
1	50–100
2	About 33
3.5	Negligible

In spite of the considerable increase in resistance close to the pile base, the ultimate resistance of the latter was in fact accurately predicted by the cone resistance value of the *undisturbed* soil by using equation 4.18. This indicates that the effect of compaction both in driven and driven and cast in-situ piles is already allowed for when using this equation.

Field trials to correlate the static cone resistance with pile loading tests are necessary in any locality where there is no previous experience to establish the relationship between the two. In the absence of such tests the base resistance should be taken as one-half of the static cone resistance with the application of a factor of safety of 2.5 to obtain the allowable unit pressure on the base of the pile. Experience has shown that if a safety factor of 2.5 is applied to the ultimate base resistance as calculated from the cone resistance the settlement at the working load is unlikely to exceed 10 mm for piles of base widths up to about 500 mm. For larger base widths it is desirable to check that pile head settlements resulting from the design end-bearing pressure are within tolerable limits. Pile head settlements can be calculated using the methods described in Section 4.6.

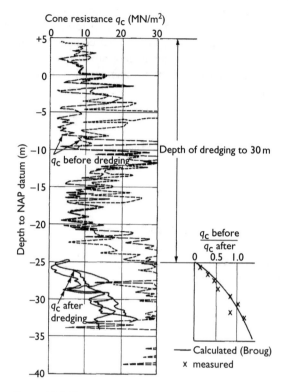

Figure 4.20 Cone resistance versus depth before and after dredging sand (after Gijt and Brassinga[4.27]).

4.3.7 Tubular steel piles driven to deep penetration into clays and sands

The research work undertaken at Imperial College, London, on the axial capacity of steel tube piles has been referred to briefly in the preceding sections. This work was undertaken on behalf of the UK Marine Technology Directorate. The design procedure adopted from the research became known as the MTD method. It is understood that MTD no longer operates but the work at Imperial College has been extended with analysis of further test data and has been published in book form by Jardine *et al.*[4.30] The design procedures which have evolved have become known as the ICP method.

The principal users of large tubular steel piles are the offshore petroleum industry and recently these piles have found increasing use as monopile foundations for offshore wind power generators. Before publication of details of the MTD method guidance for engineers designing offshore piling was available in the recommendations of the American Petroleum Institute[4.31]. Their recommendations for the shaft friction of piles in clay generally followed the αc_u relationship of Semple and Rigden[4.5]. Equation 4.16 was used for piles in sands with Brinch Hansen factors of N_q for calculating base capacity. Chow[4.32] found that the API recommendations for piles in sand were over-conservative for short piles with L/B ratios up to 30, and for dense sands with relative densities of 60% or more.

The ICP method for piles driven into clays is based on effective stresses and takes into account the effects on the interface shaft resistance of the radial displacement of the clay and the gross displacement of the clay beneath the base. To determine shaft resistance the ICP method calculates the local shear stress at failure on the interface after equalization of pore pressure changes brought about by the pile driving. The calculations are made for a succession of layers over the embedded length of the shaft. They are then integrated to give the total shaft resistance from the equation:

$$Q_s = \pi D \int \tau_f \, dZ \tag{4.20}$$

The peak local interface shear stress τ_f is obtained from the equation:

$$\tau_f = (K_f/K_o) \, \sigma'_{rc} \tan \delta_f \tag{4.21}$$

where

K_f = coefficient of radial effective stress for shaft at failure = $\sigma'_{rf}/\sigma'_{vo}$
K_o = coefficient of earth pressure at rest = $\sigma'_{rf}/\sigma'_{vo}$
σ'_{rc} = equalized radial effective stress = $K_c \, \sigma'_{vo}$
δ_f = operational interface angle of frictional failure

K_c is obtained from the equation:

$$K_c = [2.2 + 0.016\text{YSR} - 0.870 \, \Delta I_{vy}]\text{YSR}^{0.42}(h/R)^{-0.20} \tag{4.22}$$

where

I_{vy} = relative void index at yield = $\log_{10}S_t$
I_{vo} = relative void index
YSR = yield stress ratio or apparent over-consolidation ratio
S_t = clay sensitivity
h = height of soil layer above pile toe
R = pile radius
and $K_f/K_c = 0.8$

An alternative to equation 4.22 which is marginally less conservative is

$$K = [2 - 0.625 \, \Delta I_{vo}]\text{YSR}^{0.42}(h/R)^{-0.20} \tag{4.23}$$

YSR, Δ_{vy} and Δ_{vo} are obtained either from oedometer tests in the laboratory on good quality undisturbed samples or from a relationship with consolidated anisotropic undrained triaxial compression tests or by estimation from CPT or field vane tests. The clay sensitivity is determined by dividing the peak intact unconsolidated undrained shear strength by its remoulded undrained shear strength.

The operational interface angle of friction at failure δ_f lies between the peak effective shear stress angle and its ultimate or long strain value. The actual value used in equation 4.21 depends on the soil type, prior shearing history and the clay to steel interface properties. It is influenced by local slip at the interface when the blow of the hammer drives the pile downwards and at rebound when the hammer is raised at the end of the stroke. A further influence is progressive failure when the interface shear stress near the ground surface is at the ultimate state, but near the toe the relative pile–soil movement may be insufficient to reach the peak stress value.

The conditions at the interface can be simulated by determining δ in a ring shear apparatus where the remoulded clay is sheared against an annular ring fabricated from the same material and having the same roughness as the surface of the pile. Details of the apparatus and the testing technique are given in the IC publication which should be consulted for further information on the development and applications of the ICP method.

For calculating the shaft capacity of open-end piles in clay an equivalent radius R^* is substituted for R in the h/R term where

$$R^* = (R^2_{outer} - R^2_{inner})^{0.5} \tag{4.24}$$

and h/R^* is not less than 8.

Dealing with the base resistance of closed-end piles in clay, the ICP method does not accept the widely used practice of calculating the ultimate resistance from $Q_b = N_c\, c_u\, A_b$ where the bearing capacity factor N_c is assumed to be equal to 9. The data base of instrumented pile tests used in the IC research showed a wide variation in N_c which was found to be higher than 9 in all the tests analysed. However, the results did demonstrate a close correlation with the results of static cone penetration tests and led to a recommendation to adopt the relationships:

$q_b = 0.8q_c$ for undrained loading

and $\tag{4.25}$

$q_b = 1.3q_c$ for drained loading $\tag{4.26}$

The cone resistance q_c is obtained from CPT's by averaging the readings over a distance of 1.5 pile diameters above and below the toe.

For open-end piles, plugging of the pile toe with clay is defined as the stage when the plug is carried down by the pile during driving. This is deemed to occur when $[D_{inner}/D_{CPT} + 0.45q_c]/P_a$ is less than 36. The cone diameter D_{CPT} is 0.036 m and the atmospheric pressure P_a is 100 kN/m^2.

Fully plugged piles as defined above develop half the base resistance calculated by equations 4.25 and 4.26 for undrained and drained loading respectively, after a pile head displacement of D/10.

The base resistance of an unplugged open-end pile is calculated on the annular area of steel only, when q_b is taken as the average q_c at founding depth it is stated that Q_b may be increased by a factor of 1.6 for drained loading.

It is evident from the foregoing account of the application of the ICP method to piles in clay that the reliability of the method depends in the first instance on obtaining good quality undisturbed samples taken by piston samplers and using thin-wall tubes. Second, the laboratory operations involving oedometer and ring shear testing require special apparatus handled by skilled technicians. These facilities are not widely available to UK commercial site investigation contractors. Nevertheless, the accuracy of the predictions by the ICP method appears to justify its use for offshore construction where savings in estimated pile lengths are more than offset by the corresponding reduced construction costs.

The ratio of calculated to measured pile resistance derived from the IC data base of 43 piles ranging in diameter from 100 to 570 mm and in length from 3.5 to 57 m showed a statistical mean of 1.03, a standard deviation coefficient of 0.21 and a coefficient of variation of 0.20, compared with the corresponding figures of 0.99, 0.32 and 0.33 using the API (1993) recommendations.

Jardine *et al.*[4.30] recommend safety factors of 1.3 to 1.6 for the shaft resistance in compression for offshore foundations where settlements of the structures are not critical and the design is based on permissible stress methods.

In contrast, the ICP method of design for tubular piles in sands is a simple one based on the static cone penetration test. No other field work or special laboratory testing is required. The method is wholly empirical and is justified by the assumption that the penetration of the sleeved cone simulates the displacement of the soil by a closed-end or fully plugged pile.

The expression for the shaft resistance is calculated by the following sequence of equations:

Unit shaft resistance $= \tau_f = \sigma'_{rf} \tan \delta_f$ (4.27)

Radial effective stress at point of shaft failure $= \sigma'_{rf} = \sigma'_{rc} + \sigma'_{rd}$ (4.28)

Equalized radial effective stress $= \sigma'_{rc} = 0.029 q_c \, (\sigma'_{vo}/P_a)^{0.13} (h/R)^{-0.38}$ (4.29)

Dilatant increase in local radial effective stress $= \Delta\sigma'_{rd} = 2G\delta_f/R$ (4.30)

where

$\delta_f = \delta_{cr} =$ interface angle of friction at failure
$R =$ pile radius
$G =$ operational shear modulus

In equation 4.27 δ_f can be obtained either by constant volume shear box tests in the laboratory or by relating it to the pile roughness and particle size of the sand (Figure 4.21). The equalized radial stress in equation 4.29 implies that the elevated pore pressures around the shaft caused by pile driving have dissipated. The term P_a is the atmospheric pressure which is taken as 100 kN/m². Because of the difficulty in calculating or measuring the high radial stresses near the pile toe h/R is limited to 8.

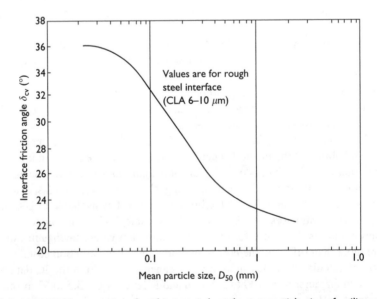

Figure 4.21 Relationship between interface friction angle and mean particle size of a silica sand (based on Jardine et al.[4.30]).

The shear modulus G in equation 4.30 can be measured in the field using a pressuremeter (Section 11.1.4) or a seismic cone penetrometer, or obtained by correlation with CPT data using the relationship established by Chow[4.32]:

$$G = q_c(A + B_\eta - C^2) \tag{4.31}$$

and

$$\eta = q_c/\sqrt{P_a\sigma'_{vo}} \tag{4.32}$$

The term δ_f in equation 4.30 is equal to twice the average roughness R_{cla} of the pile surface which is the average height of the peaks and troughs above and below the centre line. For lightly rusted steel Δ_r is 0.02 mm. σ'_{rf} is inversely proportional to the pile radius and tends to zero for large-diameter piles.

In equation 4.31:

$A = 0.0203$

$B = 0.00125$

$C = 1.216 \times 10^{-6}$

Piles driven with open ends develop a lower shaft resistance than closed-end piles because of their smaller volume displacement when a solid plug is not carried down during driving. The open unplugged end is allowed for by adopting an equivalent pile radius $R*$ (see equation 4.24). Equation 4.29 becomes

$$\sigma'_{rc} = 0.029q_c(\sigma'_{vo}/P_a)^{0.13}(h/R*)^{-0.38} \tag{4.33}$$

To use the ICP method the embedded shaft length is divided into a number of short sections of thickness h depending on the layering of the soil and the variation with depth of the CPT readings. A mean line is drawn through the plotted q_c values over the depths of the identified soil layers. A line somewhat higher than the mean is drawn when the ICP method is used to estimate pile driveability when the shaft resistance must not be underestimated.

From a data base of pile tests in calcareous sands, Jardine et al.[4.30] stated that the ICP method was viable in these materials and recommended that the density should be taken as 7.5 kN/m^3 for calculating σ'_{vo} and the interface angle δ_f as 25°. The third term in equation 4.28 is omitted ($\sigma'_{rf} = \sigma'_{rc}$) and equation 4.29 for open-end piles is modified to become $\sigma'_{rc} = 72(\sigma'_{vo}/P_a)^{0.84}(h/R*)^{-0.35}$. For closed-end piles R is substituted for $R*$.

The ICP method was used to compare the calculated distribution of interface shear stress at failure with stresses measured over the shaft depth of a well-instrumented 762 mm OD pile driven with an open end to a depth of 44 m into medium-fine silty micaceous sand in Bangladesh. The test was made as part of the trial piling for the foundations of the Jamuna River bridge at Sirajgang[4.33, 4.34] as described in Section 9.6.2. The observed and calculated distributions of stress are compared in Figure 4.22. It will be noted that the ICP method considerably over-estimated the measured stresses. This was commented on by Jardine et al.[4.30] with no conclusions as to the reasons for the over-estimate. However, he pointed out that the Jamuna piles developed very marked increases in bearing capacity with time as noted in Section 4.3.8. A study of the shaft friction measurements made on two 762 mm trial

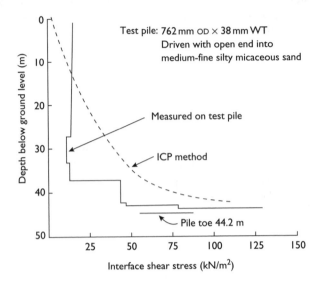

Figure 4.22 Comparison of measured and calculated interface shear stress on the shaft of a steel tube pile driven into sand.

piles showed that the distribution of interface shear stress could be represented by the relationship $\tau_f = 0.009(h/d)^{-0.5}q_f$ in compression and $0.003(h/d)^{-0.5}q_c$ in tension.

The ICP method uses CPT data to calculate the base resistance. For closed-end piles the equation is:

$$q_b = q_c \left[1 - 0.5 \log(D/D_{CPT})\right] \tag{4.34}$$

where q_c is the cone resistance averaged over 1.5 pile diameters above and below the toe, D is the pile diameter, and D_{CPT} is the cone diameter. The equation is valid provided that the variations in q_c are not extreme and the depth intervals between peaks and troughs of the q_c values are not greater than $D/2$. If these conditions are not met a q_c value below the mean should be adopted. A lower limit for q_b of $0.3q_c$ is suggested for piles having diameters greater than 0.9 m.

A rigid basal plug within an open-end pile is assumed to develop if the inner diameter in metres is less than 0.02 $(D_r - 30)$ where the relative density D_r is expressed as a percentage. Also D_{inner}/D_{CPT} should be less than $0.83q_c/P_a$ and the absolute atmospheric pressure P_a is taken as 100 kN/m².

If the above criteria are satisfied the fully plugged pile is stated to develop a base resistance of 50% of that of a closed-end pile after the head has settled by one-tenth of the diameter. A lower limit of q_b is that it should not be less than that of the unplugged pile and should not be less than $0.15q_c$ for piles having diameters greater than 0.9 m.

The base resistance of unplugged piles is taken as $0.5q_c$ multiplied by the net cross-sectional area of the pile at the toe, where q_c is the cone resistance at toe level. No contribution is allowed from the inner wall shaft friction. For a solid end pile q_b at the toe is determined from equation 4.34.

Imperial College have assessed the reliability of their method for piles in sands by comparing the predictions of shaft capacity with those of the American Petroleum Institute method[3.5] as shown below:

Method	Mean Q_c/Q_m	Standard deviation(s)	Coefficient of variation
ICP, all piles	0.99	0.28	0.28
ICP, all open-end piles	1.05	0.30	0.28
API RP2A(1993), all piles	0.87	0.58	0.60

Note
Q_c/Q_m denotes calculated/measured.

White and Bolton[4.35] re-analysed the IC data base for closed-end piles on the basis that instead of the criterion of failure being the load causing a settlement of one-tenth of the diameter they assumed that plunging settlement occurred, i.e. beyond the point D in Figure 4.1. They also made allowance for only partial embedment of some piles into the bearing stratum, and the presence in some piles of a weaker layer below base level. They found a mean of $q_b = 0.9q$ with no trend towards a reduction of q_b with increase in pile diameter. They suggested that a reduction factor to obtain the ultimate bearing capacity of a closed-end pile in sand should be linked to partial embedment and partial mobilization rather than to absolute diameter. This suggestion would appear to be part of the methodology of research based on analysis of test pile failures rather than criteria to be adopted at the design stage of piled foundations. White and Bolton noted the dearth of high-quality pile load test data in the public domain.

It was generally assumed in past years that no allowance should be made for significant changes in the bearing capacity of piles driven into coarse soils with time after installation. Neither increases nor decreases in capacity were considered although the 'set-up' or temporary increase in driving resistance about 24 hours after driving was well known. The long-term effects had not been given serious study. However, the research work at Imperial College described in the previous section did include some long-term tension tests on piles at Dunkirk[4.30]. Six 465 mm OD × 19 m long and one 465 mm OD × 10 m long steel tube piles were tested in tension at ages between 10 days and about 6 months. A progressive increase in resistance of about 150% was recorded. All the tests were 'first-time', that is, none of the piles were tested a second time.

The 762 mm OD × 44 m long test pile at the Jamuna Bridge site was referred to in the previous section[4.34]. There was an increase in tension capacity of about 270% on retest after the initial test made a few days after driving into medium-dense silty micaceous sand. Precast concrete piles on the same site showed a progressive increase of about 200% in compression at various ages up to 80 days after driving. The ultimate resistances were estimated from dynamic tests and graphical analysis of loading tests not taken to failure.

Jardine et al.[4.30] attributed the increased tension capacity at Dunkirk mainly to relaxation through creep of circumferential arching around the pile shaft leading to increase in radial effective stress.

The procedure for determining the resistance of piles driven into sand using CPT values is wholly empirical and was originally based on uninstrumented loading tests and experience. The tests were mainly made on piles of small to medium diameter. EC7 rules do not recommend any particular method of relating q_c to base or shaft resistance but state that the method adopted should have been established from pile loading tests and from comparable experience involving the same type of soil and similar structures with particular reference to local information (Clauses 7.6.2.3 and 1.5.2.2). The overall safety factor for DA1 approach is mainly influenced by the correlation factors derived from the number of in-situ tests made on the site.

Jardine et al.[4.30] do not offer any recommendations for applying EC7 procedures to their design methods.

4.3.8   Time effects for piles in coarse-grained soils

The engineer should be aware of a possible *reduction* in capacity where piles are driven into fine sands and silts. Peck et al.[4.18] stated that 'If the fine sand or silt is dense, it may be highly resistant to penetration of piles because of the tendency for dilatancy and the development of negative pore pressures during the shearing displacements associated with insertion of the piles. Analysis of the driving records by means of the wave equation may indicate high dynamic capacity but instead of freeze, large relaxations may occur.'

An example of this phenomenon was provided by the experiences of driving large diameter tubular steel piles into dense sandy clayey silts for the foundations of the new Galata Bridge in Istanbul[4.36]. The relaxation in capacity of the 2 m OD piles in terms of blows per 250 mm penetration is shown in Figure 4.23. The magnitude of the reduction in driving resistance was not related to the period of time between cessation and resumption of driving. It is likely that most of the reduction occurred within a period of 24 hours after completing a stage of driving. The widely varying time periods shown in Figure 4.23 were due to the operational movements of the piling barge from one pile location or group to another.

Correlation of blow count figures with tests made with the dynamic pile analyser (Section 7.3) showed a markedly smaller reduction in dynamic soil resistance than indicated by the reduction in blow count after the delay period.

These experiences emphasize the need to make re-driving tests after a minimum period of 24 hours has elapsed after completing the initial drive. Loading tests should not be made on piles in sands until at least seven days after driving. Where piles are driven into laminated fine sands, silts and clays, special preliminary trial piling should be undertaken to investigate time effects on driving resistance. These trials should include tests with the pile driving analyser.

Increases in shaft capacity similar to those described above are not expected with bored piles.

4.4   Piles in soils intermediate between sands and clays

Where piles are installed in sandy clays or clayey sands which are sufficiently permeable to allow dissipation of excess pore pressure caused by application of load to the pile, the base and shaft resistance can be calculated for the case of drained loading using equation 4.16. The angle of shearing resistance used for obtaining the bearing capacity factor N_q should be the effective angle ϕ' obtained from unconsolidated drained triaxial compression tests. In a uniform soil deposit, equation 4.16 gives a linear relationship for the increase of base resistance with depth. Therefore, the base resistance should not exceed the peak value of 11 MN/m^2 unless pile loading tests show that higher ultimate values can be obtained. The effective overburden pressure, σ'_{vo}, in equation 4.16 is the total overburden pressure minus the pore water pressure at the pile toe level. It is important to distinguish between uniform $c - \phi$ soils and layered c and ϕ soils, as sometimes the layering is not detected in a poorly executed soil investigation.

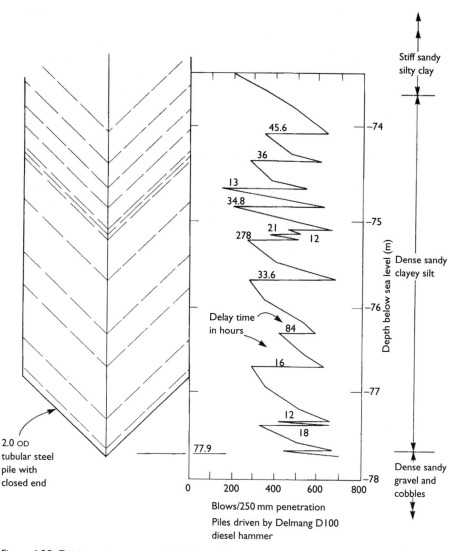

Figure 4.23 Driving resistance over final 4.5 m of penetration for 2.0 m tubular steel pile showing reduction in driving resistance after various delay periods, New Galata Bridge, Istanbul.

4.5   Piles in layered fine- and coarse-grained soils

It will be appreciated from Sections 4.2 and 4.3 that piles in fine-grained soils have a relatively high shaft friction and a low end-bearing resistance, and in coarse soils the reverse is the case. Therefore, when piles are installed in layered soils the location of the pile toe is of great importance. The first essential is to obtain a reliable picture of the depth and lateral extent of the soil layers. This can be done by making in-situ tests with static or dynamic cone test equipment (see Section 11.1.4), correlated by an adequate number of boreholes. If it is desired to utilize the potentially high end-bearing resistance provided by a dense sand or gravel layer, the variation in thickness of the layer should be determined and its continuity across the site should be reliably established. The bearing stratum should not be in the form of isolated lenses or pockets of varying thickness and lateral extent.

Where driven or driven and cast-in-place piles are to be installed, problems can arise when piles are driven to an arbitrary 'set' to a level close to the base of the bearing stratum, with the consequent risk of a breakthrough to the weaker clay layer when the piles are subjected to their working load (Figure 4.24a). In this respect the driven and cast-in-place pile with an enlarged base is advantageous, as the bulb can be hammered out close to the top of the bearing stratum (Figure 4.24b). The end-bearing resistance can be calculated conservatively on the assumption that the pile always terminates within or just above the clay layer, that is, by basing the resistance on that provided by the latter layer. This is the only possible solution for sites where the soils are thinly bedded, and there is no marked change in driving resistance through the various layers. However, this solution can be uneconomical for sites where a dense sand layer has been adequately explored to establish its thickness and continuity. A method of calculating the base resistance of a pile located in a thick stiff or dense layer underlain by a weak stratum has been established by Meyerhof[4.37]. In Figure 4.25 the unit base resistance of the pile is given by the equation:

$$q_b = q_o + \frac{q_l - q_o}{10B} H \leq q_l \tag{4.35}$$

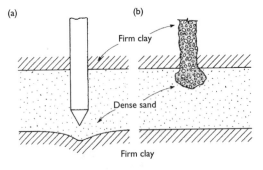

Figure 4.24 Pile driven to end bearing into relatively thin dense soil layer (a) Driven pile (b) Driven and cast-in-place pile.

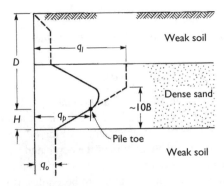

Figure 4.25 End-bearing resistance of piles in layered soils.

where

q_o = ultimate base resistance in the lower weak layer
q_l = ultimate base resistance in the upper stiff or dense stratum
H = distance from the pile toe to the base of the upper layer and
B = width of the pile at the toe

The following procedures were adopted for the piled foundations of British Coal's bulk-handling plant at Immingham, where a layer of fairly dense sandy gravel was shown to exist at a depth of about 14.6 m below ground level. The thickness of the gravel varied between 0.75 and 1.5 m and it lay between thick deposits of firm to stiff boulder clay. The end-bearing resistance in the gravel of the 508 mm diameter driven and cast in-situ piles was more than 3000 kN as derived from loading tests to obtain separate evaluations of shaft friction and base resistance. It was calculated that if the toe of the pile reached a level at which it was nearly breaking through to the underlying clay, the end-bearing resistance would then fall to 1000 kN and the safety factor of the pile would be reduced to 1.2 at the working load of 800 kN. This safety factor was inadequate, and it was then necessary to drive the pile some 3.6 m deeper to mobilize additional shaft friction so as to raise the safety factor to a satisfactory value. A record was made to compare the driving resistance of piles driven completely through the gravel to a deeper penetration and those terminating on the gravel layer (Figure 4.26). An evaluation of this record led to the establishment of the following rules:

(1) When the driving resistance in the gravel increased rapidly from 20 mm per blow to 5 mm per blow for a complete 300 mm of driving it was judged that the pile was properly seated in the gravel stratum

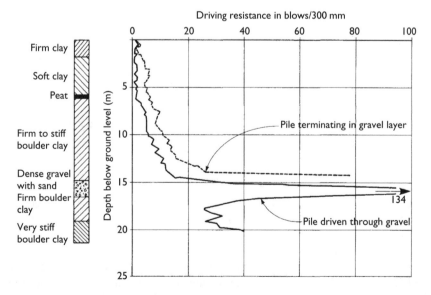

Figure 4.26 Resistance to driven and cast-in-place piles provided by a thin layer of dense sand and gravel at Immingham.

(2) The pile was then required to be driven a further 75 mm without any reduction in the driving resistance
(3) If the resistance was not maintained at 5 mm per blow, it was judged that the gravel layer was thin at that point, and the pile was liable to break through to the clay. Therefore, the pile had to be driven further to a total penetration of 20 m, which was about 3 to 4 m below the base of the gravel, to obtain the required additional frictional resistance.

The effects of driving piles in groups onto a resistant layer underlain by a weaker compressible layer must be considered in relation to the settlement of the group. This aspect is discussed in Chapter 5.

4.6   The settlement of the single pile at the working load for piles in soil

It is necessary to divide the calculated ultimate resistance of the pile (or the ultimate resistance derived from load testing) by a safety factor to obtain the design working load on the pile. A safety factor is required for the following reasons:

(1) To provide for natural variations in the strength and compressibility of the soil
(2) To provide for uncertainties in the calculation method used
(3) To ensure that the working stresses on the material forming the pile shaft are within the safe limits
(4) To ensure that the total settlement(s) of the single isolated pile or the group of piles are within tolerable limits
(5) To ensure that the differential settlements between adjacent piles or within groups of piles are within tolerable limits.

The need for a safety factor or partial factors to cover the uncertainties in the calculation methods will have been evident from the earlier part of this chapter, and in this respect they are 'factors of ignorance' rather than absolute values. With regard to reason 4 above, the load/settlement curves obtained from a very large number of loading tests in a variety of soil types, both on displacement and non-displacement piles, have shown that for piles of small to medium (up to 600 mm) diameter, the settlement under the working load will not exceed 10 mm if the safety factor is not lower than 2.5. This is reassuring and avoids the necessity of attempting to calculate settlements on individual piles that are based on the compressibility of the soils. A settlement at the working load not exceeding 10 mm is satisfactory for most building and civil engineering structures provided that the group settlement is not excessive.

However, for piles larger than 600 mm in diameter the problem of the settlement of the individual pile under the working load becomes increasingly severe with the increase in diameter, requiring a separate evaluation of the shaft friction and base load. The question of the correct safety factor then becomes entirely the consideration of the permissible settle-ment or in EC7 terms compliance with serviceability limit-state. The load/settlement rela-tionships for the two components of shaft friction and base resistance and for the total resistance of a large-diameter pile in a stiff clay are shown in Figure 4.27. The maximum shaft resistance is mobilized at a settlement of only 10 mm but the base resistance requires a settlement of nearly 150 mm for it to become fully mobilized. At this stage the pile has

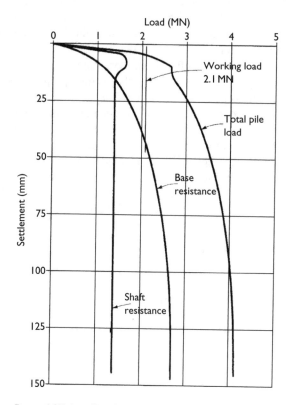

Figure 4.27 Load/settlement relationships for large-diameter bored piles in stiff clay.

reached the point of ultimate resistance at a failure load of 4.2 MN. A safety factor of 2 on this condition gives a working load of 2.1 MN, under which the settlement of the pile will be nearly 5 mm. This is well within the settlement which can be tolerated by ordinary building structures. The full shaft resistance will have been mobilized at the working load, but only 22% of the ultimate base resistance will have been brought into play. For economy in pile design the settlement at the working load should approach the limit which is acceptable to the structural designer, and this usually involves mobilizing the full shaft resistance.

By using partial safety factors on the ultimate shaft and base resistances, Burland *et al.*[4.38] have presented a simple stability criterion for bored piles in clay which states that if an overall load factor of 2 is stipulated, together with a minimum factor of safety in end bearing of 3, then the maximum safe load on the pile is the lesser of the two expressions $(\frac{1}{2}Q_p)$ and $(Q_s + \frac{1}{3}Q_b)$, where Q_p is the ultimate resistance of the whole pile, Q_s is the ultimate resistance of the shaft, and Q_b is the ultimate resistance of the base.

Burland *et al.* state that the first expression is nearly always dominant for straight-sided piles and for long piles with comparatively small under-reams, whereas the second expression often controls piles with large under-reamed bases. Satisfaction of the above criteria does not necessarily mean that the settlement at working load will be tolerable. Experience based on loading tests on piles in similar soil conditions may give a guide to the order of settlement that may be expected. If there is no such experience available, then it may be necessary to undertake loading tests on full-scale piles. This is very costly for large piles and a more economical procedure is to estimate values from the results of loading tests made on circular plates at the bottom of the pile boreholes, or in trial shafts.

However, loading tests on piles are more helpful when designing 'ductile piles' (Section 5.2.1). Instrumentation can be provided to determine the relative proportions of load carried in friction on the shaft and transmitted to the base and hence to determine the degree of settlement needed to mobilize peak friction (e.g. at a pile head settlement of about 10 mm in Figure 4.27), and to determine whether or not the lower 'long strain' value of shaft friction is operating when load distribution between piles in a group takes place.

Burland et al.[4.38] plotted the settlement of test plates divided by the plate diameter (p_i/B) against the plate bearing pressure divided by the ultimate bearing capacity for the soil beneath the plate (i.e. q/q_f) and obtained a curve of the type shown in Figure 4.28. If the safety factor on the end-bearing load is greater than 3, the expression for this curve is

$$p_i/B = K \times q/q_f \qquad (4.36)$$

When plate bearing tests are made to failure, the curve can be plotted and, provided that the base safety factor is greater than 3, the settlement of the pile base p_i can be obtained for any desired value of B.

The procedure used to estimate the settlement of a circular pile is as follows:

(1) Obtain q_f from the failure load given by the plate bearing test
(2) Check q_f against the value obtained by multiplying the shearing strength by the appropriate bearing capacity factor N_c, i.e. q_f should equal $N_c \times c_{ub}$
(3) Knowing q_f, calculate the end-bearing resistance Q_b of the pile from $Q_b = A_b \times q_f$

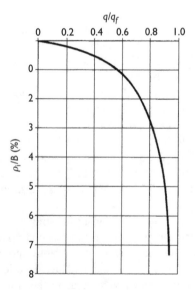

Figure 4.28 Elastic settlements of bored piles in London clay at Moorfields (after Burland et al.[4.38]).

(4) Obtain the safe end-bearing load on the pile from $W_b = Q_b/F$, where F is a safety factor greater than 3
(5) Obtain q from $q = Wb/\frac{1}{4}\pi B^2$ and hence determine q/q_f
(6) From a curve of the type shown in Figure 4.28, read off ρ_i/B for the value of q/q_f and hence obtain ρ_i (the settlement of the pile base).

Merely increasing the size of the base by providing an under-ream will not reduce the base settlement, and if the settlement is excessive it should be reduced by one or more of the following measures:

(1) Reduce the working load on the pile
(2) Reduce the load on the base by increasing the shaft resistance, i.e. by increasing the shaft diameter
(3) Increase the length of the shaft to mobilize greater shaft friction, and to take the base down to deeper and less-compressible soil.

For piles in London Clay, K in equation 4.36 has usually been found to lie between 0.01 and 0.02. If no plate bearing tests are made, the adoption of the higher value provides a conservative estimate of settlement. Having estimated the settlement of the individual pile using the above procedure it is still necessary to consider the settlement of the pile group as a whole (see Chapter 5).

The greater the length of the pile the greater is the pile head settlement. From their analyses of a large number of load/settlement curves, Weltman and Healy[4.7] established a simple relationship for the settlement of straight shaft bored and cast-in-place piles in glacial till. The relationship given below assumed a pile diameter not greater than 600 mm, a working stress on the pile shaft of about 3 MN/m², a length to diameter ratio of 10 or more, and stiff to hard glacial till with undrained shear strengths in excess of 100 kN/m². The pile head settlement is given by

$$\rho = \frac{l_m}{4} \text{ in millimetres} \tag{4.37}$$

where l_m is the length of embedment in glacial till in metres.

Precast concrete piles and some types of cast-in-place piles are designed to carry working loads with shaft stresses much higher than 3 MN/m². In such cases the settlement should be calculated from equation 4.37 assuming a stress of 3 MN/m². The settlement should then be increased *pro rata* to the designed working stress.

The above methods of Burland *et al.*, and Weltman and Healy, were developed specifically for piling in London Clay and glacial till respectively and were based on the results of field loading tests made at a standard rate of loading as specified by the Institution of Civil Engineers (Section 11.4) using the maintained loading procedure. More generally the pile settlements can be calculated if the load carried by shaft friction and the load transferred to the base at the working load can be reliably estimated. The pile head settlement is then given by the sum of the elastic shortening of the shaft and the compression of the soil beneath the base as follows:

$$\rho = \frac{(W_s + 2W_b)L}{2A_sE_p} + \frac{\pi}{4} \cdot \frac{W_b}{A_b} \cdot \frac{B(1 - v^2)I_p}{E_b} \tag{4.38}$$

where

W_s and W_b = loads on the pile shaft and base respectively
L = shaft length
A_s and A_b = cross-sectional area of the shaft and base respectively
E_p = elastic modulus of the pile material
B = pile width
v = Poisson's ratio of the soil
I_p = influence factor related to the ratio of L/R
E_b = deformation modulus of the soil beneath the pile base

For a Poisson's ratio of 0 to 0.25 and $L/B > 5$, I_p is taken as 0.5 when the last term approximates to 0.5 $W_b/(BE_b)$. Values of E_b are obtained from plate loading tests at pile base level or from empirical relationships with the results of laboratory or in-situ soil tests given in Sections 5.2 and 5.3. The value of E_b for bored piles in coarse soils should correspond to the loose state unless the original in-situ density can be maintained by drilling under bentonite or restored by base grouting.

The first term in equation 4.38 implies that load transfer from pile to soil increases linearly over the depth of the shaft. It is clear from Figure 4.22 that the increase is not linear for a deeply penetrating pile. However, with the present-day availability of computers it is possible to simulate the load transfer for wide variations in soil stratification and in cross-sectional dimensions of a pile. One of the principal programmes represents an elastic continuum model. A pile carrying an axial compression load is modelled as a system of rigid elements connected by springs and the soil resistance by external non-linear springs (Figure 4.29). The load at the pile head is resisted by frictional forces on each element. The resulting displacement of each of these is obtained from Mindlin's equation for the displacement due to a point load in a semi-infinite mass. The load/deformation behaviour is represented in the form of a t–z curve (Figure 4.29). A similar q–z curve is produced for the settlement of the pile base.

The concept of modelling a pile as a system of rigid elements and springs for the purpose of determining the stresses in a pile body caused by driving is described in Section 7.3.

It was noted at the beginning of this section that the adoption of nominal safety factors in conjunction with conventional methods of calculating pile-bearing capacity can obviate the necessity of calculating working load settlements of small-diameter piles. However, there is not the same mass of experience relating settlements to design loads obtained by EC7 methods based on partial safety factors. Hence, it is necessary to check that the design pile capacity does not endanger the serviceability limit-state of the supported structure. Equation 4.38 can be used for this check. A material factor of unity should be adopted for the design value of E_d.

EC7 (Clause 7.6.4.1) states that where piles are bearing on medium-dense to dense soils the safety requirements for ultimate limit state design are normally sufficient to prevent a serviceability limit state in the supported structure.

4.7   Piles bearing on rock

4.7.1   Driven piles

For maximum economy in the cross-sectional area of a pile it is desirable to drive the pile to virtual refusal on a strong rock stratum, thereby developing its maximum carrying capacity.

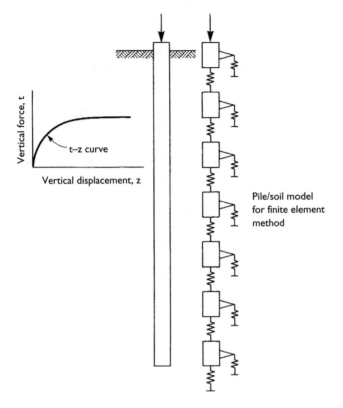

Figure 4.29 t–z curve for deformation of a pile under vertical axial loading.

Piles driven in this manner are regarded as wholly end bearing: friction on the shaft is not considered to contribute to the support of the pile. The depth of penetration required to reach virtual refusal depends on the thickness of any weak or heavily broken material overlying sound rock. If a pile can be driven to near refusal on to a strong *intact* rock the safe working load on the pile is governed by the permissible working stress on the material of the pile at the point of minimum cross-section; i.e. the pile is regarded as a short column supported against buckling by the surrounding soil. Where piles are driven through water or through very soft clays and silts of fluid consistency, then buckling as a long strut must be considered (see Section 7.5).

When steel piles are adopted, working loads based on the permissible working stress on the steel may result in concentrations of very high loading on the rock beneath the toe of the pile. The ability of the rock to sustain this loading without yielding depends partly on the compressive strength of the rock and partly on the frequency and inclination of fissures and joints in the rock mass, and whether these discontinuities are tightly closed or are open or filled with weathered material. Very high toe loads can be sustained if the rock is strong, with closed joints either in a horizontal plane or inclined at only a shallow angle to the horizontal. If the horizontal or near-horizontal joints are wide there will be some yielding of the rock mass below the pile toe but the amount of movement will not necessarily be large since the zone of rock influenced by a pile of slender cross-section does not extend very deeply below toe level. However, the temptation to continue the hard driving of slender-section piles to ensure full refusal conditions must be avoided. This is because brittle rocks may be split by the toe of the pile, thus considerably reducing the base resistance. The splitting may continue as the pile is driven down, thus requiring very deep penetration to regain the original resistance.

Where bedding planes are steeply inclined with open transverse joints there is little resistance to the downward sliding of a block of rock beneath the toe and the movement will continue until the open joints have become closed, or until the rock mass becomes crushed and locked together. This movement and crushing will take place as the pile is driven down, as indicated by a progressive tightening-up in driving resistance. Thus there should be no appreciable additional settlement when the working load is applied. However, there may be some deterioration in the end-bearing value if the piles are driven in closely spaced groups at varying toe levels. For this reason it is desirable to undertake re-driving tests whenever piles are driven to an end bearing into a heavily jointed or steeply dipping rock formation. If the re-driving tests indicate a deterioration in resistance, then loading tests must be made to ensure that the settlement under the working load is not excessive. Soil heave may also lift piles off their end bearing on a hard rock, particularly if there has been little penetration to anchor the pile into the rock stratum. Observations of the movement of the heads of piles driven in groups, together with re-driving tests indicate the occurrence of pile lifting due to soil heave. Methods of eliminating or minimizing the heave are described in Section 5.7.

Steel tubes driven with open ends, or H-section piles are helpful in achieving the penetration of layers of weak or broken rock to reach virtual refusal on a hard unweathered stratum. However, the penetration of such piles causes shattering and disruption of the weak layers to the extent that the shaft friction may be seriously reduced or virtually eliminated. This causes a high concentration of load on the relatively small area of rock beneath the steel cross-section. While the concentration of load may be satisfactory for a strong intact rock it may be excessive for a strong but closely jointed rock mass. The concentration of load can be reduced by welding stiffening rings or plates to the pile toe or, in the case of weak and heavily broken rocks, by adopting winged piles (Figure 2.19).

The H-section pile is particularly economical for structures on land where the shaft is wholly buried in the soil and thus not susceptible to significant loss of cross-sectional area due to corrosion. To achieve the maximum potential bearing capacity it is desirable to drive the H-pile in conjunction with a pile driving analyser (Section 7.3) to determine its ultimate resistance and hence the design working load, verified if necessary by pile loading tests.

The methods given below for calculating the ultimate bearing capacity assume that this is the sum of the shaft and base resistance. Both of these components are based on correlations between pile loading tests and the results of field tests in rock formations or laboratory tests on core specimens.

Where the joints are spaced widely, that is at 600 mm or more apart, or where the joints are tightly closed and remain closed after pile driving, the ultimate base resistance may be calculated from the equation:

$$q_b = 2N_\phi q_{uc} \tag{4.39}$$

where the bearing capacity factor:

$$N_\phi = \tan^2\left(45° + \frac{\phi}{2}\right)$$

The variations in N_ϕ caused by joints in the rock mass are demonstrated by the comparisons in Table 4.13 of observations of the ultimate base resistance of driven and bored piles terminated in weak mudstones, siltstones and sandstones with the corresponding N_ϕ values calculated from equation 4.40. For these rocks the ϕ values as recommended by Wyllie[4.39] are in the range of 27° to 34° giving N_ϕ values from 2.7 to 3.4.

It will be noted that the calculated N_ϕ values in Table 4.13 are considerably lower than the range of 2.7 to 3.4 established for rocks with widely spaced and tight joints. The reduction is most probably due to the jointing characteristics of the rock formation in which the tests were made. A measure of the joint spacing is the rock quality designation (RQD) determined as described in Section 11.1.4. Kulhawy and Goodman[4.40,4.41] showed that the ultimate base resistance (q_{ub}) can be related to the RQD of the rock mass as shown in Table 4.14.

Where laboratory tests can be made on undisturbed samples of weak rocks to obtain the parameters c and ϕ, Kulhawy and Goodman[4.40, 4.41] state that the ultimate bearing capacity of the jointed rock beneath the pile toe can be obtained from the equation:

$$q_{ub} = cN_c + \gamma DN_q + \gamma \frac{BN_\gamma}{2} \tag{4.40}$$

Table 4.13 Observed ultimate base resistance values of piles terminated in weak mudstones, siltstones, and sandstones

Description of rock	Pile type	Plate or pile diameter (mm)	Observed bearing pressure at failure (MN/m²)	Calculated N_ϕ
Mudstone/siltstone moderately weak	Bored	900	5.6	0.25
Mudstone, highly to moderately weathered weak	Plate test	457	9.2	1.25
Cretaceous mudstone weak, weathered, clayey	Bored	670	6.8	3.0
Weak carbonate siltstone/sandstone (coral detrital limestone)	Driven	762	5.11	1.5
Calcareous sandstone weak	Driven tube	200	3.0	1.2
Sandstone, weak to moderately weak	Driven	275	19[a]	1.75

Note
a From dynamic pile test.

where

> c = undrained shearing resistance
> B = base width
> D = base depth below the rock surface
> γ = effective density of the rock mass
> N_c, N_q, and N_γ = bearing capacity factors related to ϕ as shown in Figure 4.30.

The above equation represents wedge failure conditions beneath a strip foundation and should not be confused with Terzaghi's equation for spread foundations. Because equation 4.40 is for strip loading the value of cN_c should be multiplied by a factor of 1.25 for a square pile or 1.2 for a circular pile base. Also the term $\gamma BN_\gamma/2$ should be corrected by the factors 0.8 or 0.7 for square or circular bases respectively. The term $\gamma BN_\gamma/2$ is small compared with cN_c and is often neglected.

Where it is difficult to obtain satisfactory samples for laboratory testing to determine c or ϕ the relationship of these parameters to the uniaxial compression strength

Table 4.14 Ultimate base resistance of piles related to the uniaxial compression strength of the intact rock and the RQD of the rock mass

RQD (%)	q_{ub}	c	ϕ (°)
0–70	$0.33q_{uc}$	$0.1q_{uc}$	30
70–100	$0.33–0.8q_{uc}$	$0.1q_{uc}$	30–60

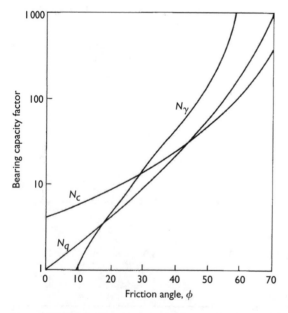

Figure 4.30 Wedge bearing capacity factors for foundations on rock (reprinted from Pells and Turner[4.42]).

and RQD of the rock as shown in Table 4.14 can be used. The q_{uc} values are determined from tests on core specimens of the intact rock to obtain its point load strength (Section 11.1.4).

It is important to note that to mobilize the maximum base resistance from equation 4.40 the settlement of the pile toe is likely to be of the order of 20% of its diameter, requiring an ample safety factor, at least 2.5, to ensure that settlements at the working load are within allowable limits (Section 4.7.4).

Driving a closed-end pile into low to medium density chalk causes blocks of the rock to be pushed aside. Crushed and remoulded chalk flows from beneath the toe, and the cellular structure of the rock is broken down releasing water trapped in the cells to form a slurry. This flows into fissures and causes an increase in pore pressure which considerably weakens the shaft resistance, although it is possible that drainage from the fissures will eventually relieve the excess pore pressure thereby increasing the shaft resistance.

Very little penetration is likely to be achieved when attempting to drive large closed-end piles into a high-density chalk formation with closed joints, but penetration is possible with open-end or H-section plies.

Because of the effects of driving piles into chalk, as described above, equations 4.39 and 4.40 cannot be used to calculate base resistance. From the results of a number of plate and pile loading tests, CIRIA Report 574[4.43] recommends that the base resistance should be related to the standard penetration test N-values (Section 11.1.4). The report gives the relationship for driven piles as

$$\text{Base resistance} = q_{ub} = 300\,N\ \text{kN/m}^2 \tag{4.41}$$

where N is the SPT resistance in blows/300 mm. A lower bound is of the order of $200\,N$ kN/m^2.

No correction should be made to the N-values for overburden pressure when using equation 4.41. Use of this equation is subject to the stress at the base of the pile not exceeding 600 to 800 kN/m^2 for low to medium density chalk, and 1000 to 1800 kN/m^2 for medium to high-density chalk. Also the allowable load on the pile should be the lesser of either:

$$P_a = \frac{Q_s}{1.0} + \frac{Q_b}{3.5} \left(P_a = \frac{Q_s}{1.5} + \frac{Q_b}{3.5} \text{ if the settlement is to be less than 10 mm} \right) \tag{4.42}$$

or,

$$P_a = \frac{Q_s + Q_b}{2} \tag{4.43}$$

Dynamic testing (Section 7.3) of trial or working piles is frequently used to determine permissible working loads in end bearing on chalk. CIRIA Report 574 states that instrumented dynamic tests using the CAPWAPC program can give a good estimate of end-bearing resistance provided that the hammer blow displaces the toe at least 6 mm during the test. Definitions of the density grades of chalk and their characteristics for use with equations 4.42 and 4.43 are given in Appendix 3.

Granite rocks are widely distributed in the territories of Hong Kong, where the fresh rock is blanketed by varying thicknesses of weathered rock in the form of a porous mass of quartz particles in a clayey matrix of decomposed feldspar and biotite[4.44]. The Geotechnical Office (GEO) of the Hong Kong Government[4.45] recommends that the end-bearing resistance of piles should be expressed in terms of a *safe* rather than an ultimate value. They recommend that piles should be driven to refusal in a fresh to moderately decomposed or partially weathered granite having a rock content greater than 50%. For these conditions the allowable load on the pile is governed by the permissible stresses on the material forming the pile. This recommendation assumes that the rock joints are widely spaced and closed. In the case of open or clay-filled joints, the yielding of the pile at the toe should be calculated using the drained elastic modulus of the rock. In Hong Kong the modulus is related to the standard penetration test N-value. The GEO publication gives an E_v' value of 3.5 to $5.5N$ (MN/m^2). It is pointed out that N may be increased by compaction during pile-driving.

The shaft friction developed on piles driven into weak weathered rocks cannot always be calculated from the results of laboratory tests on rock cores. It depends on such factors as the formation of an enlarged hole around the pile, the slurrying and degradation of rocks, the reduction in friction due to shattering of the rock by driving adjacent piles, and the presence of groundwater. Some observed values are shown in Table 4.16. In the case of brittle coarse-grained rocks such as sandstones, igneous rocks and some limestones, it can be assumed that pile driving shatters the rock around the pile shaft to the texture of a loose to medium-dense sand. The ultimate shaft friction can then be calculated from the second term in equation 4.16 using the appropriate values of K_s and δ. Where rocks such as mudstones and siltstones weather to a clayey consistency making it possible to obtain undisturbed samples from boreholes, the weathered rock can be treated as a clay and the shaft friction calculated from the methods described in Section 4.2.1.

The effects of degradation of weakly cemented carbonate soils caused by pile driving have been discussed in Section 4.3.3. Similar effects occur in carbonate rocks such as detrital coral limestones, resulting in very deep penetration of piles without any significant increase in driving resistance. An example of the low driving resistance provided by weak coral limestone to the penetration of closed-end tubular steel piles at a coastal site in Saudi Arabia is shown in Figure 4.31.

Beake and Sutcliffe[4.46] observed ultimate unit shaft resistances of 170 and 300 kN/m^2 from tension tests on 1067 and 914 mm OD tubular steel piles driven with open ends into weak carbonate siltstones and sandstones in the Arabian Gulf. The mean compression strengths of the rocks were 3.2 and 4.7 mN/m^2. The two test piles were 4.2 and 4.55 m into the rocks. The above shaft resistances were 0.04 to 0.10 of the mean unconfined compression strength of the rock.

Although a relationship was established between the base resistance and SPT N-values of piles driven into chalk as noted above, no meaningful relationship could be found with shaft resistance. The CIRIA recommendations[4.43] in Table 4.15 are the best possible estimates derived from pile loading tests. The CIRIA report recommends that whenever possible a preliminary trial pile should be tested to verify the design. It should be noted that dissipation of excess pore pressure caused by pile driving can increase the shaft resistance of piles in chalk. Therefore, as long a delay as possible

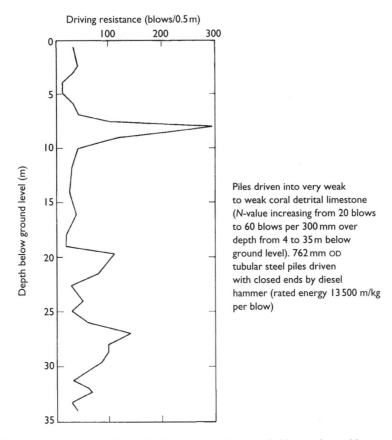

Driving resistance (blows/0.5 m)

Depth below ground level (m)

Piles driven into very weak
to weak coral detrital limestone
(N-value increasing from 20 blows
to 60 blows per 300 mm over
depth from 4 to 35 m below
ground level). 762 mm OD
tubular steel piles driven
with closed ends by diesel
hammer (rated energy 13 500 m/kg
per blow)

Figure 4.31 Low resistance to driving of tubular steel piles provided by weak coral limestone.

Table 4.15 CIRIA recommendations for the shaft resistance of displacement piles driven into chalk

Chalk classification	Type of pile	Ultimate unit shaft resistance (kN/m^2)
Low to medium density, open joints	Small displacement	20
	Small displacement, H-sections	10
	Large displacement, preformed	30
High density, closed joints	Small displacement, open end tubular	120
	Large displacement, preformed in pre-drilled holes	(100) verify by load testing

Table 4.16 Observed ultimate shaft friction values for piles driven into weak and weathered rocks

Pile type	Rock description	Ultimate shaft friction (kN/m²)	Reference
H-section	Moderately strong slightly weathered slaty mudstone	28[a]	4.47
H-section	Moderately strong slightly weathered slaty mudstone	158[b]	4.47
Steel tube	Very weak coral detrital limestone (carbonate sandstone/siltstone)	45	Unpubl.
Steel tube	Faintly to moderately weathered moderately strong to strong mudstone	127	Unpubl.
Steel tube	Weak calcareous sandstone	45	Unpubl.
Precast concrete	Very weak closely fissured argillaceous siltstone (Mercia Mudstone)	130	4.48

Notes
a Penetration 1.25 m.
b Penetration 2.2 m.

should be allowed between driving and load testing. The authors recommend a minimum period of 28 days.

Some other observed values of the shaft resistance of piles in weak rocks are shown in Table 4.16.

4.7.2   Driven and cast-in-place piles

Driven and cast-in-place piles terminated on strong rock can be regarded as end bearing. Their working load is governed by the permissible working stress on the pile shaft at the point of minimum cross-section, or by code of practice requirements. Where these piles are driven into weak or weathered rocks they should be regarded as partly friction and partly end-bearing piles.

CIRIA Report 574[4.43] recommends that the base resistance of driven and cast-in-place piles in chalk should be taken as $250\,N$ kN/m² where N is the SPT N-value. A lower bound should be $200\,N$ kN/m² with the recommendation to make a preliminary test pile whenever possible. For calculating the unit shaft resistance the effective overburden pressure σ'_{vo} should be multiplied by a factor of 0.8 where is σ'_{vo} less than 100 kN/m². If σ'_{vo} is greater than 100 kN/m² the design should be confirmed by a loading test.

4.7.3   Bored and cast-in-place piles

Where these piles are installed by drilling through soft overburden onto a strong rock the piles can be regarded as end-bearing elements and their working load is determined by the safe working stress on the pile shaft at the point of minimum cross-section, or by code of practice requirements. Bored piles drilled down for some depth into weak or weathered rocks and terminated within these rocks act partly as friction and partly as end-bearing piles. Wyllie[4.39] gives a detailed account of the factors governing the development of shaft friction over the depth of the rock socket. The factors which govern the bearing capacity and settlement of the pile are summarized as the following:

(1) The length to diameter ratio of the socket
(2) The strength and elastic modulus of the rock around and beneath the socket
(3) The condition of the side walls, that is, roughness and the presence of drill cuttings or bentonite slurry
(4) Condition of the base of the drilled hole with respect to removal of drill cuttings and other loose debris
(5) Layering of the rock with seams of differing strength and moduli
(6) Settlement of the pile in relation to the elastic limit of the side-wall strength and
(7) Creep of the material at the rock/concrete interface resulting in increasing settlement with time.

The effect of the length/diameter ratio of the socket is shown in Figure 4.32 for the condition of the rock having a higher elastic modulus than the concrete. It will be seen that if it is desired to utilize base resistance as well as socket friction the socket length should be less than four pile diameters. The high interface stress over the upper part of the socket will be noted.

The condition of the side walls is an important factor. In a weak rock such as chalk, clayey shale, or clayey weathered marl, the action of the drilling tools is to cause softening and

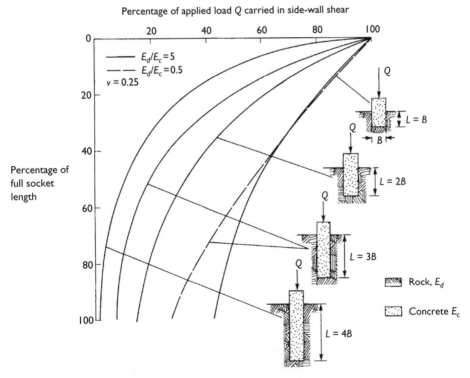

Figure 4.32 Distribution of side-wall shear stress in relation to socket length and modulus ratio (after Osterberg and Gill[4.49]).

slurrying of the walls of the borehole and, in the most adverse case, the shaft friction corresponds to that typical of a smooth-bore hole in a soft clay. In stronger and fragmented rocks the slurrying does not take place to the same extent, and there is a tendency towards the enlargement of the drill hole, resulting in better keying of the concrete to the rock. If the pile borehole is drilled through soft clay this soil may be carried down by the drilling tools to fill the cavities and smear the sides of the rock socket. This behaviour can be avoided to some extent by inserting a casing and sealing it into the rock-head before continuing the drilling to form the rock socket, but the interior of the casing is likely to be heavily smeared with clay which will be carried down by the drilling tools into the rock socket. Wyllie[4.39] suggests that if bentonite is used as a drilling fluid the rock socket shaft friction should be reduced to 25% of that of a clean socket unless tests can be made to verify the actual friction which is developed.

It is evident that the keying of the shaft concrete to the rock and hence the strength of the concrete to rock bond is dependent on the strength of the rock. Correlations between the unconfined compression strength of the rock and rock socket bond stress have been established by Horvarth[4.50], Rosenberg and Journeaux[4.51], and Williams and Pells[4.52]. The ultimate bond stress, f_s, is related to the average unconfined compression strength, $\bar{q}_{uc}$, by the equation:

$$f_s = \alpha\beta\bar{q}_{uc} \tag{4.44}$$

where

α = reduction factor relating to $\bar{q}_{uc}$ as shown in Figure 4.33
β = correction factor related to the discontinuity spacing in the rock mass as shown in Figure 4.34.

The curve of Williams and Pells in Figure 4.33 is higher than the other two, but the β factor is unity in all cases for the Horvarth and the Rosenberg and Journeaux curves. It should also be noted that the α factors for all three curves do not allow for smearing of the rock socket caused by dragdown of clay overburden or degradation of the rock.

The β factor is related to the mass factor, j, which is the ratio of the elastic modulus of the rock mass to that of the intact rock as shown in Figure 4.35. If the mass factor is not known from loading tests or seismic velocity measurements, it can be obtained approximately from the relationships with the rock quality designation (RQD) or the discontinuity spacing quoted by Hobbs[4.53] as follows:

RQD (%)	Fracture frequency per metre	Mass factor j
0–25	15	0.2
25–50	15–8	0.2
50–75	8–5	0.2–0.5
75–90	5–1	0.5–0.8
90–100	1	0.8–1

As a result of later research Horvath et al.[4.54] derived the following equation for calculating the socket shaft friction of large diameter piles in mudstones and shales:

$$\text{Unit shaft friction} = f_s = b\sqrt{\sigma'_{ucw}} \tag{4.45}$$

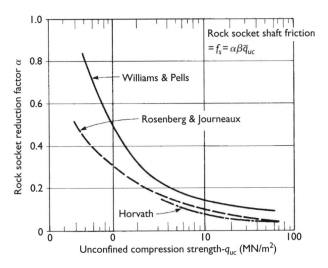

Figure 4.33 Reduction factors for rock socket shaft friction.

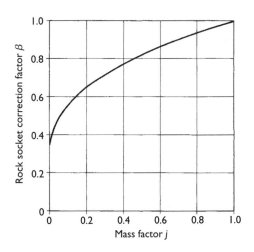

Figure 4.34 Reduction factors for discontinuities in rock mass (after Williams and Pells[4.52]).

where σ'_{ucw} is the unconfined compression strength of the weaker material (concrete or rock), and f_s and σ'_{ucw} are expressed in MN/m². The factor b is given as 0.2 to 0.3.

The shaft friction can be increased in weak or friable rocks by grooving the socket. Horvath *et al.*[4.54] described experiments in mudstones using a toothed attachment to a rotary auger. They showed that f_s was related to the depth of the groove by the equation:

$$\frac{f_s}{\sigma'_{ucw}} = 0.8(RF)^{0.45} \tag{4.46}$$

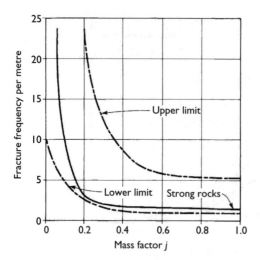

Figure 4.35 Mass factor value (after Hobbs[4.53]).

where σ'_{ucw} is the rock strength defined as above and RF is a roughness factor given as

$$RF = \frac{\bar{\Delta}_r}{r_s} \times \frac{L_t}{L_s} \qquad (4.47)$$

where

$\bar{\Delta}_r$ = average height of asperities
r_s = nominal socket radius
L_t = the total travel distance along the grooved profile
L_s = the nominal socket depth

$\bar{\Delta}_r$ is further defined as the radial distance from a socket profile to the surface of an imaginary cylinder which would fit into the grooved socket. A factor of safety of 2 is recommended to be applied to f_s calculated from equations 4.46 and 4.47. There may be practical difficulties in measuring the depth of the groove achieved by the rotary tool, particularly where direct visual or underwater television methods of inspection are used in muddy water.

CIRIA Report 570[4.55] recommends that the shaft friction of bored piles in very weak mudstones can be calculated in the same way as piles in stiff clay using either effective stress methods (equation 4.6) or undrained shear strengths (equation 4.7). However, the report points out the difficulty in obtaining satisfactory samples in weak weathered mudstones with the result that the c_u values are likely to be low and hence the calculated shaft friction will be over-conservative. When effective stress methods are used c' should be taken as zero to allow for softening and a remoulded value of ϕ' of 36° should be assumed. Laboratory tests gave K_o values of 1.5 to 1.6. Report 570 relates α in equation 4.7 and β (K in equation 4.6) to the weathering grades of mudstone shown in Table 4.17.

The allowable bearing pressure on the base of bored piles in weak rocks depends to a great extent on drilling techniques. The use of percussive drilling tools can result in the

Table 4.17 α and β values of weak mudstones related to weathering grades

Grade	α	β
IV–III	0.45	—
IV–II	0.45	0.5
II	0.3	1.71
IV–III	0.3	0.86
III	0.31–0.44	0.86–1.06
IV–II	0.45	—
IV–II	0.375	—

formation of a very soft sludge at the bottom of the drillhole which, apart from weakening the base resistance, makes it difficult to identify the true character of the rock at the design founding level. The sludge should be baled out, accompanied if necessary by flushing with air and water. Standard or cone penetration tests can be used to assess the rock quality at base level but the procedure is time consuming. It is better to judge the founding level by examination and testing of cores taken from boreholes at the site investigation stage (Section 11.1.3), with later correlation by examining drill cuttings from the pile boreholes. Assessment of pile base levels from rock cores is particularly necessary in thinly bedded strata where weak rocks are interbedded with stronger layers. In such cases the allowable bearing pressure should be governed by the strength of the weak layers, irrespective of the strength of the material on which the pile is terminated.

The introduction of powerful mechanical augers of the type described in Section 3.3 has eliminated most of the rock identification problems associated with percussion drilling.

Where pile boreholes are socketted into strong rock the allowable base pressure is governed by the permissible compression strength of the concrete forming the pile shaft (Section 2.4.2). In weak rocks which are decomposed to a soil-like consistency (complete weathering), the ultimate base resistance of the material can be determined, in the case of sandstones, by making standard or cone penetration tests in the weathered layers at the site investigation stage, with calculations from the test results as described for bored piles in coarse-grained soils in Section 4.3.6.

It should be possible to obtain undisturbed samples of completely weathered mudstones, siltstones and shales, from which shear strength tests can be made and base resistance calculated as described in Section 4.2.3. In the case of moderately weathered mudstones, siltstones and shales, uniaxial compression tests are made on rock cores, or in the case of poor core recovery, point load tests (Section 11.1.4) are made to obtain the compression strength. The base resistance is then calculated using the relationship with q_{uc} and RQD as shown in Table 4.14. Alternatively, the parameters c and ϕ can be obtained from this table and used in conjunction with equation 4.40.

In the absence of compression strength data, published relationships between the weathering grade, undrained shear strength and elastic properties of the above rocks can be used to determine the base resistance from equation 4.40. CIRIA Report 181[4.57] gives these properties as shown in Table 4.18.

High values of base resistance resulting from the calculations described above should be adopted with caution because of the risk of excessive base settlement. This can be of the order of 20% of the pile width at the toe which is required to mobilize the ultimate base

Table 4.18 Relationships between weathering grades, undrained shear strength, and elastic properties of weak rocks (see also Seedhouse and Sancters[4.56]).

Weathering grade	Undrained shear strength (c_u, kN/m^2)	Shear modulus (G, MN/m^2)	Youngs modulus (E, MN/m^2)
V–VI	250	80	115
IV	850	100	230
III	1330	350	820
III	1270	265	615
III	1230	210	490
III	1150	175	405
III	1090	150	350

resistance. If the safety factor to obtain the allowable load is too low, the resulting shaft settlement could break down the bond between the rock and concrete thus weakening the total pile resistance in cases where the design requires the load to be shared between the shaft and the base. A reduction in shaft resistance of 30% to 40% of the peak value has been observed where shear displacements of the rock socket of little more than 15 mm have occurred. It may also be difficult to remove soft or loose debris from the whole base area at the time of final clean-out before placing the concrete.

Because of the porous cellular nature of chalk and the consequent break down and softening of the material under the action of drilling tools (similar to that described in Section 4.7.1), conventional methods of calculating the base resistance and rock socket shaft friction cannot be used for bored piles in chalk. CIRIA Report 574 states that these two components of bearing capacity are best determined from relationships with the standard penetration test N-values uncorrected for overburden pressure. These give a rough indication of the weathering grade to supplement the classification based on examination of rock cores and exposures in the field. The recommendations of Report 574 are

Ultimate base resistance of bored piles: 200 N (kN/m^2)
Ultimate base resistance of continuous flight auger (CFA) piles: 200 N (kN/m^2)
Ultimate shaft resistance of bored piles in low to medium-density chalk: $0.8\bar{\sigma}'_{vo}$

The above relationships for base resistance are subject to the pile bore being certified as clean. Also where the average effective overburden pressure, σ'_{vo}, is less than 400 kN/m^2 (based on final ground levels and omitting the contribution from made ground and fill) the calculated shaft friction must be confirmed by load testing. For high-density Grade A chalk the pile should be treated as a rock socket and the shaft friction taken as 0.1 times the uniaxial compression strength. The report makes a distinction between made ground and fill. The former is regarded as an accumulation of debris resulting from the 'activities of man', whereas fill is purposefully placed.

A somewhat modified recommendation is made for continuous flight auger piles. A later CIRIA Report PR86[4.58] states:

$$\text{Average ultimate shaft resistance} = \beta\bar{\sigma}'_{vo} \tag{4.48}$$

Table 4.19 Presumed safe vertical bearing stress for foundations on horizontal ground in Hong Kong

Category	Weathering grade	Total core recovery (%)	Uniaxial compression strength (MN/m²)	Equivalent point load index strength (MN/m²)	Presumed bearing stress (MN/m²)
I(a)	≥ II	>95 of grade	≮50	≮2	7.5
I(b)	≥ II-III	>85 of grade	≮25	≮1	5
I(c)	≥III-IV	>50 of grade	—	—	3

Notes
Category I(a): Fresh to slightly decomposed, strong.
Category I(b): Slightly to moderately decomposed, moderately strong.
Category I(c): Moderately decomposed, moderately strong to moderately weak.

The factor β should be based on SPT N-values. For low values of N (less than or equal to 10) or a cone resistance q_c between 2 and 4 MN/m², and in the absence of pile test results, β should be taken as 0.45 throughout. Where N is greater than 10, or q_c greater than 4 MN/m², and in the absence of flints, β is taken as 0.8 throughout. The allowable pile load P_a should be determined by using the partial factors in equation 4.42.

Throughout reports 574 and PR86 it is emphasized that load testing is desirable at some stage as a means of confirming load capacity, and achieving economy in design. It is pointed out that a single test made to 3 times the working load is a much better aid to judgement than two tests to 1.5 times the working load.

For granites and volcanic rocks, the practice in Hong Kong is to relate the allowable base bearing pressure for bored piles to the weathering grade of the decomposed material. The recommendations of the Government Geotechnical Office[4.45] as quoted by Ng et al.[4.59] are shown in Table 4.19.

The Hong Kong Government recommends that completely weathered granite should be treated as a soil. Also the rock socket shaft friction in weak to moderately weak and strong to moderately strong granites should be determined from correlation with the uniaxial compression strength of sedimentary rocks using the method of Horvath et al.[4.50] Ng et al.[4.59] point out that observations made in loading tests in granites suggest that the value for b in equation 4.45 of 0.2 is appropriate.

4.7.4 The settlement of the single pile at the working load for piles in rocks

The effects of load transfer from shaft to base of piles on the pile head settlements have been discussed by Wyllie[4.39]. Because of the relatively short penetration into rocks which is needed to mobilize the required total pile resistance, the simpler methods of determining pile head settlement described in Section 4.6 are suitable in most cases. For piles having base diameters up to 600 mm the settlement at the working load should not exceed 10 mm if a safety factor of 2.5 has been applied to the ultimate bearing capacity.

The settlement of large diameter piles can be calculated from equation 4.38. The modulus of deformation of the rock below the pile toe can be obtained from plate bearing or pressuremeter tests or from empirical relationships developed between the modulus, the weathering grade and the unconfined compression strength of the rock given in Table 4.18 and Section 5.5.

These relationships are not applicable to high porosity chalk or weathered silty mudstone (Mercia Mudstone). The relationships given in Section 5.5 assume fairly low stress levels. Therefore calculated values based on the unconfined compression strength of the rock should take into account the high-bearing pressures beneath the base of piles.

In CIRIA Report 574[4.43] the deformation modulus of chalk is related to the weathering grade and standard penetration test N-values. For Grade A chalk where the N-value is greater than 25, the deformation modulus is 100 to 300 MN/m². For Grades B, C and D with N-values less than 25 the modulus is 25 to 100 MN/m².

Pells and Turner[4.60] have derived influence factors for calculating the settlement of a bored pile where the load is carried by rock socket shaft friction only using the equation:

$$\text{Settlement} = \rho = \frac{QI_p}{BE_d} \tag{4.49}$$

where

Q = total load carried by the pile head
I_p = influence factor
B = diameter of the socket
E_d = deformation modulus of the rock mass surrounding the shaft.

The influence factors of Pells and Turner are shown in Figure 4.36. Where the rock sockets are recessed below the ground surface or where a layer of soil or very weak rock overlies competent rock, a reduction factor is applied to equation 4.49. Values of the reduction factor are shown in Figure 4.37.

4.7.5   Eurocode recommendations for piles in rock

EC7 makes no specific recommendations for the design of piles carrying axial compression loads in rock. The design methods described in the preceding Sections 4.7.1 to 4.7.3 are based either on relationships with unconfined compression strengths or by correlation with SPT N-values.

Where the calculations are based on relationships with SPT N-values, the correlation factors based on the number of test profiles are applied as in equation 4.19 together with a calibration factor of 1.05. Where design total pile resistances are obtained from static or dynamic pile tests the correlation factors are obtained from Tables 4.7 and 4.8 respectively.

EC7(Clause 7.6.4.2) states that 'when the pile toe is placed in a medium-dense or firm layer overlying rock or very hard soil, the partial safety factors for the ultimate limit state conditions are normally sufficient to satisfy serviceability limit-state conditions'.

4.8   Piles in fill – negative skin friction

4.8.1   Estimating negative skin friction

Piles are frequently required for supporting structures that are sited in areas of deep fill. The piles are taken through the fill to a suitable bearing stratum in the underlying natural soil

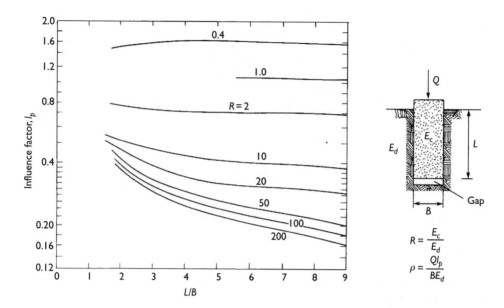

Figure 4.36 Elastic settlement influence factors for rock-socket shaft friction on piles (after Pells and Turner[4.60], courtesy of *Research Journals, National Research Council, Canada*).

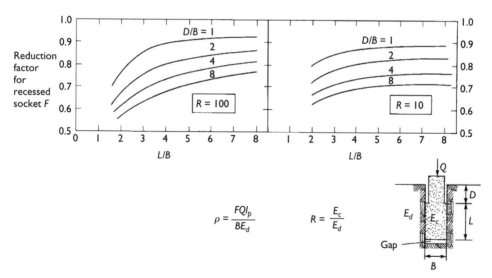

Figure 4.37 Reduction factors for calculation of settlement of recessed sockets (after Pells and Turner[4.60], courtesy of *Research Journals, National Research Council, Canada*).

or rock. No support for compressive loads from shaft friction can be assumed over the length of the pile shaft through the fill. This is because of the downward movement of the fill as it compresses under its own weight or under the weight of further soil or surcharge placed over the fill area. The downward movement results in drag-down forces, generally known as *negative skin friction*, on the pile shaft. Where fill is placed over a compressible natural soil the latter consolidates and moves downwards relative to the pile. Thus the negative skin friction occurs over the length of the shaft within the natural soil as well as within the fill.

Calculation of the magnitude of the negative skin friction is a complex problem which depends on the following factors:

(1) The relative movement between the fill and the pile shaft
(2) The relative movement between any underlying compressible soil and the pile shaft
(3) The elastic compression of the pile under the working load and
(4) The rate of consolidation of the compressible layers.

The simplest case is fill that is placed over a relatively incompressible rock with piles driven to refusal in the rock. The toe of the pile does not yield under the combined working load and drag-down forces. Thus the negative skin friction on the upper part of the pile shaft is equal to the fully mobilized value. Near the base of the fill its downward movement may be insufficiently large to mobilize the full skin friction, and immediately above rock-head the fill will not settle at all relative to the pile shaft. Thus negative skin friction cannot occur at this point. The distribution of negative skin friction on the shaft of the unloaded pile is shown in Figure 4.38a. If a heavy working load is now applied to the pile shaft, the shaft compresses elastically and the head of the pile moves downwards relative to the fill. The upper part of the fill now acts in support of the pile although this contribution is neglected in calculating the pile resistance. The distribution of negative skin friction on the shaft of the loaded pile is shown in Figure 4.38b. Where the fill has been placed at a relatively short period of time before installing the piles, continuing consolidation of the material will again cause it to slip downwards relative to the pile shaft, thus re-activating the drag-down force.

The simplified profile of negative skin friction for a loaded pile on an incompressible stratum is shown in Figure 4.38c. This diagram can be used to calculate the magnitude of the drag-down forces. The peak values for coarse soils and fill material are calculated by the method described in Section 4.3.

In the case where negative skin friction is developed in clays, the rate of loading must be considered. It was noted in Section 4.2.4 that the capacity of a clay to support a pile in skin friction is substantially reduced if the load is applied to the pile at a very slow rate. The same consideration applies to negative skin friction, but in this case it works advantageously in reducing the magnitude of the drag-down force. In most cases of negative skin friction in clays the relative movement between the soil which causes drag-down and the pile takes place at a very slow rate. The movement is due to the consolidation of the clay under its own weight or under imposed loading, and this process is very slow compared with the rate of application of the working load to the pile.

Meyerhof[4.37] advises that the negative skin friction on piles driven into soft to firm clays should be calculated in terms of effective stress from the equation:

$$\tau_{s\,neg} = \beta\sigma'_{vo} \tag{4.50}$$

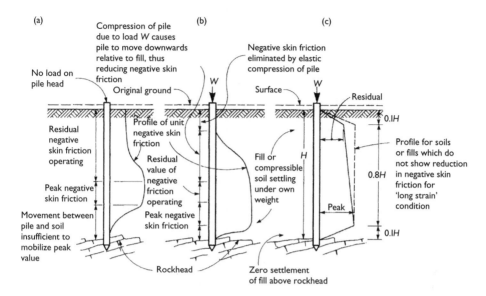

Figure 4.38 Distribution of negative skin friction on piles terminated on relatively incompressible stratum (a) No load on pile head (b) Compressive load on pile head (c) Design curve for loaded pile.

Values of the negative skin friction factor allowing for reduction of the effective angle of friction with increasing depth to the residual value δ_r are shown in Figure 4.39.

Taking the case of a pile bearing on a compressible stratum, where yielding of the pile toe occurs under the drag-down forces and the subsequently applied working load, the downward movement of the pile relative to the lower part of the fill may then be quite large, and such that negative skin friction is not developed over quite an appreciable proportion of the length of the shaft within the fill. Over the upper part of the shaft the fill moves downwards relative to the pile shaft to an extent such that the negative skin friction operates, whereas in the middle portion of the pile shaft the small relative movement between the fill and the pile may be insufficient to mobilize the peak skin friction as a drag-down force. The distribution for the unloaded pile is shown in Figure 4.40a.

When the working load is applied to the head of the pile, elastic shortening of the pile occurs, but since the load is limited by the bearing characteristics of the soil at the pile toe the movement may not be large enough to eliminate the drag-down force. The distribution of negative friction is then shown in Figure 4.40b. The diagram in Figure 4.40c can be used for design purposes, with the peak value calculated as described in Section 4.3 for coarse soils and fill and by using equation 4.50 and Figure 4.39 for soft to firm clays.

It may be seen from Figure 4.40a to c that at no time does the maximum skin friction operate as a drag-down force over the full length of the pile shaft. It is not suggested that these simplified profiles of distribution of negative skin friction represent the actual conditions in all cases where it occurs, since so much depends on the stage reached in the consolidation of the fill, and the compression of the natural soil beneath the fill. The time interval between the installation of the pile and the application of the working load is

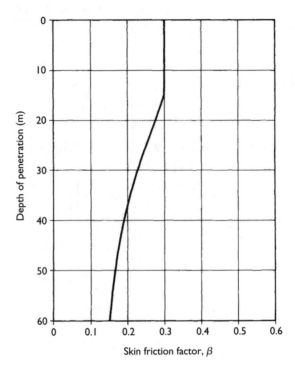

Figure 4.39 Negative skin friction factors for piles driven into soft to firm clays (after Meyerhof[(4.37)]).

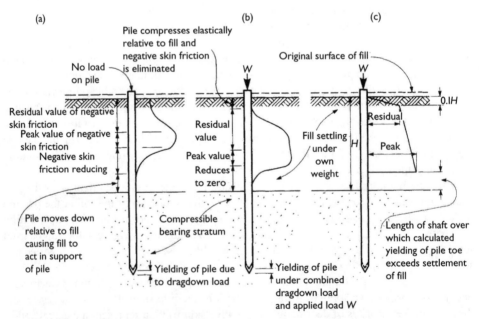

Figure 4.40 Distribution of negative skin friction on piles terminated in compressible stratum (a) No load on pile head (b) Compressive load on pile head (c) Design curve for loaded pile.

also significant. In old fill which has become fully consolidated under its own weight and where it is not proposed to impose surcharge loading the negative skin friction may be neglected, but shaft friction within the fill layer should not be allowed to help support the pile. In the case of recently placed fill it may settle by a substantial amount over a long period of years. The fill may also be causing consolidation and settlement of the natural soil, within which the pile obtains its bearing. The case of recent fill placed over a compressible soil which becomes stiffer and less compressible with depth is shown in Figure 4.41.

Modelling the load transfer by drag-down from fill and the underlying compressible soil and the distribution of resistance in positive shaft friction can be undertaken by using a pile–soil interaction analysis as described in Section 4.6. This procedure is particularly effective because the resulting t–z curves give a more accurate estimate in separate or combined form of the distribution of axial forces over the depth of the pile shaft from the compression load applied to the pile head and the shear stress on the pile surface from the drag-down loading, than is possible from semi-empirical diagrams such as shown in Figures 4.38, 4.40, and 4.41. In particular the t–z curves indicate the depth H in Figure 4.41, that is, the depth to the 'neutral point' at which the shear stress changes from negative, caused by drag-down, to positive, acting in support of the pile.

It is good practice to ignore the contribution to the support provided by friction over the length of a pile in soft clay, where the pile is driven through a soft layer to less compressible soil. This is because of the drag-down force on the pile shaft caused by heave and reconsolidation of the soft clay. The same effect occurs, of course, if a pile is driven into a stiff clay but the stiff clay continues to act in support of the pile if yielding at the toe is permitted.

Very large drag-down forces can occur on long piles. In some circumstances they may exceed the working load applied to the head of the pile. Fellenius[4.61] measured the

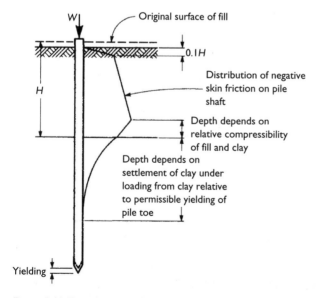

Figure 4.41 Distribution of negative skin friction on pile driven through recent fill into compressible clay stratum.

progressive increase in negative skin friction on two precast concrete piles driven through 40 m of soft compressible clay and 15 m of less-compressible silt and sand. Reconsolidation of the soft clay disturbed by pile driving contributed 300 kN to the drag-down load over a period of 5 months. Thereafter, regional settlement caused a slow increase in negative skin friction at a rate of 150 kN per year. Seventeen months after pile driving a load of 440 kN was added to each pile, followed by an additional load of 360 kN a year later. Both these loads caused yielding of the pile at the toe to such an extent that all negative skin friction was eliminated, but when the settlement of the pile ceased under the applied load the continuing regional settlement caused negative skin friction to develop again on the pile shaft. Thus with a yielding pile toe the amount of negative skin friction which can be developed depends entirely on the downward movement of the pile toe relative to the settlement of the soil or fill causing the drag-down force. If the drag-down force is caused only by the reconsolidation of the heaved soil, and if the pile can be permitted to yield by an amount greater than the settlement of the ground surface due to this reconsolidation, then negative friction need not be provided for. If, however, the negative skin friction is due to the consolidation of recent fill under its own weight or to the weight of additional fill, then the movement of the ground surface will be greater than the permissible yielding of the pile toe. Negative skin friction must then be taken into account, the distribution being as shown in Figure 4.40c or Figure 4.41.

Much greater drag-down loads occur with piles driven onto a relatively unyielding stratum. Johannessen and Bjerrum[4.62] measured the development of negative skin friction on a steel pile driven through 53 m of soft clay to rock. Sand fill was placed to a thickness of 10 m on the sea bed around the pile. The resulting consolidation of the clay produced a settlement of 1.2 m at the original sea-bed level and a drag-down force of about 1500 kN at the pile toe. It was estimated that the stress in the steel near the toe could have been about 190 N/mm^2, which probably caused the pile to punch into the rock, so relieving some of the drag-down load. The average unit negative skin friction within the soft clay was equal to 100% of the undrained shearing strength of the clay.

4.8.2   Safety factors for negative skin friction

Safety factors for piles subjected to negative skin friction require careful consideration. The concept of partial safety factors can be applied to the two components of permanent working load and negative skin friction. Thus if the negative skin friction P_n has been conservatively estimated before deciding on a value of Q_p to give a safety factor of 2.5 or more on the combined loading, it is over-conservative to add this to the working load W on the pile in order to arrive at the total allowable pile load. It is more realistic to obtain the required *ultimate* pile load Q_p by multiplying the working load only by the normal safety factor, and then to check that the safety factor given by the ultimate load divided by the working load plus the negative skin friction is still a reasonable value.

When applying the EC7 recommendations to the design of piles subjected to drag-down, the resulting axial load is treated as a permanent unfavourable action in Table 4.1 (Section 4.1.4). This is classed as a geotechnical action in Clause 7.3.2.1(3)P which can be calculated either by a pile–soil interaction analysis (Method (a)), or as an upper-bound force exerted on the pile shaft (Method (b)). As noted above, Method (a) is the more effective of the two, particularly in determining the depth to the neutral point. It is evident that if Method (b) is used the depth H over which the upper-bound force is assumed to act is critical. If the depth is over-estimated application of the action factor A1 of 1.35 in Table 4.20 will further exaggerate the drag-down force.

Frank *et al.*[1.5] give a worked example to compare the application of the three design approaches in EC7 to the case of a single pile carrying a compression load at the pile head (structural action) and a drag-down force on the shaft (geotechnical action). The magnitude and depth to the neutral point of the latter action were determined by an interaction analysis. The partial factors applied to the actions and ground resistance are shown in Table 4.20.

M2 is applied as an action factor and noted in the table because the drag-down is usually calculated by effective stress analysis using a constant, for example, β in equation 4.50, which is not directly related to the angle of shearing resistance of the soil.

The use of Method (a) requires, as a first step, a settlement analysis to determine the settlement of the fill and underlying compressible soil. Clause 7.3.2.2(5)P requires the design value of the ground in a settlement analysis to take account of the weight densities of the material. However, the partial factors for M1 and M2 sets in Table A4 of EC7 Annex A are unity (see Table 4.2 in Section 4.1.4 of this book).

When calculating drag-down on the shafts of uncased bored and cast-in-place piles, the possibility of enlargement of the pile cross-section due to overbreak should be considered as well as 'waisting' in the supporting soil layer. Clause 2.3.4.2 of EN 1992–1:2004(EC2) does not consider the possibility of enlargement, but Table 4.9 in Section 4.1.4 can be used as a guide to the required tolerance on pile diameter.

EC7 points out that drag-down and transient loading need not usually be considered to act simultaneously in load combinations.

4.8.3 Minimizing negative skin friction

The effects of drag-down can be minimized by employing slender piles (e.g. H-sections or precast concrete piles), but more positive measures may be desirable to reduce the magnitude of the drag-down forces. In the case of bored piles this can be done by placing in-situ concrete only in the lower part of the pile within the bearing stratum and using a precast concrete element surrounded by a bentonite slurry within the fill. Negative skin friction forces on precast concrete or steel tubular piles can be reduced by coating the portion of the shaft within the fill with soft bitumen.

Table 4.20 Partial factor sets for a pile axially loaded at the head and subjected to drag-down on the shaft

| Design approach | Structural action | Geotechnical action | | Resistance to compression |
		Shear strength parameter	Load	
DA1, combination 1	A1 (1.35)	M1 (1.0)	A1 (1.35)	R1 (1.0)
DA1, combination 2	A2 (1.0)	M2 (1.25)[a]	A2 (1.0)	R4 (1.3)

Note
a Applied as partial action factor.

Claessen and Horvat[4.63] describe the coating of 380 × 450 mm precast concrete piles with a 10 mm layer of bitumen having a penetration of 40 to 50 mm at 25°C. The skin friction on the 24 m piles was reduced to 750 kN compared with 1600 to 1700 kN for the uncoated piles.

Shell Composites Ltd. markets its Bitumen Compound SL for coating bearing piles. The material has the following characteristics:

Penetration at 25°C:	53 to 70 mm
Softening Point (R and B):	57°C to 63°C
Penetration Index:	Less than +2

The bitumen is heated to 180°C (maximum) and sprayed or poured onto the pile to obtain a coating thickness of 10 mm. Before coating, the pile surface should be cleaned and primed with Shell Composites Bitumen Solvent Primer applied by brush or spray at a rate of about 2 kg/10 m². Alternatively, the SL Compound can be fluxed with 29% of white spirit to provide the primer. The bitumen slip layers should not be applied over the length of the shaft which receives *support* from skin friction, and Claessen and Horvat recommend that a length at the lower end of ten times the diameter or width of the pile should remain uncoated if the full *end-bearing resistance* is to be mobilized.

Negative skin friction is a most important consideration where piles are installed in groups. The overall settlement of pile groups in fill must be analysed as described in Section 5.5.

The above measures to minimize negative skin friction can be quite costly. In most cases it will be found more economical to increase the penetration of the pile into the bearing stratum thereby increasing its capacity to carry the combined loading.

4.9   References

4.1 TOMLINSON, M. J. The adhesion of piles driven in clay soils, *Proceedings of 5th International Conference, ISSMFE*, London, Vol. 2, 1957, pp. 66–71.

4.2 TOMLINSON, M. J. The adhesion of piles in stiff clay, *Construction Industry Research and Information Association*, Research Report No. 26, London, 1970.

4.3 BOND, A. J. and JARDINE, R. J. Effects of installing displacement piles in a high OCR clay, *Geotechnique*, Vol. 41, No. 3, 1991, pp. 341–63.

4.4 RANDOLPH, M. F. and WROTH, C. P. Recent developments in understanding the axial capacity of piles in clay, *Ground Engineering*, Vol. 15, No. 7, 1982, pp. 17–25.

4.5 SEMPLE, R. M. and RIGDEN, W. J. Shaft capacity of driven pipe piles in clay, *Symposium on Analysis and Design of Pile Foundations*, American Society of Civil Engineers, San Francisco, 1984, pp. 59–79.

4.6 RIGDEN, W. J., PETTIT, J. J., ST. JOHN, H. D., and POSKITT, T. J. Developments in piling for offshore structures, *Proceedings of the Second International Conference on the Behaviour of Offshore Structures*, London, Vol. 2, 1979, pp. 276–96.

4.7 WELTMAN, A. J. and HEALY, P. R. Piling in 'boulder clay' and other glacial tills, *Construction Industry Research and Information Association (CIRIA)*, Report PG5, 1978.

4.8 TRENTER, N. A. Engineering in glacial tills, *Construction Industry Research and Information Association (CIRIA)*, Report C504, 1999.

4.9 MEYERHOF, G. G. and MURDOCK, L. J. An investigation of the bearing capacity of some bored and driven piles in London Clay, *Geotechnique*, Vol. 3, No. 7, 1953, pp. 267–82.

4.10 WHITAKER, T. and COOKE, R. W. Bored piles with enlarged bases in London Clay, *Proceedings of 6th International Conference, ISSMFE*, Montreal, Vol. 2, 1965, pp. 342–6.

4.11 SKEMPTON, A. W. Cast in-situ bored piles in London Clay, *Geotechnique*, Vol. 9, No. 4, 1959, pp. 153–73.

4.12 FLEMING, W. G. K. and SLIWINSKI, Z. The use and influence of bentonite in bored pile construction, *Construction Industry Research and Information Association (CIRIA)*, Report PG3, 1977.

4.13 KRAFT, L. M. and LYONS, C. G. Ultimate axial capacity of grouted piles, *Proceedings of the 6th Annual Offshore Technology Conference*, Houston, 1974, pp. 485–99.

4.14 JONES, D. A. and TURNER, M. J. Load tests on post-grouted micropiles in London Clay, *Ground Engineering*, September 1980, pp. 47–53.

4.15 BJERRUM, L. Problems of soil mechanics and construction in soft clay, *Proceedings of the 8th International Conference, ISSMFE*. Moscow, Vol. 3, 1973, pp. 150–7.

4.16 TAYLOR, P. T. Age effect on shaft resistance and effect of loading rate on load distribution of bored piles, PhD Thesis, University of Sheffield, October 1966.

4.17 BALLISAGER, C. C. Bearing capacity of piles in Aarhus Septarian Clay, *Danish Geotechnical Institute*, Bulletin No. 7, 1959, pp. 14–19.

4.18 PECK, R. B., HANSON, W. E., AND THORNBURN, T. H. *Foundation Engineering*, 2nd edn, John Wiley, New York, 1974.

4.19 DURGONOGLU, H. T. and MITCHELL, J. K. Static penetration resistance of soils, *Proceedings of the Conference on in-situ Measurement of Soil Properties*, American Society of Civil Engineers, Raleigh (North Carolina), Vol. 1, 1975, pp. 151–88.

4.20 VESIC, A. S. Design of pile foundations, NCHRP Synthesis 42, *Transportation Research Board*, Washington, DC, 1977.

4.21 BEREZANTSEV, V. G. *et al.* Load bearing capacity and deformation of piled foundations, *Proceedings of the 5th International Conference, ISSMFE*, Paris, Vol. 2, 1961, pp. 11–12.

4.22 KULHAWY, F. H. Limiting tip and side resistance, fact or fallacy, *Symposium on Analysis and Design of Pile Foundations*, American Society of Civil Engineers, San Francisco, 1984, *Proceedings*, pp. 80–98.

4.23 DUTT, R. N., MOORE, J. E., and REES, T. C. Behaviour of piles in granular carbonate sediments from offshore Philippines, *Proceedings of the Offshore Technology Conference*, Houston, Paper No. OTC 4849, 1985, pp. 73–82.

4.24 HIGHT, D. W., LAWRENCE, D. M., FARQUHAR G. B., and POTTS, D. M. Evidence for scale effects in the end bearing capacity of open-end piles in sand, *Proceedings of the 28th Offshore Technology Conference*, Houston, OTC 7975, 1996, pp. 181–92.

4.25 MEIGH, A. C. Cone penetration testing, *CIRIA-Butterworth*, 1987.

4.26 TE KAMP, W. C. Sondern end funderingen op palen in zand, *Fugro Sounding Symposium*, Utrecht, 1977.

4.27 DE GIJT, J. G. and BRASSINGA, H. E. Ontgraving beïnvlodet de conusweerstand, *Land and Water*, Vol. 1.2, January 1992, pp. 21–5.

4.28 BROUG, N. W. A. The effect of vertical unloading on cone resistance q_c, a theoretical analysis and a practical confirmation, *Proceedings First International Geotechnical Seminar on Deep Foundations on Bored and Auger Piles*, ed. Van Impe, Balkema, Rotterdam, 1988, pp. 523–30.

4.29 HEIJNEN, W. J. Tests on Frankipiles at Zwolle, Netherlands, *La technique des Travaux*, No. 345, January/February 1974, pp. I–XIX.

4.30 JARDINE, R., CHOW, F. C., OVERY, R., and STANDING, J. *ICP Design Methods for Driven Piles in Sands and Clays*, Thomas Telford, London, 2005.

4.31 American Petroleum Institute, RP2A-WSD, Recommended practice for planning, designing and constructing fixed offshore platforms – Working stress design, 20th ed, Washington, 1993.

4.32 CHOW, F. C. Investigations into displacement pile behaviour for offshore foundations, PhD Thesis, Imperial College, London, 1997.

4.33 TAPPIN, R. G. R., van DUIVANDIJK., and HAQUE, M. The design and construction of Jamuna Bridge, Bangladesh, *Proceedings of the Institution of Civil Engineers*, Vol. 126, 1998, pp. 162–80.

4.34 TOMLINSON, M. J. Discussion, British Geotechnical Society Meeting, 1996, *Ground Engineering*, Vol. 29, No. 10. 1996, pp. 31–33.

4.35 WHITE, D. J. and BOLTON, M. D. Comparing CPT and pile base resistance in sand, *Proceedings of the Institution of Civil Engineers, Geotechnical Engineering*, Vol. 158, (GE1), 2005, pp. 3–14.

4.36 TOGROL, E., AYDINOGLU, N., TUGCU, E. K., and BEKAROGLU. Design and Construction of large piles, *Proceedings of the 12th International Conference on Soil Mechanics*, Rio de Janeiro, Vol. 2, 1992, pp. 1067–72.

4.37 MEYERHOF, G. G. Bearing capacity and settlement of pile foundations, *Proceedings of the American Society of Civil Engineers*, GT3, 1976, pp. 197–28.

4.38 BURLAND, J. B., BUTLER, F. G., and DUNICAN, P. The behaviour and design of large diameter bored piles in stiff clay, *Proceedings of the Symposium on Large Bored Piles*, Institution of Civil Engineers and Reinforced Concrete Association, London, 1966, pp. 51–71.

4.39 WYLLIE, D. C. *Foundations on Rock*, E & FN Spon, London, 1st ed., 1991.

4.40 KULHAWY, F. H. and GOODMAN, R. E. Design of foundations on discontinuous rock, *Proceedings of the International Conference on Structural Foundations on rock*, Sydney, Vol. 1, 1980, pp. 209–20.

4.41 KULHAWY, F. H. and GOODMAN, R. E. Foundations in rock. Chapter 15, Ground Engineering Reference Book, F. G. Bell (ed.), Butterworth, London, 1987.

4.42 PELLS, P. J. N. and TURNER, R. M. End bearing on rock with particular reference to sandstone, *Proceedings of the International Conference on Structural Foundations on rock*, Sydney, Vol. 1, 1980, pp. 181–90.

4.43 LORD, J. A., CLAYTON, C. R. I., and MORTIMORE, R. N. Engineering in chalk, *Construction Industry Research and Information Association*, Report No 574, 2002.

4.44 LUMB, P. The residual soils of Hong Kong, *Geotechnique*, Vol. 15, 1965, pp. 180–94.

4.45 PILE DESIGN and CONSTRUCTION, GEO Publication No 1/96, Geotechnical Engineering Office, Civil Engineering Dept, Government of Hong Kong.

4.46 BEAKE, R. H. and SUTCLIFFE, G. Pipe pile driveability in the carbonate rocks of the Southern Arabian Gulf, *Proceedings of the International Conference on Structural Foundations on Rock*, Balkema, Rotterdam, Vol. 1, 1980, pp. 133–49.

4.47 GEORGE, A. B., SHERRELL, F. W., and TOMLINSON, M. J. The behaviour of steel H-piles in slaty mudstone, *Geotechnique*, Vol. 26, No. 1, 1976, pp. 95–104.

4.48 LEACH, B. and MALLARD, D. J. The design and installation of precast concrete piles in the Keuper Marl of the Severn Estuary, *Proceedings of the Conference on Recent Developments in the Design and Construction of Piles*, Institution of Civil Engineers, London, 1980, pp. 33–43.

4.49 OSTERBERG, J. O. and GILL, S. A. Load transfer mechanisms for piers socketted in hard soils or rock, *Proceedings of the 9th Canadian Symposium on Rock Mechanics*, Montreal, 1973, pp. 235–62.

4.50 HORVARTH, R. G. Field load test data on concrete-to-rock bond strength for drilled pier foundations, University of Toronto, publication 78–07, 1978.

4.51 ROSENBERG, P. and JOURNEAUX, N. L. Friction and end bearing tests on bedrock for high capacity socket design, *Canadian Geotechnical Journal*. Vol. 13, 1976, pp. 324–33.

4.52 WILLIAMS, A. F. and PELLS, P. J. N. Side resistance rock sockets in sandstone, mudstone and shale, *Canadian Geotechnical Journal*, Vol. 18, 1981, pp. 502–13.

4.53 HOBBS, N. B. Review paper – Rocks, *Proceedings of the Conference on Settlement of Structures*, British Geotechnical Society, Pentech Press, 1975, pp. 579–610.

4.54 HORVATH, R. G., KENNEY, T. C., and KOZICKI, P. Methods of improving the performance of drilled piles in weak rock, *Canadian Geotechnical Journal*, Vol. 20, 1983, pp. 758–72.

4.55 CHANDLER, R. J. and FORSTER, A. Engineering in Mercia mudstone, *Construction Industry Research and Information Association*, Report No. 570, 2001.

4.56 SEEDHOUSE, R. L. and SANCTERS, R. V. Investigations for cooling tower foundations in Mercia mudstone at Ratcliffe-on-Soar, Nottinghamshire, *Proceedings of the Conference on Engineering Geology of Weak Rock*, A. A. Balkema, Rotterdam, Special Publication No. 8, 1993, pp. 465–72.

4.57 GANNON, J. A., MASTERTON, G. G. T., WALLACE, W. A. and MUIR-WOOD, D. Piled foundations in weak rock, *Construction Industry Research and Information Association*, Report No. 181, 1999.

4.58 LORD, J. A., HAYWARD, T., and CLAYTON, C. R. I. Shaft friction of CFA piles in chalk, *Construction Industry Research and Information Association*, Report No. PR86, 2003.

4.59 NG, C. W. W., SIMONS, N., and MENZIES, B. A Short Course in Soil-Structure Engineering of Deep Foundations, Excavations, and Tunnels, Thomas Telford, London, 2004, p. 86.

4.60 PELLS, P. J. N. and TURNER, R. M. Elastic solutions for design and analysis of rock socketted piles, *Canadian Geotechnical Journal*, 1979, Vol. 16, pp. 481–7.

4.61 FELLENIUS, B. H. Down drag on piles in clay due to negative skin friction, *Canadian Geotechnical Journal*, Vol. 9, No. 4, 1972, pp. 323–37.

4.62 JOHANNESSEN, I. J. AND BJERRUM, L. Measurement of the compression of a steel pile to rock due to settlement of the surrounding clay, *Proceedings of the 6th International Conference, ISSMFE,* Montreal, Vol. 2, 1965, pp. 261–4.

4.63 CLAESSEN, A. I. M. and HORVAT, E. Reducing negative skin friction with bitumen slip layers, *Journal of the Geotechnical Engineering Division*, American Society of Civil Engineers, Vol. 100, No. GT8, August 1974, pp. 925–44.

4.10   Worked examples

Example 4.1

A precast concrete pile for a jetty structure is required to carry a dead load of 240 kN and an imposed load of 110 kN both in compression, together with an uplift load of 200 kN. The pile is driven through 7 m of very soft clay into a stiff boulder clay. Determine the required penetration of the 350 × 350 mm pile into the stiff clay to carry the specified loading. Undrained shear strength tests were made on samples from three boreholes with the following results:

	Mean (KN/m²)			Mean of values (KN/m²)	Characteristic all values (KN/m²)
	BH1	BH2	BH3		
c_u shaft	85	115	130	110	95
c_u base	75	80	90	95	75

Settlements are not critical to the structural design of the jetty, therefore the penetration can be calculated by permissible stress methods by applying a safety factor of 2.5 to the calculated ultimate bearing resistance. The shaft resistance within the soft clay will be ignored because of possible sea-bed erosion and lateral pile movements.

Required total pile resistance in stiff clay = (240 + 110) × 2.5 = 875 kN.

To allow for possible lower than mean c_u values at pile base level take $c_u = 0.75 \times 110$ kN/m².

Ultimate base resistance = 9 × 0.75 × 110 × 0.35² = 91 kN.
Load to be carried by shaft resistance in stiff clay = 875 – 91 = 784 kN.

Take a trial penetration of 10 m into the stiff day:

Average effective overburden pressure in layer = (0.65 × 9.81 × 7) + (1.1 × 9.81 × 5)
$$= 98 \text{ kN/m}^2.$$

For $c_u/\sigma'_{vo} = 110/98 = 1.1$, Figure 4.6 gives $\alpha_p = 0.5$, and for $L/B = 10/0.35 = 29$, the length factor $= 1.0$, giving $\alpha = 0.5$.

Ultimate shaft resistance $= 0.5 \times 110 \times 4 \times 0.35 \times 10 = 770$ kN.
Total pile resistance $= 770 + 91 = 861$ kN.
Factor of safety on compression load $= 861/(240 + 110) = 2.5$ which is satisfactory.
Factor of safety on uplift load $= 770 / 200 = 3.8$ (satisfactory).

The above calculations using permissible stress methods can be checked for compliance with the EC7 recommendations. Considering first the actions using the factors in Table 4.1:

Set A1, design value of actions $= F_d = 1.35 \times 240 \times 1.5 \times 110 = 489$ kN.
Set A2, design value of actions $= F_d = 1.0 \times 240 \times 1.3 \times 110 = 383$ kN.

The pile resistances are calculated from equation 4.10 for each borehole as follows:

BH1: $R_{cal} = 9 \times 75 \times 0.35 + 0.5 \times 85 \times 4 \times 0.35 \times 10 = 83 + 595 = 678$ kN.
BH2: $R_{cal} = 83 \times 80/75 + 595 \times 115/85 = 893$ kN.
BH3: $R_{cal} = 83 \times 95/75 + 595 \times 130/85 = 1015$ kN.

The mean and minimum values of R_{cal} are

R_{cal} (mean) $= (687 + 893 + 1015)/3 = 862$ kN.
R_{cal} (min) $= 678$ kN (BH1).

The above values are divided by correction factors 1.33 (mean) and 1.23 (min) for three boreholes (Table 4.6) to give the characteristic values:

R_{ck} (mean) $= 862/1.33 = 648$ kN,   and
R_{ck} (min) $= 672/1.23 = 551$ kN.

The least value of R_{ck} is taken as 551 kN to calculate the design value R_{cd}. Because the jetty structure is relatively flexible R_{ck} is not multiplied by the factor of 1.1.

Using design approach DA1 and combination 1 (sets A1 + M1 + R1), for a driven pile in compression, the factors γ_b and γ_s for M1 and R1 sets are unity (Tables 4.2 and 4.3).

Therefore design $R_{cd} = (83 + 595) / 1.23 = 551$ kN which is greater than F_d for the A1 set.

For DA1 and combination 2 (sets A2 + M1 + R4), M1 (structural actions) is unity and R4 is 1.4. Hence design $R_{cd} = (83 + 595) / 1.23 \times 1.4 = 393$ kN, whcih is greater than F_d for the A2 set.

From the above check it can be concluded that the required penetration depth of 10 m into the stiff clay as calculated by permissible stress methods with a safety factor of 2.5 complies with the recommendations of EC7.

Example 4.2

A steel tubular pile 1.220 m in outside diameter forming part of a berthing structure is required to carry a working load in compression of 16 MN and a uplift of 8 MN. The pile is driven with a closed end into a deep deposit of normally consolidated marine clay. The undrained shearing strength–depth profile of the clay is shown in Figure 4.42. Determine the depth to which the pile must be driven to carry the working load with a safety factor of 2.0.

In dealing with problems of this kind it is a good practice to plot the calculated values of ultimate shaft friction, end bearing and total resistance for various depths of penetration. The required pile length can then be read off from the graph. This is a convenient procedure for a marine structure where the piles may have to carry quite a wide range of loading.

Outside perimeter of pile $= \pi \times 1.220 = 3.83$ m.
Overall base area of pile $= \frac{1}{4} \times \pi \times 1.220^2 = 1.169$ m^2.

From Figure 4.42, at 160 m,

$$\frac{c_u}{\sigma'_{vo}} = \frac{260}{0.65 \times 9.81 \times 160} = 0.25.$$

From Figure 4.6a, the adhesion factor, α_p is 1.0 over the full depth.

At 50 m below the sea bed:

Average shearing strength along shaft $= \frac{1}{2} \times 80 = 40$ kN/m^2.
From Figure 4.6b, adhesion factor for L/B value of $50/1.22 = 41$ is 1.0.
From equation 4.8, total shaft friction on outside of shaft

$$= \frac{1.0 \times 1.0 \times 40 \times 3.83 \times 50}{1000} = 7.66 \text{ MN}.$$

From equation 4.4, end-bearing resistance $= \dfrac{9 \times 80 \times 1.169}{1000} = 0.84$ MN.

Thus total pile resistance $= 8.50$ MN

At 75 m below the sea bed:

Average shearing strength along shaft $= \frac{1}{2} \times 120 = 60$ kN/m^2.

Length factor for L/B value of 61 is 0.9.

Thus total skin friction on outside of shaft $= \dfrac{1 \times 0.9 \times 60 \times 3.83 \times 75}{1000} = 15.51$ MN

End-bearing resistance $= \dfrac{9 \times 120 \times 1.169}{1000} = 1.26$ MN
Thus total pile resistance $= 16.77$ MN.

Similarly the total pile resistances at depths of 100, 125, and 150 m below the sea bed are 26.19, 38.01, and 50.78 MN respectively.

The calculated values of pile resistance are plotted in Figure 4.42, from which it may be seen that a penetration depth of 113 m is required to develop an ultimate resistance of 32 MN, which is the value required to support a compressive load of 16 MN with a safety

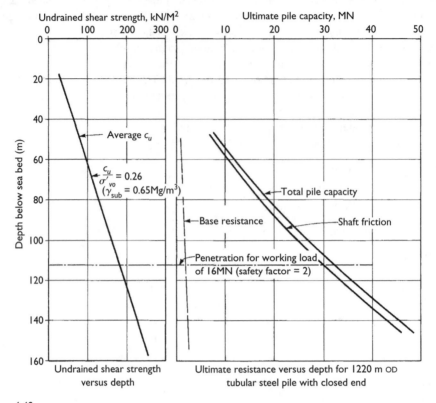

Figure 4.42

factor of 2. At a depth of 113 m the shaft friction is 30 MN. Therefore, the safety factor on uplift is 30/8, that is, nearly 4.

Checking the working stress on the steel,

$$\text{For a wall thickness of 25 mm max. compressive stress} = \frac{16 \times 10^6}{\frac{1}{4}\pi(1270^2 - 1234^2)}$$

$$= 170 \text{ Nmm}^2$$

This is 50% of the yield stress for high-tensile steel which is satisfactory for the easy driving conditions. The thickness can be decreased to 16 mm in the lower 50 m of the pile.

Example 4.3

A building column carrying a dead load of 1100 kN and an imposed load of 300 kN is to be supported by a single bored pile installed in firm to stiff fissured London Clay (Figure 4.43). Select suitable dimensions and penetration depth to obtain a safety factor of 2 in total pile resistance, or safety factors of 3 in end bearing and unity in shaft friction. Calculate the immediate settlement at the working load.

A pile of 1 m diameter is suitable. A penetration depth of 13 m (12 m below cut-off level) will be tried.

Average shearing strength along pile shaft (from Figure 4.43)
$$= \tfrac{1}{2}(35 + 182) = 108.5 \text{ kN/m}^2.$$

Take an adhesion factor of 0.45. Then from equation 4.7,

total shaft friction on pile shaft $= 0.45 \times 108.5 \times \pi \times 1 \times 12 = 1841$ kN.

Because the clay is fissured it is desirable to reduce the average shearing strength at pile base level to obtain the end-bearing resistance. Thus

end-bearing resistance $= 9 \times 0.75 \times 182 \times \tfrac{1}{4}\pi \times 1^2 = 965$ kN.

For a safety factor of 2 on the total pile resistance,

allowable working load $= \tfrac{1}{2}(1\,841 + 965) = 1403$ kN.

For a safety factor of 3 on the base resistance and 1 on the skin friction,

allowable working load $= (\tfrac{1}{3} \times 965) + 1841 = 2163$ kN.

The alternative of a bored pile with an enlarged base is now considered. Take the same shaft diameter and a base diameter of 2 m, a base level for the pile being at 10 m below ground level. The top of the base enlargement will be at 9 m and the base resistance must be ignored for 2 m above this level (Figure 4.9).

Average shearing strength from cut-off level at 1 to 7 m $= \tfrac{1}{2}(35 + 111) = 73$ kN/m^2

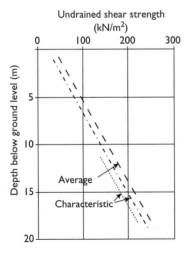

Figure 4.43 Characteristic and average shear strength values from four boreholes.

An adhesion factor of 0.3 should be used to allow for delays and softening effects while drilling the enlarged base. Thus

total shaft friction on pile shaft $= 0.3 \times 73 \times \pi \times 1 \times 6 = 413$ kN.

As before a reduction factor of 0.75 should be used on the shearing strength of 145 kN/m^2 at a depth of 10 m. Thus from equation 4.4,

end-bearing resistance $= 9 \times 0.75 \times 145 \times \frac{1}{4}\pi \times 2^2 = 3075$ kN.

For a safety factor of 2 on the total pile resistance,

allowable working load $= \frac{1}{2}(413 \times 3075) = 1744$ kN.

For a safety factor of 3 on base resistance and 1 on skin friction,

allowable working load $= (\frac{1}{3} \times 3075) + 413 = 1438$ kN.

Therefore a 1 m diameter bored pile with an enlarged base of 2 m diameter at a depth of 10 m below ground level is satisfactory for a working load of 1400 kN.

Considering the settlement of the straight-sided pile, the ultimate shaft friction of 1841 kN exceeds the working load, and therefore the settlement of the pile will be no more than that required to mobilize the ultimate resistance; i.e. a settlement of about 10 mm may be expected.

For the full mobilization of shaft resistance at the working load,

load on base $= 1400 - 413 = 987$ kN,

Unit pressure on base $= \dfrac{987}{\frac{1}{4}\pi \times 2^2} = 314$ kN/m^2,

Unit ultimate base resistance $= 9 \times 0.75 \times 145 = 979$ kN/m^2.

In the absence of plate bearing test data, take K in equation 4.36 as 0.02. Thus

$$\rho_i = 2 \times 0.02 \times \frac{314}{979} \times 1000 = 13 \text{ mm}.$$

This is an acceptable value.

However, in view of the efficiency of modern pile drilling equipment it is likely that the cost of under-reaming and concreting to form an enlarged base at 10 m will exceed the cost drilling the extra 3 m for a straight-sided pile.

Checking the design of the straight-sided pile for compliance with the EC7 requirements, from Table 4.9:

Factored pile diameter $= 0.95 \times 1.0 = 0.95$ m, giving area of base $= 0.71$ m^2 and area of shaft for 12 m penetration $= 35.81$ m^2.

It is assumed that the shear strength/depth realtionships in Figure 4.43 were based on an adequate number of boreholes and soil samples, and that the straight line graphs are a cautions estimate derived from the plotted data. Hence Figure 4.43 represents characteristic values and the correlation factors need not be applied.

Calculating the mean and minimum pile resistances:

$$R_{ck} \text{ (mean)} = 9 \times 148 \times 0.71 + 0.45 \times 97.5 \times 35.81$$
$$= 946 + 1572 = 2518 \text{ kN}$$
$$R_{ck} \text{ (min)} = 946 \times (135 / 148) + 1572 = 2435 \text{ kN}$$

Taking the minimum value and allowing for the stiffness of the building structure:

$$\text{Design } R_{ck} = 2435 \times 1.1 = 2678 \text{ kN}$$

For DA1 combination 1, the partial factor sets are A1 + M1 + R1; from Table 4.1 the A1 factors are 1.35 (permanent unfavourable) and 1.5 (variable unfavourable). Therefore design value of actions:

$$F_d = 1.35 \times 1100 + 1.5 \times 300 = 1935 \text{ kN}$$

From Table 4.2, the M1 factors are all unity and R1 for combined compression is 1.15. Therefore design value of resistance:

$$R_{cd} = 2678 / 1.0 \times 1.15 = 2329 \text{ kN which is greater than } F_d \text{ of 1935 kN}$$

For DA1 combination 2 (sets A2 + M1 + R4), the factors for permanent and variable actions are unity and 1.3 respectively. Therefore

$$F_d = 1.0 \times 1100 + 1.3 \times 300 = 1490 \text{ kN}$$

M1 (structural action) is unity and R4 is 1.6, therefore

$$R_{cd} = 2678 / 1.0 \times 1.6 = 1674 \text{ kN which is greater than } F_d \text{ of 1490 kN}$$

Checking from equation 4.39, taking the drained deformation modulus of the clay at pile base level as $140c_u$ (for long-term loading), we get

$$\rho = \frac{(413 + 2 \times 987) \times 8\,000}{2 \times 0.7854 \times 30 \times 10^6} + \frac{0.5 \times 987 \times 1\,000}{2 \times 140 \times 145}$$
$$= 0.4 + 12.2$$
$$= 13 \text{ mm}$$

Example 4.4

A precast concrete pile 450 mm square forming part of a jetty structure is driven into a medium dense over-consolidated sand. Standard penetration tests made in the sand gave an average value of N of 15 blows/300 mm. The pile is required to carry a compressive load of 250 kN and an uplift load of 180 kN. Determine the required penetration depth of the pile for a safety factor of at least 2.5 on both compressive and uplift loads.

The unit shaft friction developed on a pile in sand is rather low, and thus the penetration depth in this case is likely to be governed by the requirements for uplift resistance.

Take a trial penetration depth of 7 m below sea bed. From Figure 4.10 for $N = 15$, $\phi = 31°$. The submerged density of the sand may be taken as 1.2 Mg/m³. For an over-consolidated sand we can take $K_o = 1$. Table 4.10 gives $K_s/K_o =$ say 1.5, giving $K_s = 1.5$. From Table 4.11, take $\delta = 0.8\phi = 0.8 \times 31° = 24.8°$.

$$\text{Total shaft friction} = \tfrac{1}{2} \times 1.5 \times 1.2 \times 9.81 \times 7 \times \tan 24.8° \times 7 \times 4 \times 0.45$$

$$= 360 \text{ kN}$$

$$\text{Factor of safety on uplift load} = \frac{360}{180} = 2 \text{ which is satisfactory.}$$

Checking the base resistance using Berezantsev's value of N_q in equation 4.16, from Figure 4.13 with $\phi = 31°$, $N_q = 25$ ($D/B = 15$). Thus

$Q_b = 25 \times 1.2 \times 9.81 \times 7 \times 0.45^2 = 417$ kN.
Total pile resistance $= 417 + 360 = 777$ kN.

$$\text{Safety factor on compressive load} = \frac{777}{250} = 3.1,$$

which is satisfactory.

Example 4.5

Isolated piles are required to carry a working load of 900 kN on a site where borings and static cone penetration tests recorded the soil profile shown in Figure 4.44. Select suitable types of pile and determine their required penetration depth to carry the working load with a safety factor of 2.5. Previous tests in the area have shown that the ultimate base resistance of piles driven into the dense sand stratum is equal to the static cone resistance.

The piles will attain their bearing within the sand stratum. Any type of bored and cast-in-place pile will be uneconomical compared with the driven type. A driven and cast-in-place pile is suitable.

Driven and cast-in-place pile

For 25-grade concrete and a working stress of 25% of the works cube strength (BS 8004), working stress on concrete $= 0.25 \times 25 = 6.25$ N/mm². Therefore

$$\text{required diameter of pile shaft} = \sqrt{\frac{900 \times 1\,000 \times 4}{\pi \times 6.25}} = 428 \text{ mm}$$

Say nominal shaft diameter of 450 mm.

$$\text{Base pressure at ultimate load} = 900 \times 2.5/(\pi/4 \times 0.45^2 \times 10^3) = 14 \text{ MN/m}^2$$

For this value a penetration depth of about 29 m is required. Taking into account the available resistance in shaft friction, the likely penetration depth will be about 28 m. From Figure 4.44 and Table 4.12 take unit shaft friction, $f_s = 0.012\ q_c$

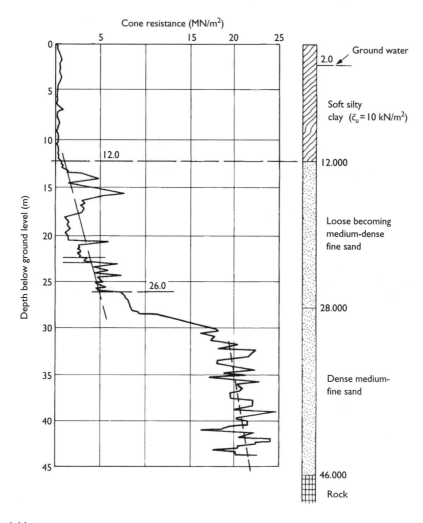

Figure 4.44

At 12 m, ultimate unit shaft friction resistance $= f_s = 0.012 \times 1 \times 10^3 = 12$ kN/m²
At 26 m, $f_s = 0.012 \times 5 \times 10^3 = 60$ kN/m²
At 28 m, $f_s = 0.012 \times 8 \times 10^3 = 96$ kN/m²

$$\text{Total shaft friction} = \frac{(12 + 60) \times 14 + (60 + 96) \times 2}{2} \times \pi \times 0.45 = 933 \text{ kN}$$

$$\text{Required ultimate base resistance} = \frac{(900 \times 2.5 - 933)}{\pi/4 \times 0.45^2 \times 10^3} = 8 \text{ MN/m}^2$$

This resistance is likely to be available at 28 m as an average over a distance of 8 pile diameters above the toe and 4 diameters below the toe. Over 4 diameters below the toe

$q_{c-1} = (7 + 15) \times 0.5 = 11$ MN/m^2, and over 8 diameters above the toe $q_{c-2} = (5 + 8) \times 0.5 = 6.5$ MN/m^2 as equation 4.19, giving average q_c at 28m of 8.75 MN/m^2. Therefore

Base resistance = $8.75 \times 0.45^2 \times 10^3 = 1\,392$ kN
Total pile resistance = $933 + 1392 = 2225$ kN.
Factor of safety = $2225 / 900 = 2.5$ (satisfactory).

Example 4.6

Calculate the ultimate bearing capacity of a 914 mm OD $\times$ 19 mm wall thickness tubular steel pile driven with a closed end to a depth of 17 m below ground level in the soil conditions shown in Figure 4.18a and compare the capacity with that of an open-end pile driven to the same depth.

Pile characteristics are: External perimeter 2.87 m
Internal perimeter 2.75 m
For 38 mm shoe, net cross-sectional area at toe = 0.1046 m^2.
Gross base area = 0.656 m^2.
    The shaft friction on the pile shaft is calculated from the characteristic curve shown in Figure 4.18a. Ignore any shaft friction in the soft clay. The coefficient from Table 4.12 is 0.008.
At 6.5 m ultimate unit shaft friction = $0.008 \times 4 = 0.032$ MN/m^2
At 14.5 m ultimate unit shaft friction = $0.008 \times 9 = 0.072$ MN/m^2
At 17.0 m ultimate unit shaft friction = $0.008 \times 9.9 = 0.079$ MN/m^2

Total external shaft friction 6.5 to 17.0 m
$= [(0.032 + 0.072) \times 8 + (0.072 + 0.079) \times 2.5] \times 0.5 \times 2.87 = 1.74$ MN

From average curve in Figure 4.18a, base resistance = $9.9 \times 0.656 = 6.49$ MN

Total ultimate pile resistance = $1.74 + 6.49 = 8.23$ MN

Figure 4.19 shows that the limiting value of about 10 MN/m^2 for base resistance in fine to coarse sand is not exceeded.
    The total pile resistance can be checked by converting the cone resistance to the equivalent angle of shearing resistance of the sand. Assume ground water level at ground level.

Effective overburden pressure at 17 m = $8 \times 6.5 + 10 \times 10.5 = 157$ kN/m^2

From Figure 4.11, for $q_c = 9.9$ MN/m^2, $\theta = 37°$. From Figure 4.13, $N_q = 70$.
    Therefore, ultimate unit bearing pressure = $70 \times 157 \times 10^{-3} = 11$ MN/m^2, which is close to the value taken directly from Figure 4.18a.
    Adopting a factor of safety of 2.5, the allowable load by permissible stress methods is: $8.23/2.5 = 3.29$ MN.

The allowable load can be checked for compliance with the EC7 recommendations. For an allowable load of 3.29 MN, the design actions are for set A1, $1.35 \times 3.29 = 4.44$ MN, and for set A2, $1.0 \times 3.29 = 3.29$ MN.

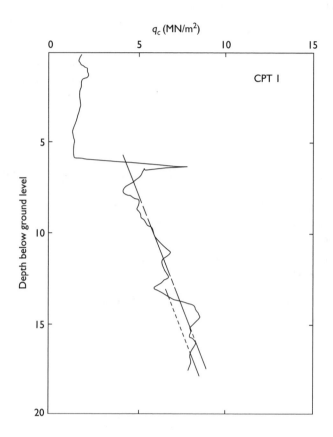

Figure 4.45

From plots of individual cone readings, shown typically for CPT 1 in Figure 4.45 the average and characteristic cone resistances over the length of the shaft and at the base are

	CPT1	CPT2	CPT3	CPT4	Average	Characteristic
Shaft, average q_c MN/m²	6.0	7.5	8.4	8.5	7.6	7.35
Base, min q_c MN/m²	8.2	9.7	10.1	10.7	9.7	9.3

For CPT1, Shaft friction $= 0.008 \times 6.0 \times 2.87 \times 10.5 = 1.45$ MN
  Base resistance $= 8.2 \times 0.656 = 5.38$ MN
  $R_{cal} = 5.38 + 1.45 = 6.83$ MN
  Similarly for CPT2: $R_{cal} = 6.36 + 1.81 = 8.17$ MN
        CPT3: $R_{cal} = 6.63 + 2.02 = 8.65$ MN
        CPT4: $R_{cal} = 7.02 + 2.05 = 9.07$ MN

The mean and minimum values of R_{cal} are

R_{cal} (mean) = (6.82 + 8.17 + 8.65 + 9.07)/4 = 8.18 MN
R_{cal} (min) = 6.82 for CPT1

The above values are divided by the correlation factors 1.31 (mean) and 1.20 (min) for four cone tests (Table 4.6) to give the characteristic values:

R_{ck} (mean) = 8.18/1.31 = 6.24 MN
R_{ck} (min) = 6.82/1.20 = 5.68 MN

The design value R_{cd} will be calculated from the minimum R_{ck} of 5.68 MN. A 'stiffness factor' will not be applied.

Using design approach DA1 combination 1 (sets A1 + M1 + R1) for a driven pile in compression the factors γ_b and γ_s for M1 and R1 sets are unity (Tables 4.2 and 4.5).

Therefore, R_{cd} = 5.68 MN which is greater than the design action for A1 of 4.44 MN.

For DA1 and combination 2 (sets A2 + M2 + R1) γ_b and γ_s are 1.25 (the value for ϕ') and R4 is 1.3. Therefore design R_{cd} = 5.68 / 1.25 × 1.3 = 3.49 MN, which is greater than the design action for A2 of 3.29 MN.

The above calculations show that the allowable load of 3.29 MN as calculated by permissible stress methods complies with the recommendations of EC7.

The ultimate resistance of the open end pile can be calculated by the ICP method. Assuming that the sand has a D_{50} size of 0.03 mm, Figure 4.21 gives a value of 27° for the interface angle of friction, δ_{cv}. From equation 4.24, the equivalent radius of the open-end pile is

$$R^* = (0.457^2 - 0.438^2)^{0.5} = 0.130 \text{ m}$$

The shear modulus G in equation 4.31 can be calculated from Figure 5.24 and equation 6.49. Figure 5.24 gives E_{50} = 30 MN/m². Take Poisson's ratio as 0.2, giving G = 30/2(1 + 0.2) = 12.5 MN/m²

Take the average roughness as 2 × 1 × 10⁻⁵ mm. From equation 4.30:

$$\Delta\sigma'_{rd} = 2 \times 12.5 \times 2 \times 10^{-5}/0.457 = 0.001 \text{ MN/m}^2$$

This is small in relation to σ'_{rc} as calculated below and can be neglected.

The 10.5 m of embedment into the sand is divided into 9 × 1.0 m segments and an uppermost segment of 1.5 m (the limiting height to the lowermost segment of 8.0 × 0.13 m = 1.04 m is not exceeded).

Calculating σ'_{rc} for the lowermost layer, the effective overburden pressure at the centre of the layer is (8 × 6.50 + 10 × 10.0) = 152 kN/m², and the average q_c is 9.5 MN/m². Take P_a = 100 kN/m². From equation 4.29, σ'_{rc} = 0.029 × 9.5 × (152/100)^{0.13} × (0.5/0.13)^{-0.38} = 0.174 MN/m². From equation 4.27:

Unit shaft resistance = 0.174 × tan 27° = 0.087 MN/m²
Shaft resistance on segment = 0.087 × 2.87 = 0.250 MN

The resistance of the remaining segments to the top of the sand layer is calculated in the same way as shown in the following table.

Depth of Segment (m bql)	$h(m)$	$\left(\dfrac{h}{R^*}\right)^{-0.38}$	$\left(\dfrac{\sigma'_{vo}}{p_o}\right)^{0.13}$	q_c (MN/m^2)	σ'_{rc} (MN/m^2)	$\tau_f = \sigma'_{rc} \tan \delta_{cv}$ (MN/m^2)	Q_s (MN)
17–16	0.5	0.599	1.055	9.5	0.174	0.087	0.250
16–15	1.5	0.395	1.047	8.9	0.107	0.053	0.152
15–14	2.5	0.325	1.037	8.4	0.082	0.041	0.118
14–13	3.5	0.286	1.026	7.9	0.067	0.033	0.095
13–12	4.5	0.260	1.015	7.4	0.057	0.028	0.080
12–11	5.5	0.241	1.003	6.8	0.048	0.024	0.069
11–10	6.5	0.226	0.989	6.3	0.041	0.020	0.057
10–9	7.5	0.214	0.974	5.8	0.035	0.017	0.049
9–8	8.5	0.204	0.958	5.2	0.029	0.015	0.043
8–6.5	7.25	0.217	0.935	4.4	0.026	0.013	0.037

Total Q_s = 0.95 MN

Calculating the base resistance, q_c at base = 9.7 MN/m^2 D_{inner}/D_{CPT} = 0.438/0.036 = 12.2 which is less than 0.83 × 9.7 × 10^3/100 = 80.5. Therefore a rigid basal plug will develop.

From equation 4.35, q_b = 9.7(0.5 − 0.25 log 914/36) = 1.45 MN/m^2 which is not less than 0.15q_c. Therefore Q_b = 1.45 × 0.656 = 0.95 MN, which is less than that of the unplugged pile (Q_b = 9.7 × 0.1046 = 1.01 MN.

Total pile resistance = 0.95 + 1.01 = 1.96 MN

Example 4.7

A bored and cast-in-place pile is required to carry a working load of 9000 kN at a site where 4 m of loose sand overlies a weak jointed cemented mudstone. Core drilling into the mudstone showed partly open joints and RQD values increased from an average of 15% at rockhead to 35% at a depth of 10 m. Tests on rock cores gave an average unconfined compression strength of 4.5 MN/m^2. Determine the required depth of the pile below rockhead and calculate the settlement of the pile at the working load.

The stress on the shaft of a 1.5 m pile is $\dfrac{9}{(\pi/4) \times 1.5^2}$ = 5.1 MN/m^2

This is satisfactory for concrete with a cube crushing strength of 25 MN/m^2 (allowable stress = 6 MN/m^2).

Load carried in shaft friction in the loose sand will be negligible.

From Figure 4.33 for q_{uc} = 4.5 MN/m^2, α = 0.2. The mass factor, j, for RQD from 15% to 35% is 0.2. Therefore, β, from Figure 4.34 is 0.65.

From equation 4.44, rock socket shaft friction = 0.2 × 0.65 × 4500

= 585 kN/m^2

Taking a 7 m socket length, ultimate shaft friction = 585 × π × 1.5 × 7 = 19297 kN

Therefore, factor of safety on shaft friction = $\dfrac{19297}{9000}$ = 2.1

Because of the open joints in the rock it will be advisable to assume that the base resistance does not exceed the unconfined compression strength of the rock.

Total base resistance = $\frac{\pi}{4} \times 1.5^2 \times 4500 = 7952$ kN

Total pile resistance $= 19297 + 7952 = 27249$ kN

Factor of safety $= \dfrac{27249}{9000} = 3.0$ which is satisfactory

If the rock socket skin friction were to be only half the calculated value, no load would be transferred to the pile base. Therefore, the pile head settlement will be caused by compression in the rock socket only.

From Section 5.5 the modulus ratio of a cemented mudstone is 150, and for a mass factor of 0.2 the deformation modulus of the rock mass is $0.2 \times 150 \times 4.5 = 135$ MN/m^2. In Figure 4.36 the modulus ratio E_c/E_d is $20 \times 10^3/135 = 148$ and for $L/B = 7/1.5 = 4.7$ the influence factor I is 0.25. The ratio D/B for a recessed socket is $4/1.5 = 2.7$. There the reduction factor from Figure 4.37 is about 0.8. Hence from equation 4.49:

$$\text{Pile head settlement} = \frac{0.8 \times 9 \times 10^3 \times 0.25}{1.5 \times 135} = 9 \text{ mm}$$

Checking the calculated shaft friction from equation 4.45 and take b as 0.25,

$$f_s = 0.25 \sqrt{4.5} = 0.53 \text{ MN/m}^2 \text{ which agrees closely with the value from equation 4.44.}$$

If the socket is grooved to an average depth of 25 mm over shortened socket length of 5.0 m with the grooves at vertical intervals of 0.75 m then:

In equation 4.47, $\Delta_r = 0.775 - 0.75 = 0.025$ m and total length of travel $= 4.7 \times 6.67 = 31.35$ m

From equation 4.47, RF $= 0.025 \times 31.33/0.75 \times 5.0 = 0.21$

From equation 4.46, $f_s = 0.8(0.21)^{0.45} \times 4.5 = 1.78$ MN/m^2

Total shaft friction on 5 m socket length $= 1.78 \times \pi \times 1.5 \times 5 = 42$ MN

Factor of safety in shaft friction $= 42/9 = 4.7$, therefore grooving the socket would theoretically provide a much shorter socket length than the 7 m required for an ungrooved shaft.

Example 4.8

A tubular steel pile with an outside diameter of 1067 mm is driven with a closed end to near refusal in a moderately strong sandstone (average $q_{uc} = 20$ MN/m^2) overlain by 15 m of soft clay. Core drilling in the rock showed a fracture frequency of 5 joints per metre. Calculate the maximum working load which can be applied to the pile and the settlement at this load.

Only a small penetration below rockhead will be possible with sandstone of this quality, and the rock will be shattered by the impact. Hence, frictional support both in the soft clay and the rock will be negligible compared with the base resistance.

Pile driving impact is likely to open joints in the rock hence the base resistance should not exceed the unconfined compression strength of the intact rock.

Total ultimate pile resistance = base resistance$= \frac{\pi}{4} \times 1.067^2 \times 20 = 17.9$ MN

For a safety factor of 2.5 allowed load $= \dfrac{17.9}{2.5} = 7.2$ MN

For a wall thickness of 19 mm in mild steel ($f_y = 240$ MN/m^2)
Allowed load $= 0.5 \times 240 \times 0.0626 = 7.5$ MN

Therefore take a maximum working load of 7.2 MN.

Pile driving impact may increase the fracture frequency from 5 to 10, say, fractures per metre giving a mass factor of 0.2. From Section 5.4 the modulus ratio of sandstone is 300.

Deformation modulus $= 0.2 \times 300 \times 20 = 1200$ MN/m^2

From equation 4.38:

$$\text{Settlement of pile head} = \frac{7.2 \times 15 \times 1000}{0.0626 \times 2 \times 10^5} + \frac{0.5 \times 7.2 \times 1000}{1.067 \times 1200}$$
$$= 8.6 + 2.8$$
$$= 11.4 \text{ mm } (say \text{ 10 to 15 mm})$$

Example 4.9

A 5 m layer of hydraulic fill consisting of sand is pumped into place over the ground shown in Figure 4.44. The calculated time/settlement curve for the surface of the hydraulic fill is shown in Figure 4.46. Two years after the completion of filling a closed-end steel cased pile with an outside diameter of 517 mm is driven to a penetration of 27 m to carry a working load of 900 kN. Calculate the negative friction which is developed on the pile shaft and assess whether or not any deeper penetration is required to carry the combined working load and negative skin friction.

It can be seen from the time/settlement curve that about 120 mm of settlement will take place from the time of driving the pile until the clay beneath the fill layer is fully consolidated. This movement is considerably larger than the compression of the pile head under the working load (about 10 mm of settlement would be expected under the working load of 900 kN). Therefore negative skin friction will be developed over the whole depth of the pile within the hydraulic fill. Considering now the negative shaft friction within the soft clay, if it is assumed that drag-down will not occur if the clay settles relatively to the pile by less than 5 mm, then adding the settlement of the pile toe (10 mm at the working load) negative

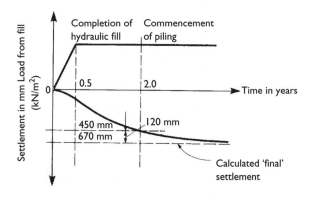

Figure 4.46

skin friction will not be developed below the point where the clay settles by less than 15 mm relative to site datum. After pile driving, the full thickness of the clay settles by 120 mm at the surface of the layer. By simple proportion, a settlement of 5mm occurs at a point $12 \times 15/120 = 1.5$ m above the base of the layer. This assumes uniform compressibility in the clay, but there is decreasing compressibility with increasing depth such that the settlement decreases to less than 15 mm at a point not less than 2 m above the base of the layer. A closer estimate could be obtained by a t–z analysis. However, the above approximate assessment will be adequate for the present case.

Adopting Meyerhof's factor from Figure 4.39 for the negative skin friction:

Unit negative skin friction 2 m above the base of clay layer
$= 0.3\sigma'_{vo} = 0.3 \times 9.81[(5 \times 2) + (2 \times 1.9) + (8 \times 0.9)]$
$= 62 \text{ kN/m}^2$

Unit negative skin friction at top of clay stratum
$= 0.3 \times 9.81 \times 5 \times 2 = 29 \text{ kN/m}^2$

Unit negative skin friction 2 m below top of clay stratum (at groundwater level)
$= 0.3 \times 9.81 \times [(5 \times 2) + (2 \times 1.9)] = 41 \text{ kN/m}^2$

Total negative skin friction in clay
$= \pi \times 0.517[\frac{1}{2}(29 + 41)2 + \frac{1}{2}(41 + 62) \times 8] = 783 \text{ kN}$

Drainage of the fill will produce a medium-dense state of compaction for which K_o is 0.45 and K_s in equation 4.16 is 0.67 (Table 4.10) and $\delta = 0.7 \times 30° = 20°$. Therefore

Negative skin friction
$= \frac{1}{2} \times 0.67 \times 9.81 \times 2 \times 5 \times \tan 21° \times \pi \times 0.517 \times 5$
$= 102 \text{ kN}$

Total negative skin friction on pile $= 102 + 783 = 885 \text{ kN}$
Total load on pile $= 885 + 900 = 1785 \text{ kN}$

From Example 4.5 the average CPT resistance at a penetration of 28 m is about 8.75 MN/m and the shaft frictional resistance from 12 to 28 m is $933 \times 517/450 = 1072 \text{ kN}$

Ultimate base resistance at 28 m $= \pi/4 \times 0.517^2 \times 8.75 \times 10^3 = 1837 \text{ kN}$
Total pile resistance $= 1072 + 1837 = 2909 \text{ kN}$

Safety factor on combined working load and downdrag $= 2909/1785 = 1.6$ (satisfactory).

The working stress on the pile shaft must be checked under the combined loading. For a wall thickness of 4.47 mm the steel area is 7193 mm^2.

Permissible load on steel at 30% yield stress
$= \dfrac{0.3 \times 255 \times 7193}{1000} = 550 \text{ kN}$

If the pile is filled with concrete with a working stress of 25% of the strength:

Allowable load on concrete $= \frac{1}{4}\pi \times 0.508^2 \times 0.25 \times 25 \times 1000$
$= 1267 \text{ kN}$

Total allowed load $= 1267 + 550 = 1817 \text{ kN}$.

If there is concern about long-term corrosion of the steel section in the hydraulic fill, the cube strength of the concrete filling could be increased so that the whole of the load is carried by the concrete.

Checking for compliance with the EC7 requirements, the pile is a concrete filled casing therefore dimensional tolerances are not applied. The shaft resistance is fully mobilized over the depth below the base of the fill. The partial action and resistance factors are selected from Table 4.20.

For design approach DA1, combination 1, the partial factors are: Structural action, A1 = 1.35. Drag-down action, A1 = 1.35, M1 = 1.0, A1 = 1.35. Resistance, R1 = 1.0.

Design value of actions = (900 + 885) × 1.35 = 2410 kN
Design resistance = 2909 × 1.0 = 2909 kN (greater than 2410 kN)

For design approach DA1, combination 2, Structural, A2 = 1.0. Drag-down, M2 = 1.25, A2 = 1.0. Resistance, R4 = 1.3.

Design value of actions = 900 × 1.0 + 885 × 1.25 = 2006 kN
Design resistance = 2909/1.3 = 2273 kN (greater than 2006 kN)

The above calculations show that the penetration of 28 m for the pile satisfies the requirements of EC7 in respect of the ultimate limit state GEO.

Chapter 5

Pile groups under compressive loading

5.1 Group action in piled foundations

The supporting capacity of a group of vertically loaded piles can, in many situations, be considerably less than the sum of the capacities of the individual piles comprising the group. In all cases the elastic and consolidation settlements of the group are greater than those of a single pile carrying the same working load as that on each pile within the group. This is because the zone of soil or rock which is stressed by the entire group extends to a much greater width and depth than the zone beneath the single pile (Figure 5.1). Even when a pile group is bearing on rock the elastic deformation of the body of rock within the stressed zone can be quite appreciable if the piles are loaded to their maximum safe capacity.

Group action in piled foundations has resulted in many recorded cases of failure or excessive settlement, even though loading tests made on a single pile have indicated satisfactory performance. A typical case of foundation failure is the single pile driven to a satisfactory set in a compact or stiff soil layer underlain by soft compressible clay. The latter formation is not stressed to any significant extent when the single pile is loaded (Figure 5.2a) but when the load from the superstructure is applied to the whole group, the stressed zone extends down into the soft clay. Excessive settlement or complete general shear failure of the group can then occur (Figure 5.2b).

The allowable loading on pile groups is sometimes determined by the so-called efficiency formulae, in which the efficiency of the group is defined as the ratio of the

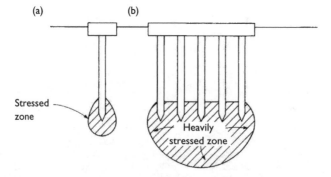

Figure 5.1 Comparison of stressed zones beneath single pile and pile group (a) Single pile (b) Pile group.

average load per pile when failure of the complete group occurs, to the load at failure of a single comparable pile. The various efficiency ratios are based simply on experience without any relationship to soil mechanics principles. For this reason the authors do not consider this to be a desirable or logical approach to the problem and prefer to base design methods on the assumption that the pile group behaves as a block foundation with a degree of flexibility which depends on the rigidity of the capping system and the superimposed structure. By treating the foundation in this manner, normal soil mechanics practice can be followed in the calculations to determine the ultimate bearing capacity and settlement. Load transfer in shaft friction from the pile shaft to the surrounding soil is allowed for by assuming that the load is spread from the shafts of friction piles at an angle of 1 in 4 from the vertical. Three cases of load transfer are shown in Figure 5.3a to c.

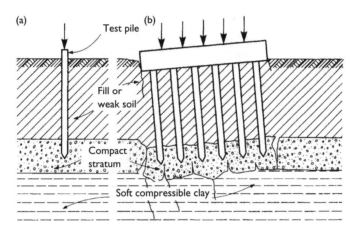

Figure 5.2 Shear failure of pile group (a) Test load on single isolated pile when soft clay is not stressed significantly (b) Load applied to group of piles when soft clay is stressed heavily.

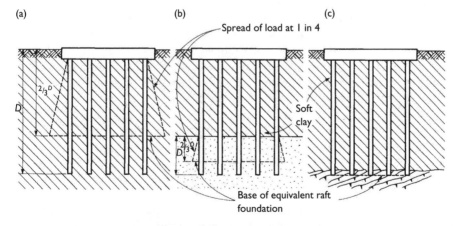

Figure 5.3 Load transfer to soil from pile group (a) Group of piles supported predominantly by shaft friction (b) Group of piles driven through soft clay to combined shaft friction and end bearing in stratum of dense granular soil (c) Group of piles supported in end bearing on hard rock stratum.

An important point to note in the application of soil mechanics methods to the design of pile groups is that, whereas in the case of the single pile the installation method has a very significant effect on the selection of design parameters for shaft friction and end bearing, the installation procedure is of lesser importance when considering group behaviour. This is because the zone of disturbance of the soil occurs only within a radius of a few pile diameters around and beneath the individual pile, whereas the soil is significantly stressed to a depth to or greater than the width of the group (Figure 5.1). The greater part of this zone is well below the ground which has been disturbed by the pile installation.

Computer programs have been established to model pile–soil interaction behaviour from which the settlement of pile groups and the loads on individual piles within the group can be determined.

Some of the programs are

DEFPIG   Non-linear continuum analysis using interaction factors
GAPFIX   Non-linear continuum analysis, complete solution
M-PILE   Simplified continuum analysis using interaction factors
PGROUP   Complete linear continuum analysis

In the above programs soil behaviour is modelled on the basis of the theory of elasticity. Poulos[5.1] states, 'Despite the gross simplification which this model involves when applied to real soil, it provides a useful basis for the prediction of pile behaviour provided that appropriate elastic parameters are selected for the soil. A significant advantage of using an elastic model for soil is that it provides a rational means of analysis of pile groups and evaluation of immediate and final movement of a pile. In determining immediate movements, the undrained elastic parameters of the soil are used in the theory, whereas for final movements the drained parameters are used'. A useful comparison of the M-PILE and PGROUP programs is given in the UK Department of Transport Publication BD 25/88. The interaction factors depend on the geometry, stiffness and spacing of the piles and the elastic modulus of the soil between them.

In view of the above reservations and the difficulties of obtaining representative values of the undrained and drained deformation parameters (particularly the latter) from field or laboratory testing of soils and rock, the authors believe that the equivalent raft method is sufficiently reliable for most day-to-day settlement predictions. Nevertheless, it could be convenient and time saving to use an available computer program particularly when making studies to determine the effect of varying parameters such as pile diameter, length and spacing.

In most practical problems piles are taken down to a stratum of relatively low compressibility and the resulting total and differential settlements are quite small such that an error of plus or minus 50% due to deficiencies in theory or unrepresentative deformation parameters need not necessarily be detrimental to the structure carried by the pile group (see also Section 11.1.4).

As an example of the relative accuracy of the methods Figure 5.4 shows a 4×4 pile group where the piles spaced at 3 diameters centre to centre are taken down to a depth of 24 m into a firm becoming stiff normally consolidated clay where the undrained shear strength and compressibility vary linearly with depth. The group settlements calculated by the equivalent raft method used the influence factors of Butler (Figure 5.19).

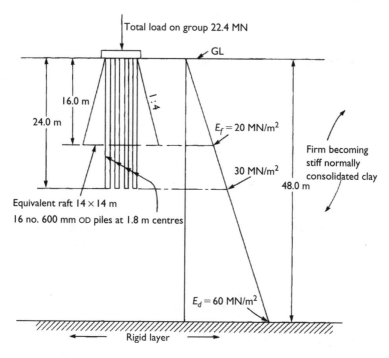

Figure 5.4 Pile group settlement by equivalent raft method.

The comparative group settlements were

DEFPIG   42 mm
PGROUP   31 mm
Equivalent raft   30 mm

The principal problems concerned with pile groups are constructional effects such as ground heave, the interference of closely spaced piles which have deviated from line during driving, and the possibilities of damage to adjacent structures and services. It is, of course, necessary to calculate the total and differential settlements of pile groups and overall piled areas to ensure that these are within limits acceptable to the design of the superstructure. The criteria of relative deflections, angular distortion, and horizontal strain which can be tolerated by structures of various types have been reviewed by Burland and Wroth[5.2].

When checking group settlement calculations to verify compliance with serviceability limit criteria, EC7 recommends a partial factor of 1.0 for actions and ground properties unless otherwise specified.

5.2   Pile groups in fine-grained soils

5.2.1   Ultimate bearing capacity

It is sometimes the practice for regulatory authorities to require that each pile in a group should be designed to carry a working load which has a conventional safety factor on its ultimate bearing capacity as determined for a given penetration by calculation, static load testing, or an empirical method. Burland[5.3] has stated his strong opinion that this requirement can result in grossly uneconomic foundation design, because it ignores the capability of a raft to redistribute loads from the superstructure on to the piles forming the group. Redistribution of loading can be permitted provided that

(1) The raft has sufficient flexibility (ductility) to perform this function without failure as a structural unit
(2) The superstructure has sufficient flexibility to accommodate any resulting movements in the raft
(3) The pile group has an adequate safety factor against failure or excessive settlement when considered as an equivalent block foundation and
(4) Account is taken of the effects of ground heave or subsidence of the mass of soil encompassed by the pile group during the construction stage (Section 5.7).

Burland recommends that redistribution should be effected by permitting piles carrying the heavier loading to mobilize their ultimate resistance in shaft friction, thereby yielding and transferring some of their load to surrounding piles within the group. This concept of 'ductile foundations' is discussed further in Section 5.10.

In all cases where piles are designed to transmit loading as a group terminating in a clay or sand stratum, whether or not some of the piles are permitted to yield, it is essential to consider the risks of general shear failure or excessive total and differential settlement of the equivalent block foundation taking the form shown in Figure 5.5.

The ultimate bearing capacity (ultimate limit state) of the block foundation as shown in Figure 5.5 can be calculated by using Brinch Hansen's general equation[5.4]. This was referred to in Section 4.3 with reference to the bearing capacity factor N_q in equation 4.16. The complete Brinch Hansen equation as applied to a shallow spread foundation embedded in soil with a level ground surface is

$$Q/A \quad \text{or} \quad q_u = cN_c\, s_c\, d_c\, i_c\, b_c + p_o\, N_q\, s_q\, d_q\, i_q\, b_q + 0.5\ \gamma B\, N_\gamma\, s_\gamma\, d_\gamma\, i_\gamma\, b_\gamma \tag{5.1}$$

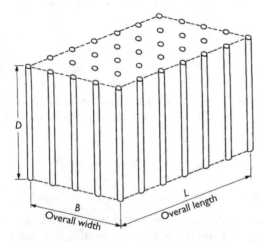

Figure 5.5 Pile group acting as block foundation.

where c = cohesion intercept of soil

N_c, N_q and N_γ = bearing capacity factors
s_c, s_q and s_γ = shape factors
d_c, d_q and d_γ = depth factors
i_c, i_q and i_γ = load inclination factors
b_c, b_q and b_γ = base inclination factors
γ = density of the soil
p_o = pressure of the overburden soil at foundation level.

For undrained conditions ($\phi = 0°$) the second term of the equation is omitted and c_u is substituted for c. For drained conditions, c' (the cohesion intercept in terms of effective stress) is used instead of c. Values of the factors in equation 5.1 are shown in Figures 5.6 to 5.10.

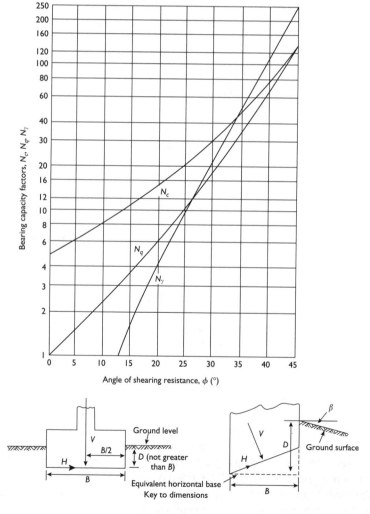

Figure 5.6 Bearing capacity factors N_c, N_q and N_γ (after Brinch Hansen[5.4]).

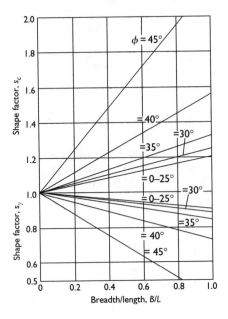

Figure 5.7 Shape factors s_c and s_γ (after Brinch Hansen[5.4]).

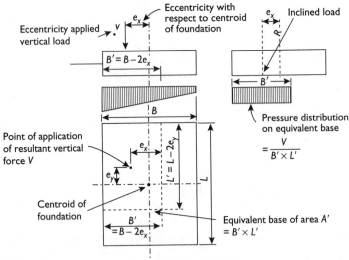

Figure 5.8 Transformation of eccentrically-loaded foundation to equivalent rectangular area carring uniformly distributed load.

The equation in similar form is given in EC7, Annex D, but the third term and the depth factors have been omitted, in the latter case because the application of the Eurocode is essentially for shallow spread foundations (D not greater than B). When applied to the block foundation equivalent to a pile group, the depth of the group could be large relative to its width. Thus by omitting the depth factors the value of q_u could be over-conservative. The use of equation 5.1 in checking for compliance with the EC7 recommendations is described in Section 5.4.

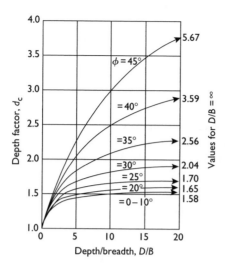

Figure 5.9 Depth factor d_c (after Brinch Hansen [5.4]).

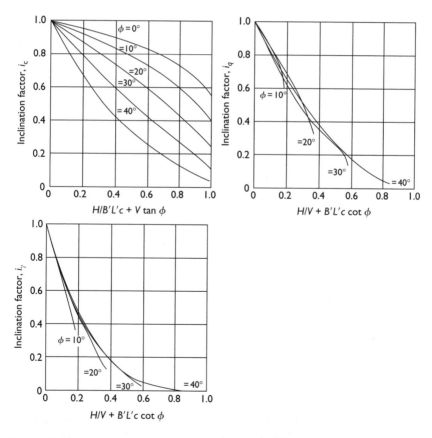

Figure 5.10 Inclination factors i_c, i_q and i_γ (after Brinch Hansen[5.4]).

Values of the shape factors s_c and s_γ for centrally applied vertical loading are obtained from Figure 5.7, and s_q from the equation:

$$s_q = \frac{s_c(s_c - 1)}{N_q} \tag{5.2}$$

Inclined loading is considered in relation to the effective breadth B' and the effective length L' of the equivalent block foundation. The plan dimensions of the block, as derived by Meyerhof[5.5], are shown in Figure 5.8. Thus for loading in the direction of the breadth:

$$B' = B - 2e_x \tag{5.3a}$$

where e_x is the eccentricity of loading in relation to the centroid of the base.

Similarly,

$$L' = L - 2e_y \tag{5.3b}$$

The shape factors s are modified for inclined loading by the equations:

$$s_{CB} = 1 + 0.2i_{CB}\, B'/L' \tag{5.4}$$

$$s_{CL} = 1 + 0.2i_{CL}\, L'/B' \tag{5.5}$$

$$s_{qB} = 1 + \sin\phi i_{qB}\, B'/L' \tag{5.6}$$

$$s_{qL} = 1 + \sin\phi i_{qL}\, L'/B' \tag{5.7}$$

$$s_{\gamma B} = 1 - 0.4i_{\gamma B}\, B'/L' \tag{5.8}$$

$$s_{\gamma L} = 1 - 0.4i_{\gamma L}\, L'/B' \tag{5.9}$$

Where B' is less than L', approximate values of the shape factors for centrally applied vertical loading which are sufficiently accurate for most practical purposes are

Shape of base	s_c	s_q	s_γ
Continuous strip	1.0	1.0	1.0
Rectangle	1 + 0.2 B/L	1 + 0.2 B/L	1 − 0.4 B/L
Square	1.3	1.2	0.8
Circle (diameter B)	1.3	1.2	0.6

Values of the depth factor d_c are obtained from Figure 5.9. The values on the right-hand side of the figure are for $D = \infty$. d_q is obtained from

$$d_q = \frac{d_c - 1}{N_q} \tag{5.10}$$

The depth factor d_γ can be taken as unity in all cases, also when $\phi = 0°$, $d_q = 1.0$. Where ϕ is greater than $25°$ d_q can be taken as equal to d_c. A simplified value of d_c and d_q where ϕ is less than $25°$ is $1 + 0.35\, D/B$. The use of the depth factors assumes that the soil above foundation level is not significantly weaker in shear strength than that of the soil below this level. However, in the case of pile groups, the piles are usually taken down through weak soils into stronger material, when either the depth factors should not be used or the depth

D should be taken as the penetration depth of the piles into the bearing stratum. Values of the load inclination factors i_c, i_q, and i_γ are shown in Figure 5.10 in relation to ϕ and the effective breadth B' and length L' of the foundation. Simplified values where the horizontal load H is not greater than $V \tan \delta + cB'L'$, and where c and δ are the parameters for cohesion and friction respectively of the soil beneath the base are given by the following equations:

$$i_c = 1 - \frac{H}{2cB'L'} \qquad\qquad (5.11)$$

$$i_q = 1 - \frac{1.5H}{V} \qquad\qquad (5.12)$$

$$i_\gamma = i_q^2 \qquad\qquad (5.13)$$

Equation 5.13 is strictly applicable only for $c = 0$ and $\phi = 30°$ but Brinch Hansen advises that it can be used for other value of ϕ.

The base of an equivalent block foundation, i.e. pile toe level, is usually horizontal but where piles are terminated on a sloping bearing stratum, the base of the block can be treated as horizontal at a depth equal to that of the lowest edge and bounded by vertical planes through the other three edges (Figure 5.6). The base factors b_c, b_q and b_γ are unity for a horizontal base.

It is evident from the foregoing account of the application of the Brinch Hansen equation that it is not readily adaptable from its original use in the design of relatively shallow spread foundations to deep pile groups subjected to appreciable transverse loading. In such cases it is preferable to use a computer program which can simulate interaction between the piles and the surrounding soil and can give a visual display of the extent of any overstressed zones in the soil below the group. Further aspects of group behaviour under transverse loading are discussed in Section 6.4.

Equation 5.1 ignores friction on the sides of the block foundation. The contribution of side shear is only a small proportion of the total where piles are taken down through a weak soil into a stronger stratum. In cases of marginal stability side shear resistance can be calculated as the shear resistance on a soil to soil interface on the sides of the group.

Where piles are installed in relatively small numbers there is a possibility of excessive base settlement if two or more piles deviate from line and come into near or close contact at the toe and the toe loads are concentrated over a small area. While failure would not occur if the safety factor in end bearing was adequate, the settlement would be higher than that which would occur when the piles were at their design spacing. This would lead to differential settlement between the piles in the group. A safeguard against this occurrence is the adoption of a centre-to-centre spacing of piles in clay of at least three pile diameters, with a minimum of 1 m. BS 8004 recommends a centre-to-centre spacing for friction piles of not less than the perimeter of the pile or for circular piles three times the diameter. Closer spacing can be adopted for piles carrying their load mainly in end bearing but the space between adjacent piles must not be less than their least width. Special consideration must be given to the spacing of piles with enlarged bases, including a study of interaction of stresses and the effect of construction tolerances. German practices for driven piles and for bored piles are shown in Figures 5.11 and 5.12 respectively, (see also BS EN 1536 and EN 12699).

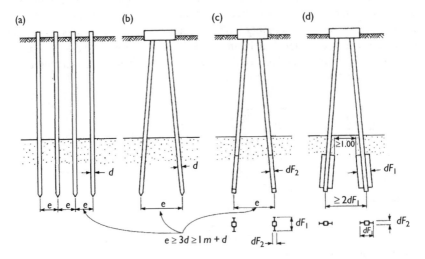

Figure 5.11 German practice for spacing of driven piles in groups (a) Vertical piles (b) Raking piles (c) and (d) Raking winged piles.

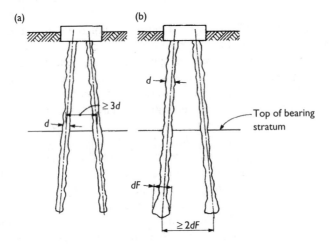

Figure 5.12 German practice for spacing of bored and cast in-situ piles in groups (a) Straight-sided piles (b) Piles with enlarged bases.

5.2.2 Settlement

The first step in the settlement analysis is to determine the vertical stress distribution below the base of the equivalent block foundation (Figure 5.3) using the curves shown in Figure 5.13, where the stress at any depth z below this level is related to its length/breath ratio. The curves assume that the foundation is rigid, but it is sufficiently accurate to assume that the superstructure, pile cap, piles and soil surrounding them have the required degree of rigidity.

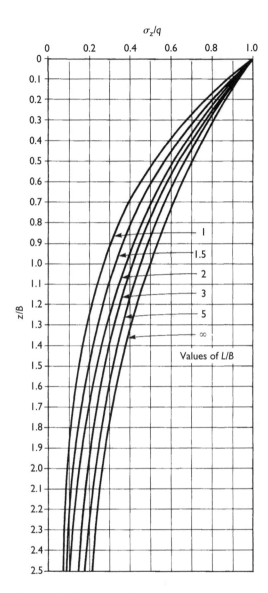

Figure 5.13 Calculation of mean vertical stress (σ_z) at depth z beneath rectangular area $a \times b$ on surface loaded at uniform pressure q.

The second step is to determine the depth of soil over which the stresses transmitted by the block foundation are significant. This is usually taken as the depth at which the vertical stress resulting from the net pressure at foundation level has decreased to 20% of the net overburden pressure at that level (Figure 5.14). A deeper level should be considered for soft highly compressible alluvial clays and peats.

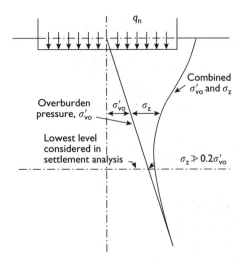

Figure 5.14 Vertical pressure and stress distribution for deep clay layer.

The third step is to calculate the settlement of the foundation which takes place in two phases. The first is immediate settlement (ρ_i) caused by elastic compression of the soil without dissipation of pore pressure. It is followed by consolidation settlement (ρ_c) which takes place over the period of pore pressure dissipation at a rate which depends upon the permeability of the soil. There is also the possibility of very long-term secondary settlement (ρ_∞) or creep of the soil. In the case of the very soft soils referred to in the previous paragraph, secondary settlement could be a significant proportion of the total. The equation for calculating it is given at the end of Section 5.2.

The net immediate settlement of foundations on clays is calculated from the equation:

$$\rho_i = \frac{q_n \times B \times (1 - v^2) \times I_p}{E_u} \tag{5.14}$$

where q_n = net foundation pressure
   B = foundation width
   v = Poisson's ratio
   E_u = undrained deformation modulus
   I_p = influence factor

E_v (or for drained conditions E'_v) can be obtained by one or more of the following methods:

(1) From the stress–strain curves established in the field by plate-bearing tests (Figure 5.15)
(2) From drained triaxial compression tests on good quality samples (to obtain E'_v)
(3) From oedometer tests to obtain the modulus of volume compressibility (m_v), when E'_v is the reciprocal of m_v and

(4) From relationships with the shear modulus (G) obtained in the field by pressuremeter tests:

$$E = 2G(1 + v_u)$$

$E = 2G(1 + v')$, where v_u and v' are the undrained and drained values of Poisson's ratio respectively.

With regard to method (1) a typical stress/strain curve obtained by a plate-bearing test in undrained conditions is shown in Figure 5.15. Purely elastic behaviour occurs only at low stress levels (line AB in Figure 5.15). Adoption of a modulus of elasticity (Young's modulus) corresponding to AB could result in under-estimating the settlement. The usual procedure is to draw a secant AC to the curve corresponding to a compressive stress equal to the net foundation pressure at the base of the equivalent block foundation. More conservatively the secant AD can be drawn at a compressive stress of 1.5 times or some other suitable multiple of the foundation pressure. The deformation modulus E_u is then obtained as shown in Figure 5.15.

As an alternative to direct determination of E_u from field tests, it can be obtained from a relationship with the undrained shear strength c_u, the plasticity index and over-consolidation ratio of the clay established by Jamiolkowski et al.[5.6] (Figure 5.16). The latter value is derived from oedometer tests or from a knowledge of the geological history of the deposit.[5.7] These tests are used to calculate the long-term consolidation settlement of the foundation as described below. Knowing the oedometer settlement (ρ_{oed}) provides another way of determining the immediate, consolidation and final settlements using the following relationships established by Burland et al.[5.8]

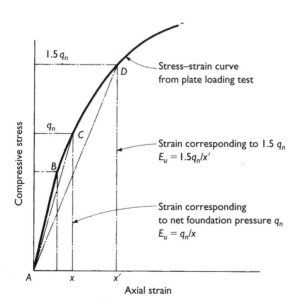

Figure 5.15 Determining deformation modulus E_u from stress/strain curve.

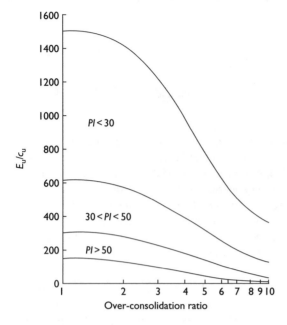

Figure 5.16 Relationship between E_u/c_u ratio for clays with plasticity index and degree of over-consolidation (after Jamiolkowski et al.[5.9]).

For stiff over-consolidated clays:

$$\text{immediate settlement} = \rho_i = 0.5 \text{ to } 0.6\rho_{oed} \tag{5.15}$$

$$\text{consolidation settlement} = \rho_i = 0.4 \text{ to } 0.5\rho_{oed} \tag{5.16}$$

$$\text{final settlement} = \rho_{oed} \tag{5.17}$$

For soft normally consolidated clays:

$$\text{immediate settlement} = 0.1 = \rho_{oed} \tag{5.18}$$

$$\text{consolidation settlement} = \rho_{oed} \tag{5.19}$$

$$\text{final settlement} = \rho_{oed} \tag{5.20}$$

The E_u/c_u ratio is also strain dependent showing a reduction in the ratio with increasing strain. Jardine *et al.*[5.9] showed this effect in London Clay from the results of undrained triaxial tests on good quality samples (Figure 5.17). Normally loaded foundations, including pile groups, usually exhibit a strain of 0.01% to 0.1%, which validates the frequently used relationship $E_u = 400c_u$ for the deformation modulus of intact blue London Clay.

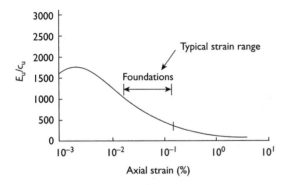

Figure 5.17 Relationship between E_u/c_u and axial strain (after Jardine et al.[5.9]).

Table 5.1 Poisson's ratio for various
soils and rocks

Clays (undrained)	0.5
Clays (stiff, drained)	0.1–0.3
Silt	0.3
Sands	0.1–0.3
Rocks	0.2

Marsland [5.10] obtained E_u/c_u equal to 348 for an upper glacial till, and 540 for a laminated glacial clay at Redcar, North Yorks.

The influence factor I_p in equation 5.14 is obtained from Steinbrenner's curves (Figure 5.18) using the method developed by Terzaghi [5.11]. Values of F_1 and F_2 in Figure 5.18 are related to the Poisson's ratio (v) of the foundation soil. For a ratio of 0.5, $I_p = F_1$. When the ratio is zero, $I_p = F_1 + F_2$. Some values of Poisson's ratio are shown in Table 5.1.

When using the curves in Figure 5.18 to calculate the immediate settlement of a flexible pile group, the square or rectangular area in Figure 5.5 is divided into four equal rectangles. Equation 5.14 then gives the settlement at the corner of each rectangle. The settlement at the centre is then equal to 4 times the corner settlement. In the case of a rigid pile group such as a group with a rigid cap or supporting a rigid superstructure, the settlement at the centre of the longest edge (twice the corner settlement) is obtained and the average settlement of the group obtained from the equation:

$$\rho_{average} = (\rho_{centre} + \rho_{corner} + \rho_{centre\ long\ edge})/3 \tag{5.20a}$$

These calculations can be performed by computer using a program such as VDISP in the OASYs GEO suite.

The curves in Figure 5.18 assume that E_u is constant with depth. Calculations based on a constant value can over-estimate the settlement. Usually the deformation modulus in soils and rocks

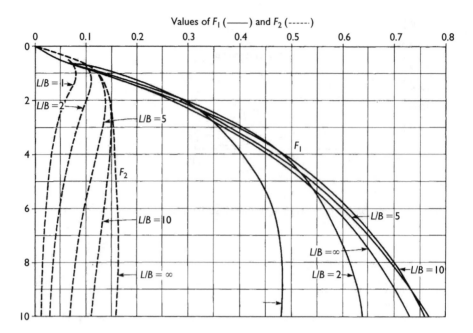

Figure 5.18 Values of Steinbrenner's influence factor I_p (for v of 0.5, $I_p = F_1$, for $v = 0, I_p = F_1 + F_2$.

Note

When using this diagram to calculate p_i at the centre of a rectangular area take B as half the foundation width to obtain H/B and L/B.

increases with depth. For materials with a linear increase, Butler[5.12] developed a method based on the research of Brown and Gibson[5.13], for calculating settlements where E_u or E_v' increases linearly with depth through a layer of finite thickness. The value of the modulus at any depth z below the base of the equivalent block foundation is given by the equation:

$$E = E_f(1 + kz/B) \qquad (5.21)$$

where E_f is the modulus at the base of the equivalent foundation.

To obtain k, values of E_u or E_v' obtained by one or more of the methods listed above are plotted against depth and a straight is drawn through the plotted points. The value of k is then obtained using Figure 5.19 which also shows the values of the influence factor I_p. The curves in this figure are based on normally consolidated clays having a Poisson's ratio of 0.5 and are appropriate to a compressible layer of thickness not greater than 9 times the breadth of the foundation. For a rigid pile group the immediate settlement as calculated for a *flexible* pile group is multiplied by a factor of 0.8 to obtain the *average* settlement of the rigid group, and a depth factor is applied using the curves in Figure 5.20.

Where a piled foundation consists of a number of small clusters of piles or individual piles connected by ground beams or a flexible ground floor slab the foundation arrangement can be considered as flexible.

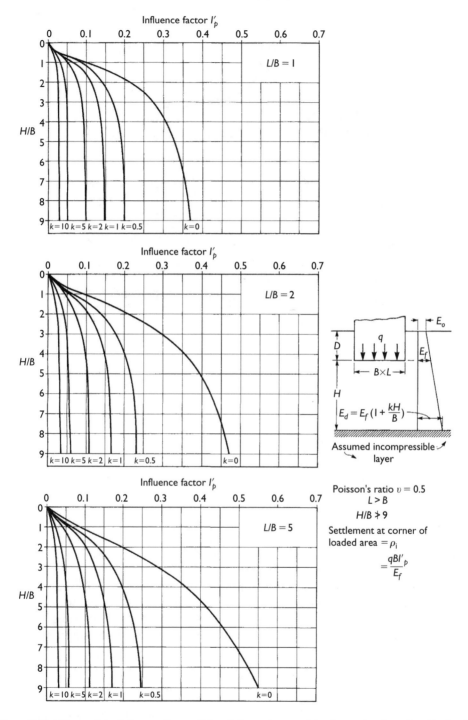

Figure 5.19 Values of the influence factor I'_p for deformation modulus increasing linearly with depth and modular ratio of 0.5 for normally consolidated days (after Butler[5.12]).

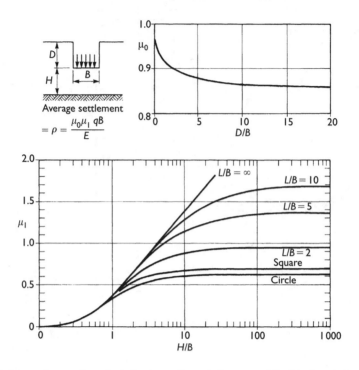

Figure 5.20 Influence factors for calculating immediate settlements of flexible foundations of width B at depth D below ground surface (after Christian and Carrier[5.14]).

When making a settlement analysis for a pile group underlain by a layered soil strata with different, but progressively increasing modulus values with depth, the strata are divided into a number of representative horizontal layers. An average modulus value is assigned to each layer. The dimensions L and B in Figure 5.20 are determined for each layer on the assumption that the vertical stress is spread to the surface of each layer at an angle of 30° from the edges of the equivalent block foundation (Figure 5.21). The total settlement of the piled foundation is then the sum of the average settlements calculated for each layer.

The foregoing procedure is referred to in EC7, Annex F, as the *Stress–Strain* method. The other procedure described in Annex F is the *Adjusted Elasticity* method. A typical example of the latter is the use of the Christian and Carrier[5.14] influence factors shown in Figure 5.20. These give the average settlement of the pile group from the equation:

$$\text{Average settlement} = \rho_i = \mu_1 \mu_0 q_n B / E_u \qquad (5.22)$$

In the above equation Poisson's ratio is taken as 0.5. The influence factors μ_1, and μ_0 are related to the depth and the length/breadth ratio of the equivalent block foundation, and the thickness of the compressible layer as shown in Figure 5.20. E_u is obtained by means of one or more of the methods listed above.

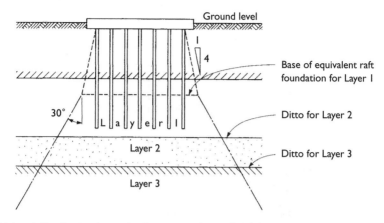

Figure 5.21 Load distribution beneath pile group in layered soil formation.

The *consolidation* settlement ρ_c is calculated from the results of oedometer tests made on clay samples in the laboratory. The curves for the pressure/voids ratio obtained from these tests are used to establish the coefficient of volume compressibility m_v.

In hard glacial tills or weak highly weathered rock it may be difficult to obtain satisfactory undisturbed samples for oedometer tests. If the results of standard penetration tests are available, values of m_v (and also c_u) can be obtained from empirical relationships established by Stroud[5.7] shown in Figure 5.22.

Having obtained a representative value of m_v for each soil layer stressed by the pile group, the *oedometer settlement* ρ_{oed} for this layer at the centre of the loaded area is calculated from the equation:

$$\rho_{oed} = \mu_d m_v \times \sigma_z \times H \tag{5.23}$$

where μ_d is a depth factor, σ_z is the average effective vertical stress imposed on the soil layer due to the net foundation pressure q_n at the base of the equivalent raft foundation and H is the thickness of the soil layer. The depth factor μ_d is obtained from Fox's correction curves[5.15] shown in Figure 5.23. To obtain the average vertical stress σ_z at the centre of each soil layer the coefficients in Figure 5.13 should be used. The oedometer settlement must now be corrected to obtain the field value of the consolidation settlement. The correction is made by applying a 'geological factor' μ_g to the oedometer settlement, where

$$\rho_c = \mu_g \times \rho_{oed} \tag{5.24}$$

Published values of μ_g have been based on comparisons of the settlement of actual structures with computations made from laboratory oedometer tests. Values established by Skempton and Bjerrum[5.16] are shown in Table 5.2.

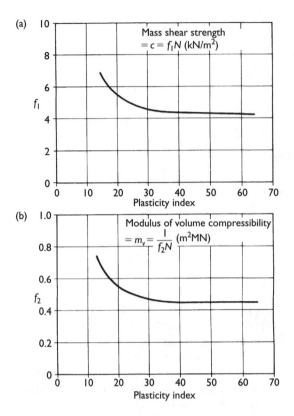

Figure 5.22 Relationship between mass shear strength, modulus of volume compressibility, plasticity index, and standard penetration test N-values (after Stroud[5.7]) (a) N-value versus undrained shear strength (b) N-value versus modulus of volume compressibility.

The total settlement of the pile group is then the sum of the immediate and consolidation settlements calculated for each separate layer. A typical case is a gradual decrease in compressibility with depth. In such a case the stressed zone beneath the pile group is divided into a number of separate horizontal layers, the value of m_v for each layer being obtained by plotting m_v against the depth as determined from the laboratory oedometer tests. The base of the lowermost layer is taken as the level at which the vertical stress has decreased to one-tenth of q_n. The depth factor μ_d is applied to the sum of the *consolidation* settlements calculated for each layer. It is not applied to the immediate settlement if the latter has been calculated from the factors in Figure 5.20.

Another method of estimating the *total* settlement of a structure on an *over-consolidated* clay is to use equation 5.14, making the substitution of a deformation modulus obtained for loading under drained conditions. This modulus is designated by the term $E_v{}'$, which is substituted for E_u in the equation. It is approximately equal to $1/m_v$. The equation implies a homogeneous and elastic material and thus it is not strictly valid when used to calculate consolidation settlements. However, when applied to over-consolidated clays for which the settlements are relatively small, the method has been found by experience to give reasonably reliable predictions. Success in using the method depends on the collection of sufficient data

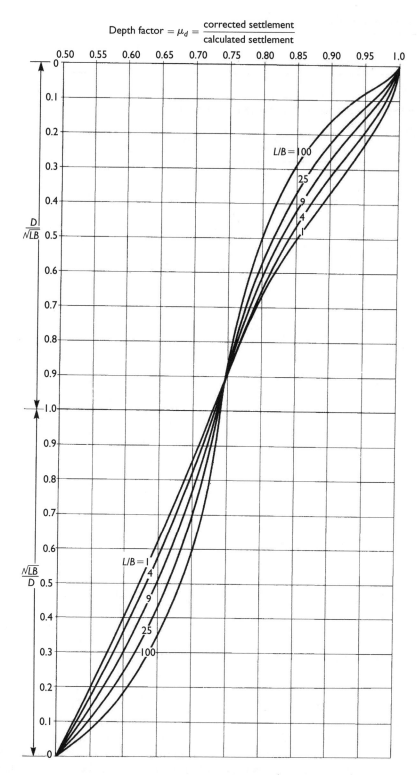

Figure 5.23 Depth factor μ_d for calculating oedometer settlements (after Fox[5.15]).

Table 5.2 Value of geological factor μ_g

Type of clay	μ_g value
Very sensitive clays (soft alluvial, estuarine, and marine clays)	1.0–1.2
Normally consolidated clays	0.7–1.0
Over-consolidated clays (London Clay, Weald, Kimmeridge, Oxford, and Lias Clays)	0.5–0.7
Heavily over-consolidated clays (unweathered glacial till, Mercia Mudstone)	0.2–0.5

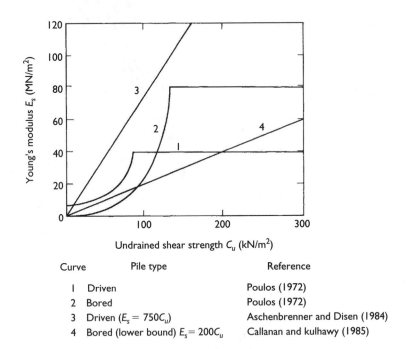

Curve	Pile type	Reference
1	Driven	Poulos (1972)
2	Bored	Poulos (1972)
3	Driven ($E_s = 750C_u$)	Aschenbrenner and Disen (1984)
4	Bored (lower bound) $E_s = 200C_u$	Callanan and kulhawy (1985)

Figure 5.24 Correlations for soil modulus for piles in clay (after Callanan and Kulhawy, for references see Poulos[5.1]).

correlating the observed settlements of structures with the determinations of E'_v from plate loading tests and laboratory tests on good undisturbed samples of clay. Butler[5.12] in his review of the settlement of structures on over-consolidated clays has related E'_v to the undrained cohesion c_u and arrived at the relationship $E'_v = 130c$ for London Clay.

Various correlations between the soil modulus and the undrained shear strength of clays for piles with a length to diameter ratio equal to or greater than 15 are shown in Figure 5.24. In commenting on these data Poulos[5.1] stated that they should be taken as representing values of the undrained modulus. He commented on the wide spread of the data suggesting that this could be due to differences in the method of measuring c_u and the soil modulus, differences in the level of loading at which the modulus was measured, and differences between the type and over-consolidation ratio of the various clays. Where the undrained shear strength increases linearly with depth, equation 5.21 can be used to obtain E'_v and hence the total settlements from Figure 5.19. From an extensive review of published and unpublished data, Burland and Kalra[5.17] established the relationship for London Clay: $E'_v = 7.5 + 3.9z$ (MN/m²) where z is the depth below ground level.

Generally, the authors prefer the more logical method of considering immediate and consolidation settlements separately. This properly takes into account time effects and the geological history of the site. Provided that a sufficient number of good undisturbed samples have been obtained at the site investigation stage, the prediction of consolidation settlements from oedometer tests made in the laboratory has been found to lead to reasonably accurate results. The adoption of the method based on the total settlement deformation modulus depends on the collection of adequate observational data, first regarding the relationship between the undrained shearing strength and the deformation modulus, and secondly regarding the actual settlement of structures from which the relationships can be checked. Any attempt to obtain a deformation modulus from triaxial compression tests in the laboratory is likely to result in serious error. The modulus is best obtained from the E_u/c_u and E_u/E'_v relationships, which must be established from well-conducted plate bearing tests and field observations of settlement.

The steps in making a settlement analysis of a pile group in, or transmitting stress to, a fine-grained soil can be summarized as follows:

(1) For the required length of pile, and form of pile bearing (i.e. friction pile or end-bearing pile), draw the equivalent flexible raft foundation represented by the group (see Figure 5.3)
(2) From the results of field or laboratory tests assign values to E_u and m_v for each soil layer significantly stressed by the equivalent raft
(3) Calculate the immediate settlement of ρ_i of each soil layer using equation 5.22, and assuming a spread of load of 30° from the vertical to obtain q_n at the surface of each layer (Figure 5.21). Alternatively calculate on the assumption of a linearly increasing modulus
(4) Calculate the consolidation settlement ρ_c for each soil layer from equations 5.23 and 5.24, using Figure 5.13 to obtain the vertical stress at the centre of each layer
(5) Apply a rigidity factor to obtain the average settlement for a rigid pile group.

The consolidation settlement calculated as described above is the final settlement after a period of some months or years after the completion of loading. It is rarely necessary to calculate the movement at intermediate times, i.e. to establish the time/settlement curve, since in most cases the movement is virtually complete after a period of a very few years and it is only the final settlement which is of interest to the structural engineer. If time effects are of significance, however, the procedure for obtaining the time/settlement curve can be obtained from standard works of reference on soil mechanics.

5.3   Pile groups in coarse-grained soils

5.3.1   Estimating settlements from standard penetration tests

Where piles are driven in groups to near-refusal into a dense sand or gravel it is unlikely that there will be sufficient yielding of individual piles under working load to permit redistribution of superstructure loading to surrounding piles as described in Section 5.2.1. Sufficient yielding to allow redistribution may occur where bored pile groups are terminated in sand, or where piles are driven to a set pre-determined from loading tests to allow a specified amount of settlement under working loads.

Provided that the individual pile has an adequate safety factor against failure under compressive loading there can be no risk of the block failure of a pile group terminated in and applying stress to a coarse soil. As in the case of piles terminated in a clay, there is a risk of differential settlement between adjacent piles in small groups if the toe loads of a small group become concentrated in a small area when the piles deviate from their intended line. The best safeguard against this occurrence is to adopt a reasonably wide spacing between the piles. Methods of checking the deviation of piles caused by the installation method are described in Chapter 11.

The immediate settlement of the pile group due to 'elastic' deformation of the coarse soil beneath the equivalent flexible raft foundation must be calculated. Equation 5.22 is applicable to this case and the deformation modulus E_v' is substituted for E_u as obtained from plate loading tests in trial pits, or from standard penetration, pressuremeter or Camkometer tests, made in boreholes. Schultze and Sherif[5.18] used case histories to establish a method for predicting foundation settlements from the results of standard penetration tests using the equation:

$$\rho = \frac{s \cdot p}{N^{0.87}(1 + 0.4D/B)} \tag{5.25}$$

where s is a settlement coefficient

p is the applied stress at foundation level

N is the average SPT N-value over a depth of $2B$ below foundation level, or d_s if the depth of cohesion-less soil is less than $2B$

D and B are the foundation depth and width respectively

Values of the coefficient s and d_s are obtained from Figure 5.25.

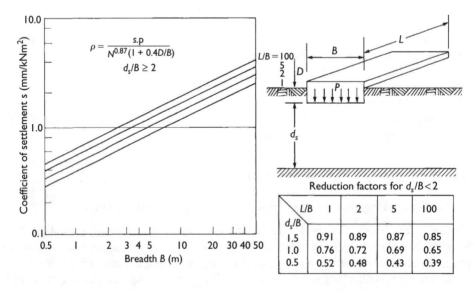

Figure 5.25 Determining foundation settlements from results of standard penetration tests (after Schultze and Sherif[5.18]).

Burland and Burbidge[5.19] have developed an empirical relationship between standard penetration test N-values and a term they have called the foundation subgrade compressibility, a_f. This term is used in the equations:

$$I_c = \frac{a_f}{B^{0.7}} \tag{5.26}$$

and

$$a_f = \frac{\Delta_{pi}}{\Delta_q} \quad \text{(in mm per kN/m}^2) \tag{5.27}$$

where I_c is a compressibility index
      B is the foundation width
      Δ_{pi} is the immediate settlement in mm
      Δ_q is the increment of foundation pressure in kN/m²

I_c and a_f are related to the standard penetration test results shown in Figure 5.26 for normally consolidated granular soils. In very fine and silty sands below the water table where N is greater than 15 the Terzaghi and Peck correction factor should be applied, giving

$$N \text{ (corrected)} = 15 + 0.5(N - 15) \tag{5.28}$$

Where the material is gravel or sandy gravel, Burland and Burbidge recommend a correction:

$$N \text{ (corrected)} = 1.25 \, N \tag{5.29}$$

It should be noted that the I_c values in Figure 5.26 are based on the average N-values over the depth of influence, z_I, of the foundation pressure. The depth of influence is related to the width of the loaded breadth B in Figure 5.27 for cases where N increases or is constant with depth. Where N shows consistent decrease with depth, z_f is taken as equal to $2B$ or the base of the compressive layer, whichever is the lesser. The average N in Figure 5.26 is the arithmetic mean of the N-values over the depth of influence.

In a normally consolidated sand the immediate average settlement, ρ_i, corresponding to the average net applied pressure, q', is given by

$$\rho_i = q' \times B^{0.7} \times I_c \text{ (in mm)} \tag{5.30}$$

In an over-consolidated sand or for loading at the base of an excavation for which the maximum previous overburden pressure was σ_{vo} and where q' is greater than σ_{vo}, the immediate settlement is given by

$$\rho_i = \left(q' - \frac{2}{3}\sigma_{vo} \right) B^{0.7} \times I_c \text{ (in mm)} \tag{5.31a}$$

where q' is less than σ_{vo} equation 5.31a becomes

$$\rho_i = q' \times B^{0.7} \times \frac{I_c}{3} \text{ (in mm)} \tag{5.31b}$$

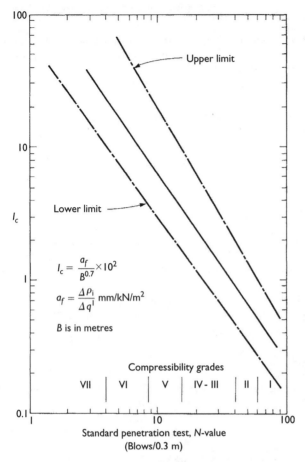

Figure 5.26 Relationship between compressibility index and average N-value over depth of influence (after Burland and Burbidge[5.19]).

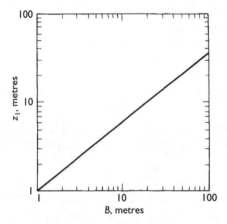

Figure 5.27 Relationship between breadth of loaded area and depth of influence z_I (after Burland and Burbidge[5.19]).

In the case of pile groups the width B is the width at the base of the equivalent raft as shown in Figure 5.3. The Burland and Burbidge method was developed essentially for shallow foundations and correlations with published settlement records given in their paper were mainly confined to foundations where their depth was not greater than their width. They state that the depth to width ratio did not influence the settlements to any significant degree and hence a depth factor of the type shown in Figure 5.20 should not be applied. However, a correction should be applied to allow for the foundation shape and for the thickness of the compressible layer beneath the foundation where this is less than the depth of influence, z_I.

The correction factors are

$$\text{Shape factor} = f_s = \left(\frac{1.25 \, L/B}{L/B + 0.25} \right)^2 \tag{5.32a}$$

$$\text{Thickness factor} = f_l = \frac{H_s}{z_I} \left(2 - \frac{H_s}{z_I} \right) \tag{5.32b}$$

where L is the length of the loaded area $(L > B)$
       B is the width of the loaded area
       H_s is the thickness of the compressible layer $(H_s < z_I)$

Burland and Burbidge state that most settlements on granular soils are time-dependent, i.e. they show a long-term creep settlement and a further time correction factor is applied using the equation:

$$f_t = \frac{\rho_t}{\rho_i} = \left(1 + R_3 + R\log\frac{t}{3} \right) \tag{5.33}$$

where t is equal to or greater than 3 years
       R_3 is the proportion of the immediate settlement which takes place in the loaded area
       R is the creep ratio expressed as the proportion of the immediate settlement that takes place per log cycle of time

Burland and Burbidge give conservative values of R and R_3 as 0.2 and 0.3 respectively for static loading and 0.8 and 0.7 respectively for fluctuating loads.

Summarizing all the above corrections, the *average consolidation settlement* is given by

$$\rho_c = f_s f_l f_t [(q' - \tfrac{2}{3}\sigma'_{vo}) \times B^{0.7} \times I_c] \quad \text{(in mm)} \tag{5.34}$$

The wide range of I_c values between the upper and lower limit shown in Figure 5.26 can cause difficulty in obtaining a reasonably close estimate of pile group settlements, particularly where the group is underlain by medium-dense sands. For example, the average I_c value for a sand with an N-value of 10 is 6 compared with upper and lower limit values of 20 and 3 respectively, giving an upper limit of settlement of three times that calculated from the average curve. However, in most cases piles are taken down to dense sands to obtain the maximum end-bearing resistance, where the settlement calculated from the upper limit curve is likely to be relatively small.

5.3.2   Estimating settlements from static cone penetration tests

Where total and differential settlements are shown to be large and critical to the superstructure design, it is desirable to make static cone penetration tests (Section 11.1.4) from which the soil modulus E'_v values can be derived and then to use the Steinbrenner (Figure 5.18) or Christian and Carrier (Figure 5.20) charts to obtain the group settlement. Relationships between the cone-resistance (q_c) values and the drained Young's modulus for normally consolidated quartz sands are shown in Figure 5.28. The E_{25} and E_{50} values represent the drained modulus at a stress level of 25% and 50% respectively of the failure stress. In a general review of the application of cone penetration testing to foundation design, Meigh[5.23] stated that the E_{25} values are appropriate for most foundation problems but the E_{50} values may be more relevant to calculating settlements of the single pile.

The E values in Figure 5.28 greatly overestimate settlements in over-consolidated sands. Lunne and Christoffersen[5.24] established a relationship between initial tangent constrained modulus (the reciprocal of the modulus of volume compressibility m_v) and q_c for normally and over-consolidated sands as shown in Figure 5.29.

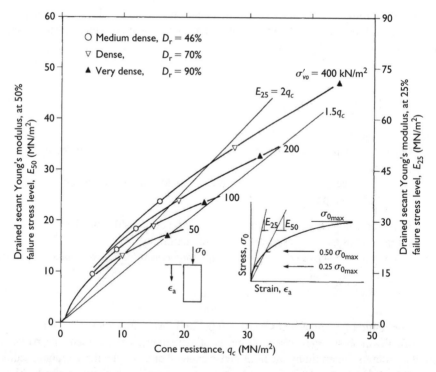

Figure 5.28 Drained deformation modulus values (E_d) for uncemented normally consolidated quartz sands in relation to cone resistance (after Meigh[5.20]), Robertson and Campanella[5.21]), Baldi et al.[5.22]).

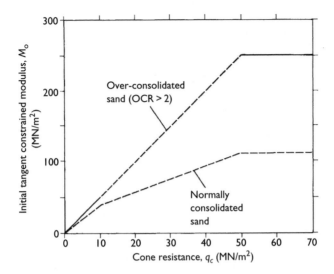

Figure 5.29 Initial tangent constrained modulus for normally consolidated and over-consolidated sand related to cone resistance (after Lunne and Christoffersen[5.24]).

Another method of estimating the settlements of pile groups in granular soils based on static cone penetration test values has been developed by Schmertmann[5.25] and Schmertmann et al.[5.26] Their basic equation for the settlement of a loaded area is

$$r = C_1 C_2 \Delta_p \sum_0^{2B} \frac{I_z}{E_v'} \Delta_z \tag{5.35}$$

where C_1 is a depth correction factor (see below), C_2 is a creep factor (see below), Δ_p is the net increase of load on the soil at the base of the foundation due to the applied loading, B is the width of the loaded area, I_z is the vertical-strain influence factor (see Figure 5.30), E_v' is the deformation modulus, and Δ_z is the thickness of the soil layer.

The value of the depth correction factor is given by

$$C_1 = 1 - 0.5 \left(\frac{\sigma'_{vo}}{\Delta_p} \right) \tag{5.36}$$

where σ'_{vo} is the effective overburden pressure at foundation level (i.e. at the base of the equivalent raft).

Schmertmann[5.25] states that while the settlement of foundations on granular soils is usually regarded as immediate, i.e. the settlement is complete within a short time after the completion of the application of load, observations have frequently shown long-continuing secondary settlement or creep. He gives the value of the creep factor as

$$C_2 = 1 + 0.2 \log_{10} \left(\frac{\text{time in years}}{0.1} \right) \tag{5.37}$$

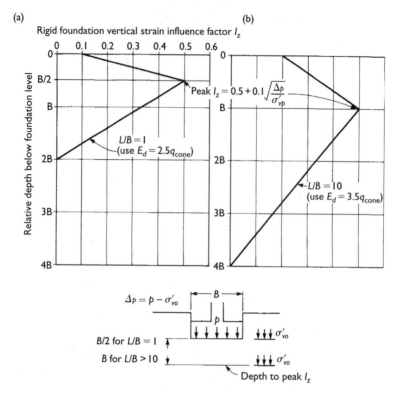

Figure 5.30 Schmertmann's influence factors for calculating immediate settlements of foundations on sands (after Schmertmann et al.[5.26]).

Schmertmann et al.[5.26] have established an improved curve for obtaining the vertical-strain influence factor based on elastic half-space theory where the factor I_z is related to the foundation width, as shown in Figure 5.30.

The vertical strain influence factor is obtained from one of the two curves shown in Figure 5.30. For square pile groups (axisymmetric loading) the curve in Figure 5.30a should be used. For long pile groups (the plane strain case) where the length is more than 10 times the breadth, use the curve in Figure 5.30b. Values for rectangular foundations for L/B of less than 10 can be obtained by interpolation.

The deformation modulus for square and long pile groups in normally consolidated sands is obtained by multiplying the static cone resistance, q_c, by a factor of 2.5 and 3.5 respectively. The deformation modulus applicable for a stress increase of Δ_p above the effective overburden pressure, σ'_{vo}, is given by the equation:

$$E'_v = E\sqrt{\frac{\sigma'_{vo} + (\Delta_p/2)}{\sigma'_{vo}}} \qquad (5.38)$$

Where standard penetration tests only are available the static cone resistance (in MN/m²) can be obtained by multiplying the SPT N-values (in blows/300 mm) by an empirical factor N for which Schmertmann suggests the following values:

Silts, sandy silts, and slightly cohesive silty sands:	$N = 0.2$
Clean fine to medium sands, slightly silty sands:	$N = 0.35$
Coarse sands and sands with a little gravel:	$N = 0.5$
Sandy gravel and gravels:	$N = 0.6$

Where static cone resistance data are available the relationships in Figures 5.28 or 5.29 can be used to obtain values of E_v' for substitution in equation 5.35.

The procedure for estimating settlements by the Schmertmann method is first to divide the static cone resistance diagram into layers of approximately equal or representative values of q_c in a manner shown in Figure 5.31. The base of the equivalent raft representing the pile group is then drawn to scale on this diagram and the influence curve is superimposed beneath the base of the raft. The settlements in each layer resulting from the loading Δ_p at the base of the equivalent raft are then calculated using the values of E_v' and I_2 appropriate to each of the representative layers. The sum of these settlements is corrected for depth and creep from equations 5.36 to 5.38. The various steps in the calculation are made in tabular form as illustrated in Example 5.3.

Where piles are terminated in a coarse soil stratum underlain by compressible clay, the settlements within the zone of clay stressed by the pile group are calculated by the methods described in Section 5.2.2. The form of load distribution to be used in this analysis to obtain the dimensions of the equivalent raft on the surface of the clay layer is shown in Figure 5.21.

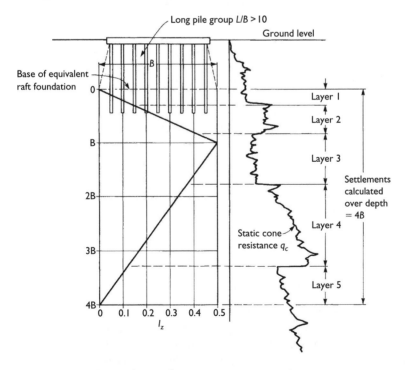

Figure 5.31 Establishing the vertical strain from static cone penetration tests.

5.4   Eurocode 7 recommendations for pile groups

Clause 7.6.2.1 of EC7 requires the stability of a pile group to be considered both in relation to the risk of failure of an individual pile in the group and to the failure of the group considered as an equivalent block foundation. Sub-clause (4) states that the block foundation can be considered to act as a single large-diameter pile. However, no guidance is given as to relationship between the diameter and depth of this pile to the shape, base area and depth of the group. If it is assumed that the plan area of the large-diameter pile is equal to the gross area of the group, then in the case of square (or rectangular) groups the resulting calculations could give an over-conservative value of the design load. This is because the perimeter and shape factors for rectangular bases are larger than those for circular sections. Also it is reasonable to assume that the shaft friction of the pile should be calculated on the basis of a soil-to-soil interface using the undisturbed shear strength of the surrounding soil. Whereas when calculating the shaft friction on an individual pile the installation method has an important influence on the resistance of a pile to soil interface. Where a group of piles are driven into a clay the surrounding soil is strengthened by expulsion of pore water, and a sand is strengthened by densification. Conversely, drilling for a group of bored piles could cause weakening of a clay due to relaxation of a fissured structure or drilling in sand could result in loss of resistance in friction.

If as an alternative to the large-diameter pile assumption the pile group is treated as an equivalent block foundation, the partial factors for actions and material properties are the same as used for piled foundations (Tables 4.1 and 4.2 in Section 4.1.4). The base resistance factor for spread foundations, γ_{Rv}, and the factor for sliding γ_{Rh}, are both unity for set 1 in DA1 (Table A5 in Annex A of EC7). There are no R4 resistance factors for spread foundations.

Worked Example 5.1 at the end of this chapter shows that the assumption of a single large-diameter pile under-estimates the resistance of a rectangular group in clay, compared with calculations assuming a block foundation of the same dimensions as the prototype, when using the calculation method described in Section 5.2.1.

Clause 7.6.4.2(2)P states that the assessment of settlement of pile groups should take into account the settlement of the individual piles as well as that of the group, but it does not make it clear whether the settlement analysis should assume that the group acts as an equivalent large-diameter pile or as a block foundation. Presumably the latter is the case, for which Clause 6.6.2, considering the settlement of spread foundations, requires the depth of the compressible soil layer to be taken normally as the depth at which the effective vertical stress due to the foundation load is 20% of that of the effective overburden stress, which may in many cases be roughly estimated as one to two times the foundation width or less for lightly loaded foundation rafts. In the case of pile groups the authors assume this to be the depth below the base of the equivalent rafts shown in Figure 5.3.

5.5   Pile groups terminating in rock

The stability of a pile group bearing on a rock formation is governed by that of the individual pile. For example, one or more of the piles might yield due to the presence of a pocket of weathered rock beneath the toe. There is no risk of block failure unless the piles are terminated on a sloping rock formation, when sliding on a weak clay-filled bedding plane might occur if the bedding is unfavourably inclined to the direction of loading (Figure 5.32). The possibility of such occurrences must be studied in the light of the information available on the geology of the site.

The settlement of a pile group may be of significance if the piles are heavily loaded. Immediate settlements can be calculated as described in Section 5.2.2, and equations 5.14 and 5.22 are applicable, where E_d is reasonably constant with depth.

It is possible to obtain a rough estimate of the deformation modulus of a jointed rock mass from empirical relationships with the unconfined compression strength of the intact rock. BS 8004 gives a relationship as $E_d = j \times M_r \times q_c$ where j is the mass factor (see Section 4.7.3 for values) and M_r is the ratio of the elastic modulus of the intact rock to its unconfined compression strength. BS 8004 gives the following values for M_r:

		Values for M_r
Group 1	Pure limestones and dolomites	600
	Carbonate sandstones of low porosity	
Group 2	Igneous	300
	Oolitic and marly limestones	
	Well-cemented sandstones	
	Indurated carbonate mudstones	
	Metamorphic rocks including slates and schists (flat cleavage/foliation)	
Group 3	Very marly limestones	150
	Poorly cemented sandstones	
	Cemented mudstones and shales	
	Slates and schists (steep cleavage/foliation)	
Group 4	Uncemented mudstones and shales	75

Chalk and Mercia Mudstone (Keuper Marl) are excluded from the above groups. Some observed values of E_d for chalk are given in Table 5.3 and for Mercia Mudstone in Table 5.4.

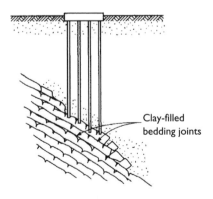

Clay-filled bedding joints

Figure 5.32 Instability of pile group bearing on sloping rock surface.

Table 5.3 Values of deformation modulus of Chalk (after Lord et al.[4.43])

Density	Grade	'Yield' stress (MN/m²)	Ultimate bearing capacity (MN/m²)	Secant modulus at applied stress of 200 kN/m² (MN/m²)	'Yield' modulus (MN/m²)
Medium/high	A	—	16	1500–3000	—
	B	0.3–0.5	4.0–7.7	1500–2000	35–80
	C	0.3–0.5	4.0–7.7	300–1500	35–80
Low	B and C	0.25–0.5	1.5–2.0	200–700	15–35
(Low)	D_c	0.25–0.5	—	200	20–30
	D_m	—	—	6	—

Table 5.4 Values of deformation modulus of Mercia Mudstone (Keuper Marl) at low stress levels (after Chandler and Davis[5.27])

Zone	Deformation modulus (MN/m²)
I	26–250
II	9–70
III	2–48
IV	2–13

It is likely that weathered rocks will show an increase in E_d with depth as the state of weathering decreases from complete at rockhead to the unweathered condition. If it is possible to draw a straight line through the increasing values the influence factors in Figure 5.19 can be used in conjunction with equation 5.21 to obtain the settlement at the centre of the loaded area. These curves were established by Butler[5.12] for a Poisson's ratio of 0.5, but most rock formations have lower ratios. Meigh[5.23] stated that the Poisson's ratio of Triassic rocks is about 0.1 to 0.3.

Meigh[5.23] derived curves for the influence factors shown in Figure 5.33 for various values of the constant k in equation 5.39 where

$$k = \frac{(E_d - E_f)B/H}{E_f} \tag{5.39}$$

and for a Poisson's ratio of 0.2. He applied further corrections to the calculation of the settlement at the *corner* of the foundation where

$$\text{Settlement at corner} = \rho_i = \frac{q_n B I_p'}{E_f} \tag{5.40}$$

The corrected settlement is given by

$$\rho_c \text{ (corrected)} = \frac{q_n B I_p'}{E_f} \times F_B \times F_D \tag{5.41}$$

where

F_B = correction factor for roughness of base (Figure 5.34)
F_D = correction factor for depth of embedment (Figure 5.35)

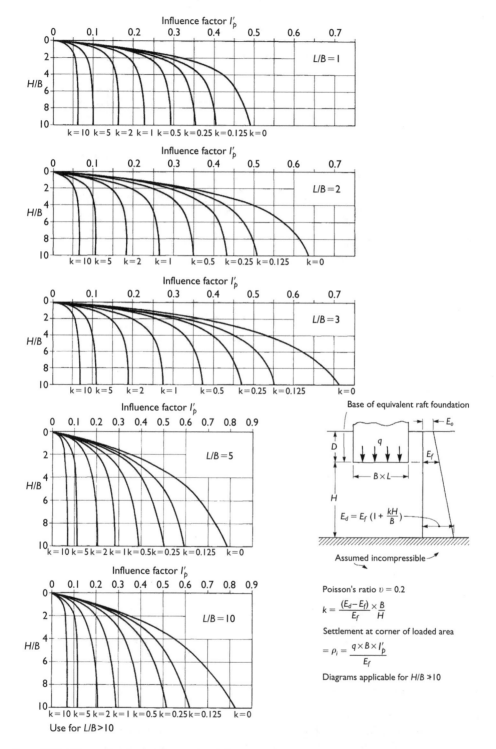

Figure 5.33 Values of influence factor for deformation modulus increasing linearly with depth and modular ratio of 0.2 in rock (after Meigh[5.23]).

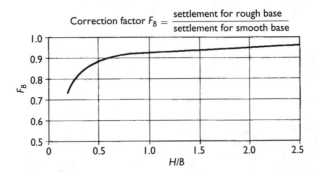

Figure 5.34 Correction factors for roughness of base of foundation (after Meigh[5.23]).

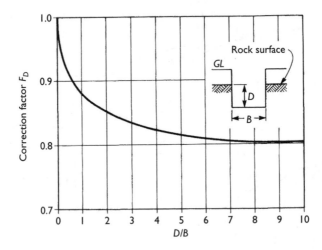

Figure 5.35 Correction factors for depth of embedment of foundation below surface of rock (after Meigh[5.23]).

The equivalent raft is assumed to have a rough base and is divided into four equal rectangles and the settlement computed for the corner of each rectangle from equation 5.41. The settlement at the centre of the pile group is then four times the corner settlement.

5.6 Pile groups in filled ground

The problem of negative skin friction or drag-down on the shafts of isolated piles embedded in fill was discussed in Section 4.8. This drag-down is caused by the consolidation of the fill under its own weight, or under the weight of additional imposed fill. If the fill is underlain by a compressible clay the consolidation of the clay under the weight of the fill also causes negative skin friction in the portion of the shaft within this clay. Negative skin friction also occurs on piles installed in groups but the addition to the working load on each of the piles in the group is not necessarily more severe than that calculated for the isolated pile. The basis for calculating the negative skin friction as described in Section 4.8.1, is that the ultimate skin friction on the pile shaft is assumed to act on that length of pile over which the fill and any underlying compressible clay move downwards relative to the shaft. The magnitude of this skin friction cannot increase as a result of grouping the piles at close centres, and the total negative skin friction acting on the group cannot exceed the total weight of fill enclosed by the piles. Thus in Figure 5.36a

$$\text{total load on pile group} = \text{working load} + (B \times L \times \gamma'D') \tag{5.42}$$

where γ' is the unit weight of fill, and D' is the depth over which the fill is moving downwards relative to the piles.

Where the fill is underlain by a compressible clay, as in Figure 5.36b

$$\text{total load on pile group} = \text{working load} + B \times L(\gamma'D' + \gamma''D'') \tag{5.43}$$

where D'' is the total thickness of fill, γ'' is the unit weight of compressible clay, and D''' is the thickness of compressible clay moving downwards relative to the piles.

It should also be noted that the negative skin friction acting on the piles in the group does not increase the *settlement* of the group caused by the working load on the piles. If the filling has been in place for a long period of years any underlying compressible soil will have been fully consolidated and the only additional load on the compressible soil causing settlement of the group is that from the working load on the piles. However, if the fill is to be placed only a short time before driving the piles, then any compressible soil below the fill will consolidate. The amount of this consolidation can be calculated separately and added to the

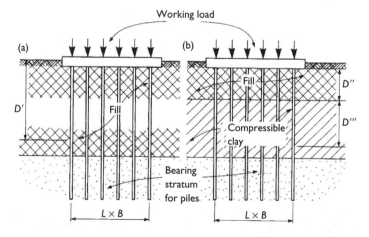

Figure 5.36 Negative skin friction on pile groups in filled ground (a) Fill overlying relatively incompressible bearing stratum (b) Fill placed on compressible clay layer.

settlement caused by the working load on the piles. The negative skin friction on the piles is not included in the working load for the latter analysis.

EC 7 gives no specific guidance for the design of pile groups carrying compression loading in filled ground. As in the case of the single pile calculation, the load distribution on individual piles in the group is best undertaken by an interaction analysis as described in Section 4.3.8. It is evident that treatment of the group as a single large-diameter pile as recommended in Clause 7.6.2.1(4) for the determination of group stability is not valid for application to an interaction analysis.

Clause 7.3.2.2(P) requires account to be taken of the weight density of materials in a settlement analysis for piles in filled ground. As noted in the case of the single pile the partial factors for weight in Table A.4 of EC 7 Annex A are unity (see Table 4.2 Section 4.1.4).

5.7   Effects on pile groups of installation methods

When piles are driven in groups into clay the mass of soil within the ground heaves and also expands laterally, the volume of this expansive movement being approximately equal to the volume occupied by the piles. High pore pressures are developed in the soil mass, but in the course of a few days or weeks these pore pressures dissipate and the heaving directly caused by pore pressure subsides. In soft clays the subsidence of the heaved soil can cause negative skin friction to develop. It is not usual to add this negative skin friction to the working load since it is of relatively short duration, but its effect can be allowed for by ignoring any *support* provided in shaft friction to the portion of the pile shaft within the soft clay. Methods of calculating the surface heave within a pile group have been discussed by Hagerty and Peck[5.28]. Chow and Teh[5.29] have established a theoretical model relating the pile head heave/diameter ratio to the pile spacing/diameter ratio for a range of length/diameter ratios in soft, firm and stiff clays.

It is not good practice to terminate pile groups within a soft clay since the reconsolidation of the heaved and remoulded soil can result in the substantial settlement of a pile group, and neighbouring structures can be affected. It may be seen from Figure 5.37 that there is little difference between the extent of the stressed zone around and beneath a surface raft and a group of short friction piles. The soil beneath the raft is not disturbed during construction and hence the settlement of the raft may be much less than that of a pile group carrying the same overall loading. This was illustrated by Bjerrum[5.30], who compared the settlement of buildings erected on the two types of foundation construction on the deep soft and sensitive clays of Drammen near Oslo.

A building where the gross loading of 65 kN/m² was reduced by excavation for a basement to a net loading of 25 kN/m² was supported on 300 timber friction piles 23 m long. In 10 years the building had settled by 110 mm and the surrounding ground surface had settled by 80 mm. A nearby building with a gross loading of 55 kN/m² had a fully compensated unpiled foundation, i.e. the weight of the soil removed in excavating for the basement balanced the superstructure and substructure giving a net intensity of loading of zero on the soil. Nearly 30 mm of heave occurred in the base of excavation and thus the settlement of the building was limited to the re-consolidation of the heaved soil. The net settlement 9 years after completing the building was only 5 mm.

Lateral movement of a clay soil and the development of high pore pressures can damage structures or buried services close to a pile group. Adams and Hanna[5.31] measured the pore

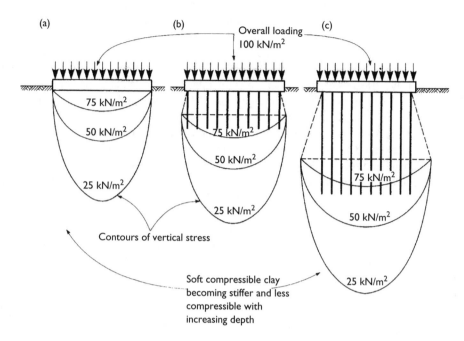

Figure 5.37 Comparison of stress distribution beneath shallow raft foundation and beneath pile groups (a) Shallow raft (b) Short friction piles (c) Long friction piles.

pressures developed within the centre of a large group of driven piles at Pickering Nuclear Power Station, Ontario. The horizontal ground strains were also measured at various radial distances from the centre. The group consisted of 750 piles driven within a circle about 46 m in diameter. Steel H-section piles were selected to give a minimum of displacement of the 15 m of firm to very stiff and dense glacial till, through which the piles were driven to reach bedrock. From measurements of the change in the distance between adjacent surface markers it was calculated that the horizontal earth pressure at a point 1.5 m from the edge of the group was 84 kN/m² while at 18.8 m from the edge the calculated pressure was only 1 kN/m². Earth pressure cells mounted behind a retaining wall 9 m from the group showed no increase in earth pressure due to the pile driving. Very high pore pressures were developed at the centre of the piled area, the increase being 138 kN/m² at a depth of 6 m, dissipating to 41 kN/m², 80 days after completing driving of the instrumented pile, when all pile driving in the group had been completed.

The average ground heave of 114 mm measured over the piled area represented a volume of soil displacement greater than the volume of steel piles which had been driven into the soil, for which the theoretical ground heave was 108 mm.

Substantial heave accompanied by the lifting of piles already driven can occur with large displacement piles. Brzezinski et al.[5.32] made measurements of the heave of 270 driven and cast-in-place piles in a group supporting a 14-storey building in Quebec. The piles had a shaft diameter of 406 mm and the bases were expanded by driving. The piles were driven through 6.7 to 11 m of stiff clay to a very dense glacial till. Precautions against uplift were taken by providing a permanent casing to the piles and the concrete was not placed in the shafts until the pile bases had been re-driven by tapping with a drop hammer to the extent necessary to overcome the effects of uplift. The measured heave of a cross-section of the piled area is shown in Figure 5.38. It was found that the soil heave caused the permanent casing to become detached from the bases, as much as 300 mm of separation being observed. Heave effects were not observed if the piles were driven at a spacing wider than 12 diameters. This agrees with the curves established by Chow and Teh[5.29] which show a pile head heave of only about 1 mm for a spacing of 12 diameters.

Similar effects were observed by Cole[5.33]. At three sites the heave was negligible at pile spacings wider than 8 to 10 diameters. Cole observed that uplift was more a function of the pile diameter and spacing than of the soil type or pile length. Where piles carry their load mainly in end bearing, the effect of uplift is most damaging to their performance and on all sites where soil displacement is liable to cause uplift, precautions must be taken as described in Section 5.8. Heave is not necessarily detrimental where piles are carried by shaft friction in firm to stiff clays in which there will be no appreciable subsidence of the heaved soil to cause negative skin friction to develop on the pile shaft. On a site where a 12-storey block of flats was supported by driven and cast in-situ piles installed in 5 m of firm London Clay to terminate at the base of a 4 m layer of stiff London clay, about 0.5 m of heave was observed in the ground surface after 70 piles had been driven within the 24 × 20 m area of the block. A pile was tested in an area where 220 mm of heave had occurred. The settlement at 1300 kN (i.e. twice the working load) was 23 mm, while the settlement at the working load was only 2.5 mm.

Heaving and the development of high pore pressures do not occur when bored and cast-in-place piles are installed in groups. However, general subsidence around the piled area can be caused by the 'draw' or relaxation of the ground during boring. In soft sensitive clays the bottom of a pile borehole can heave up due to 'piping', with a considerable loss of ground. These effects can be minimized by keeping the pile borehole full of water or bentonite slurry during drilling and by placing the concrete within a casing which is only withdrawn after all concrete placing is completed.

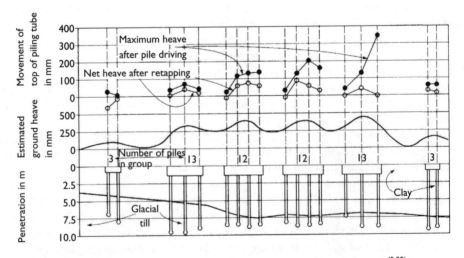

Figure 5.38 Observations of heave due to pile driving in clay (after Brzezinski et al.[5.32]).

Detrimental effects from heave are not usually experienced when driving piles in groups in coarse soils. A loose soil is densified, thus requiring imported filling to make up the subsided ground surface within and around the group. Adjacent structures may be damaged if they are within the area of subsidence. A problem can arise when the first piles to be installed drive easily through a loose sand but, as more piles are driven, the sand becomes denser thus preventing the full penetration of all the remaining piles. This problem can be avoided by paying attention to the order of driving, as described in Section 5.8.

Subsidence due to the loss of ground within and around a group in a coarse soil can be quite severe when bored and cast-in-place piles are installed, particularly when 'shelling' is used as the boring method (see Section 3.3.7). The subsidence can be very much reduced, if not entirely eliminated, by the use of rotary drilling with the assistance of a bentonite slurry (see Section 3.3.8).

5.8   Precautions against heave effects in pile groups

It will have been noted from Section 5.7 that the principal problems with soil heave and the uplift of piles occur when large displacement piles are driven into clay. In coarse soils the problems can be overcome to a great extent by using small displacement piles such as H-sections or open-ended steel tubes. To adopt a spacing between piles of 10 or more diameters is not usually practical if pile group dimensions are to be kept within economical limits. Pre-boring the pile shaft is not always effective unless the pre-bored hole is taken down to the pile base, in which case the shaft friction will be substantially reduced if not entirely eliminated. Jetting piles is only effective in a coarse soil and the problems associated with this method are described in Section 3.1.9. The most effective method is to re-drive any risen piles, after driving all the piles in a cluster that are separated from adjacent piles by at least 12 diameters has been completed.

In the case of driven and cast-in-place piles, a permanent casing should be used and the re-driving of the risen casing and pile base should be effected by tapping the permanent casing with a 3-tonne hammer, as described by Brzezinski *et al.*[5.32] Alternatively, the 'Multitube' method described by Cole[5.33] can be used. This consists of providing sufficient lengths of withdrawable casing to enable all the piling tubes to be driven to their full depth and all the pile bases to be formed before the pile shafts in any given cluster are concreted. An individual cluster dealt with in this way must be separated from a neighbouring cluster by a sufficient distance to prevent the uplift of neighbouring piles or to reduce this to an acceptable amount. On the three sites described by Cole it was found possible to drive piles to within 6.5 diameters of adjacent clusters without causing an uplift of more than 3 mm to the latter. This movement was not regarded as detrimental to the load/settlement behaviour. Cole stated that, although the 'Multitube' system required eight driving tubes to each piling rig, the cost did not exceed that of an additional 2 m on each pile.

Curtis[3.26] states that it is possible to re-drive risen driven and cast in-situ piles using a 3-to 4-tonne hammer with a drop not exceeding 1.5 m. The head of the pile should be protected by casting on a 0.6 m capping cube in rapid-hardening cement concrete.

Cole[5.33] stated that the order of driving piles did not affect the incidence of risen piles but it did change the degree of uplift on any given pile in a group. Generally, the aim should be to work progressively outwards or across a group, and in the case of an elongated group from end to end or from the middle outwards in both directions. This procedure is particularly important when driving piles in coarse soils. If piles are driven from the perimeter towards the centre of a group, a coarse-grained soil will 'tighten-up' so much due to ground vibrations that it will be found impossible to drive the interior piles.

It is desirable to adopt systematic monitoring of the behaviour of all piles installed in groups by taking check levels on the pile heads, by carrying out re-driving tests, and by making loading tests on working piles selected at random from within the groups. Loading tests undertaken on isolated piles before the main pile driving commences give no indication of the possible detrimental effects of heave. Lateral movements should also be monitored as necessary.

5.9   Pile groups beneath basements

Basements may be required beneath a building for their functional purpose, for example, as an underground car park or for storage. The provision of a basement can be advantageous in reducing the loading which is applied to the soil by the building. For example, if a basement is constructed in an excavation 7 m deep the soil at foundation level is relieved of a pressure equivalent to 7 m of overburden, and the gross loading imposed by the building is reduced by this amount of pressure relief. It is thus possible to relieve completely the net loading on the soil. An approximate guide to the required depth of excavation is the fact that a multistorey dwelling block in reinforced concrete with brick and concrete external walls, lightweight concrete partition walls, and plastered finishes weighs about 12.5 kN/m² per storey. This loading is inclusive of 100% of the dead load and 60% of the design imposed load. Thus a 20-storey building would weigh 250 kN/m² at ground level, requiring a basement to be excavated to a depth of about 20 m to balance the loading (assuming the groundwater level to be 3 m below ground level and taking the submerged density of the soil below water level).

Deep basement excavations in soft compressible soils can cause considerable constructional problems due to heave, instability and the settlement of the surrounding ground surface. Because of this it may be desirable to adopt only a partial relief of loading by excavating a basement to a moderate depth and then carrying the net loading on piles taken down to soil having a lesser compressibility.

In all cases where piles are installed to support structures it is necessary to consider the effects of soil swelling and heave on the transfer of load from the basement floor slab to the piles. Four cases can be considered as described below and shown in Figure 5.39.

Piles wholly in compressible clay (Figure 5.39a)

In this case the soil initially heaves due to swelling consequent on excavating the foundation, and further heave results from pile driving. The heaved soil is then trimmed off to the correct level and the basement slab concreted. If the concreting is undertaken within a few days or a week after the pile driving there is a tendency for the heaved soil to slump down, particularly in a soft clay which developed high pore pressures. A space may tend to open between the underside of the concrete and the soil surface. When the superstructure is erected the piles will carry their working load and if correctly designed they will settle to an acceptable degree. This will in turn cause the basement slab to settle but pressure will not develop on its underside because the soil within and beneath the settling piles will move down with them. Thus the maximum pressure on the underside of the basement slab is due to the soil swelling at an early stage before partial slumping of the heaved soil takes place and before

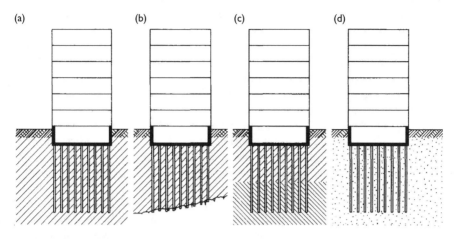

Figure 5.39 Piled basements in various ground conditions (a) Wholly in compressible clay (b) Compressible clay over bedrock (c) Soft clay over stiff clay (d) Loose sand becoming denser with depth.

the piles carry any of their designed loading. The uplift pressure on the basement slab will be greater if bored piles are used since no heaving of the soil is caused by installing the piles, and if the basement slab is completed and attached to the piles soon after completing the excavation, the swelling pressures on the underside of the slab will cause tension to be developed in the piles. This is particularly liable to happen where bored piles are installed from the ground surface before the excavation for the basement commences. Concreting of the pile shaft is terminated at the level of the underside of the basement slab and the construction of the basement slab usually takes place immediately after the completion of excavation and before any heave of the excavation can take place to relieve the swelling pressure. Generally, in any piled basement where *bored piles* are installed wholly in compressible clay, the basement slab should be designed to withstand an uplift pressure equal at least to one-half of the dead and sustained imposed load of the superstructure. Alternatively, a void can be provided beneath the basement slab by means of collapsible cardboard or plastics formers. The piles can be designed to be anchored against uplift or they can be sleeved over the zone of swelling. Anchoring the piles against uplift by increasing the shaft length to increase shaft friction below the swelling zone is often the most economical solution to the problem. Where void formers made of cardboard or plastics are used to eliminate swelling pressure beneath the basement slab, there is a risk of bio-degradation of the organic materials causing an accumulation of methane gas in the void. Venting the underside of the slab can be difficult and costly.

Providing an increased shaft length can be made more economical than sleeving the pile shaft within the swelling zone. Fleming and Powderham[5.34] recommended that where piles are reinforced to restrain uplift the friction forces should not be underestimated and they suggest that if the forces are estimated conservatively it would be appropriate to reduce the load factors on the steel, perhaps to about 1.1.

Hydrostatic pressure will, of course, act on the basement slab in water-bearing soil. The piles must be designed to carry the net full weight of the structure (i.e. the total weight less the weight of soil and soil water excavated from the basement).

When installing piles for 'top-down' construction the empty borehole from the pile head below the lowest basement floor slab up to the ground surface is supported by temporary casing. Steel columns are set on the pile heads and are surrounded with sand before the casing is extracted. Particular care is necessary in establishing the position of the pile borehole and maintaining verticality in drilling. If this is not done there could be considerable error in the position of the pile head and difficulties in locating the column in the design position.

The tolerance in pile position permitted by the Institution of Civil Engineers' Specification for Piling[2.5] is 75 mm and the permitted deviation from the vertical of a bored pile is 1 in 75. (see also Section 3.4.12) Taking as an example a 3-storey basement with an overall depth from ground surface to pile head level (beneath the lowest floor slab) of 15 m the centre of the pile could be 275 mm from the design position. Consequently, either the column set in its intended position in the superstructure would be off-plumb or the column would apply an eccentric load to the pile head. The permitted tolerances should be kept in mind when considering the pile diameter and the design of reinforcement to provide for eccentric loading.

Piles driven through compressible clay to bedrock (Figure 5.39b)

In this case soil swelling takes place at the base of the excavation followed by heave if driven piles are employed. As before, the heaved soil tends to slump away from the underside of the basement slab if the latter is concreted soon after pile driving. Any gap which might form will be permanent since the piles will not settle except due to a very small elastic shortening of the shaft. If bored piles are adopted, with a long delay between concreting the base slab and applying the superstructure loading to the piles, the pressure of the underside of the slab due to long-term soil swelling might be sufficient to cause the piles to lift from their seating on the rock. The remedy then is to provide a void beneath the slab, and to anchor the piles to rock or to sleeve them through the swelling zone.

Piles driven through soft clay into stiff clay (Figure 5.39c)

This case is intermediate between the first two. There is a continuing tendency for the heaved soft clay to settle away from the underside of the basement slab, because the settlement of the piles taking their bearing in the stiff clay is less than that caused by the reconsolidation of the heaved and disturbed soft clay. Uplift pressure occurs on the underside of the base slab if bored piles are used, and a design value equal at least to one-half of the combined dead and imposed load of the superstructure should be considered. Alternatively, the effects of heave should be eliminated as described above.

Piles driven into loose sand (Figure 5.39d)

In this case it is presumed that the piles are driven through loose sand to an end bearing in deeper and denser sand. The slight heave of the soil caused by excavating the basement is an instantaneous elastic movement. No heave occurs because either pile driving causes some settlement of the ground surface due to densification, or a loss of ground results due to pile boring. When the superstructure load is applied to the piles they compress, but the soil follows the pile movement, and any soil pressures developed on the underside of the basement slab are relatively small. Hydrostatic pressure occurs in a water-bearing soil.

In all cases when designing piled basements the full design working load should be considered as acting on the piles and, in the case of piles bearing on rock or granular soils of low compressibility, the load on the underside of the basement slab can be limited to that caused by the soil pressure (i.e. the overburden pressure measured from the ground surface around the basement) and hydrostatic pressure. Sometimes a tall building is constructed close to a low-rise podium (Figure 5.40) and both structures are provided with a piled basement. Piling beneath the podium is required to reduce differential movement between the heavily loaded tower block and the podium. Uplift of the latter may occur if the weight of the superstructure is less that that of the soil removed in excavating for the basement. In such a case the piles must be anchored below the zone of soil swelling and designed to take or eliminate tension. The pressure on the underside of the podium basement slab will be equal to the swelling pressure exerted by the soil unless a void former is used to eliminate the pressure. A vertical movement joint passing completely through the basement and superstructure should be provided between the tower and podium to allow freedom of movement.

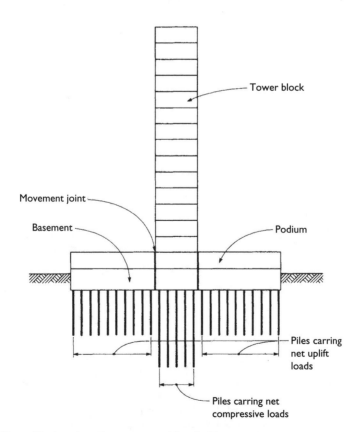

Figure 5.40 Tower block and podium supported by piled basement.

Measurements of the relative loads carried by the piles and the underside of the slab of a piled basement raft were described by Hooper[5.35]. The measurements were made during and subsequently to the construction of the 31-storey building of the Hyde Park Cavalry Barracks in London. The 90 m high building was constructed on the piled raft 8.8 m below ground level. The 51 bored and cast-in-place piles supporting the raft had a shaft diameter of 910 mm and an enlarged base 2400 mm in diameter (Figure 5.41a). The piles were installed by drilling from ground level and concreting the shaft up to raft level before commencing the bulk excavation.

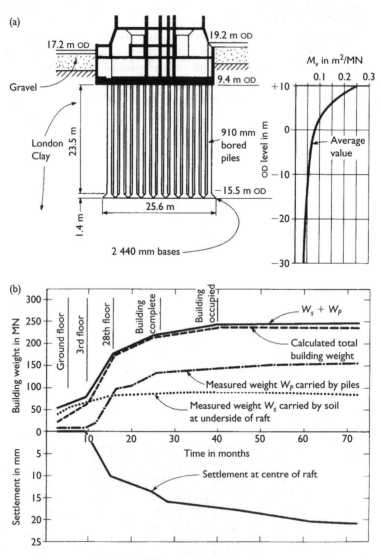

Figure 5.41 Piled raft foundations for Hyde Park Cavalry Barracks, London (a) Foundation arrangements and soil characteristics (b) Distribution of loading between raft and piles.

The weight of the building (including imposed load but excluding wind load) was calculated to be 228 MN. The weight of soil removed when excavating through gravel on to the stiff London clay at raft level was 107 MN, giving a net load to be transferred by the raft and piles to the London clay of 121 MN, or a net bearing pressure at raft level of 196 kN/m^2. Load cells were installed in three of the piles to measure the load transferred from the raft to the pile shaft, and three earth pressure cells were placed between the raft and the soil to measure the contact pressures developed at this interface. Settlements of the raft at various points were also measured by means of levelling points installed at ground level.

The observations of pile loadings and contact pressures were used to estimate the proportion of the total load carried by the piles and the basement raft from the initial stages of construction up to 3 years after completing the building. The results of these calculations are shown in Figure 5.41b and are compared with the calculated total weight of the building at the various stages of construction. Hooper[5.35] estimated that at the end of construction 60% of the building load was carried by the piles and 40% by the underside of the raft. In the post-construction period there was a continuing trend towards the slow transfer of more load to the piles, about 6% of the total downward structural load being transferred to the piles in the three-year period.

5.10   The optimization of pile groups to reduce differential settlements in clay

Cooke et al.[5.36] measured the proportion of load shared between the piles and raft and also the distribution of load to selected piles in different parts of a 43.3 m by 19.2 m piled raft supporting a 16-storey building in London Clay at Stonebridge Park. There were 351 piles in the group with a diameter of 0.45 m and a length of 13 m. The piles were uniformly spaced on a 1.6 m square grid. The overall loading on the pile group was about 200 kN/m^2

At the end of construction the piles carried 78% of the total building load, the remainder being carried by the raft. The distribution of the load to selected piles near the centre, at the edges, and at the corners of the group is shown in Figure 5.42. It will be seen that the loads carried by the corner and edge piles were much higher than those on the centre piles. The loading was distributed in the ratio 2.2:1.4:1 for the corner, edge, and centre respectively.

Advantages can be taken of the load sharing between raft and piles and between various piles in a group to optimize the load sharing whereby differential settlement is minimized and economies obtained in the design of the structural frame and in the penetration depth and/or diameter of the piles (Section 5.3). The procedure in optimization is described by Padfield and Sharrock[5.37]. Central piles are influenced by a larger number of adjacent piles than those at the edges. Hence, they settle to a greater extent and produce the characteristic dished settlement. Therefore, if longer stiffer piles are provided at the centre they will attract a higher proportion of the load. The outer piles are shorter and thus less stiff and will yield and settle more, thus reducing the differential settlement across the group. The alternative method of varying the settlement response to load is to vary the cross-sectional dimensions. The centre piles are made long with straight shafts and mobilize the whole of their bearing capacity in shaft friction at a settlement of between 10 and 15 mm. The shorter outer piles can be provided with enlarged bases which require a greater settlement to mobilize the total ultimate bearing capacity (see Section 4.6). An example of this is given by Burland and Kalra[5.17].

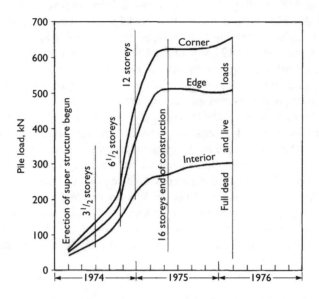

Figure 5.42 Load distribution on piled raft in London Clay (after Cooke *et al.*[5.36]).

Randolph[1.1] pointed out that where the ratio of the width of a pile group to the pile length is greater than unity the pile cap contributes significantly to the load transfer from the superstructure to the soil. Hence, the stiffness of a piled raft where the piles are arranged to cover the whole foundation area will be similar to that of the raft structure without the piles. Thus by concentrating the piles in the central area and using shorter piles (or no piles) around the edges, the bending moments due to dishing of the raft are considerably reduced. In the case of a uniformly loaded foundation area analyses show that piles of length greater than 70% of the foundation width situated over the central 25% to 40% of the raft area are required. Hence, instead of conventionally spreading the piles uniformly over the whole foundation area, as little as 30% to 50% of the cumulative length of all the piles is needed.

Load distribution between the piles is achieved through the continuous pile cap which must be designed to be stiff enough to achieve this. With perfect optimization differential settlement can be reduced to zero. The analysis to achieve optimization is complex and involves interaction factors discussed in Section 5.4. A computer is required to perform the necessary calculations. It is also necessary to check that the stress is not excessive on the shafts of the central piles which are designed to carry a high proportion of the load.

5.11   References

5.1   POULOS, H. G. Pile behaviour – theory and application, *Geotechnique*, Vol. 39, No. 3, 1989, pp. 365–415.

5.2   BURLAND, J. B. and WROTH, C. P. General report on Session 5: Allowable and differential settlements of structures, including settlement damage and soil structure interaction, *Proceedings of the Conference on Settlement of Structures*, Cambridge, 1974, Pentech Press, London, 1975, pp. 611–54.

5.3 BURLAND, J. B. Interaction between structural engineers and geotechnical engineers, *The Structural Engineer*, Vol. 84, No. 8, 2006, pp. 29–37.

5.4 HANSEN, J. B. A general formula for bearing capacity, Danish Geotechnical Institute, Bulletin No. 11, 1961; also, A revised and extended formula for bearing capacity, Danish Geotechnical Institute, Bulletin No. 28, 1968, and Code of Practice for Foundation Engineering, Danish Geotechnical Institute, Bulletin No. 32, 1978.

5.5 MEYERHOF, G. G. Some recent research on bearing capacity, *Canadian Geotechnical Journal*, Vol. 1, 1963, pp. 16–26.

5.6 JAMIOLKOWSKI, M. *et al.* Design parameters for soft clays, *Proceedings of the 7th European Conference on Soil Mechanics*, Brighton, 1979, pp. 21–57.

5.7 STROUD, M. A. The standard penetration test in insensitive clays, *Proceedings of the European Symposium on Penetration Testing*, Stockholm, 1975, Vol. 2, pp. 367–75.

5.8 BURLAND, J. B., BROMS, B. B., and DE MELLO, V. Behaviour of foundations and structures, *Proceedings of the 9th International Conference on Soil Mechanics*, Tokyo, Session 2, 1977.

5.9 JARDINE, R., FOURIE, A., MASWOSE, J., and BURLAND, J. B. Field and laboratory measurements of soil stiffness, *Proceedings of the 11th International Conference on Soil Mechanics*, San Francisco, Vol. 2, 1985, pp. 511–14.

5.10 MARSLAND, A. In-situ and laboratory tests on glacial clays at Redcar, *Proceedings of a Symposium on Behaviour of Glacial Materials*, Midlands Geotechnical Society, 1975, pp. 164–80.

5.11 TERZAGHI, K. *Theoretical Soil Mechanics*, John Wiley, New York, 1943, p. 425.

5.12 BUTLER, F. G. General Report and state-of-the-art review, Session 3, *Proceedings of the Conference on Settlement of Structures*, Cambridge, 1974, Pentech Press, London, 1975, pp. 531–78.

5.13 BROWN, P. T. and GIBSON, R. E. Rectangular loads on inhomogeneous soil, *Proceedings of the American Society of Civil Engineers*, Vol. 99 (SM10), 1973, pp. 917–20.

5.14 CHRISTIAN, J. T. and CARRIER, W. D. Janbu, Bjerrum and Kjaernsli's chart reinterpreted, *Canadian Geotechnical Journal*, Vol. 15, 1978, pp. 123–8.

5.15 FOX, E. N. The mean elastic settlement of a uniformly-loaded area at a depth below the ground surface, *Proceedings of the 2nd International Conference, ISSMFE*, Rotterdam, Vol. 1, 1948, pp. 129–32.

5.16 SKEMPTON, A. W. and BJERRUM, L. A contribution to the settlement analysis of foundations on clay, *Geotechnique*, Vol. 7, No. 4, 1957, pp. 168–78.

5.17 BURLAND, J. B. and KALRA, J. C. Queen Elizabeth Conference Centre: geotechnical aspects, *Proceedings of the Institution of Civil Engineers*, Vol. 80, No. 1, 1986, pp. 1479–503.

5.18 SCHULTZE, E. and SHERIF, G. Predictions of settlements from evaluated settlement observations for sand, *Proceedings of the 8th International Conference, ISSMFE*, Moscow, Vol. 1, 1973, p. 225.

5.19 BURLAND, J. B. and BURBIDGE, M. C. Settlement of foundations on sand and gravel, *Proceedings of the Institution of Civil Engineers*, Vol. 78, No. 1, 1985, pp. 1325–37.

5.20 MEIGH, A. C. *Cone Penetration Testing–Methods and Interpretation*, CIRIA/Butterworth, 1987.

5.21 ROBERTSON, P. K. and CAMPANELLA, R. G. Interpretation of cone penetration tests, parts 1 and 2, *Canadian Geotechnical Journal*, Vol. 20, 1983, pp. 718–45.

5.22 BALDI, G., BELLOTI, R., GHIONNA, V., JAMIOLKOWSKI, M., and PASQUALINIE. Cone resistance of dry medium sand, *Proceedings of the 10th International Conference, ISSMFE*, Stockholm, Vol. 2, 1981, pp. 427–32.

5.23 MEIGH, A. C. The Triassic rocks, with particular reference to predicted and observed performance of some major structures, *Geotechnique*, Vol. 26, 1976, pp. 393–451.

5.24 LUNNE, T. and CHRISTOFFERSEN, H. P. Interpretation of cone penetration data for offshore sands, *Proceedings of the Offshore Technology Conference* 15, Houston, 1983, Vol. 1, pp. 181–92.

5.25 SCHMERTMANN, J. H. Static cone to compute static settlement over sand, *Journal of the Soil Mechanics and Foundations Division*, American Society of Civil Engineers, Vol. 96, No. SM3, May 1970, pp. 1011–43.

5.26 SCHMERTMANN, J. H., HARTMAN, J. P., and BROWN, P. R. Improved strain influence diagrams, *Proceedings of the American Society of Civil Engineers*, Vol. GT8, 1978, pp. 1131–5.

5.27 CHANDLER, F. W. and DAVIS, A. G. Further work on the engineering properties of Keuper Marl *Construction Industry Research and Information Association (CIRIA)*, Report 47, 1973.

5.28 HAGERTY, A. and PECK, R. B. Heave and lateral movements due to pile driving, *Journal of the Soil Mechanics and Foundation Division*, American Society of Civil Engineers, No. SM11, November 1971, pp. 1513–32.

5.29 CHOW, Y. K. and TEH, C. I. A theoretical study of pile heave, *Geotechnique*, Vol. 40, No. 1, 1990, pp. 1–14.

5.30 BJERRUM, L. Engineering geology of normally-consolidated marine clays as related to the settlement of buildings, *Geotechnique*, Vol. 17, No. 2, 1967, pp. 83–117.

5.31 ADAMS, J. I. and HANNA, T. H. Ground movements due to pile driving, *Proceedings of the Conference on the Behaviour of Piles*, Institution of Civil Engineers, London, 1970, pp. 127–33.

5.32 BRZEZINSKI, L. S., SHECTOR, L., MACPHIE, H. L., and VAN DER NOOT, H. J. An experience with heave of cast-in-situ expanded base piles, *Canadian Geotechnical Journal*, Vol. 10, No. 2, May 1973, pp. 246–60.

5.33 COLE, K. W. Uplift of piles due to driving displacement, *Civil Engineering and Public Works Review*, March 1972, pp. 263–9.

5.34 FLEMING, W. G. K. and POWDERHAM, A. J. Soil down-drag and heave on piles, Institution of Civil Engineers, *Ground Engineering Group*, notes for meeting on 25/10/89.

5.35 HOOPER, J. A. Observations on the behaviour of a piled-raft foundation on London Clay, *Proceedings of the Institution of Civil Engineers*, Vol. 55, No. 2, December 1973, pp. 855–77.

5.36 COOKE, R. W., BRYDEN SMITH, D. W., GOOCH, M. N., and SILLETT, D. F. Some observations on the foundation loading and settlement of a multi-storey building on a piled raft foundation in London Clay, *Proceedings of the Institution of Civil Engineers*, Vol. 7, No. 1, 1981, pp. 433–60.

5.37 PADFIELD, C. J. and SHARROCK, M. J. Settlement of structures on clay soils, *Construction Industry Research and Information Association (CIRIA)*, Special Publication 27, 1983.

5.12 Worked examples

Example 5.1

Bored piles 500 mm in diameter drilled to a depth of 13.9 m below ground level into a firm to stiff clay are arranged in a group consisting of 10 rows each of seven piles, each carrying a dead load of 250 kN and an imposed load of 110 kN. From the results of tests on samples from three boreholes, the characteristic undrained shear strength of the clay increases from 60 kN/m^2 at 1.5 m below ground surface to 110 kN/m^2 at the base of the pile group. The fissured strength of the clay at pile toe level is 80 kN/m^2. Profiles of the undrained deformation modulus E_u and the coefficient of compressibility m_v are shown in Figure 5.43. Determine the overall stability and settlement of the pile group.

The first step is to calculate the factor of safety of the individual pile under the combined dead and imposed loads, from equations 4.4 and 4.7:

$$\text{Ultimate bearing capacity} = 9 \times 80 \times \pi/4 \times 0.5^2 + 0.45 \times (60 + 110)/2 \times \pi \times 0.5 \times 12.4$$
$$= 141 + 745 = 886 \text{ kN}$$

Factor of safety = 886/360 = 2.5 which is satisfactory, and because of the increasing strength of the clay below toe level, block failure of the group should not occur. However, for the purpose of comparison with the recommendation in EC7 Clause 7.6.2.1 to assume that the pile group acts as a single large-diameter pile, the stability of the group will be checked for compliance with the EC7 rules.

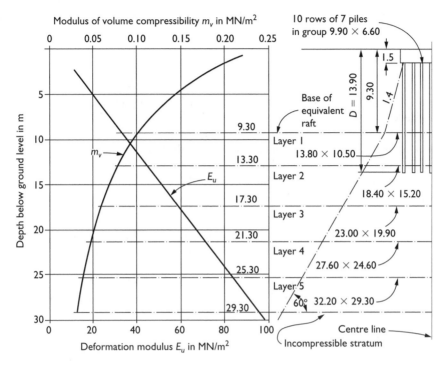

Figure 5.43 Profiles of the undrained deformation modulus E_u and the coefficient of compressibility m_v.

Determining the stability of the group assuming that it acts as a block foundation, and using the procedure, in Sections 5.1 and 5.2.1, the shaft resistance of the block is calculated on the fissured strength of the clay as a result of general relaxation of the mass of soil surrounding the group. Because there is a relatively small pile-to-soil contact around the periphery of the group an adhesion factor will not be applied.

Unit shear resistance on sides of block $= 0.75 \times (60 + 110)/2 = 63.75$ kN/m^2

Total shear resistance $= 2 \times 12.4 \times (9.9 + 6.6) \times 63.75 = 26\,086$ kN

For calculating the base resistance from equation 5.1, N_c from Figure 5.6 is 5.14 (the classic value for a shallow foundation on clay in undrained shear), s_c is 1.3, d_c for $D/B = 1.5$ and $\phi = 0°$ is 1.0 (Figure 5.9), i_c is 1.0 for a centrally applied vertical load, and b_c is 1.0. Taking the intact shear strength of the clay at base level as 110 kN/m^2:

Base resistance of foundation $= 110 \times 5.14 \times 1.3 \times 1.5 \times 1.0 \times 1.0 \times 65.34 = 72\,039$ kN

Total resistance of foundation $= 26\,086 + 72\,039 = 98\,125$ kN

Factor of safety against general shear failure $= 98\,125/70 \times 360 = 4.0$

Checking for compliance with EC7, the stability of the individual pile is calculated using the procedure described in Section 4.2.3. The calculations are not included in this example. The same procedure is used for checking the equivalent large-diameter pile for compliance. The partial factors are as follows:

For actions (Table 4.1: Set A1, permanent unfavourable = 1.35, variable unfavourable = 1.5
                    Set A2, permanent unfavourable = 1.0, variable unfavourable = 1.3
Design value of actions Set A1, $F_d = (1.35 \times 250 + 1 = 5 \times 110) \times 70 = 35\,175$ kN
                    for Set A2 = $(1.0 \times 250 + 1.3 \times 110) \times 70 = 27\,510$ kN
Soil parameter resistances for undrained shear strengths (Table 4.2): M1 = 1.0, M2 = 1.4
Base resistances for bored piles (Table 4.4), R1 = 1.25, R2 = 1.1, R3 = 1.0
Shaft resistances for bored piles (Table 4.4), R1 = 1.0, R2 = 1.1, R3 = 1.3
Combined resistances (Table 4.4), R1 = 1.15, R2 = 1.1, R3 = 1.0

The correlation factor ξ_3 in Table 4.6 for mean values of c_u and three test profiles is divided by 1.1 because the pile loads on the group will be redistributed by a stiff pile cap. Therefore correlation factor for mean strengths $\xi_3 = 1.33/1.1 = 1.21$.

Calculating for design approach DA1, Combination1 (Sets A1 + M1 + R1), an adhesion factor of 0.45 will be used for the concrete–soil interface of the equivalent pile.

Mean group resistance of pile = $9 \times 80 \times 65.34 + 0.45 \times 85 \times 177.6$
$$= 47\,045 + 6793 = 53\,838 \text{ kN}$$

Characteristic group resistance of pile = $R_{ck} = 53\,838/1.21 = 44\,494$ kN

Design value of combined resistance = $R_{cd} = 44\,494/1.0 \times 1.15 = 38\,690$ kN
which is greater than the design action of 35 125 kN.

For design approach DA1, Combination 2 (Sets A2 + M2 + R1): $R_{cd} = 44\,494/1.4 \times 1.0$
$$= 31\,781 \text{ kN}$$
$$(>27\,510 \text{ kN})$$

Checking the equivalent pile for separate shaft and base resistances:

For DA1 Combination 1, $R_{cd} = 47\,045/1.21 \times 1.0 \times 1.25 + 6793/1.21 \times 1.0 \times 1.0$
$$= 31\,104 + 5\,614 = 36\,718 \text{ kN } (>35175 \text{ kN})$$
For DA1 Combination 2, $R_{cd} = 47\,045/1.21 \times 1.4 \times 1.25 + 6793/1.21 \times 1.4 \times 1.0$
$$= 26\,227 \text{ kN } (<27\,510 \text{ kN})$$

The lowest design resistance of the equivalent large-diameter pile from the above calculations is 26 227 kN; this can be compared with the ultimate bearing capacity of the block foundation calculated by the method in Sections 5.1 and 5.21 of 98 125 kN when divided by a nominal safety factor of 2.5 gives an allowable bearing capacity of 39 250 kN. This shows that for the particular group dimensions of Example 5.1, the EC7 concept of an equivalent large-diameter pile gives an over-conservative design.

For the arrangement of the piles shown in Figure 5.43, the overall dimensions of the group are $9 \times 1.1 = 9.9$ m, $6 \times 1.1 = 6.6$ m. The spread of the load shown in Figure 5.3a applies.

Depth to centre of equivalent raft $= \frac{2}{3} \times 13.90 = 9.3$ m

Dimensions of equivalent raft $= 6.60 + (\frac{1}{4} \times 7.80 \times 2) = 10.5$ m

and $9.90 + (\frac{1}{4} \times 7.80 \times 2) = 13.8$ m

Pressure at level of equivalent raft $q_n = \dfrac{70 \times 360}{10.5 \times 13.8} = 174$ kN/m^2

The settlements are calculated over the zone of soil down to the level of the incompressible stratum, that is at a depth of 20 m below the base of the equivalent raft. It is convenient to divide the soil into five 4 m layers commencing at 9.30 m and extending to 29.30 m. The immediate and consolidation settlements are then calculated for each layer.

Immediate settlement in Layer 1

From Figure 5.43, average $E_u = 39$ MN/m^2. From Figure 5.20, for $H/B = 4/10.5 = 0.38$, and $L/B = 13.8/10.5 = 1.3$, $\mu_1 = 0.15$, and for $D/B = 9.3/10.5 = 0.9$, and $L/B = 1.3$, $\mu_0 = 0.93$. Therefore from equation 5.22:

$$\text{Immediate settlement} = \rho_i = \frac{0.15 \times 0.93 \times 174 \times 10.5 \times 1\,000}{39 \times 1\,000} = 6.5 \text{ mm}$$

The settlements in the underlying four layers are calculated in a similar manner, the calculations for all five layers being tabulated thus

Layer	B (m)	L (m)	q_n (kN/m^2)	μ_1	μ_0	E_u (MN/mv^2)	ρ_i (mm)
1	10.5	13.8	174	0.15	0.93	39	6.5
2	15.1	18.4	90	0.06	0.93	52	1.5
3	19.7	23.0	55	0.03	0.92	64	0.5
4	24.3	27.6	37	0.02	0.92	76	0.2
5	28.9	32.2	27	0.01	0.93	88	0.1
Total immediate settlement							8.8

The immediate settlement can be checked from equation 5.21 because the deformation modulus increases linearly with depth. At the level of equivalent raft E_u is 32 MN/m^2 and at 20 m below this level it is 97 MN/m^2. Therefore from equation 5.21:

$97 = 32(1 + 20k/10.5)$

$k = 1.1$

Dividing equivalent raft into four rectangles, each 6.9×5.25 m. From Figure 5.20 for $L/B = 6.9/5.25 = 1.3$, $H/B = 20/5.25 = 3.8$ and $k = 1.1$, I_p' is 0.13. From equation in Figure 5.19:

$$\text{Settlement at corner of rectangle} = \frac{174 \times 5.25 \times 0.13 \times 1000}{32 \times 1000} = 3.7 \text{ mm}$$

Settlement at centre of equivalent raft $= 4 + 3.7 = 14.8$ mm

Oedometer settlement for Layer I

Depth to centre of layer $= 9.3 + 2.0 = 11.3$ mm

From Figure 5.13 with $L/B = 13.8/10.5 = 1.3$ and $z/B = 2/10.5 = 0.19$, stress at the centre of layer $= 0.8 \times 174$ kN/m^2. From Figure 5.43 average m_v at centre of layer $= 0.09$ MN/m^2. Therefore

$$\text{oedometer settlement from equation 5.23} = \rho_{oed} = \frac{0.09 \times 0.80 \times 174 \times 4 \times 1000}{1000}$$

$$= 50.1 \text{ mm}$$

The oedometer settlements for all five layers are calculated in a similar manner and are tabulated thus.

Layer	Depth to centre of layer (m)	z (m)	z/B	σ_z		m_v (MN/m^2)	ρ_{oed} (mm)
I	11.3	2	0.19	0.80×174		0.09	50.1
2	15.3	6	0.57	0.51×174		0.07	24.8
3	19.3	10	0.95	0.33×174		0.05	11.5
4	23.3	14	0.33	0.22×174		0.04	6.1
5	27.3	18	0.71	0.15×174		0.04	4.2
Total oedometer settlement							96.7

From Figure 5.23 the depth factor μ_d for $D/\sqrt{LB} = 9.30/\sqrt{13.8 \times 10.5} = 0.77$, is 0.78 and for London Clay the geological factor μ_g is about 0.5. Therefore

corrected consolidation settlement $= \rho_c = 0.5 \times 0.78 \times 96.7 = 37.7$ mm.
Total settlement of pile group $= \rho_i + \rho_c = 8.8 + 37.7 = 46.5$ mm.

In practice a settlement between 30 and 60 mm would be expected.

Example 5.2

Part of the jetty structure referred to in Example 4.4 carries bulk handling equipment with a total dead and imposed load of 6 MN. Design a suitable pile group to carry this equipment and calculate the settlement under the dead and imposed loading.

It has been calculated in Example 4.4 that a 450×450 mm precast concrete pile driven to 7 m below the sea bed could carry a working load of 250 kN in compression with a safety factor on the ultimate resistance of a single pile of 3.1. For uniformity in design and construction it is desirable to adopt a pile of the same dimensions to carry the bulk handling plant. A group of forty-two piles arranged in seven rows of six piles should be satisfactory.

Spacing the piles at centres equal to three times the width the dimensions of the group are $6 \times 1.35 = 8.10$ m by $5 \times 1.35 = 6.75$ m. A suitable pile cap in the form of a thick slab would be $10.5 \times 9.0 \times 1.25$ m deep. Take a depth of water of 12 m and a height of 4 m from water level to the underside of the pile cap.

Weight of pile group above sea-bed level
$$= 9.81([10.5 \times 9.0 \times 1.25 \times 2.5] + \{42 \times 0.45^2[(12 \times 1.5) + (4 \times 2.5)]\}) = 5233 \text{kN}$$

$$\text{Working load on each pile} = \frac{(6 + 5.233)}{42} \times 1\,000 = 267 \text{ kN}$$

Safety factor on ultimate load of 777 kN (Example 4.4) $= 777/267 = 2.9$, which is satisfactory.

Because the piles are driven into a uniform sand carrying their load partly in skin friction and partly in end bearing the distribution of load shown in Figure 5.3a applies.

Depth below sea bed to equivalent raft $= \frac{2}{3} \times 7 = 4.67$ m

Thus the dimensions of the equivalent raft are

$$L = 8.1 + (\tfrac{1}{4} \times 2 \times 4.67) = 10.4 \text{ m}$$
$$B = 6.75 + (\tfrac{1}{4} \times 2 \times 4.67) = 9.1 \text{ m}$$

In calculating settlements it is only necessary to consider the dead and imposed loading from the bulk handling plant. The piles and pile cap settle immediately as they are constructed and the pile cap is finished to a level surface.

$$\text{Pressure on sand below raft due to weight of plant} = \frac{6 \times 1000}{10.4 \times 9.1} = 63 \text{ kN/m}^2$$

At level of raft, effective overburden pressure $= 1.2 \times 9.81 \times 4.67 = 55 \text{ kN/m}^2$

From Figure 5.26 for a standard penetration test N-value of 15 blows/300 mm, I_c is 4×10^{-2}.

Assume for the purposes of illustration that the previous overburden pressure was 75 kN/m^2. Then from equation 5.31b the immediate settlement for an effective pressure increase of 63 kN/m^2 is:

$$\rho_i = 63 \times 9.1^{0.7} \times \frac{4 \times 10^{-2}}{3} = 3.9 \text{ mm}$$

From Figure 5.27 the depth of influence z_I for B of 9.1 m is 5 m. This is less than the thickness of the compressible layer. Hence the thickness factor, f_s, is unity. From equation 5.32a:

$$\text{Shape factor, } f_s = \left(\frac{1.25 \times 10.4/9.1}{10.4/0.1 + 0.25} \right)^2 = 1.05$$

The time factor for settlement at 30 years and static loading condition from equation 5.33 is

$$f_t = 1 + 0.3 + 0.2\frac{\log 30}{3} = 2.5$$

Therefore from equation 5.34, consolidation settlement $= 1.05 \times 1 \times 2.5 \times 3.9$
$$= 10.2 \text{ mm}$$

The imposed loading would be intermittent in operation.

Checking from equation 5.25, for d_s greater than $2B$ and $L/B = 1.1$, Figure 5.25 gives $s = 1.1$.

$$\text{Immediate settlement} = \frac{1.1 \times 63}{15^{0.87}\left(1 + 0.4 \times \dfrac{4.67}{9.1}\right)} = 5 \text{ mm}$$

Therefore the pile group would be expected to settle between 5 and 10 mm under the dead and imposed loading from the bulk handling equipment.

Example 5.3

The driven and cast-in-place piles in Example 4.5 each carry a working load of 900 kN and are arranged in a group of 20 rows of 15 piles spaced at 1.60 m centres in both directions. Calculate the settlement of the pile group using the static cone resistance diagram in Figure 4.44. Length of pile group $= 19 \times 1.6 = 30.4$ m. Width of pile group $= 14 \times 1.6 = 22.4$ m.

The transfer of load from the piles to the soft clay in skin friction is relatively small, and therefore the distribution of load shown in Figure 5.3b applies.

Depth to equivalent raft foundation $= \frac{2}{3} \times 15 = 10$ m below the surface of the sand stratum or 22 m below ground level, as shown in Figure 4.44.

Length of equivalent raft $L = 30.4 + (2 \times 10 \times \frac{1}{4}) = 35.4$ m

Width of equivalent raft $B = 22.4 + (2 \times 10 \times \frac{1}{4}) = 27.4$ m

Pressure on soil beneath raft $= \dfrac{270 \times 1\,000}{35.4 \times 27.4} = 278 \text{ kN/m}^2$

The settlement can be calculated by the Schmertmann method. It is convenient to divide the cone resistance diagram shown in Figure 4.44 into three layers between the base of the equivalent raft and rock head. The sub-division of these layers and the superimposition of the Schmertmann curves beneath the base of the raft are shown in Figure 5.44. The settlement is calculated over a period of 25 years.

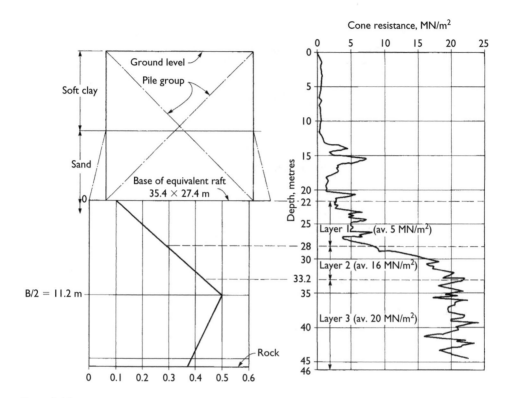

Figure 5.44

From Figure 5.44,

$$I_z$$

Layer	For $L/B = 1$	For $L/B = 10$
1	0.20	0.24
2	0.36	0.30
3a	0.46	
3b	0.40	
0.39		

Layer	For $L/B = 1$	E_d	For $L/B = 10$
1	$5 \times 2.5 = 12.5$ MN/m^2		$5 \times 3.5 = 17.5$ MN/m^2
2	$16 \times 2.5 = 40$ MN/m^2		$16 \times 3.5 = 56$ MN/m^2
3	$20 \times 2.5 = 50$ MN/m^2		$20 \times 3.5 = 70$ MN/m^2

For axisymmetric loading ($L/B = 1$) from equation 5.35:
Uncorrected settlement is given by:

$$\text{Layer 1} = \frac{278 \times 0.20 \times 6 \times 1000}{12.5 \times 1000} = 27 \text{ mm}$$

$$\text{Layer 2} = \frac{278 \times 0.36 \times 5.2 \times 1\,000}{40 \times 1\,000} = 13 \text{ mm}$$

$$\text{Layer 3a} = \frac{278 \times 0.46 \times 2.5 \times 1\,000}{50 \times 1\,000} = 6 \text{ mm}$$

$$\text{Layer 3b} = \frac{278 \times 0.4 \times 10.3 \times 1\,000}{50 \times 1\,000} = 23 \text{ mm}$$

$$\text{Total} = 69 \text{ mm}$$

Similarly, for $L/B > 10$ the uncorrected settlements are

Layer 1 = 19 mm
Layer 2 = 9 mm
Layer 3 = 21 mm
    Total = 49 mm

By interpolation the settlement for $L/B = 1.3$ is 66 mm

Effective overburden pressure at base of raft

$$= 9.81[(2 \times 1.9) + (10 \times 0.9) + (10 \times 0.9)] = 214 \text{ kN/m}^2$$

From equation 5.36 $C_1 = 1 - 0.5 \times \dfrac{214}{278} = 0.62$

From equation 5.37 $C_2 = 1 + 0.2lg\dfrac{25}{0.1} = 1.48$

Corrected settlement at 25 years = $0.62 \times 1.48 \times 66 = 61$ mm say between 50 and 70 mm

Example 5.4

Nuclear reactors, their containment structures, and ancillary units weighing 900 MN are to be constructed on a base 70×32 m sited on 8 m loose to medium-dense sand overlying a moderately strong sandstone. Rotary cored boreholes showed that below a thin zone of weak weathered rock the RQD value of the sandstone was 85% and the average unconfined compression strength was 14 MN/m². Design a suitable piled foundation and calculate the settlement.

$$\text{Overall loading on base} = \frac{900}{70 \times 32} = 0.402 \text{ MN/m}^2$$

Under this loading the settlement of the sand will be excessive. A piled foundation is required and relatively few large-diameter piles will be more economical than a large number of lightly loaded piles.

Adopt piles 1.5 m in diameter taken 2 m below weak weathered rock on to the moderately strong sandstone. It will be possible to permit loading of the concrete forming the bored piles to the maximum working stress on the concrete of the pile shaft. For 25-grade concrete, max. working stress $= 0.25 \times 25 = 6.25$ MN/m^2 (factor of safety on $q_{uc} = 2.2$) and since max. working load on pile $\frac{1}{4} \times \pi + 1.5^2 \times 6.25 = 11$ MN, required number of piles $= 900/11 = 82$.

A suitable arrangement of piles consists of fourteen rows of six piles placed at 5 m centres in both directions.

Length of pile group $L = 13 \times 5 = 65$ m
Width of pile group $B = 5 \times 5 = 25$ m

The transfer of load in skin friction to the sand is relatively small and the piles can be regarded as end bearing on the rock. The base of the equivalent raft will be shown in Figure 5.3c.

$$\text{Overall loading at base of raft} = \frac{900}{65 \times 25} = 0.55 \text{ MN/m}^2$$

From Section 4.7.3 for RQD of 85%, mass factor $= 0.7$ and from Section 5.5 the modulus ratio of a well-cemented sandstone is 300, deformation modulus of sandstone $= E_d = 0.7 \times 300 \times 14 = 2940$ MN/m^2, say 3000 MN/m^2.

From Figure 5.20 with $H/B = \infty$, and $L/B = 65/25 = 2.6$, $\mu_1 = 1.1$, and with $D/B = (8 + 2)/25 = 0.4$ and $L/B = 2.6$, $\mu_0 = 0.95$. From equation 5.22:

$$\text{settlement of foundation} = \frac{1.1 \times 0.95 \times 0.55 \times 25 \times 1000}{3000} = 5 \text{ mm}.$$

Example 5.5

A site, where the ground conditions consist of 5.5 m of soft organic silty clay overlying 35 m of stiff to very stiff over-consolidated clay followed by rock, is reclaimed by placing and compacting 4 m of sand fill covering the entire site area. Six months after completing the reclamation a 12-storey building imposing an overall dead load and sustained imposed load of 160 kN/m^2 on a ground floor area of 48 m by 21 m is to be constructed on the site. The average undrained shearing strength of the stiff clay stratum is 90 kN/m^2 at the surface of the stratum, increasing to 430 kN/m^2 at rockhead. Measurements of the deformation modulus and modulus of volume compressibility show a linear variation, with average values at the top and bottom of the stiff clay stratum as follows:

At top: $E_u = 40$ MN/m^2, $m_v = 0.08$ m^2/MN
At bottom: $E_u = 120$ MN/m^2, $m_v = 0.04$ m^2/MN

Design suitable piled foundations and estimate the settlement of the completed building.

Because of the heavy loading it is economical to provide large-diameter bored and cast in-situ piled foundations. A suitable arrangement consists of fourteen rows of six piles (Figure 5.45). Trial-and-adjustment calculations show that a pile diameter of 1200 mm is suitable. The pile spacing must be a minimum of 3 diameters, giving a spacing of at least 3.6 m.

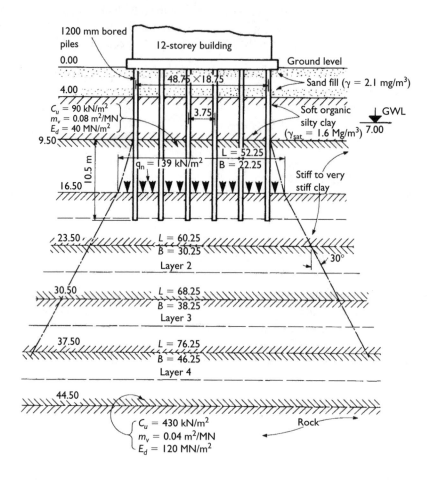

Figure 5.45

Adopt a spacing of say 3.75 m in both directions. Thus the dimensions of the pile group are $5 \times 3.75 = 18.75$ m and $13 \times 3.75 = 48.75$ m.

$$\text{Average load carried by piles} = \frac{48 \times 21 \times 160}{14 \times 6} = 1920 \text{ kN per pile}$$

The central two rows of piles carry higher loads than the outer two rows on each side. A likely loading for the centre rows is $2\,200$ kN per pile. The required penetration of the piles will be calculated on this loading. The exterior piles will be taken to the same depth but adopting a reduced diameter as required by the lesser loading.

The piles carry negative skin friction due to the consolidation of the soft clay under the imposed loading of the sand fill. At 6 months, settlement of the soft clay will be continuing at a very slow rate and it is appropriate to use Figure 4.39 to calculate the negative skin friction in this layer.

Unit negative skin friction at top of layer

$$= 0.30\sigma'_{vo} = 0.30 \times 9.81 \times 2.1 \times 4 = 4.7\,\text{kN/m}^2$$

Unit negative skin friction at groundwater level (see Figure 5.45)

$$= 0.30 \times 9.81[(2.1 \times 4) + (1.6 \times 3)] = 8.8\,\text{kN/m}^2$$

Unit negative skin friction at bottom of layer

$$= 0.30 \times 9.81[(2.1 \times 4) + (1.6 \times 3) + (0.6 \times 2.5)] = 43.3\,\text{kN/m}^2$$

Therefore total negative skin friction in soft clay:

$$= \pi \times 1.2 \times [\tfrac{1}{2}(4.7 + 8.8) \times 3 + \tfrac{1}{2}(8.8 + 43.3) \times 2.5] = 746\,\text{kN}$$

Because the pile will settle due to yielding of the stiff clay when the full working load is applied, the pile will move downwards relative to the lower part of the soft clay. Thus negative skin friction will be developed only over about 80% of the length within the soft clay. Thus

approximate total negative skin friction in soft clay $= 0.8 \times 746 = 597\,\text{kN}$

The negative skin friction in the sand can be calculated using the coefficients for K_s in Table 4.10. Although the compacted sand fill is dense it will be loosened by pile boring to give a coefficient K_s of 1 and a value of ϕ of 30°. From equation 4.16:

negative skin friction on pile in sand fill

$$= \tfrac{1}{2} \times 1 \times 9.81 \times 2.1 \times 4 \times \tan 30° \times \pi \times 1.2 \times 4 = 359\,\text{kN}$$

Total negative skin friction on pile $= 359 + 597 = 956\,\text{kN}$

Total applied load on piles in centre rows $= 956 + 2200 = 3156\,\text{kN}$

The required pile penetration depth is calculated on the basis of the building loading, with a check being made to ensure that the safety factor on the combined building load and negative skin friction is adequate.

Required ultimate pile resistance for overall safety factor of 2 (Section 4.6) $=$ $2 \times 2200 = 4400\,\text{kN}$

Take a trial penetration depth of 10 m into the stiff clay stratum. At the pile base level $c_{ub} = 190\,\text{kN/m}^2$ and the average value of c_u on the shaft is 140 kN/m². Thus

Ultimate base resistance $= \tfrac{1}{4} \times \pi \times 1.2^2 \times 9 \times 190 = 1935\,\text{kN}$

Load to be carried in skin friction $= 4400 - 1935 = 2465\,\text{kN}$

The adhesion factor for a straight-sided pile can be taken as 0.45. Therefore from equation 4.7:

total load $= 2465 = 0.45 \times 140 \times \pi \times 1.2 \times l$

from which $l = 10.4$ m (say 10.5 m) and the trial depth is satisfactory.

Checking the criterion of a safety factor of 3 in end bearing and unity in skin friction, allowable load $= (\frac{1}{3} \times 1935) + 2\,465 = 3110$ kN, which roughly equals the building load plus the negative skin friction. Checking the overall safety factor on the combined loading,

Safety factor $= (1\,935 + 2\,465)/3\,156 = 1.4$

This is satisfactory since the negative skin friction on the piles will not contribute to the settlement of the pile group.

The transfer of load from the pile group to the soil will be as shown in Figure 5.3b.

Depth below ground level to base of equivalent raft
$= 4 + 5.5 + (\frac{2}{3} \times 10.5) = 16.5$ m

The dimensions of the equivalent raft are:

$L = 48.75 + (\frac{2}{3} \times 10.5 \times 2 \times \frac{1}{4}) = 52.25$ m

$B = 18.75 + (\frac{2}{3} \times 10.5 \times 2 \times \frac{1}{4}) = 22.25$ m

Pressure on base of equivalent raft due to building load $= \dfrac{48 \times 21 \times 160}{52.25 \times 22.25} = 139$ kN/m^2

Calculating the immediate settlement
At a level of equivalent raft, $E_u = E_f = 65$ MN/m^2
At rockhead, $E_u = 120$ MN/m^2
From equation 5.21, $120 = 65(1 + 28\,k/22.25)$
$$k = 0.7$$
Divide equivalent raft into four rectangles, each 26.1×11.1 m
From Figure 5.19 for $L/B = 26.1/11.1 = 2.3$, $H/B = 28/11.1 = 2.5$, and $k = 0.7$, I_p' is 0.14.

Settlement at corner of rectangle $= \dfrac{139 \times 11.1 \times 0.14 \times 1\,000}{65 \times 1\,000} = 3.3$ mm

Settlement at centre of equivalent raft $= 4 \times 3.3 = 13.2$ mm

Calculating the consolidation settlement
To calculate the settlement of the pile group due to the building loads only, the 28 m layer of clay between the equivalent raft and rockhead is divided into four 7 m layers.

Oedometer settlement in Layer 1.
From Figure 5.13 with and $z/B = 3.5/22.2 = 0.16$ and $L/B = 52.2/22.2 = 2.3$, stress at centre of rectangle $= 0.83 \times 139 = 118$ kN/m^2. Modulus of volume compressibility $= 0.07$ m^2/MN. Then from equation 5.23:

oedometer settlement uncorrected by depth factor
$$= \dfrac{0.07 \times 118 \times 7 \times 1\,000}{1\,000} = 57.8 \text{ mm}$$

The settlements in the remaining layers are calculated similarly, the results for the four layers are tabulated as follows:

Layer	Depth to centre of layer (m)	z (m)	z/B	σ_z (kN/m²)	m_v (MN/m²)	ρ_{oed} (mm)
1	20.00	3.5	0.16	118	0.07	57.8
2	27.00	10.5	0.47	88	0.06	37.0
3	34.00	17.5	0.79	64	0.05	22.4
4	41.00	24.5	1.10	50	0.04	14.0
Total uncorrected oedometer settlement						131.2

The above summation must be corrected by a depth factor which is given by Figure 5.23, with $D/\sqrt{LB} = 16.5/\sqrt{52.25 \times 22.25} = 0.48$ and $L/B = 2.35$, as $\mu_d = 0.85$.

To obtain the consolidation settlement ρ_c the summation is also multiplied by the geological factor μ_g, which is 0.5 for an overconsolidated clay. Therefore

total consolidation settlement $= 0.85 \times 0.5 \times 131.2 = 55.8$ mm

Total settlement of pile group due to building load only $= \rho_i + \rho_c = 13.2 + 55.8 = 69.0$ mm

To this figure must be added the consolidation settlement of the stiff clay due to the sand filling. The immediate settlement is not taken into account since this will have taken place before commencing the construction of the building.

Oedometer settle due to 4 m of sand fill for an average m_v of 0.06 m²/MN in clay layer

$$= \frac{0.06 \times 9.81 \times 2.1 \times 4 \times 1000}{1000} = 4.9 \text{ mm}$$

Correcting for the geological factor as equation 5.24:

$\rho_c = 0.5 \times 4.9 = 2.4$ mm

A time-settlement calculation would show that about one-third of this settlement would be complete before completing the pile installation. Thus

settlement of 12-storey building due to combined loading from building and sand layer
$= 69.0 + (\tfrac{2}{3} \times 2.4) = 70.6$ mm

Therefore it would be reasonable to assume that the final settlement of the building would be between 50 and 100 mm.

It will be noted that the negative skin friction on the piles was not added to the loading on the equivalent raft when calculating the settlement of the building. However, it is necessary to check that the individual piles will not settle excessively under the combined building load and negative skin friction.

Maximum load on pile = 3156 kN. If shaft friction on pile is fully developed, the end-bearing load is $3156 - 2465 = 691$ kN, and thus

$$\text{End-bearing pressure} = \frac{691}{\frac{1}{4}\pi \times 1.2^2} = 611 \text{ kN/m}^2$$

Ultimate unit base resistance $= 9 \times 190 = 1\,710$ kN/m^2

From equation 4.36, with $K = 0.01$,

$$\rho_i = 0.01 \times \frac{611}{1\,710} \times 1\,200 = 4 \text{ mm}$$

Therefore individual piles will not settle excessively and the critical factor is the overall settlement of the complete pile group, for which a movement of 50 to 100 mm over a long period of years is by no means excessive.

Chapter 6

The design of piled foundations to resist uplift and lateral loading

6.1 The occurrence of uplift and lateral loading

When vertical piles are installed beneath buoyant structures such as drydocks, basements and pumping stations they are required to resist uplift loads. Where the hydrostatic pressure always exceeds the downward loading, as in the case of some underground tanks and pumping stations, the anchorages are permanently under tension and cable anchors may be preferred to piles. However, in the case of the shipbuilding dock floor in Figure 6.1, for example, the anchorages may be under tension only when the dock is unwatered before the commencement of shipbuilding. As the loading on the floor from ship construction increases to the stage at which the uplift pressure is exceeded, the anchor piles are required to carry compressive loads. Cable anchors might not then be suitable if the dock floor was underlain by soft or loose soil.

Vertical piles are used to restrain buildings against uplift caused by the swelling of clay soils. Swelling can occur, for example, when mature trees are removed from a building site. The desiccated soil in the root zone of the trees gradually absorbs water from the surrounding clay, and the consequent swelling of the clay may amount to an uplift of 50 to 100 mm of the ground surface, causing severe damage to buildings sited over the root zone. In sub-tropical countries where there is a wide difference in seasonal climatic conditions, i.e. a hot dry summer and a cool wet winter, the soil zone affected by seasonal moisture changes can extend to a depth of several metres below the ground surface. In clay soils these changes cause the ground surface to alternately rise and fall with a differential movement of 50 mm or more. The depth to which these swelling (or alternate swelling and shrinkage) movements can occur, usually makes the use of piled foundations taken below the zone of soil movements more economical and technically more suitable than deep strip or pad foundations.

Vertical piles must have a sufficient depth of penetration to resist uplift forces by the development of shaft friction in the soil beneath the zone of soil movements (Figure 6.2). Uplift on bored piles can be reduced by casting the concrete in the upper part of the pile within a smooth polyvinylchloride (pvc) sleeve, or by coating a precast concrete or steel tubular pile with soft bitumen (see Section 4.8.3). Uplift can be further reduced by supporting the superstructure clear of the ground surface, or by providing a compressible layer beneath pile caps and ground beams (see Figure 7.15). Piles in large groups may also be lifted due to ground heave, as described in Section 5.7.

In countries where frost penetrates deeply below the ground surface, frost expansion of the soil can cause uplift on piles, resulting in severe effects in 'permafrost' regions, as

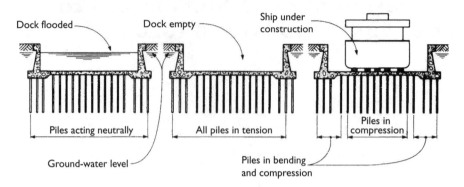

Figure 6.1 Tension/compression piles beneath floor of shipbuilding dock.

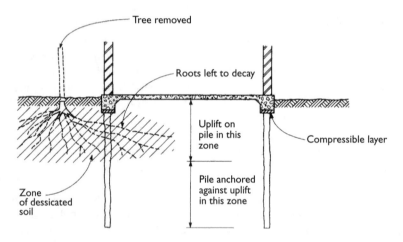

Figure 6.2 Uplift on pile due to swelling of soil after removal of mature tree.

described in Section 9.4. Floating ice on lakes and rivers can jam between piles in groups causing them to lift when water levels rise or when the ice sheet buckles.

The most frequent situation necessitating design against lateral and uplift forces occurs when the piles are required to restrain forces causing the sliding or overturning of structures. Lateral forces may be imposed by earth pressure (Figure 6.3a), by the wind (Figure 6.3b), by earthquakes, or by the traction of braking vehicles (Figure 6.3c). In marine structures lateral forces are caused by the impact of berthing ships (Figure 6.4), by the pull from mooring ropes, and by the pressure of winds, currents, waves, and floating ice. A vertical pile has a very low resistance to lateral loads and, for economy, substantial loadings are designed to be resisted by groups of inclined or raking piles (sometimes referred to as 'batter' piles). Thus in Figure 6.5 the horizontal force can be resolved into two components, producing an axial compressive force in pile A and a tensile force in pile B. It is usual to ignore the restraint offered by the pile cap; thus the magnitude of each component is

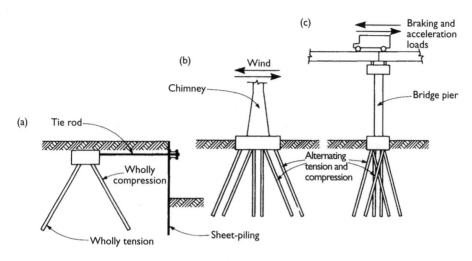

Figure 6.3 Raking piles to resist overturning forces. (a) Piled anchorage to tie rods restraining sheet-piled retaining wall (b) Raking piles to withstand wind forces on chimney (c) Raking piles to withstand traction forces from vehicles on bridge.

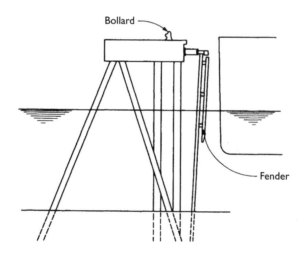

Figure 6.4 Raking and vertical piles in breasting dolphin.

obtained from a simple triangle of forces as shown. Where lateral forces are transient in character, for example, for wind loadings, they may be permitted to be carried wholly or partly by the pile cap where this is bearing on the ground (see Section 7.8). If raking piles are installed in fill or compressible soil which is settling under its own weight or under a surcharge pressure, considerable bending stresses can be induced in the piles, requiring a high moment of resistance to withstand the combined axial and bending stresses as discussed in Section 6.4.

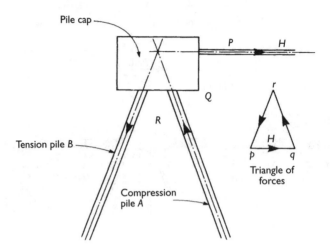

Figure 6.5 Restraint of horizontal force by raking piles.

6.2   Uplift resistance of piles

6.2.1 General

The simplest method of restraining piles against uplift is to employ a pile shaft that is sufficiently long to take the whole of the uplift load in shaft friction. However, where there is rock beneath a shallow soil overburden it may not be possible to drive the piles deeply enough to mobilize the required frictional resistance. In such cases the shaft resistance must be augmented by adding dead weight to the pile to overcome the uplift load, or by anchoring the pile to the rock.

Adding dead weight to counteract uplift loading is not usually feasible or economical. The piles may be required to carry alternating uplift and compressive loads, in which case the added dead weight would result in a large increase in the compressive loading. In the case of shipbuilding dock floors (Figure 6.1), dead weight in the form of a thick floor would add considerably to the construction costs, and in piled dolphins (Figure 6.4) the provision of a massive pile cap could make a substantial addition to the load on the compression rakers. Experience has shown that anchors in the form of grouted-in bars, tubes or cables are the most economical means of providing the required uplift resistance for piles taken down to a shallow rock layer.

6.2.2 The uplift resistance of friction piles

The resistance of straight-sided piles in shaft friction to statically applied uplift loads is calculated in exactly the same way as the shaft friction on compression piles, and the calculation methods given in Sections 4.2 to 4.5 can be used. However, for cyclic loading the frictional resistance is influenced by the rate of application of the load and the degree of degradation of the soil particles at the interface with the pile wall. In the short term, the uplift resistance of a bored pile in clay is likely to be equal to its frictional resistance in compression; however, Radhakrishna and Adams[6.1] noted a 50% reduction in the uplift resistance of cylindrical augered footings and a 30% to 50% reduction in belled footings in clay when sustained loads were carried over a period of 3 to 4 months. It was considered that the reduction in uplift was due to a loss of suction beneath the pile base and the dissipation of negative pore pressures set up at the initial loading stage. These authors pointed out that such reductions are unlikely for piles where the depth/width ratio is greater than 5.

The ICP method[4.30] can be used to determine the tension capacity of driven piles carrying tension loading. For piles in clay the method does not differentiate between shaft resistance in compression or tension, i.e. equations 4.20 to 4.24 can be used without modification for either type of loading. Conditions are different for piles in sands where the degradation of the soil particles at the pile–soil interface has a greater effect on stability. Also in the case of tubular steel piles the radial contraction across the diameter under tension loads is a further weakening effect on frictional resistance, particularly for open end piles. Accordingly, equation 4.27 is modified to become

$$\tau_f = (0.8\,\sigma'_{rc} + \Delta\sigma'_{rd})\tan\delta'_{cv} \tag{6.1}$$

where σ'_{rc} and $\Delta\sigma'_{rd}$ are calculated as described for compression loading in Section 4.3.7. For open end piles in tension τ_f as calculated by equation 6.1 is reduced by a factor of 0.9.

Cyclic loading generally results in a weakening of shaft capacity. The reduction can be significant for offshore structures where piles are subjected to repetitive loading from wave action. The degree of reduction depends on the amplitude of shear strain at the pile–soil interface, the susceptibility of the soil grains to attrition, and the number and direction of the load-cycles, i.e. one-way or two-way loading. The amplitude of the shear strain depends in turn on the ratio of the applied load to the ultimate shaft capacity. In clays the repeated load applications increase the tendency for the soil particles to become re-aligned in a direction parallel to the pile axis at the interface which may eventually result in residual shear conditions with a correspondingly low value of δ_{cv}. In sands, it is evident that the greater the number of load-cycles the greater the degree of degradation, although the residual silt-sized particles produced by a silica sand will have an appreciable frictional resistance.

Degradation, both in sands and clays, takes place initially in the region of the soil-line where the amplitude of the tensile strain is a maximum; it then decreases progressively down the shaft but may not reach the pile toe if the applied load is a relatively small proportion of the ultimate shaft capacity.

Jardine et al.[4.30] recommend cyclic shear tests in the laboratory using the site-specific materials as a means of quantifying the reduction in friction capacity. In clays the interface shear is likely to occur in undrained conditions; accordingly, the laboratory testing programme should provide for simple cyclic undrained shear tests. An alternative to laboratory testing suggested by Jardine et al. is to simulate the relative movement between pile and soil under repetitive loading by finite element or t–z analyses (Section 4.6).

EC7 adopts a criterion for avoiding the ultimate limit state for single piles or pile groups in tension by the expression similar to that for compression loading, that is

$$F_{td} \leq R_{td} \tag{6.2}$$

where F_{td} is the design value for actions in tension on a pile or pile group and R_{td} is the design value of resistance in tension of the pile or the foundation. Partial factors for actions are as shown for compression piles in Table 4.1.

Two modes of failure are to be examined:

(1) The pull-out of the pile from the ground mass and
(2) Uplift of a block of ground containing the piles.

For condition (1) the risk of pull-out of a cone of soil adhering to the pile is to be considered. The adverse effects of cyclic loading as described above are to be taken into account.

EC7 permits the ultimate tensile resistance to be determined by pile loading tests. It is recommended that more than one test should be made, and in the case of a large number of piles at least 2% should be tested. Correlation factors (see Table 4.7) are applied to the test results to obtain the characteristic tension resistance R_{tk}.

Analytical methods as described in Chapter 4 for compression loading are permitted by EC7 to be used for calculating resistance to tension loading. The correlation factors shown in Table 4.6 are applied to the results of the calculations to obtain characteristic values (R_{stk}). The factors depend on the number of ground test results used to provide the basis for the calculations. The partial factor for shaft resistance, γ_{st}, is then applied to obtain R_{std}. The partial factors shown in Tables 4.3–4.5 depend on the type of pile. It will be noted that the factors are generally higher than those for shaft resistance in compression reflecting the potentially more damaging effects of failure of a foundation in uplift.

The 'model pile' and the design approaches, as described in Section 4.1.4, are used in conjunction with analytical methods of determining tension resistance. Determination by dynamic pile testing appears to be excluded by EC7. Presumably this is because the method measures shaft resistance in compression requiring an empirical correction to obtain the tension value and because the direction, speed and duration of loading are such that predictions could be misleading.

Where vertical piles are arranged in closely spaced groups the uplift resistance of the complete group may not be equal to the sum of the resistances of the individual piles. This is because, at ultimate-load conditions, the block of soil enclosed by the pile group is lifted. The manner in which the load is transferred from the pile to the soil is complex and depends on the elasticity of the pile, the layering of the soil and the disturbance to the ground caused by installing the pile. A spread of load of 1 in 4 from the pile to the soil provides a simplified and conservative estimate of the volume of a coarse soil available to be lifted by the pile group, as shown in Figure 6.6. For simplicity in calculation, the weight of the pile embedded

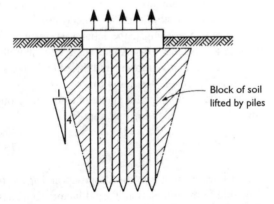

Figure 6.6 Uplift of group of closely spaced piles in fine-grained soils.

in the ground is assumed to be equal to that of the volume of soil it displaces. If the weight of the block of soil is calculated by using a diagram of the type shown in Figure 6.6, then the safety factor against uplift can be taken as unity, since frictional resistance around the periphery of the group is ignored in the calculation. The submerged weight of the soil should be taken below groundwater level.

In the case of fine-grained soils the uplift resistance of the block of soil in undrained shear enclosed by the pile group in Figure 6.7 is given by the equation:

$$Q_u = (2LH + 2BH)\bar{c}_u + W \tag{6.3}$$

where Q_u is the total uplift resistance of the pile group, L and B, are the overall length and width of the group, respectively, H is the depth of the block of soil below pile cap level, $\bar{c}_u$ is the value of average undisturbed undrained shear strength of the soil around the sides of the group, and W is the combined weight of the block of soil enclosed by the pile group plus the weight of the piles and pile cap. Submerged densities are used for the soil and portion of the structure below groundwater level when calculating W.

A safety factor of 2 should be used with equation 6.3 to allow for the possible weakening of the soil around the pile group caused by the method of installation. For long-term sustained loading a safety factor of 2.5 to 3 would be appropriate.

If either of the above two methods is used to calculate the combined uplift resistance of a pile group, the allowable resistance must not be greater than that provided by the sum of the skin-frictional resistance of the individual piles in the group divided by the appropriate safety factor.

EC7 (Clause 7.6.3.1) recommends calculating the uplift resistance of a block of soil surrounding the pile group in a manner similar to that described above. The design value of the uplift load combined with the uplift force from buoyancy on the underside of the soil block, V_{dstd}, are resisted by the design values of the friction on the vertical outer surfaces of the block, T_d, and the stabilizing forces, G_{stbd}, of the mass of soil composing the block, the pile cap, or other substructures supported by the piles, and the weight of any soil overburden above these structures. The resistances of the piles to pull-out are not included in the stabilizing forces but are considered separately since they provide no resistance if failure is by lifting of the mass of soil.

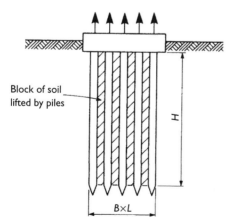

Block of soil lifted by piles

H

$B \times L$

Figure 6.7 Uplift of group of piles in coarse-grained soils.

Because buoyancy is a destabilizing factor, EC7 (Clause 2.4.7.4) requires verification of stability by the UPL criteria as given by the equation:

$$V_{dstd} \leq G_{stbd} + R_d \tag{6.3a}$$

where

$$V_{dstd} = G_{dstd} + Q_{dstd} \tag{6.4}$$

and

V_{dstd} = design value of the permanent destabilizing vertical action on the substructure
G_{stbd} = design value of the permanent stabilizing vertical actions
G_{dstd} = design value of the permanent destabilizing actions
Q_{dstd} = design values of the variable actions
R_d = any additional resistance to uplift
R_d = considered as a permanent vertical action.

The partial factors for actions for the ultimate limit state (UPL) verification are set out in Annex A of EC7 as shown in Table 6.1. For verification of the uplift resistances of the soil surrounding the block, and of the pull-out resistances of the piles in the group, where derived by calculations using soil parameters the partial factors shown in Table 6.2 are used.

6.2.3 Piles with base enlargements

When bored piles are constructed in clay soils, base enlargements can be formed to anchor the piles against uplift. The enlargements are made by the belling tools described in

Table 6.1 Partial factors for actions (γ_f) for uplift limit-state (UPL) verifications

Action	Symbol	Value
Permanent		
Unfavourable[a]	γ_{Gdst}	1.0
Favourable[b]	γ_{Gstb}	0.9
Variable		
Unfavourable[a]	γ_{Qdst}	1.5

Notes
a denotes destabilizing.
b denotes stabilizing.

Table 6.2 Partial factors for soil parameters and resistances

Soil parameter	Symbol	Values
Angle of shearing resistance[a]	γ_ϕ'	1.25
Effective cohesion	γ_c'	1.25
Undrained shear strength	γ_{cu}	1.40
Tensile pile resistance	γ_{st}	1.40
Anchorage resistance	γ_a	1.40

Note
a This value is applied to tan ϕ'.
The factors in Tables 6.1 and 6.2 may be modified when the British National Annex is published.

Section 3.3.1. Enlargements cannot be formed in coarse soils unless the borehole is drilled with the support of a bentonite slurry. The size and stability of an enlargement made in this way is problematical. Full-scale loading tests are essential to prove the reliability of the bentonite method for any particular site. Reliable predictions cannot be made of the size and shape of base enlargements formed by hammering out a bulb of concrete at the bottom of a driven and cast in-situ pile as described in Section 2.3.2. End enlargements formed on precast concrete or steel piles, although providing a substantial increase in compressive resistance when driven to a dense or hard stratum, do not offer much uplift resistance since a gap of loosened soil is formed around the shaft as the pile is driven down.

In the case of bored piles in fine-grained soils installed using belling tools, resistance to uplift loading provided by the straight-sided portion of the shaft is calculated over the depth H in Figure 6.8 minus the overall depth of the under-ream. Failure under short-term loading takes place in undrained shear on the pile to clay interface. The mobilized resistance should take into account the effects of installation as described in Section 4.2.3. Uplift resistance of the projecting portion of the enlarged base is assumed to be provided by compression resistance of the soil overburden.

Resistance to long-term uplift loading on piles in fine-grained soils is calculated by effective stress methods as described for clayey sands in the following paragraphs.

Meyerhof and Adams[6.2] investigated the uplift resistance of a circular plate embedded in a partly clayey $(c - \phi)$ soil and established the equation

$$Q_u = \pi c B H + s \times \frac{1}{2}\pi \times \gamma \times B(2D - H)\, H K_u \tan \phi + W \qquad (6.5)$$

where Q_u is the uplift resistance of the plate, B is the diameter of the plate, H is the height of the block of soil lifted by the pile (Figure 6.9), c is the cohesive strength of the soil, s is a shape factor (see below), γ is the density of the soil (the submerged density being taken below groundwater level), D is the depth of the plate, K_u is a coefficient obtained from Figure 6.9, ϕ is the angle of shearing resistance of the soil, and W is the weight of the soil resisting uplift by the plate.

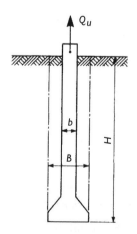

Figure 6.8 Uplift of single pile with base enlargement in fine-grained soil $(\phi = 0)$.

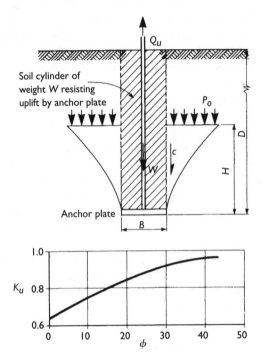

Figure 6.9 Uplift of circular plate in partly clayey $(c - \phi)$ or sandy $(c = 0)$ soil (after Meyerhof and Adams[6.2]).

If equation 6.5 is adapted to a pile with an enlarged base the weight of the pile is taken in conjunction with the weight of the soil when calculating W.

It will be noted that for deeply embedded plates or pile enlargements, H does not extend up to ground level and its value can be obtained from tests made by Meyerhof which gave the following results:

ϕ	20°	25°	30°	35°	40°	45°	48°
H/B	2.5	3	4	5	7	9	11

The shape factor s for deep foundations (including piles) is equal to $1 + mH/B$, where m depends on the angle of shearing resistance ϕ of the soil. Meyerhof's values of m and the maximum permissible values of the shape factor are as follows:

ϕ	20°	25°	30°	35°	40°	45°	48°
m	0.05	0.1	0.15	0.25	0.35	0.50	0.60
max. s	1.12	1.30	1.60	2.25	3.45	5.50	7.60

The value of Q_u calculated from equation 6.5 must not exceed the combined resistance of the enlarged base (considered as a buried deep foundation) and the pile shaft friction. These components are calculated as described in Chapter 4.

The shaft length is taken as the overall depth of the pile, from which the depth of the enlargement and any allowance made for the shrinkage of the soil away from the pile at the ground surface are deducted. Where piles in clay have to carry long-term sustained uplift loading, and the ratio of the depth of these piles to the width of the enlarged base is less than 5, the uplift resistance, as calculated by equation 6.5 or the methods in Chapter 4, should be reduced by one-half.

Where piles with base enlargements are installed in groups the uplift resistance of the group can be calculated as described in Section 5.2.1.

6.2.4 Anchoring piles to rock

Rock anchors are provided for tension piles when the depth of soil overburden is insufficient to develop the required uplift resistance on the pile in shaft friction. In weak rocks such as chalk or marl it is possible to drive piles into the rock, or to drill holes for bored piles so that the frictional resistance can be obtained on the pile shaft at its contact surface with the rock. However, driving piles into a strong rock achieves only a small penetration and so shatters the rock that no worthwhile resistance can be obtained. The cost of drilling into a strong rock to form a bored pile is not usually economical compared with that of drilling smaller and deeper holes for anchors as described below, although drilling-in large-diameter piles to carry ship berthing forces in marine structures is sometimes practised (see Section 8.2).

Anchorages in rock are formed after driving an open-ended tubular pile to seat the toe of the pile into the rock surface. The pile must not be driven too hard at this stage as otherwise the toe will buckle, thus preventing the entry of the cleaning-out tools and the anchor drilling assembly. The soil plug within the pile is removed by baling, washing or 'airlifting'. If a bored pile is to be anchored, the borehole casing is drilled below rock level to seal off the overburden. All the soil within the piling tube is cleaned out, and drilling pipes with centralizers are lowered down to the rock level. The anchor hole is then drilled to the required depth. The cuttings washed out of the hole are removed by reverse circulation up the drilling pipe or through a conductor tube up to the surface. The anchor, which can consist of a high-tensile steel bar or a stranded cable, is fed down the hole. A small-bore nylon tube is taped to the anchor and used to inject the grout at the bottom of the drilling hole (Figure 6.10). At this level the bar or cable is provided with a compression fitting to ensure that the full bonded length of the anchor acts in resisting uplift. Stranded cables are parted (after removing the sheath) and the strands are degreased over the lower part which is bonded to the grout.

Grout is injected through the nylon tube to fill the annulus completely, and it is also allowed to fill the piling tube to the required level. Where the anchors are stressed, the bar or cable is carried up to the top of the pile or pile cap to which the stress from the anchor is transferred by a stressing head and jack. The anchor is greased and sheathed in a plastic tube supplemented by wrapping with waterproof tape to protect the unbonded length from corrosion. The space surrounding the sheathed length can be filled with grout or concrete, or left as a void. The latter is usually required in the case of piles in marine structures to allow them to flex under lateral loading.

Unstressed or 'dead' anchors can consist of steel tubes installed by drilling them down into rock. On reaching the required depth, grout is pumped down the drilling pipe where it emerges at the drilling bit and fills the annulus between the anchor tube and the rock. A sealing plate prevents the grout from entering the space between the anchor tube and the drilling pipe, as shown in Figure 6.11. The grout is allowed to fill the pile to the height necessary to cover the top of the anchor tube, so as to protect it from corrosion and to serve as the

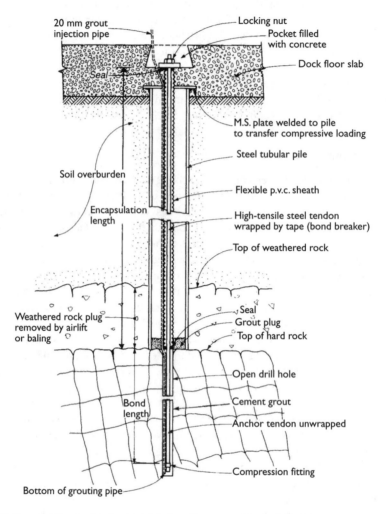

Figure 6.10 Stressed bar tendon in steel tubular pile supporting dock floor.

medium transferring the uplift load from the pile on to the anchor. Where large uplift loads are carried, the transfer of load is effected by welding a mild steel strip on to the interior surface of the pile and the exterior of the anchor tube to act as a shear key, as described in the following section. The drilling bit is left in place at the bottom of the tube where it acts as a compression fitting, but the drilling rods are disconnected at a special coupling.

6.2.5 The uplift resistance of drilled-in rock anchors

The resistance to pull-out of anchors drilled and grouted into rock depends on five factors, each of which must be separately evaluated. They are as follows:

(1) The safe working stress on the steel forming the anchor

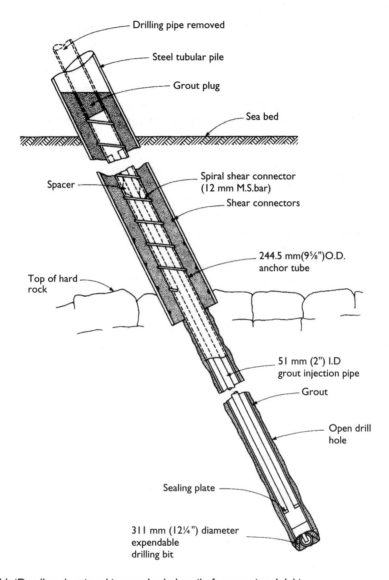

Drilling pipe removed

Steel tubular pile

Grout plug

Sea bed

Spacer

Spiral shear connector
(12 mm M.S.bar)

Shear connectors

244.5 mm(9⅝")O.D.
anchor tube

Top of hard
rock

51 mm (2") I.D
grout injection pipe

Grout

Open drill
hole

Sealing plate

311 mm (12¼") diameter
expendable
drilling bit

Figure 6.11 'Dead' anchor in raking steel tubular pile for mooring dolphin.

(2) The allowable bond stress between the anchor and the grout
(3) The allowable bond stress between the grout and the rock
(4) The dead weight of the mass of rock and any overlying soil which is lifted by the anchor, if prior failure does not occur due to the preceding three factors and
(5) The dead weight of the mass of rock and any soil overburden which is lifted by a group of closely spaced anchors.

The allowable bond stress between the anchor and the grout depends on the compressive strength of the grout, the amount of keying or roughening given to the steel surface, the diameter of the anchor and the influence of the bottom compression fitting in short anchors. The anchor diameter is of significance since with large-diameter high-capacity anchors there is an appreciable diminution of diameter caused by the inward radial strain that occurs under the tensile load. This creates a tendency to weaken the bond between the steel and the grout.

Specifications for anchorage materials and grouting cements and recommendations for the bond strength at the grout to tendon interface are given in British Standard 8081, Code of Practice for Ground Anchors. This code provides a wealth of useful and practical information on the design, installation, and testing of anchors in soils and rock. The code recommendations for ultimate bond strength are

Plain bar	not greater than 1 N/mm^2
Clean strand or deformed bar	not greater than 2 N/mm^2
Locally noded strand	not greater than 3 N/mm^2

Mixes of pumpable normal Portland cement grout have compressive strengths in the range of 14 to 21 N/mm^2 at 3 days. Special grouts are formulated for injection into the annulus between an anchor and a tubular pile or between a pile and a surrounding sleeve. The grouts incorporate plasticizers, expanding agents and fibrous bonding materials. By adopting a water/cement ratio of about 0.5 compressive strengths of the order of 24 N/mm^2 are attainable at 3 days. Alternatively for marine work a mix consisting of 100 parts of API Oilwell B cement to 34 parts of seawater will develop a characteristic cube strength of about 22 N/mm^2 at 3 days. However, when such special grouts are used to transfer the load between large-diameter piles and a surrounding sleeve, correspondingly high bond stresses cannot be achieved. This is because of the shrinkage of the grout in the relatively wide annulus and the diminution in the diameter of the inner member due to the inward radial strain when under tensile load.

The transfer of load from a pile to the sleeve can be effected wholly through shear keys formed on the inner surface of the sleeve, and outer surface of the pile, and these should be in the form of beads of weld metal or welded-on steel strips.

The ultimate grout to steel bond strength on the surface of tubular piles on pile sleeves either with or without mechanical shear connectors can be calculated by an equation recommended by the UK Department of Energy[6.3] as follows:

$$f_{buc} = K \cdot C_L(9C_s + 1100h/s) \times (f_{cu})^{0.5} \tag{6.6}$$

where f_{buc} = *characteristic* bond strength in N/mm^2
f_{cu} = *characteristic* grout compressive strength in N/mm^2
K = stiffness factor (see below)
C_L = coefficient for grout length to pile diameter ratio
C_s = surface condition factor
h = minimum shear connector outstand in mm
s = nominal shear connector spacing in mm

The stiffness factor is given by

$$K = \frac{1}{m}\left(\frac{D}{t}\right)_g^{-1} + \left[\left(\frac{D}{t}\right)_p + \left(\frac{D}{t}\right)_s\right]^{-1} \tag{6.7}$$

where m is the modular ratio of steel to grout (take conservatively as 18 in absence of other data for long term, i.e. 28 days or more), D is the outside diameter, t is the wall thickness. The suffixes g, p, and s refer to the grout, pile and sleeve respectively.

The basis of equation 6.6 was a research programme described in references 6.4 (a) to (c). This covered a limited range of geometry of the pile connections. For this reason the Department of Energy have applied limits to the values of C_L as follows:

L/D_p	C_L
2	1.0
4	0.9
8	0.8
≥ 12	0.7

Where L is the nominal grouted connection length, referred to in EC7 (Section 8) as the tendon bond length. Obtain intermediate values for $L/D_p < 12$ by linear interpolation.

The surface condition factor C_s should be taken according to the following:

(i) If shear connectors are present and satisfy requirement $h/s > 0.005$ take C_s as 1.0.
(ii) For plain pipe connections and for connections with shear connectors but with $h/s < 0.005$, in the absence of test data take $C_s = 0.6$.

The values in (i) and (ii) refer to shotblasted or lightly rusted surface conditions. Other conditions (e.g. painted surfaces) should receive special consideration.

Equation 6.6 should be applied only to connections which satisfy the geometrical limits given below, and where mechanical shear connectors are used they are present both on pile and sleeve surfaces in contact with grout; the connectors are at uniform spacing along the length of the connection; outstand and spacing of shear connectors on pile and sleeve are the same; shear connectors on driven piles should cover sufficient length to ensure contact with grout after driving; shear connector cross-section and welds on each grout/steel interface should be designed to transmit total load applied to the grouted connection. The geometrical limits are

Sleeve geometry $50 \leq (D/t)s \leq 140$ Pile geometry $24 \leq (D/t)p \leq 40$
Grout annulus geometry $10 \leq (D/t)g \leq 45$ $L/D_p \geq 2$
Shear connector height ratio $0^* \leq h/D_p \leq 0.006$
Shear connector spacing ratio $0^* \leq D_p/s \leq 8$
Shear connector ratio $0^* \leq h/s \leq 0.04$
Shear connector shape factor $1.5 \leq w/h \leq 3$ (consistent with welded square bar or hemispherical weldbeads)
where w is the nominal width of the shear connector including welds.

The American Petroleum Institute recommendations[3.5] are much simpler. The equations are:
For operating and environmental conditions combined with dead loads and maximum or minimum live loads appropriate to normal operations of the platform:

$$f_{ba} = 0.138 + 0.5 f_{cu} \frac{h}{s} \text{ (MN/m}^2) \tag{6.8}$$

For design environmental conditions with dead loads and maximum or minimum live loads appropriate for combining with extreme conditions:

$$f_{ba} = 0.184 + 0.67 f_{cu} \frac{h}{s} \; (MN/m^2) \tag{6.9}$$

where f_{ba} is the *allowable* axial load transfer strength (steel to grout bond strength) and other terms are as previously defined.

It is evident from equations 6.8 and 6.9 that the first term represents the bond strength value for the plain pipe. The equations are again subject to limitations as follows:

Grout compression strength to be greater than 17.25 N/mm² but less than 110 N/mm²
Ratio D_p/t_p not more than 40
Ratio D_s/t_s not more than 80
Ratio D_g/t_g greater than 7, not more than 45
Ratio D_p/s greater than 2.5, not more than 8
Ratio h/s not greater than 0.10
Ratio w/h greater than 1.5, not less than 3
Product $f_{cu} \times h/s$ equal to or not less than 5.5 MN/m²

The allowable bond stress between grout and rock depends on the compressive strength of the intact rock, the size and spacing of joints and fissures in the rock, the keying of the rock effected by the drilling bit, and the cleanliness of the rock surface obtained by the flushing water. The size of the drill hole and the size of the annular space between the anchor and the wall of the hole are also important. Usually the diameter of the drill hole is taken as 1.7 to 2.5 times the diameter of the anchor. The lower end of the range is used in a strong massive rock and the higher end in a weak fractured rock. A large-diameter hole or a thick annulus gives rise to problems due to the shrinkage of the grout and the consequent weakening of the grout-to-rock bond. These difficulties can be overcome to some extent by using a special bonding grout as described above. Also, by using a compression fitting at the bottom of the anchor, part of the grout column is put in compression. The smaller the annulus and the shorter the bonded length, the higher is the compressive stress on the grout and hence its ability to lock into the surrounding rock. The value of the bond between grout and rock will be small if the rock softens to a slurry under the action of drilling and flushing. This occurs with chalk, weathered marl, and weathered clayey shales. Some observed values of bond stress at failure for drill holes of up to 75 mm in diameter are given in Table 6.3.

If the bond stress between the grout and the rock is a critical factor in designing the anchors, the allowable value should be obtained by increasing the length of the anchor rather than by increasing the diameter of the drill hole, for the reasons already stated. However, in certain conditions it is possible that the bond stress will not be reduced in direct proportion to the increase in bond length. This is because of the possibility of progressive failure in a hard rock. The maximum stretch in the anchor occurs at the top of the bonded length, and this may cause local bond failure with the rock or the pulling out of a small cone of rock (Figure 6.12a). Progressive failure then extends down to the bottom of the anchor. By limiting the bond length and sheathing the upper part of the anchor, referred to in EC7 (Section 8) as the tendon free length, within the rock (Figure 6.12b), the pulling-out of a cone of rock is prevented and the whole column of grout is compressed and acts in bond resistance with the rock.

Table 6.3 Examples of bond stress between grout and rock

Type of rock	Bond stress between grout and rock at failure N/mm²	Reference
Chalk (Grade I)	0.21	Littlejohn[6.5]
Chalk (Grade III)	0.80	Littlejohn[6.5]
Keuper Marl (Zones I and II)	0.17–0.25	Littlejohn[6.5]
Chalk	1.0	Hutchinson[6.6]
Weathered shaley slate	0.27	Unpublished[6.7]
Hard shaley slate	1.0–1.7	Unpublished[6.7]
Billings shale (Ottawa)	3.0	Freeman et al.[6.8]
Sandstone	>0.6	Unpublished[6.7]

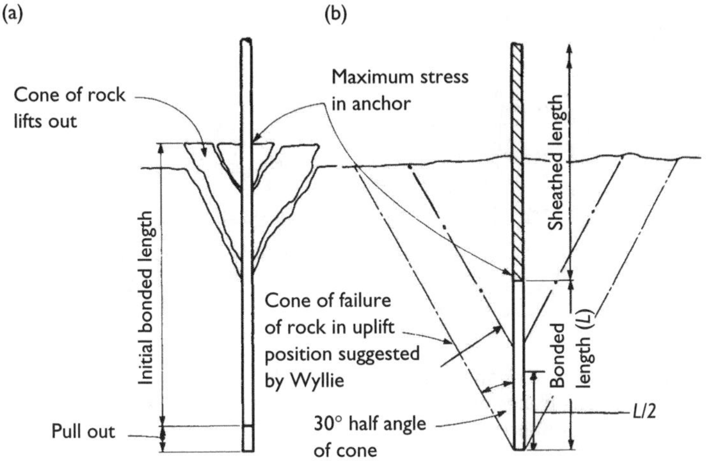

Figure 6.12 Pull-out of cone of rock (a) Fully bonded anchor (b) Upper part sheathed, lower part bonded.

The pull-out resistance of the mass of rock (as shown in Figure 6.12b) is the final criterion of the performance of an individual anchor. The actual shape of the mass of rock lifted (if failure does not occur due to the failure of the steel-to-grout or rock-to-grout bond) depends on the degree of jointing and fissuring of the rock and the inclination of the bedding planes. Various forms of failure are sketched in Figure 6.13. A cone with a half angle of 30° gives a conservative value for the pull-out resistance and represents conditions for a heavily jointed or shattered rock. Wyllie[4.39] suggests that the base of the cone should be taken at the mid-point of the bonded length (Figure 6.12b), but this arrangement would not apply for the case of a compression fitting at the bottom of the anchor. Because shear at the interface between the surface of the cone and the surrounding rock is neglected, a safety factor of unity can be taken on the weight of the rock cone, where the rock is bedded horizontally or at moderate angles from the vertical (Figure 6.13a). Where the bedding planes or other joint systems are steeply inclined, as shown in Figure 6.13b to d, either an increased factor should be allowed, or an attempt should be made to calculate the uplift resistance of the rock mass by rock mechanics' principles. The submerged weight should be taken for rock below groundwater level or below the sea. The uplift resistance of the cylinder or cone of soil overburden above the rock cone can be calculated as described in Section 6.2.3. The dimensions B and H in equations 6.3 and 6.5 are as shown in Figure 6.14. Shaft friction on the pile does not operate to resist uplift for this mode of failure. The mode of failure of a group of anchors, assuming no failure occurs in the bond between grout and steel or grout and rock, is shown in Figure 6.15. The anchors can be splayed out as shown in Figure 6.16 to increase the volume of rock bounded by the group.

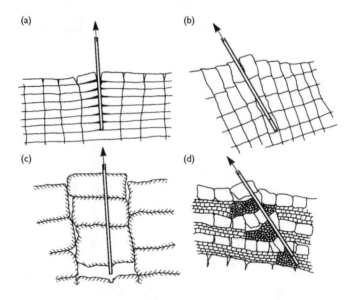

Figure 6.13 Pull-out failure in rock anchors (a) Horizontally bedded rock (thinly bedded) (b) Steeply inclined bedding planes with anchor raked in direction of bedding joints (c) Horizontally bedded rock (d) Alternating thinly and thickly bedded rocks.

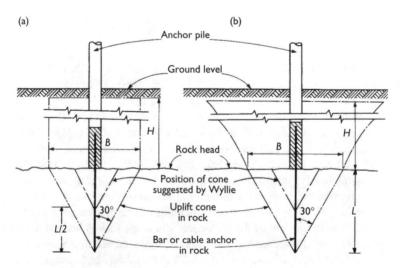

Figure 6.14 Approximate method of calculating ultimate uplift resistance of rock anchors with soil overburden (a) Clay overburden (b) Granular soil overburden.

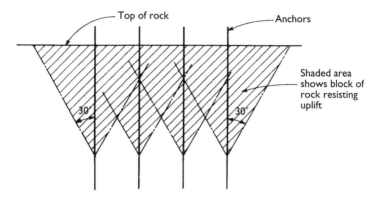

Figure 6.15 Failure condition at group of anchors in rock.

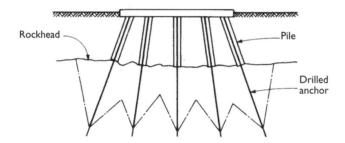

Figure 6.16 Splaying anchors in group to increase uplift resistance.

The calculation of the volume of rock V_c in a single cone with a half-angle of 30° at various angles of inclination θ to a horizontal rock surface can be performed with the aid of the curve for V_c/L^3 in Figure 6.17a. The effect of overlapping cones of rock in groups of vertical or raking anchors can be calculated by reference to Figures 6.17a and b. These charts enable the overlapping volumes ΔV_m and ΔV_n to be calculated for a group of anchors arranged on a rectangular grid. They are not applicable to a diagonal (i.e. 'staggered') pattern. All the anchors in the group are assumed to be arranged at the same angle of inclination to the horizontal and the charts are based on a cone with a 30° half-angle. The charts are not valid if the sum of $(P/n)^2$ and $(S/m)^2$ is less than 4 when composite overlapping occurs. In such a case the total volume acting against uplift should be estimated from the geometry of the system.

Because of the various uncertainties in the design of rock anchors as described above, it is evident that it is desirable to adopt stressed anchors with every anchor individually stressed and hence checked for pull-out resistance at a proof load of 1.5 times the working load. However, it should be noted that the technique of stressing anchors by jacking against the reaction provided by the pile does not check the pull-out resistance of the cone of rock: this is clear from Figure 6.10. The resistance offered by the mass of rock can be tested only by providing a reaction beam with bearers sited beyond the influence of the conjectural rock cone. Tests of this description are very expensive to perform and it is usual to avoid them by adopting conservative assumptions for the dimensions of the cone, and applying a safety factor to the calculated weight if required.

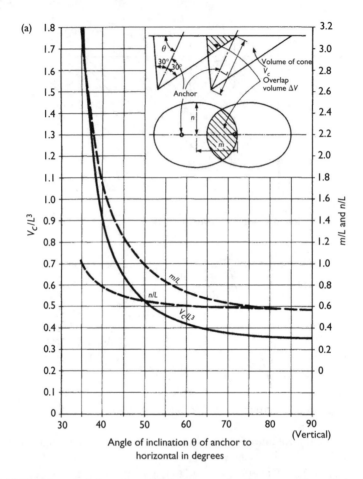

Figure 6.17 (a) Chart for calculating volumes of single or over-lapping cones with vertical or inclined axes.

In addition to dealing with the use of anchorages for restraining uplift on structures, EC7 (Section 8) makes recommendations covering anchorages for retaining walls, slopes, cuttings and tunnels. Only their applications relevant to uplift on structures will be described below. EC7 requires tension piles, where used as anchors, to be designed as described for compression piles in Section 7 of the Code. In the case of anchorages formed by grouting tendons into drilled holes, EC7 requires their design and installation to be in accordance with BS EN 1537: 1999, Ground anchors, which could eventually supersede the more comprehensive BS 8081 referred to above.

BS EN 1537 defines temporary anchors as those with a design life of less than two years, and permanent anchors as those with a design life of two years or more. The ultimate limit-states to be considered are the same as those listed at the beginning of this section. In addition, BS EC7 and EN 1537 require design measures against the following:

(1) Structural failure of the anchor head
(2) Distortion or corrosion of the anchor head

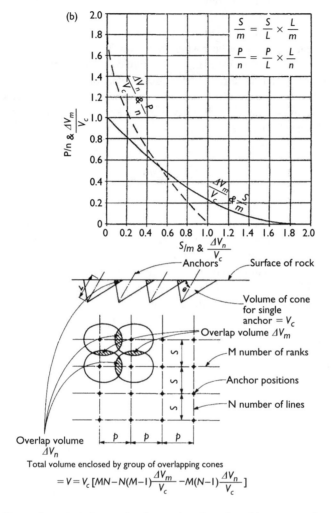

Figure 6.17 (b) Charts for calculating total volume of rock enclosed by groups of anchors arranged on rectangular grid pattern.

(3) loss of anchorage force by excessive displacement of the anchor head or by creep and relaxation

(4) failure or excessive distortion on parts of the structure due to the applied anchorage force and

(5) interaction of groups of anchorages with the ground and adjoining structures.

BS EN 1537 requires construction steel in anchors to be in accordance with EC2 and EC3. Prestressing steel used for tendons is to comply with pr EN 10138, Design of prestressing steel, and EC2 part 1–5, Use of unbonded and external prestressing tendons. Cement grouts are to comply with BS EN 445, 446 and 447; resin grouts may be used subject to appropriate tests for the particular application. Admixtures and inert fillers are permitted to be used in grout mixes provided that they do not contain materials liable to cause corrosion of the tendons. Resin grouts are permitted.

BS EN 1537 gives detailed consideration to the selection, design, fabrication, and installation of plastics sheathing used for corrosion protection. Temporary anchors are not required to be sheathed provided that they have protection from corrosion suitable for their design life.

Drilling for anchorages is required to be within a deviation limit of not more than 1/30 of the anchor length. The procedure for making permeability tests in the drilled holes using water and grout to investigate the possibility of grout loss is described.

BS EN 1537 defines three types of test on anchorages:

(1) Investigation test
(2) Suitability test
(3) Acceptance test

The investigation test is made to establish the ultimate resistance of the anchor at the grout/ground interface and to determine the characteristics of the anchorage in the working load range. The suitability test is made to confirm that a particular anchorage system will be adequate for the ground conditions on the project site. In the case of permanent anchorages, the test is made with sheathed tendons and is required to establish acceptable limits of creep or load loss at the proof and lock-off loads. In cases where no investigation tests are made the suitability test is undertaken to demonstrate anchorage characteristics and to provide criteria for acceptance creep and load loss.

The acceptance test is made at the project construction stage on each working anchor with the following requirements:

(1) To demonstrate that the proof load can be sustained
(2) To determine the apparent tension free length
(3) To ensure that the lock-off load is at the design load level, excluding friction and
(4) To determine creep or load loss characteristics at the serviceability limit-state where necessary.

The acceptance tests are to be made after lock-off and before the anchorage becomes operational.

BS EN 1537 gives detailed information on the procedure for conducting anchorage tests, the interpretation of the results, monitoring of behaviour, and record keeping. Items such as health and safety, and environmental matters including air and water pollution, noise and vibration are dealt with.

For the purpose of design verification characteristic values of anchorage resistance R_{ak} obtained from pull-out tests are divided by the partial factor γ_a to determine the design resistance R_{ad}.

$$R_{ad} = R_{ad} / \gamma_a \tag{6.10}$$

values of γ_a related to the Ground Resistance R series (see Section 4.1.4) are shown in Table 6.4. Correlation factors can be applied to obtain R_{ad} from suitability tests. Figures for these factors are not given in Annex A of EC7, but it is specified that at least three tests should be made for each distinct condition of ground and structure.

Table 6.4 Partial resistance factors (γ_a) for prestressed anchorages

Resistance	Symbol	Set	
		R I	R4
Temporary	γ_{at}	1.1	1.1
Permanent	γ_{ap}	1.1	1.1

Where R_{ad} is derived by calculation, the design approach DAI as described in Section 4.1.4 is to be used, with verification of stability against uplift of the structure by application of the UPL partial factors as described in Section 6.2.2 for friction piles.

To verify the serviceability limit-state of a structure restrained against uplift by prestressed anchorages, the tendons are regarded as elastic prestressed springs. The analysis is required to consider the most adverse combinations of minimum and maximum anchorage stiffness, and minimum and maximum prestress. To prevent damaging effects of interaction between close-spaced groups of anchors EC7 and BS EN 1537 require tendons to be spaced at least 1.5 m apart.

6.3 Single vertical piles subjected to lateral loads

The ultimate resistance of a vertical pile to a lateral load and the deflection of the pile as the load builds up to its ultimate value are complex matters involving the interaction between a semi-rigid structural element and the soil, which deforms partly elastically and partly plastically. Taking the case of a vertical pile unrestrained at the head, the lateral loading on the pile head is initially carried by the soil close to the ground surface. At a low loading the soil compresses elastically but the movement is sufficient to transfer some pressure from the pile to the soil at a greater depth. At a further stage of loading the soil yields plastically and transfers its load to greater depths. A short rigid pile unrestrained at the top and having a length to width ratio of less than 10 to 12 (Figure 6.18a) tends to rotate, and passive resistance develops above the toe on the opposite face to add to the resistance of the soil near the ground surface. Eventually the rigid pile will fail by rotation when the passive resistance of the soil at the head and toe are exceeded. The short rigid pile restrained at the head by a cap or bracing will fail by translation in a similar manner to an anchor block which fails to restrain the movement of a retaining wall transmitted through a horizontal tied rod (Figure 6.18b).

The failure mechanism of an infinitely long pile is different. The passive resistance of the lower part of the pile is infinite, and thus rotation of the pile cannot occur, the lower part remaining vertical while the upper part deforms to a shape shown in Figure 6.19a. Failure takes place when the pile fractures at the point of maximum bending moment, and for the purpose of analysis a plastic hinge capable of transmitting shear is assumed to develop at the point of fracture. In the case of a long pile restrained at the head, high bending stresses develop at the point of restraint, for example, just beneath the pile cap, and the pile may fracture at this point (Figure 6.19b).

The pile head may move horizontally over an appreciable distance before rotation or failure of the pile occurs, to such an extent that the movement of the structure supported by the pile or pile group exceeds tolerable limits. Therefore, having calculated the ultimate load

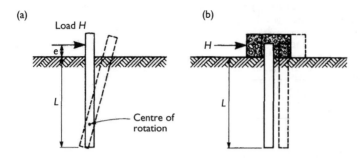

Figure 6.18 Short vertical pile under horizontal load.

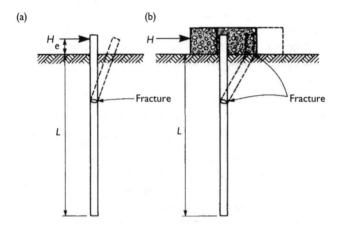

Figure 6.19 Long vertical pile under horizontal load (a) Free head (b) Fixed head.

and divided it by the appropriate safety factor, it is still necessary to check that the permissible deflection of the pile is not exceeded.

There are many inter-related factors which govern the behaviour of laterally loaded piles. The dominant one is the pile stiffness, which influences the deflection and determines whether the failure mechanism is one of the rotation of a short rigid element, or is due to flexure followed by the failure in bending of a long pile. The type of loading, whether sustained (as in the case of earth pressure transmitted by a retaining wall) or alternating (say, from reciprocating machinery) or pulsating (as from the traffic loading on a bridge pier), influences the degree of yielding of the soil. External influences such as scouring around piles at sea-bed level, or the seasonal shrinkage of clay soils away from the upper part of the pile shaft, affect the resistance of the soil at a shallow depth.

Methods of calculating ultimate resistance and deflection under lateral loads are presented in the following sections of this chapter. No attempt is made to give their complete theoretical basis. Various simplifications have been necessary in order to provide simple solutions to complex problems of soil–structure interaction, and the limitations of the methods are stated where these are particularly relevant. Most practical calculations are processes of trial and adjustment, starting with a very simple approach to obtain an approximate measure of the required stiffness, and embedment depth of the pile. The process can then be elaborated to some degree to narrow the margin of error, and to provide the essential data for calculating bending moments, shearing forces and deflections at the working load. Very elaborate calculation processes are not justified, because of the non-homogeneity of most natural soil deposits and the disturbance to the soil caused by installing piles. None of these significant factors can be reproduced in their entirety by the calculation methods.

EC7, Section 7.7 requires the design of transversely loaded piles to be consistent with the design rules previously described in Chapter 4 for piles under compression loading. Failure mechanisms to be considered are failure of a short rigid pile by rotation or translation, and failure of a long slender pile in bending with local fracture and displacement of the soil near the pile head.

Pile load tests, when undertaken as a means of determining the transverse resistance, need not necessarily be taken to the stage of failure, but the magnitude and line of action of the test load should conform to the design requirements. The effects of interaction between piles in groups and fixity at the pile head are required to be considered.

Where transverse resistance is determined by calculation, the method based on the concept of a modulus of horizontal subgrade reaction as described in Section 6.3.1 is permitted. The structural rigidity of the connection of the piles to the pile cap or substructure is to be considered as well as the effects of load reversals and cyclic loading.

For any important foundation structure which has to carry high or sustained lateral loading, it is advisable to make field loading tests on trial piles having at least three different shaft lengths, in order to assess the effects of embedment depth and structural stiffness. For less important structures, or where there is previous experience of pile behaviour to guide the engineer, it may be sufficient to make lateral loading tests on pairs of working piles by jacking or pulling them apart. These tests are rapid and economical to perform (see Section 11.4.4) and provide a reliable check that the design requirements have been met.

6.3.1 Calculating the ultimate resistance of short rigid piles to lateral loads

The first step is to determine whether the pile will behave as a short rigid unit or as an infinitely long flexible member. This is done by calculating the stiffness factors R and T for the particular combination of pile and soil. The stiffness factors are governed by the stiffness (EI value) of the pile and the compressibility of the soil. The latter is expressed in terms of a 'soil modulus', which is not constant for any soil type but depends on the width of the pile B and the depth of the particular loaded area of soil being considered. The soil modulus k has been related to Terzaghi's concept of a modulus of horizontal subgrade reaction[6.9]. In the case of a stiff over-consolidated clay, the soil modulus is generally assumed to be constant with depth. For this case

$$\text{stiffness factor } R = \sqrt[4]{\frac{EI}{kB}} \text{ (in units of length)} \qquad (6.11)$$

For short rigid piles it is sufficient to take k in the above equation as equal to the Terzaghi modulus k_1, as obtained from load/deflection measurements on a 305 mm square plate. It is related to the undrained shearing strength of the clay, as shown in Table 6.5.

Table 6.5 Relationship of modulus of subgrade reaction (k_1) to undrained shearing strength of stiff over-consolidated clay

Consistency	Firm to stiff	Stiff to very stiff	Hard
Undrained shear strength (c_u) kN/m²	50–100	100–200	>200
Range of k_1 MN/m³	15–30	30–60	>60

For most normally consolidated clays and for granular soils the soil modulus is assumed to increase linearly with depth, for which

$$\text{stiffness factor } T = \sqrt[5]{\frac{EI}{n_h}} \text{ (in units of length)} \tag{6.12}$$

where soil modulus $K = n_h \times x/B$ $\tag{6.13}$

Values of the coefficient of modulus variation n_h were obtained directly from lateral loading tests on instrumented piles in submerged sand at Mustang Island, Texas. The tests were made for both static and cyclic loading conditions and the values obtained, as quoted by Reese et al.[6.10], were considerably higher than those of Terzaghi[6.9]. The investigators recommend that the Mustang Island values should be used for pile design and these are shown together with the Terzaghi values in Figure 6.20.

Other observed values of n_h are as follows:

Soft normally-consolidated clays: 350 to 700 kN/m³
Soft organic silts: 150 kN/m³

Having calculated the stiffness factors R or T, the criteria for behaviour as a short rigid pile or as a long elastic pile are related to the embedded length L as follows:

Pile type	Soil modulus	
	Linearly increasing	Constant
Rigid (free head)	$L \leq 2T$	$L \leq 2R$
Elastic (free head)	$L \geq 4T$	$L \geq 3.5R$

Brinch Hansen's method[6.12] can be used to calculate the ultimate lateral resistance of short rigid piles. The method is a simple one which can be applied both to uniform and layered soils. It can also be applied to longer semi-rigid piles to obtain a first approximation of the required stiffness and embedment length to meet the design requirements before undertaking the more rigorous methods of analysis for long slender piles described in Sections 6.3.4 and 6.3.5. The resistance of the rigid unit to rotation about point X in Figure 6.21a is given by the sum of the moments of the soil resistance above and below this point. The passive resistance diagram is divided into a convenient number n of horizontal elements of depth L/n. The unit passive resistance of an element at a depth z below the

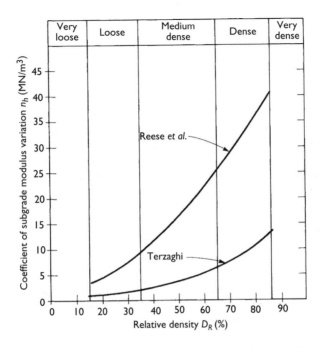

Figure 6.20 Relationship between coefficient of modulus variation and relative density of sands (after Garassino *et al.*[(6.11)]).

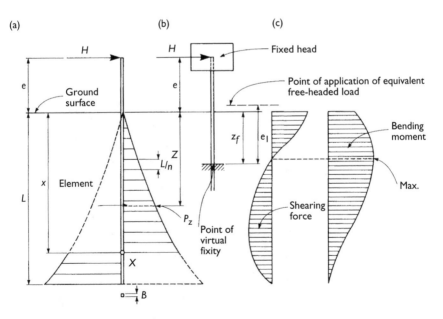

Figure 6.21 Brinch Hansen's method for calculating ultimate lateral resistance of short piles (a) Soil reactions (b) Shearing force diagram (c) Bending moment diagram.

ground surface is then given by

$$p_z = p_{oz}K_{qz} + cK_{cz} \tag{6.14}$$

where p_{oz} is the effective overburden pressure at depth z, c is the cohesion of the soil at depth z, and K_{qz} and K_{cz} are the passive pressure coefficients for the frictional and cohesive components respectively at depth z. Brinch Hansen[6.12] has established values of K_q and K_c in relation to the depth z and the width of the pile B in the direction of rotation, as shown in Figure 6.22.

The total passive resistance on each horizontal element is $p_z \times \frac{L}{n} \times B$ and, by taking moments about the point of application of the horizontal load:

$$\Sigma M = \sum_{z=0}^{z=x} p_z \frac{L}{n}(e + z)B - \sum_{z=x}^{z=L} p_z \frac{L}{n}(e + z)B \tag{6.15}$$

The point of rotation at depth x is correctly chosen when $\Sigma M = 0$, that is, when the passive resistance of the soil above the point of rotation balances that below it. Point X is thus determined by a process of trial and adjustment. If the head of the pile carries a moment M instead of a horizontal force, the moment can be replaced by a horizontal force H at a distance e above the ground surface where M is equal to $H \times e$.

Where the head of the pile is fixed against rotation, the equivalent height e_1 above ground level of a force H acting on a pile with a free head is given by

$$e_1 = \tfrac{1}{2}(e + z_f) \tag{6.16}$$

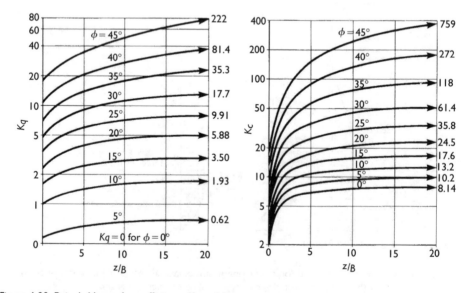

Figure 6.22 Brinch Hansen's coefficients K_q and K_c.

where e is the height from the ground surface to the point of application of the load at the fixed head of the pile (Figure 6.21a), and z_f is the depth from the ground surface to the point of virtual fixity. The depth z_f is not known at this stage but for practical design purposes it can be taken as 1.5 m for a compact granular soil or stiff clay (below the zone of soil shrinkage in the latter case), and 3 m for a soft clay or silt. The American Concrete Institute[6.13] recommend that z_f should be taken as $1.4R$ for stiff, over-consolidated clays and $1.8T$ for normally consolidated clays, granular soils and silt, and peat (see equations 6.11 and 6.12).

Having obtained the depth to the centre of rotation from equation 6.15, the ultimate lateral resistance of the pile to the horizontal force H_u can be obtained by taking moments about the point of rotation, when

$$H_u(e + x) = \sum_0^x p_z \frac{L}{n}B(x - z) + \sum_x^{x+L} p_z \frac{L}{n} + B(z - x) \tag{6.17}$$

The final steps in Brinch Hansen's method are to construct the shearing force and bending moment diagrams (Figure 6.21b and c). The ultimate bending moment, which occurs at the point of zero shear, should not exceed the ultimate moment of resistance M_u of the pile shaft. The appropriate load factors are applied to the horizontal design force to obtain the ultimate force H_u.

When applying the method to layered soils, assumptions must be made concerning the depth z to obtain K_q and K_c for the soft clay layer, but z is measured from the top of the stiff clay stratum to obtain K_c for this layer, as shown in Figure 6.23.

The undrained shearing strength c_u is used in equation 6.14 for short-term loadings such as wave or ship-berthing forces on a jetty, but the drained effective shearing strength values (c' and ϕ') are used for long-term sustained loadings such as those on retaining walls. A check should be made to ensure that there is an adequate safety factor for undrained conditions in the early stages of loading. The step-by-step procedure using the Brinch Hansen method is illustrated in worked Example 6.4.

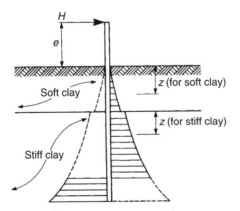

Figure 6.23 Reactions in layered soil on vertical pile under horizontal load.

6.3.2 Calculating the ultimate resistance of long piles

The passive resistance provided by the soil to the yielding of an infinitely long pile is infinite. Thus the ultimate lateral load which can be carried by the pile is determined solely from the ultimate moment of resistance M_u of the pile shaft.

A simple method of calculating the ultimate load, which may be sufficiently accurate for cases of light loading on short or long piles of small to medium width, for which the cross-sectional area is governed by considerations of the relatively higher compressive loading, is to assume an arbitrary depth z_f to the point of virtual fixity. Then from Figure 6.24:

ultimate lateral load on free-headed pile $H_u = M_u/(e + z_f)$ (6.18)

ultimate lateral load on fixed-headed pile $H_u = 2M_u/(e + z_f)$ (6.19)

Arbitrary values for z_f which are commonly used are given in the reference to the Brinch Hansen method.

It has already been stated that vertical piles offer poor resistance to lateral loads. However, in some circumstances it may be justifiable to add the resistance provided by the passive resistance of the soil at the end of the pile cap and the friction or cohesion on the embedded sides of the cap. The pile cap resistance can be taken into account when the external loads are transient in character, such as wind gusts and traffic loads, but the resulting elastic deformation of the soil must not be so great as to cause excessive deflection and hence overstressing of the piles. The design of pile caps to resist lateral loading is discussed in Section 7.9.

6.3.3 The deflection of vertical piles carrying lateral loads

A simple method which can be used to check that the deflections due to small lateral loads are within tolerable limits and as an approximate check on the more-rigorous methods described below, is to assume that the pile is fixed at an arbitrary depth below the ground surface and then to calculate the deflection as for a simple cantilever either free at the head, or fixed at the head but with freedom to translate.

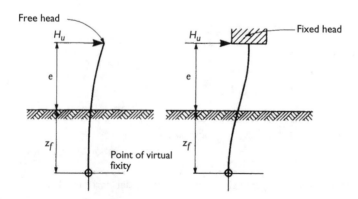

Figure 6.24 Piles under horizontal load considered as simple cantilever.

Thus from Figure 6.24

$$\text{deflection at head of free-headed pile } y = \frac{H(e + z_f)^3}{3EI} \qquad (6.20)$$

and

$$\text{deflection at head of fixed-headed pile } y = \frac{H(e + z_f)^3}{12EI} \qquad (6.21)$$

where E is the elastic modulus of the material forming the pile shaft, and I is the moment of inertia of the cross-section of the pile shaft. Depths which may be arbitrarily assumed for z_f are noted in Section 6.3.1.

6.3.4   Elastic analysis of laterally loaded vertical piles

The suggested procedure for using this section and Section 6.3.2 is first to calculate the ultimate load H_u for a pile of given cross-section (or to determine the required cross-sections for a given ultimate load) and then to divide H_u by an arbitrary safety factor to obtain trial working load H. The alternative procedure is to calculate the deflection y_0 at the ground surface for a range of progressively increasing loads H up to the value of H_u. The working load is then taken as the load at which y_0 is within the allowable limits. As a first approximation, H_u can be obtained by the Brinch Hansen method (Section 6.3.1) or from equations 6.18 and 6.19. A preliminary indication of the likely order of pile head deflection under this load can be obtained from equations 6.20 or 6.21 depending on the fixity conditions at the head.

It may be necessary to determine the bending moments, shearing forces, and deformed shape of a pile over its full depth at a selected working load. These can be obtained for

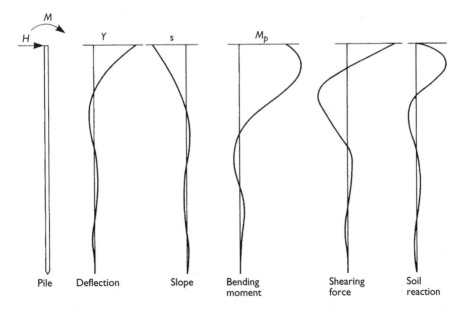

Figure 6.25 Deflections, slopes, bending moments, shearing forces, and soil reactions for elastic conditions (after Reese and Matlock[6.14]).

working-load conditions on the assumption that the pile behaves as an elastic beam on a soil behaving as a series of elastic springs. Calculations for the bending moments, shearing forces, deflections, and slopes of laterally loaded piles are necessary when considering their behaviour as energy absorbing members resisting the berthing impact of ships (see Section 8.1.1), or the wave forces in offshore platform structures (see Section 8.2).

Reese and Matlock[6.14] have established a series of curves for normally consolidated and cohesion-less soils for which the elastic modulus of the soil E_s is assumed to increase from zero at the ground surface in direct proportion to the depth. The deformed shape of the pile and the corresponding bending moments, shearing forces, and soil reactions are shown in Figure 6.25.

Coefficients for obtaining these values are shown for a lateral load H on a free pile head in Figure 6.26a to e, and for a moment applied to a pile head in Figure 6.27a to e. The coefficients for a fixed pile head are shown in Figure 6.28a to c. For combined lateral loads and applied moments the basic equations for use in conjunction with Figures 6.26 and 6.27 are as follows:

$$\text{Deflection } y \quad = y_A + y_B = \frac{A_y H T^3}{EI} + \frac{B_y M_t T^2}{EI} \tag{6.22}$$

$$\text{Slope} \quad = s_A + s_B = \frac{A_s H T^2}{EI} + \frac{B_s M_t T}{EI} \tag{6.23}$$

$$\text{Bending moment} = M_A + M_B = A_m H T + B_m M_t \tag{6.24}$$

$$\text{Shearing force} \quad = V_A + V_B = A_v H + \frac{B_v M_t}{T} \tag{6.25}$$

$$\text{Soil reaction} \quad = P_A + P_B = \frac{A_p H}{T} + \frac{B_p M_t}{T^2} \tag{6.26}$$

For a fixed pile head the basic equations are as follows:

$$\text{Deflection} \quad = y_F = \frac{F_y H T^3}{EI} \tag{6.27}$$

$$\text{Bending moment} = M_F = F_m H T \tag{6.28}$$

$$\text{Soil reaction} \quad = P_F = F_p \frac{H}{T} \tag{6.29}$$

In equations 6.22 to 6.29, H is the horizontal load applied to the ground surface, T (a stiffness factor) $= \sqrt[5]{EI/n_h}$ (as equation 6.12), M_t is the moment applied to the head of the pile, A_y and B_y are deflection coefficients (Figures 6.26a and 6.27a), A_s and B_s are slope coefficients (Figures 6.26b and 6.27b), A_m and B_m are bending-moment coefficients (Figures 6.26c and 6.27c), A_v and B_v are shearing-force coefficients (Figures 6.26d and 6.27d), A_p and B_p are soil resistance coefficients (Figures 6.26e and 6.27e), F_y is the deflection coefficient for a fixed pile head (Figure 6.28a), F_m is the moment coefficient for a fixed pile head (Figure 6.28b), and F_p is the soil resistance coefficient for a fixed pile head (Figure 6.28c).

In Figures 6.26 to 6.28 the above coefficients are related to a depth coefficient Z for various values of Z_{max}, where Z is equal to the depth x at any point divided by T (i.e. $Z = x/T$) and Z_{max} is equal to L/T. The use of curves in Figure 6.28 is illustrated in Example 6.6.

The case of a load H applied at a distance e above the ground surface can be simulated by assuming this to produce a bending moment M_t equal to $H \times e$, this value of M_t being used

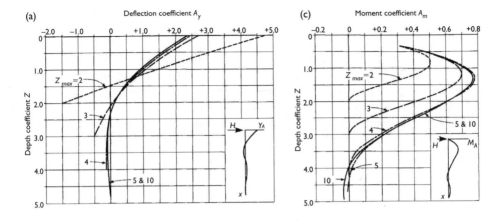

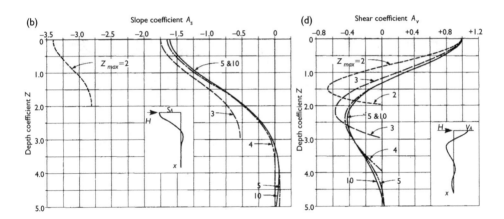

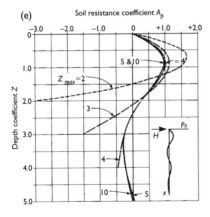

Figure 6.26 Coefficients for laterally loaded free-headed piles in soil with linearly increasing modulus (after Reese and Matlock[6.14]) (a) Coefficients for deflection (b) Coefficients for slope (c) Coefficients for bending moment (d) Coefficients for shearing force (e) Coefficients for soil resistance.

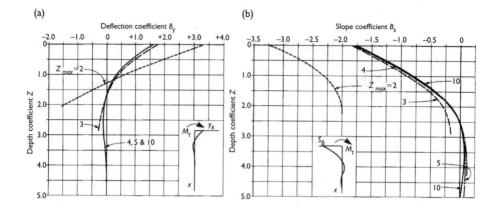

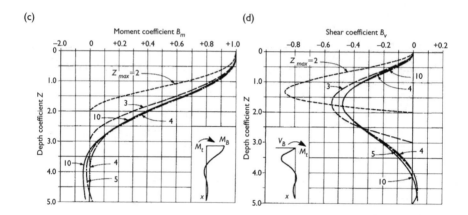

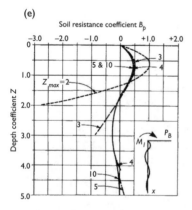

Figure 6.27 Coefficients for piles with moment at free head in soil with linearly increasing modulus (after Reese and Matlock[6.14]) (a) Coefficients for deflection (b) Coefficients for slope (c) Coefficients for bending moment (d) Coefficients for shearing force (e) Coefficients for soil resistance.

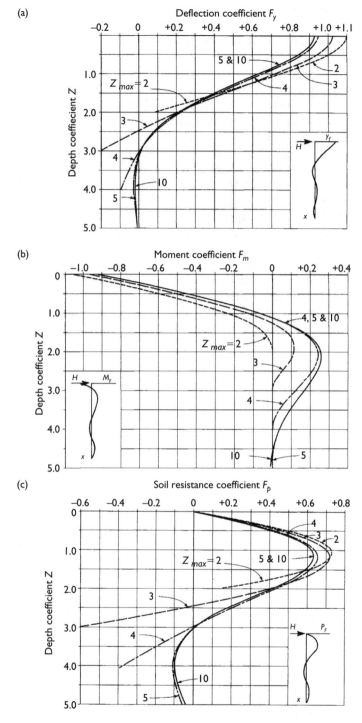

Figure 6.28 Coefficients for fixed headed piles with lateral load in soil with linearly increasing modulus (after Reese and Matlock[6.14]) (a) Coefficients for deflection (b) Coefficients for bending moment (c) Coefficients for soil resistance.

in equations 6.22 to 6.29. The moments M_a produced by load H applied at the soil surface are added arithmetically to the moments M_b produced by moment M_t applied to the pile at the ground surface. This yields the relationship between the total moment and the depth below the soil surface over the embedded length of the pile. The deflection of a pile due to a lateral load H at some distance above the soil surface is calculated in the same manner. The deflections of the pile and the corresponding slopes due to the load H at the soil surface are calculated and added to the values calculated for moment M_t applied to the pile at the surface. To obtain the deflection at the head of the pile, the deflection as for a free-standing cantilever fixed at the soil surface is calculated and added to the deflection produced at the soil surface by load H and moment M_t, together with the deflection corresponding to the calculated slope of the pile at the soil surface. This procedure is illustrated in Example 8.2.

Davisson and Gill[6.15] have analysed the case of elastic piles in an elastic soil of constant modulus. The bending moments and deflections are related to the stiffness coefficient R (equation 6.11) but in this case the value of K is taken as Terzaghi's subgrade modulus k_1, using the values shown in Table 6.5. The dimensionless depth coefficient Z in Figure 6.29 is equal to x/R. From these curves, deflection and bending moment coefficients are obtained for free-headed piles carrying a moment at the pile head and zero lateral load (Figure 6.29a) and for free-headed piles with zero moment at the pile head and carrying a horizontal load (Figure 6.29b). These curves are valid for piles having an embedded length L greater than $2R$ and different moment and deflection curves are shown for values of $Z_{max} = L/R$ of 2, 3, 4, and 5. Piles longer than $5R$ should be analysed for $Z_{max} = 5$. The equations to be used in conjunction with the curves in Figure 6.29 are as follows:

Load on pile head   For free-headed pile:

Moment M;   Bending moment $= MM_m$ (6.30)

Moment M;   Deflection $= My_m\dfrac{R^2}{EI}$ (6.31)

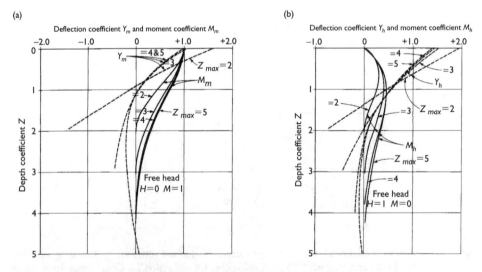

Figure 6.29 Coefficients for free headed piles carrying lateral load or moment at pile head in soil of constant modulus (after Davisson and Gill[6.15]) (a) Coefficients for deflection and bending moment for piles carrying moment at head and zero lateral (b) Coefficients for deflection and bending moment for piles carrying horizontal load at head and zero moment.

$$\text{Horizontal load } H; \quad \text{Bending moment} = HM_hR \tag{6.32}$$

$$\text{Horizontal load } H; \quad \text{Deflection} \quad = Hy_h\frac{R^3}{EI} \tag{6.33}$$

The effect of fixity at the pile head can be allowed for by plotting the deflected shape of the pile from the algebraic sum of the deflections (equations 6.31 and 6.33) and then applying a moment to the head which results in zero slope for complete fixity, or the required angle of slope for a given degree of fixity. The deflection for this moment is then deducted from the calculated value for the free-headed pile. The use of the curves in Figure 6.29 is illustrated in Example 8.2. Conditions of partial fixity occur in jacket-type offshore platform structures where the tubular jacket member only offers partial restraint to the pile that extends through it to below sea-bed level.

Where marine structures are supported by long piles $(L \geq 4T)$, Matlock and Reese[6.16] have simplified the process of calculating deflections by re-arranging equation 6.27 to incorporate a deflection coefficient C_y. Then

$$y = C_y\frac{HT^3}{EI} \tag{6.34}$$

where

$$C_y = A_y + \frac{M_tB_y}{HT} \tag{6.35}$$

Values of C_y are plotted in terms of the dimensionless depth factor $Z(= x/T)$ for various values of M_t/HT in Figure 6.30. Included in these curves are the fixed-headed case (i.e. $M_t/HT = -0.93$) and the free-headed case (i.e. $M_t = 0$).

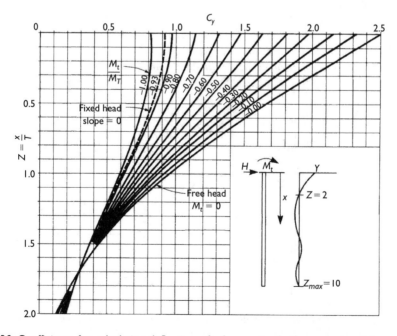

Figure 6.30 Coefficients for calculating deflection of pile carrying both moment and lateral load (after Matlock and Reese[6.16]).

The elastic deflections of piles in layered soils, each soil layer having its individual constant modulus, have been analysed by Davisson and Gill[6.15] who have produced design charts for this condition.

6.3.5 The use of p–y curves

The analytical methods of Reese and Matlock[6.14] and Davisson and Gill[6.15] that are described in the previous section are applicable only to the deflections of piles which are within the range of the elastic compression of the soil caused by the lateral loading on the piles. However, these analytical methods can be extended beyond the elastic range to analyse movements where the soil yields plastically up to and beyond the stage of shear failure. This can be done by employing the artifice of 'p–y' curves, which represent the deformation of the soil at any given depth below the soil surface for a range of horizontally applied pressures from zero to the stage of yielding of the soil in ultimate shear, when the deformation increases without any further increase of load. The p–y curves are independent of the shape and stiffness of the pile and represent the deformation of a discrete vertical area of soil that is unaffected by loading above and below it.

The form of a p–y curve is shown in Figure 6.31a. The individual curves may be plotted on a common pair of axes to give a family of curves for the selected depths below the soil surface, as shown in Figure 6.31b. Thus for the deformed shape of the pile (and also the induced bending moments and shearing forces) to be predicted correctly using the elastic analytical method described above, the deflections resulting from these analyses must be compatible with those obtained by the p–y curves for the given soil conditions. The deflections obtained by the initial elastic analysis are based on an assumed modulus of subgrade variation n_h and this must be compared with the modulus obtained from the pressures corresponding to these deflections, as obtained from the p–y curve for each particular depth analysed. If the soil moduli, expressed in terms of the stiffness factor T, do not correspond, the stiffness factor must be modified by making an appropriate adjustment to the soil modulus E_s and from this to a new value of n_h and hence to the new stiffness factor T. The deflections are then recalculated from the Reese and Matlock curves, and the corresponding pressures again obtained from the p–y curves. This procedure results in a new value of the soil modulus which is again compared with the second trial value, and the process repeated until reasonable agreement is obtained.

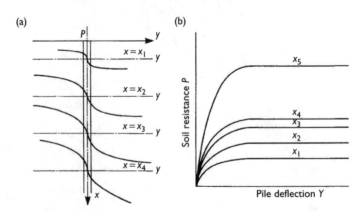

Figure 6.31 p–y curves for laterally loaded piles (a) Shape of curves at various depths x below soil surface (b) Curves plotted on common axes.

Methods of drawing sets of p–y curves have been established for soils which have a linearly increasing modulus, that is, soft to firm normally consolidated clays and coarse soils. Empirical factors were obtained by applying lateral loads to steel tubular piles driven into soft to firm clays and sands. The piles were instrumented to obtain soil reactions and deflections over their full embedded depth.

The method of establishing p–y curves for soft to firm clays is described by Matlock[6.17]. The first step is to calculate the ultimate resistance of the clay to lateral loading. Matlock's method is similar in concept to those described in Section 6.3.1. but the bearing-capacity factor N_c is obtained on a somewhat different basis.

Below a critical depth x_r the coefficient is taken conventionally as 9 as in Section 6.3.1. Above this depth it is given by the equation:

$$N_c = 3 + \frac{\gamma x}{c_u} + \frac{Jx}{B}$$ (6.36)

where γ is the density of the overburden soil, x is the depth below ground level, c_u is the undrained cohesion value of the clay, J is an empirical factor, and B is the width of the pile.

The experimental work of Matlock yielded values of J of from 0.5 for a soft clay to 0.25 for a stiffer clay. The critical depth is given by the equation:

$$x_r = \frac{6B}{\frac{\gamma B}{c_u} + J}$$ (6.37)

The ultimate resistance above and below the critical depth is expressed in the p–y curves as a force p_u per unit length of pile, where p_u is given by the pile width multiplied by the undrained shear strength c_u and a bearing capacity factor N_c, usually taken as 9.

Up to the point a in Figure 6.32 the shape of the p–y curve is derived from that of the stress/strain curve obtained by testing a soil specimen in undrained triaxial compression, or from the load/settlement curve in a plate loading test (Figure 5.15). The shape of the curve is defined by the equation:

$$\frac{p}{p_u} = 0.5 \sqrt[3]{\frac{y}{y_c}}$$ (6.38)

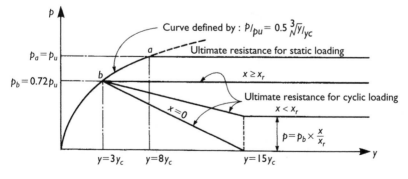

Figure 6.32 Determining shape of p–y curve in soft to firm clay (after Matlock[6.17]).

where y_c is the deflection corresponding to the strain ε_c at a stress equal to the maximum stress resulting from the laboratory stress/strain curve. The strain ε_c can also be obtained from the established relationship between c_u and the deformation modulus E_u (see Section 5.2.2). Matlock[6.17] quotes values of ε_c of 0.005 for 'brittle and sensitive clays' and 0.020 for 'disturbed or remoulded clays or unconsolidated sediments'. These values of ε_c have been based on the established range of E_u/c_u of 50 to 200 for most clays, and they can be applied to stiff over-consolidated clays, for example the value of E_u/c_u for stiff London Clay is 400. Matlock[6.17] recommends an average value of 0.010 for normally consolidated clays for use in the equation:

$$y_c = 2.5\varepsilon_c B \tag{6.39}$$

The effect of cyclic loading at depths equal to or greater than x_r can be allowed for by cutting off the p–y curve by a horizontal line representing the ultimate resistance p_b of the clay under cyclically applied loads. From the experimental work of Matlock[6.17], the point of intersection of this line with the p–y curve (shown in Figure 6.32 as point b) is given by

$$p_b/p_u = 0.72 \tag{6.40}$$

The p–y curves for cyclic loading with values of y/y_c from 3 to 15 and for depths of less than x_r, at $x = 0$ are shown in Figure 6.32.

There are little published data on values of p_b for various types of clay. The application of a static horizontal load after a period of cyclic loading, say in a deep-sea structure where a berthing ship strikes a dolphin after a period of wave loading, produces a more complex shape in the p–y curve and a method of establishing the curve for this loading condition has been described by Matlock[6.17].

The shape of a p–y curve for a pile in sand as established by Reese et al. is shown in Figure 6.33. It is in the form of a three-part curve up to the stage of the ultimate failure p_u.

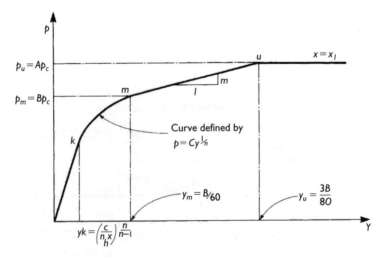

Figure 6.33 Determining shape of p–y curve in sand (after Reese et al.[6.10]).

Calculations to determine the ultimate resistance per unit depth of the pile shaft at a given depth x are obtained by using the angle of shearing resistance and density of the sand as determined by field or laboratory tests. The procedure for obtaining the shape of the curve, and the trial and adjustment process using various assumed values of the coefficient of subgrade modulus variation n_h to obtain the stiffness factor T are more complex than those described above for piles in normally consolidated clays. The reader is referred to the recommendations of the American Petroleum Institute[4.31] and reference 6.18 for guidance.

It will be evident from the foregoing account of the construction and use of p–y curves for laterally loaded piles in clays and sands that the procedure using longhand methods is extremely time-consuming (see Worked Example 8.2). However, computer programs have been established from which the required data on pile deflections, bending moments and soil resistances can be readily determined. The programs can deal with cyclic loading and predetermined variations in pile width or wall thickness over the depth of the shaft. A widely used program Lpile plus[6.19] was developed by the work of Reese and others at the University of Texas at Austin, and the ALP program[6.20] is available from OASYS Limited.

The use of p–y curves as described above is strictly applicable to piles in soils having a linearly increasing modulus (i.e. coarse soils and normally consolidated clays). In the case of stiff clays having a constant modulus of subgrade reaction k_1 equation 6.36 can be used to obtain values of N_c above the critical depth. The latter can be calculated from equation 6.37 using a value of 0.25 for coefficient J. Values of n_h are obtained by plotting the soil modulus E_s against the depth, but the trial line is a vertical one passing through the plotted points, again with weight being given to depths of $0.5R$ or less. Cyclic loading can be a critical factor in stiff clays. The relationship in equation 6.40 should preferably be established for the particular site by laboratory and field tests, but the factor of 0.72 may be used if results of such studies are not available.

Instead of relating the deflection y_c to the strain ε_c at a stress corresponding to the maximum stress obtained in the laboratory stress–strain curve for use in equation 6.38, Reese and Welch[6.18] adopted the following relationship for stiff clays:

$$\frac{p}{p_u} = 0.5 \sqrt[4]{\frac{y}{y_{50}}} \tag{6.41}$$

where p and p_u are as previously defined, and y_{50} is the deflection corresponding to the strain ε_{50} at one-half of the maximum principal stress difference in the laboratory stress–strain curve.

If no value of ε_{50} is available from laboratory tests a figure of between 0.005 and 0.010 can be used in equation 6.39 but substituting y_{50} for y_c and ε_{50} for ε_c. The larger of these two values is the more conservative. Reese and Welch[6.18] have described a method for establishing p–y curves for cyclic loading on piles in stiff clay.

6.3.6   Effect of method of pile installation on behaviour under lateral loads and moments applied to pile head

The method of installing a pile, whether driven, driven-and-cast-in-situ or bored and cast-in-situ, has not been considered in Sections 6.3.1 to 6.3.4. The effect of the installation method on the behaviour under lateral load, can be allowed for by appropriate adjustments to the soil parameters. For example, when considering the resistance to lateral loads, of piles driven into a soft sensitive clay the remoulded shearing strength could be in conjunction with the Brinch Hansen method (Section 6.3.1), to obtain the ultimate resistance over a period of a few days or weeks after driving. If the piles are not to be subjected to loading for a few months after driving, the full 'undisturbed' shearing strength can be used. There is unlikely to be much difference between the ultimate lateral resistance of short rigid piles driven into stiff over-consolidated clays and bored piles in the same type of soil. The softening effects for bored piles mentioned in Section 4.2.3 occur over a very short radial distance from the pile and the principal resistance to lateral loads is provided by the undisturbed soil beyond the softened zone.

In the case of piles installed in coarse soils the effect of loosening due to the installation of bored piles can be allowed for by assuming a low value of ϕ when determining K_q from Figure 6.22. When considering the deflection of bored piles in coarse soils the value of the soil modulus n_h in Figure 6.20 should be appropriate to the degree of loosening which is judged to be caused by the method of installing the piles.

p–y curves were developed primarily for their application to the design of long driven piles, mainly for offshore structures. Because such piles are required to have sufficient strength to cope with driving stresses, they have a corresponding resistance to bending stresses from lateral loading. On the other hand, bored and cast-in-place piles are required to have only nominal reinforcement, unless they are designed to act as columns above ground level, or to carry uplift or lateral loading. Nip and Ng[6.21] investigated the behaviour of laterally loaded bored piles. They noted that while allowance can be made, arbitrarily, by assuming that the stiffness of a cracked reinforced pile section is 50% of that of an uncracked pile, this assumption can result in over-predicting the deflections and under-predicting the bending moments. By comparing the deflections measured in lateral load tests with predictions made by calculations using p–y curves they concluded that the latter can be used to predict deflections, bending moments, and soil reactions of laterally loaded bored piles with varying EI values corresponding to uncracked, partially cracked, and fully cracked sections.

6.3.7   The use of the pressuremeter test to establish p–y curves

The pressuremeter test (see Section 11.1.4) made in a borehole (or in a hole drilled by the pressuremeter device) is particularly suitable for use in establishing p–y curves for laterally loaded piles. The test produces a curve of the type shown in Figure 6.34a. The initial portion represents a linear relationship between pressure and volume change, that is the radial expansion of the walls of the borehole. At the creep pressure p_f the pressure/volume relationship becomes non-linear indicating plastic yielding of the soil; at the limit pressure p_l the volume increases rapidly without increase of pressure as represented by the horizontal portion of the p–y curve.

Menard used a Poisson's ratio of 0.33 to derive an expression for determining the pressuremeter modulus of the soil from the initial portion of the curve in Figure 6.34a. This equation as given by Baguelin et al.[6.22] is

$$E_m = 2.66 V_m \frac{\Delta_p}{\Delta_v} \tag{6.42}$$

where

$$\frac{\Delta_p}{\Delta_v} = \text{slope of the curve between } V_0 \text{ and } V_f$$

V_m = midpoint volume

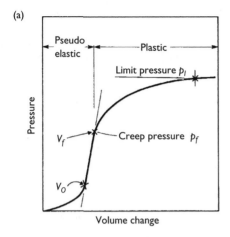

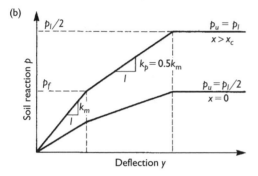

Figure 6.34 Obtaining soil reaction values from pressuremeter test (a) pressure/volume change curve (b) design reaction curve (after Baguelin et al.[6.22]).

Baguelin et al.[6.22] give two sets of curves relating the response of the soil to lateral loading for the two stages in the pressuremeter tests as shown in Figure 6.34b. The upper curve is for depths below the ground surface equal to or greater than the critical depth, x_c, at which surface heave affects the validity of the calculation method. In fine-grained soils x_c is taken as twice the pile width and in coarse soils it is four times the width. Where there is a pile cap there is no surface heave, x_c is zero and the lower curve in Figure 6.34b applies. The value of the coefficient of subgrade reaction, k_m in Figure 6.34b, is given by

for pile widths greater than 600 mm

$$\frac{1}{k_m} = \frac{2}{9E_m}B_0\left(\frac{B}{B_0} \cdot 2.65\right)^\alpha + \frac{\alpha B}{6E_m} \tag{6.43a}$$

for pile widths less than 600 mm

$$\frac{1}{k_m} = \frac{B}{E_m}\left(\frac{4(2.65)^\alpha + 3\alpha}{18}\right) \tag{6.43b}$$

where

E_m = mean value of the pressuremeter modulus over the characteristic length of the pile

B_0 = pile width

α = rheological factor varying from 1.0 to 0.5 for clays, 0.67 and 0.33 for silts, and 0.5 to 0.33 for sands

Between the ground surface and the critical depth, X_c, the value of k_m should be reduced by the coefficient λ_z, given by

$$\lambda_z = \frac{1 + \dfrac{X}{X_c}}{2} \qquad\qquad (6.44)$$

A simplified procedure in a homogeneous soil is to assume that there will be no lateral soil reaction between the ground surface and a depth equivalent to 0.5 to 0.75B then to use the full reaction given by the upper curve in Figure 6.34b.

Baguelin et al.[6.22] give the following equations for calculating deflections, bending moments and shears at any depth z below the ground surface for conditions of a constant value of the pressuremeter modulus with depth:

Deflection $y(z) = \dfrac{2H}{Rk_mB} \cdot F_1 + \dfrac{2M_t}{R^2k_mB} \cdot F_4$ \qquad (6.45a)

Moment $M(z) = \dfrac{H.R.F_3}{R} + M_tF_2$ \qquad (6.45b)

Shear $T(z) = HF_4 - \dfrac{2M_t}{R}F_3$ \qquad (6.45c)

where

R = stiffness coefficient given by equation 6.11 (Baguelin refers to this as the transfer length, l_0)

H = horizontal load applied to the pile head

M_t = bending moment at the pile head

z = dimensionless coefficient equal to $\dfrac{X}{R}$

Values of the coefficients F_1 to F_4 are given in Figure 6.35.

At the ground surface the deflection becomes

$$y_0 = \frac{2H}{Rk_mB} + \frac{2M_t}{R^2k_mB} \qquad\qquad (6.46)$$

and slope $= y_0' = \dfrac{-2H}{R^2k_mB} - \dfrac{4M_t}{R^1k_mB}$ radians \qquad (6.47)

If the head of the pile is fixed so that it does not rotate ($y_0 = 0$) equations 6.45 to 6.47 become

$$y(z) = \frac{H}{Rk_mB}F_2 \qquad\qquad (6.48a)$$

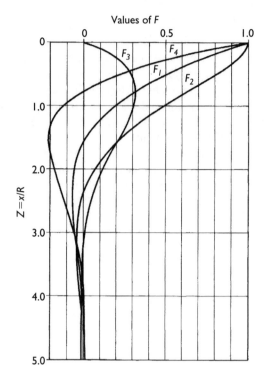

Figure 6.35 Values of the coefficients F_1 to F_4 (after Baguelin et al.[6.22]).

$$M(z) = \frac{-HR}{2} F_4 \tag{6.48b}$$

$$T(z) = H \cdot F_3 \tag{6.48c}$$

$$y_0 = \frac{H}{Rk_m B} \tag{6.48d}$$

$$M_t = \frac{-HR}{2} \tag{6.48e}$$

To draw the pile load/deflection curve, the deflections corresponding to soil reactions equal to the creep pressure, p_f, and the limit pressure, p_l, are calculated from the relationship $p = k_m y$. The lateral pile loads then follow from equations 6.45a, 6.46, 6.48a or 6.48d. For soil reactions between the limit pressure and creep pressure the value of k_m is halved as shown in Figure 6.34. The procedure is illustrated in Worked Example 6.8 where the pressuremeter tests show a linearly increasing soil modulus. The values of n_h can be calculated from equation 6.13 taking K as k_{mB}. Deflections are calculated from the Matlock and Reese curves (Figures 6.26 to 6.28).

6.3.8   Calculation of lateral deflections and bending moments by elastic continuum methods

The method of preparing p–y curves described in Section 6.3.5 was based on the assumption that the laterally loaded pile could be modelled as a beam supported by discrete springs. The springs would be considered as possessing linear or non-linear behaviour. In the latter case the method could be used to model pile behaviour in strain conditions beyond the elastic range.

In many cases where lateral forces are relatively low and piles are stiff the pile head movements are within the elastic range and it may be convenient to use the elastic continuum model to calculate deflections and bending moments.

Randolph[6.23] used finite element analyses to establish relationships between pile deflections and bending moments with depth for lateral force and moment loading as shown in Figure 6.36. The following notation applies to the parameters in this figure:

y_0 = lateral displacement at ground surface
z = depth below ground level
H_0 = lateral load applied at ground surface
M = bending moment in the pile
M_0 = bending moment at ground surface
r_0 = radius of the pile
E_p' = effective Young's modulus of a solid circular pile of radius r_0 (i.e. $4E_pI_p/\pi r_0{}^4$)
G_c = characteristic modulus of the soil, that is, the average value of G^* over depths less than l_c

$$G^* = G(1 + \tfrac{3}{4}v)$$

where

G = shear modulus of the soil
v = Poisson's ratio
l_c = critical length of the pile
$l_c = 2r_0(E_p'/G^*)^{2/7}$ for homogeneous soil
$= 2r_0(E_p'/m^*r_0)^{2/9}$ for soil increasing linearly in stiffness with depth

$$m^* = m\left(1 + \frac{3v}{4}\right)$$

$m = \dfrac{G}{z}$ where G varies with depth as $G = mz$

ρ_c is a homogeneity factor

where

$$\rho_c = \frac{G^* \text{ at } l_c/4}{G^* \text{ at } l_c/2}$$

The use of the Randolph curves is illustrated in worked Example 8.2.

The Randolph method is useful where the shear modulus is obtained directly in the field using the pressuremeter. If Young's modulus values only are available the shear modulus for an isotropic soil can be obtained from the equations:

$$E_u = 2G(1 + v_u) \quad \text{and} \quad E' = 2G(1 + v') \tag{6.49}$$

where v_u and v' are the undrained and drained Poisson's ratios respectively.

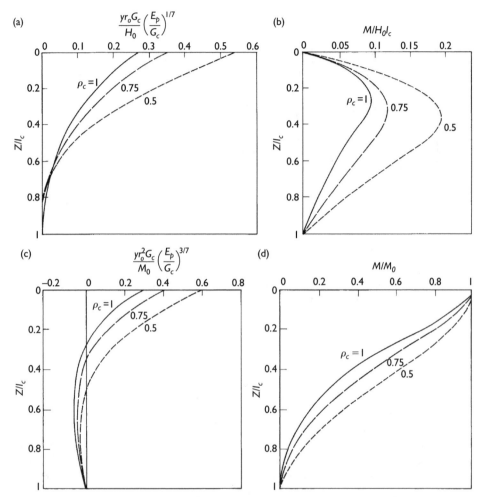

Figure 6.36 Generalized curves giving deflected pile shape and bending moment profile for lateral force and bending moment applied to pile head (after Randolph[6.23]) (a) Deflected pile shape for lateral force loading (b) Bending moment profile for lateral force loading; (c) Deflected pile shape for moment loading (d) Bending moment profile for moment loading.

6.3.9   Bending and buckling of partly embedded single vertical piles

A partly embedded vertical pile may be required to carry a vertical load in addition to a lateral load and a bending moment at its head. The stiffness factors R and T as calculated from equations 6.11 and 6.12 have been used by Davisson and Robinson[6.24] to obtain the equivalent length of a free-standing pile with a fixed base, from which the factor of safety against failure due to buckling can be calculated using conventional structural design methods.

A partly embedded pile carrying a vertical load P, a horizontal load H, and a moment M at a height e above the ground surface is shown in Figure 6.37a. The equivalent height L_e of the fixed-base pile is shown in Figure 6.37b.

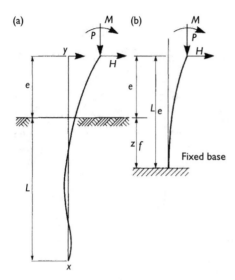

Figure 6.37 Bending of pile carrying vertical and horizontal loads at head (a) Partly embedded pile (b) Equivalent fixed base pile or column.

For soils having a constant modulus:

Depth to point of fixity $z_f = 1.4R$ (6.50)

For soils having a linearly increasing modulus:

$z_f = 1.8T$ (6.51)

The relationships 6.50 and 6.51 are only approximate, but Davisson and Robinson state that they are valid for structural design purposes provided that l_{max}, which is equal to L/R, is greater than 4 for soils having a constant modulus and provided that z_{max}, which is equal to L/T, is greater than 4 for soils having a linearly-increasing modulus. From the above equations the equivalent length L_e of the fixed-base pile (or column) is equal to $e+z_f$ and the critical load for buckling is

$$P_{cr} = \frac{\pi^2 EI}{4(e+z_f)^2} \text{ for free-headed conditions}$$ (6.52)

and

$$P_{cr} = \frac{\pi^2 EI}{(e+z_f)^2} \text{ for fixed- (and translating-) headed conditions}$$ (6.53)

6.4 Lateral loads on raking piles

The most effective way of arranging piles to resist lateral loads is to have pairs of piles raking in opposite directions as shown in Figure 6.5. The simple graphical method of

(a)   (b)

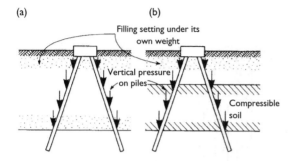

Figure 6.38 Bending of slender raking piles due to loading from soil subsidence (a) Fill settling under own weight (b) Fill overlying compressible soil.

determining the compressive and tensile forces in the piles by a triangle of forces assumes that the piles are hinged at their point of intersection and that the lateral loads are carried only in an axial direction by the piles. The tension pile will develop its maximum pull-out resistance with negligible movement, and the yielding of a properly designed compression pile of small to medium diameter is unlikely to exceed 10 mm at the working load. Thus the horizontal deflections of the pile cap will be quite small.

For economy, the raking piles should be installed at the largest possible angle from the vertical. This depends on the type of pile used (see Section 3.4.11). Where raking piles are embedded in fill which is settling under its own weight (Figure 6.38a) or in a compressible clay subjected to a surcharge load or to superimposed fill (Figure 6.38b) the vertical loading on the upper surface of the rakers may induce high bending moments in the pile shaft. Because of this, raking piles may not be an appropriate form of construction in deep fill or compressible layers.

6.5   Lateral loads on groups of piles

Loads on individual piles forming a group of vertical piles that is subject to horizontal loading or to combined vertical and horizontal loading can be determined quite simply (for cases where the resultant cuts the underside of the pile cap) by taking moments about the neutral axis of the pile group. Thus in Figure 6.39 the vertical component V of the load on any pile produced by an inclined thrust R, where R is the resultant of a horizontal load H and a vertical load W is given by

$$V = \frac{W}{n} + \frac{We\bar{x}}{\sum \bar{x}^2} \qquad (6.54)$$

where W is the total vertical load on the pile group, n is the number of piles in the group, e is the distance between the point of intersection of R with the underside of the pile cap and the neutral axis of the pile group, and $\bar{x}$ is the distance between the pile and the neutral axis of the pile group ($\bar{x}$ is positive when measured in same direction as e and negative when in the opposite direction).

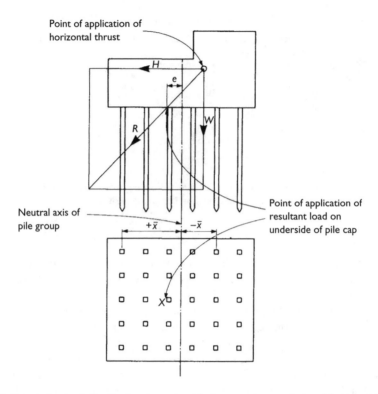

Point of application of
horizontal thrust

Neutral axis of
pile group

Point of application of
resultant load on
underside of pile cap

Figure 6.39 Calculating load distribution on group of piles carrying vertical and horizontal loading.

Determination of the individual loads on groups of raking or combined raking and vertical piles is a complex matter if there are more than three rows of piles in the group. The latter case can be analysed by static methods if it is assumed that the piles are hinged at their upper ends, that horizontal loads are carried only by axial forces in the inclined piles and that vertical piles do not carry any horizontal loading. The forces in the piles are resolved graphically as shown in Figure 6.40. The same method can be used if pairs of piles or individual groups of three closely spaced piles are arranged in not more than three rows, as shown in Figure 6.41. To produce the parallelogram of forces the line of action of the forces in the piles is taken as the centre-line of each individual group.

The determination of the individual loads on piles installed in groups comprising multiple rows of raking or combined raking and vertical piles is a highly complex process which involves the analysis of movements in three dimensions; i.e. movements in vertical and horizontal translation and in rotational modes. The analysis of loadings on piles subjected to these movements requires the solution of six simultaneous equations, necessitating the use of a computer for practical design problems.

The reader is referred to the work of Poulos and Davis[6.25], and Poulos[6.26] for an account of their research into the behaviour of laterally loaded pile groups in an elastic medium. Randolph[6.23] gives expressions to determine the interaction factor between adjacent piles in groups carrying compression and lateral loading and compares them with values derived

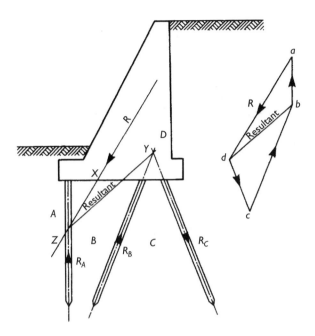

Figure 6.40 Graphical method for determining forces on groups of vertical and raking piles under inclined loading.

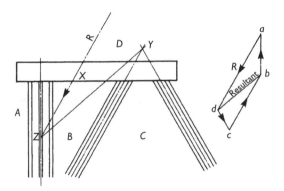

Figure 6.41 Graphical method for determining forces on groups of closely spaced vertical and raking piles under inclined loading.

by Poulos and with results of model tests. Computer programs have been developed to aid the design of pile groups for these loading conditions. They include the ALP program previously referred to[6.20]. Also DEFPIG developed by Poulos at the University of Sydney Engineering Laboratory, and PGROUP, available from various bureaux.

The case of closely spaced groups of piles acting as a single unit when subject to lateral loads must also be considered. Prakash[6.27] showed that piles behave as individual units if they are spaced at more than three pile widths in a direction normal to the direction of

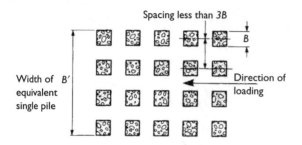

Figure 6.42 Piles at close spacing considered as single unit (after Prakash[6.27]).

loading and at more than 6 to 8 diameters parallel to this direction. Piles at a closer spacing can be considered to act as a single unit in order to calculate the ultimate resistance and deflections under lateral loads (Figure 6.42). In soft clays and sands the effect of driving piles in groups at close spacing is to stiffen the soil enclosed by the group thus increasing its capability as a single unit to resist movement when carrying horizontal loading.

Calculations to determine the ultimate bearing capacity of pile groups carrying vertical and horizontal or inclined loading can be performed using the Brinch Hansen general equation (Section 5.2). The calculations assume that the pile group takes the form of an equivalent block foundation. EC7 in Clause 7.6.2.1(4) states that the resistance of the group to compression loading can be calculated by assuming that the group acts as a single large-diameter pile. However, Clause 7.6.3.1 requires the resistance of a group subjected to tension loading to be provided by the frictional resistance of the soil enclosing a block foundation. No guidance is given in respect of pile groups carrying transverse loading. Clause 7.7.1(4)P merely requires group action 'to be considered'.

6.6 References

6.1 RADHARKRISHNA, H. S. and ADAMS, J. I. Long-term uplift capacity of augered footings in fissured clay. *Canadian Geotechnical Journal*, Vol. 10, No. 4, November 1973, pp. 647–52.

6.2 MEYERHOF, G. G. and ADAMS, J. I. The ultimate uplift capacity of foundations, *Canadian Geotechnical Journal*, Vol. 5, No. 4, November 1968, pp. 225–44.

6.3 UK DEPARTMENT OF ENERGY. *Offshore installations: Guidance on design and construction*, HMSO, London, 1984.

6.4 (a) *The strength of grouted pile-sleeve connections for offshore structures: Static tests relating to sleeve buckling.* Wimpey Offshore Engineers and Constructors Ltd. HMSO Offshore Technology Series, OTH.85.223, 1986.
(b) *The strength of grouted pile-sleeve connections.* A composite report for DoE. HMSO Offshore Technology Services, OTH.86.210, 1986.
(c) *A study of length, longitudinal stiffening and size effects on grouted pile-sleeve connections.* Wimpey Offshore Engineers and Constructors Ltd. HMSO Offshore Technology Series, OTH.86.230, 1987.

6.5 LITTLEJOHN, G. S. Soil Anchors, *Proceedings of the Conference on Ground Engineering*, Institution of Civil Engineers, London, 1970, pp. 41–4.

6.6 HUTCHINSON, J. N. Discussion on ref. 6.5, Conference proceedings, p. 85.

6.7 WIMPEY LABORATORIES LTD. unpublished report.

6.8   FREEMAN, C. F., KLAJNERMAN, D., and PRASAD, G.D. Design of deep socketted caissons into shale bedrock, *Canadian Geotechnical Journal*, Vol. 9, No. 1, February 1972, pp. 105–14.

6.9   TERZAGHI, K. Evaluation of coefficients of subgrade reaction, *Geotechnique*, Vol. 5, No. 4, 1955, pp. 297–326.

6.10 REESE, L. C., COX, W. R., and KOOP, F. B. Analysis of laterally loaded piles in sand, *Proceedings of the Offshore Technology Conference*, Houston, Texas, 1974, Paper No. OTC 2080.

6.11 GARASSINO, A., JAMIOLKOWSKI, M., and PASQUALINI, E. Soil modulus for laterally-loaded piles in sands and NC clays. *Proceedings of the 6th European Conference, ISSMFE*, Vienna, 1976, Vol. I, No. 2, pp. 429–34.

6.12 HANSEN, J. B. The ultimate resistance of rigid piles against transversal forces, *Danish Geotechnical Institute*, Bulletin No. 12, Copenhagen 1961, pp. 5–9.

6.13 *Recommendations for the Design, Manufacture and Installation of Concrete Piles*, American Concrete Institute, Report ACI 543R-74 (re-affirmed 1960).

6.14 REESE, L. C. and MATLOCK, H. Non-dimensional solutions for laterally-loaded piles with soil modulus assumed proportional to depth, *Proceedings of the 8th Texas Conference on Soil Mechanics and Foundation Engineering*, Austin, Texas, 1956, pp. 1–41.

6.15 DAVISSON, M. T. and GILL, H. L. Laterally-loaded piles in a layered soil system, *Journal of the Soil Mechanics Division*, American Society of Civil Engineers, Vol. 89, No. SM3, May 1963, pp. 63–94.

6.16 MATLOCK, H. and REESE, L. C. Foundation analysis of offshore pile-supported structures, *Proceedings of the 5th International Conference, ISSMFE*, Paris, Vol. 2, 1961, pp. 91–7.

6.17 MATLOCK, H. Correlations for design of laterally loaded piles in soft clay, *Proceedings of the Offshore Technology Conference*, Houston, Texas, 1970, Paper OTC 1204.

6.18 REESE, L. C. and WELCH, R. C. Lateral loading of deep foundations in stiff clay, *Journal of the Geotechnical Engineering Division*, American Society of Civil Engineers, Vol. 101, No. GT7, July 1975, pp. 633–49.

6.19 Ensoft Inc, Lpile Plus4, *A Program for the Analysis of Piles and Drilled Shafts Under Lateral Loads*, Ensoft Inc, Austin, Texas, 2000.

6.20 *Oasys ALP Laterally-Loaded Pile Analysis Manual*, Ove Arup and Partners, London, 1997.

6.21 NIP, D. C. N. and NG, C. W. W. Back-analysis of laterally-loaded bored piles, *Geotechnical Engineering*, Vol. 158, No. GE2, 2005, pp. 63–73.

6.22 BAGUELIN, F., JEZEQUEL, J. F., and SHIELDS, D. G. *The Pressuremeter and Foundation Engineering*, Trans Tech Publications, Clausthal, 1978.

6.23 RANDOLPH, M. F. The response of flexible piles to lateral loading, *Geotechnique*, Vol. 31, No. 2, June 1981, pp. 247–59.

6.24 DAVISSON, M. T. and ROBINSON, K. E. Bending and buckling of partially embedded piles, *Proceedings of the 6th International Conference, ISSMFE*, Montreal, Vol. 2, 1965, pp. 243–6.

6.25 POULOS, H. G. and DAVIS, E. H. *Pile Foundation Analysis and Design*, John Wiley, New York, 1980.

6.26 POULOS, H. G. Behavior of laterally loaded piles: I – Single piles, *Proceedings of the American Society of Civil Engineers*, Vol. SM5, May 1971, pp. 711–31.

6.27 PRAKASH, S. Behaviour of pile groups subjected to lateral loads, Thesis University of Illinois, Urbana, Illinois, 1962.

6.7 Worked examples

Example 6.1

A bored and cast in-situ pile having a shaft diameter of 0.6 m and a 2 m diameter enlarged base is installed to a depth of 11 m in a clay with an undrained shear strength of 40 kN/m^2 and an angle of shearing resistance of 25°. The groundwater level is below 11 m. Calculate the ultimate resistance in uplift of the pile for sustained loading conditions. In Figure 6.9 the failure surface does not extend up to the ground surface.

For $\phi = 25°$, $H/B = 3$, giving $H = 3B = 3 \times 2 = 6$ m. Also from Figure 6.9, with $\phi = 25°$, $K_u = 0.88$ and the value of the shape factor is the maximum of 1.3.

The weight W can be taken conservatively as the weight of a cylinder of soil 2 m in diameter extending up to ground level (i.e. the weight of the soil displaced by the pile is assumed to be equal to the weight of the concrete). Then from equation 6.5

$$Q_u = (\pi \times 40 \times 2 \times 6) + \{1.3 \times \tfrac{1}{2}\pi \times 2.1 \times 9.81 \times 2[(2 \times 11) - 6]$$

$$\times 6 \times 0.88 \times \tan 25°\} + (\tfrac{1}{4}\pi \times 2^2 \times 2.1 \times 11 \times 9.81) = 5534 \text{ kN}$$

This value of the uplift resistance must not be greater than that given by the uplift resistance Q_{ub} of the projecting part of the enlarged base plus the skin friction Q_{us} on the pile shaft. The latter value is calculated on the overall depth minus the depth of the enlarged base (i.e. 1.5 m) and a zone 1.5 m deep at the ground surface where the soil can shrink away from the shaft. From the Brinch Hansen factors from Figure 5.6 for $\phi = 25°$ are

$N_c = 21$, $N_q = 10.5$, $s_c = 1.2$, $d_c = 1.6$, $d_q = 1.6 - (1.6 - 1)/10.5 =$ approximate 1.6, $s_q = 1.2 - (1.2 - 1)/10.5 =$ approximate 1.2, and $i_c = i_q = 1.0$.

The third term in equation 5.1 is small and can be neglected.

$$Q_{ub} = \tfrac{\pi}{4}(2^2 - 0.6^2) [40 \times 21 \times 1.2 \times 1.6 \times 1.0 + (2.1 \times 9.81 \times 11$$

$$\times 10.5 \times 1.2 \times 1.6 \times 1.0)] = 17\,671 \text{ kN}$$

From equation 4.7 the shaft friction within the stiff clay from the undrained shear strength component only is

$$Q_{us} = 0.3 \times 40 \times \pi \times 0.6 [11 - (2 \times 1.5)] = 181 \text{ kN}$$
Total uplift resistance $= Q_{ub} + Q_{us} = 17\,671 + 181 = 17\,832 \text{ kN}$

Although the length/diameter ratio is 5.5 it is advisable to take only one-half of this value as the ultimate resistance; i.e. a value of 8926. Therefore the ultimate resistance is governed by equation 6.5, i.e. has a value of 5534 kN.

Example 6.2

The floor of a shipbuilding dock covers an area of 210×60 m. The 0.8 mm floor is restrained against uplift by precast concrete shell piles having an overall diameter of 450 mm which are driven through 8 m of soft clay ($\bar{c}_u = 16 \text{ kN/m}^2$) on to a strong shale ($\gamma = 2.3 \text{ Mg/m}^3$). The piles are spaced on a 3 m square grid and each pile carries an uplift load of 1 100 kN. Design a suitable anchorage system for the dock floor using stressed cable anchors.

From Figure 4.6, for $c_u/\sigma'_{vo} = 16/9.81 \times 0.8 \times 8 = 0.25$, $\alpha = 1.0$ and length factor F, for $L/B = 8/0.45 = 18$, of 1.0, Equation 4.8 gives:

$$Q_s = 1 \times 16 \times \pi \times 0.45 \times 8 = 181 \text{ kN}$$

For a safety factor of 2.5:

allowable uplift resistance of the pile in soft clay $= 181 / 2.5 = 72 \text{ kN}$

Thus the load to be carried by anchorage in shale $= 1100 - 72 = 1028$ kN. This load can be resisted by an anchor cable made up from seven 15.2 mm 7-wire Dyform strands. The breaking load of each strand is 300 kN. Therefore

$$\text{working load} = \frac{1\,028}{7 \times 300} = 0.49 \times \text{breaking load of strand}$$

which is satisfactory. The approximate overall diameter of the 7-strand cable is 45 mm. Therefore for an *allowable* bond stress between steel and grout of 1.0 N/mm²:

$$\text{required bond length of cable} = \frac{1\,028 \times 1\,000}{\pi \times 45 \times 1.0 \times 1\,000} = 7.3 \text{ m}$$

Drill the cable hole to 9 m and provide an unwrapped and cleaned bond length of 7 m with compression fittings swaged on to the lower end. The cable can be fed down a 150 mm borehole for which,

$$\text{working bond stress between rock and grout} = \frac{1028 \times 1000}{\pi \times 150 \times 7.0 \times 1000} = 0.31 \text{ N/mm}^2$$

which is satisfactory for a strong shale (Table 6.3). The stress is not excessive if the anchors are stressed to 1.5 times the working load during installation.

From Figure 6.17a, the volume of a rock cone with a 30° half angle lifted by single anchor cable is $0.35 \times 9^3 = 255$ m³. The submerged weight of the rock cone $= 1.3 \times 9.81 \times 255/1\,000 = 3.25$ MN.

$$\text{Factor of safety against uplift} = \frac{3.25}{1.03} = 3.1 \text{ (which is satisfactory)}$$

The anchorage of the whole dock floor requires 70 lines of anchors (at right-angles to the centre-line of the dock) and 20 ranks of anchors (parallel to the centre-line of the dock) to form the 3 m square grid. Therefore in Figure 6.17b, $N = 70$, $M = 20$, and $P = S = 3$ m. From Figure 6.17a, $m/L = n/L = 0.57$, and therefore, $m = n = 0.57 \times 9 = 5.1$ m. Then $P/n = s/m = 0.59$ so that, from Figure 6.17b, $\Delta V_n/V_c = \Delta V_m/V_c = 0.45$. Because $(P/n)^2 + (S/m)^2 = 2 \times 0.59^2 = 0.7$ is less than 4, there is composite overlapping of the rock cones, and the charts are not valid. The intersecting cones represent a rock volume roughly estimated to be $69 \times 3 \times 19 \times 3 \times 6 = 70\,794$ m³.

The sum giving the total force resisting uplift is thus as follows:

$$\text{Weight of dock floor} = \frac{210 \times 60 \times 0.8 \times 2.4 \times 9.81}{1000} \qquad = 237.3 \text{ MN}$$

$$\text{Submerged weight of soft clay} = \frac{210 \times 60 \times 8.0 \times 0.8 \times 9.81}{1000} \qquad = 791.1 \text{ MN}$$

$$\text{Submerged weight of anchored rock} = \frac{70\,794 \times 1.3 \times 9.81}{1000} \qquad = 902.8 \text{ MN}$$

$$\text{Total} = 1931.2 \text{ MN}$$

$$\text{Total uplift on underside of floor} = \frac{70 \times 20 \times 1100}{1000} = 1\,540 \text{ MN}.$$

Therefore

$$\text{factor of safety against uplift} = \frac{1931.2}{1540} = 1.25 \text{ (which is satisfactory and a closer estimate of the rock volume is not needed).}$$

Therefore the anchorage should consist of a 3 m square grid of 450 mm shell piles driven on to rockhead, each pile being anchored 9 m below rockhead with a seven 15.2 mm/7-wire Dyform strand cable having the bottom 7 m bonded to the rock.

Example 6.3

A piled dolphin carrying a horizontal pull of 1800 kN consists of a pair of compression piles and a pair of tension piles, raked at angles of 1 horizontal to 3 vertical. Design 'dead' anchors for the tension piles, which are driven through 3 m of weak weathered chalk to near-refusal on strong rock chalk (having an average submerged density of 0.5 mg/m³).

From the triangle of forces (Figure 6.43) the uplift load on a pair of tension piles is 2800 kN. The load to be carried by a single pile is thus $0.5 \times 2800 = 1400$kN

From Table 4.15, the ultimate shaft friction in weathered chalk is 30 kN/m². For a safety factor of 2.5 in uplift, in weathered chalk on a 600 mm steel tubular pile:

$$\text{allowable load} = \frac{\pi \times 0.6 \times \sqrt{(3^2 + 1^2)} \times 30}{2.5} = 71 \text{ kN}$$

The load to be carried by the dead anchor is thus $1400 - 71 = 1329$ kN . Use a steel tube having an overall diameter of $6\frac{5}{8}$ in (168.3 mm). A wall thickness of $\frac{5}{8}$ in (16 mm) provides a cross-sectional area of 7600 mm². Thus

$$\text{Working tensile stress in anchor} = \frac{1329 \times 1000}{7600} = 175 \text{ N/mm}^2$$

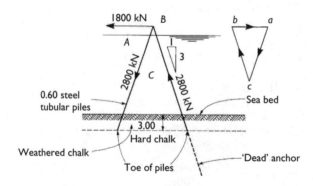

Figure 6.43

If high-tensile steel (which has a yield stress of 347 N/mm^2) is used

factor of safety at yield $= \dfrac{347}{175} = 2.0$ (which is satisfactory).

The anchor will be installed in a 215 mm diameter drill-hole. Thus for an ultimate bond stress between grout and strong rock of 0.8 N/mm^2 (Table 6.5), allowable bond stress $= 0.8/2.5 = 0.3$ MN/m^2.

required bond length $= \dfrac{1320 \times 1000}{\pi \times 215 \times 0.3 \times 1000} = 6.6$ m.

Drill the hole to say 7 m below the toe of the pile and grout the annulus fully. Then

working stress between steel and grout $= \dfrac{1329 \times 1000}{\pi \times 168.3 \times 7 \times 1000} = 0.36$ N/mm^2
(which is satisfactory)

The bond length should be increased by approximately $L/2$ to comply with Figure 6.12b to give a bond length over the cone of 10 m below the surface of the weathered chalk and space the piles at 4 m centres. Then in Figure 6.17a, $V_c = 0.35 \times 10^3 = 350$ m^3. Since $m/L = 0.61$, $m = 10 \times 0.61 = 6.1$ m . In Figure 6.17b, $S = 4$ m, so that $S/m = 4/7 = 0.66$, and thus $\Delta V_m/V_c = 0$. $M = 2$, $N = 1$, and $P = 0$, and therefore $\Delta V_n/V_c = 0$.

Rock volume anchored by pair of anchors $= 350[(2 \times 1) - (1 \times 0.40)] = 560$ m^3.
Weight of rock resisting uplift $= 560 \times 0.5 \times 9.81 = 2747$ kN.
Factor of safety against uplift $= 2747/2800 = 0.98$.

This is insufficient for compliance with the EC7 recommendations (see below) but the frictional resistance on the sloping surfaces of the overlapping cones can be taken into account. As a rough approximation, assume that the two cones act as a rectangular block having a volume of 560 m^3, say $10 \times 8 \times 7$ m deep, and take the angle of shearing resistance of the chalk as 30° and take K_o as 1.5.

Average unit frictional resistance on the vertical surfaces of the block
$= 1.5 \times \tan 30° \times 9.81 \times 0.5 \times 3.5 = 14.9$ kN/m^2
Frictional resistance to uplift $= 2 (10 + 8) \times 7 \times 14.9 = 3755$ kN
Total resistance $= 2747 + 3755 = 6502$ kN
Factor of safety against uplift $= 6502/2800 = 2.3$

Checking the design of the anchorage for compliance with the EC7 requirements, the axial uplift load on a single anchor pile is a variable unfavourable action. From Table 4.1 take $\gamma_G = 1.5$.
Therefore design value of uplift $= F_{dt} = 1.5 \times 1400 = 2100$ kN.
The value of 71.5 kN as the uplift resistance of the 600 mm pile within the weathered chalk was based on a number of tests in compression and uplift (ref. 4.43). Take the results of three tests as applicable to the selected unit resistance of 30 kN/m^2 to give a correlation factor of 1.2 (Table 4.7).
Design resistance in weak chalk $= 71.5 \times 2.5/1.2 = 148$ kN.
The grout to strong chalk ultimate bond stress of 0.8 N/mm^2 (Table 6.3) was based on pull-out tests, for which the standard practice of cycling the load would have been adopted. Table 6.2 gives an anchorage resistance partial factor of 1.4, therefore for a 10 m bond length:

Design resistance of deadman anchor in strong chalk $= 0.8 \times 1000 \times \pi \times 0.215 \times 10/1.4 = 3860$ kN.

Total design resistance of pile and anchor $= 148 + 3860 = 4008$ kN, which is greater than the design uplift of 2100 kN.

Checking the uplift resistance of the overlapping cones using the UPL limit-state factors, Table 6.2 gives a partial factor of 1.25 on the angle of shearing resistance. Therefore the characteristic resistance due to the submerged weight of the cones and the frictional resistance surrounding them is equal to $2747 + 3755/1.25 = 5751$ kN. From Table 6.1 the partial factor to obtain the design value of the favourable stabilizing action is 0.9.

Design value $= 0.9 \times 5751 = 5175$ kN.

Vertical component of variable destabilizing action $= 2800 \times \sin 71.5° = 2655$ kN.

Table 6.1 gives a partial factor of 1.5 for variable unfavourable actions. From equation 6.4:

$V_{dstd} = 1.5 \times 2655$ kN $= 3983$ which is less than the design resistance of 5175 kN.

If shear connectors are to be provided equation 6.6 can be used to calculate the required bond length. It is not strictly valid for the geometry of the connection but this example will illustrate the use of the equations. Assume a characteristic grout compression strength of 25 N/mm^2 at 3 days and a modular ratio of 18. For a shear key upstand height of 10 mm and a spacing of 150 mm, the ratio h/s of 0.067 gives $C_s = 1.0$. Assume a bond length to anchor tube diameter greater than 12 to give $C_L = 0.7$.

From equation 6.7, stiffness factor:

$$K = \frac{1}{18}\left(\frac{568}{200}\right)^{-1} + \left(\frac{168}{16} + \frac{600}{16}\right)^{-1}$$

$$= 0.04$$

From equation 6.6, characteristic bond strength:

$$= 0.04 \times 0.7 \left(9 \times 1 + 1100 \times \frac{10}{150}\right)\sqrt{25}$$

$$= 11.5 \text{ N/mm}^2$$

For work of this kind special grout monitoring would not be used. Hence a safety factor of 6 should be used giving,

Allowable bond strength $= \dfrac{11.5}{6} = 1.9$ N/mm^2

Required bond length over anchor $= \dfrac{1329 \times 1000}{\pi \times 168.3 \times 1.9} = 1323$ mm

Therefore provide $1323/150 = 8.8$, say 9 shear keys spaced at 150 mm centre over a distance of 1.3 m over anchor tube and pile.

Example 6.4

A vertical bored and cast in-situ pile 900 mm in diameter is installed to a depth of 6 m in a stiff over-consolidated clay ($\bar{c}_u = 120$ kN/m^2, $c' = 10$ kN/m^2, $\phi' = 25°$). Find the ultimate sustained horizontal load which can be applied at a point 4 m above ground level. Also find the maximum working load if the lateral deflection of the pile at ground level is limited to not more than 25 mm.

Consider first the ultimate horizontal load. For conditions of immediate application, that is, using the undrained shearing strength, from Table 6.5 for $\bar{c}_u = 120$ kN/m², the soil modulus k is 7.5 mN/m². If the elastic modulus of concrete is 26×10^3 MN/m² and the moment of inertia of the pile is $0.0491 \times (900)^4$ mm⁴, from equation 6.11:

$$R = \sqrt[4]{\frac{26 \times 10^3 \times 0.0490 \times 0.9^4}{7.5 \times 0.9}} = 3.3 \text{ m}$$

L is 6 m which is less than $2R$, therefore the pile will behave as a short rigid unit, and Brinch Hansen's method can be used. Brinch Hansen's coefficients, as shown in Figure 6.22 with $c = c_u = 120$ kN/m² and $\phi = 0$ are as tabulated below:

z (m)	0	1	2	3	4	5	6
$\frac{z}{B}$	0	1.1	2.2	3.3	4.4	5.5	6.6
K_c	2.2	5.5	6.2	6.7	7.0	7.2	7.3
$c_u K_c$	264	660	744	804	840	864	876

The soil resistance of each element 1 m wide by 1 m deep is plotted in Figure 6.44a. As a trial assume the point of rotation X is at 4.0 m below ground level. Then taking moments about point of application of H_u:

$$\sum M = (462 \times 1 \times 4.5) + (702 \times 1 \times 5.5) + (774 \times 1 \times 6.5) + (822 \times 1 \times 7.5)$$
$$- [(852 \times 1 \times 8.5) + (870 \times 1 \times 9.5)] = +16\,291 \text{ Nm per metre width of pile}$$

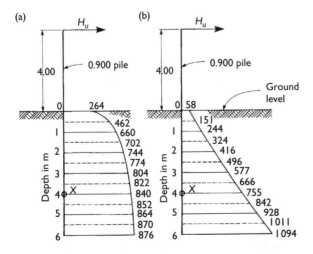

Figure 6.44

If the point of rotation is raised to 3.9 m below ground level, $\sum M = +297$ kNm, which is sufficiently close to zero for the purpose of this example.

Taking moments about the centre of rotation,

$$H_u \times 7.9 = (462 \times 3.4) + (702 \times 2.4) + (774 \times 1.4) + (820.2 \times 0.9 \times 0.45)$$
$$+ (838.2 \times 0.1 \times 0.05) + (852 \times 0.6) + (870 \times 1.6),$$

and thus $H_u = 828$ kN per metre width. For a pile 0.9 m wide, $H_u = 0.9 \times 828 = 745$ kN

Now consider the long-term stability under sustained loading, when the drained shearing strength parameters $c' = 10$ kN/m^2 and $\phi' = 25°$ apply. From Figure 6.22, Brinch Hansen's coefficients for K_c and K_q are tabulated below:

z (m)	0	1	2	3	4	5	6
$\frac{z}{B}$	0	1.1	2.2	3.3	4.4	5.5	6.6
K_c	5.8	16	20	23	26	27	28
$c'K_c$	58	160	200	230	260	270	280
K_q	3.3	5.0	5.5	5.9	6.3	6.7	6.9
p_0(kN/m^2)	0	18.6	39.3	58.8	78.5	98.2	118
p_0K_q(kN/m^2)	0	93	216	347	495	658	814
$c'K_c + p_0K_q$(kN/m^2)	58	253	416	577	755	928	1094

The soil resistance of each element 1 m deep for a pile 1 m wide is plotted in Figure 6.44b. As a trial, consider the point of rotation X to be 4.0 m below ground level. Taking moments about the point of application of H_u:

$$\sum M = (155 \times 1 \times 4.5) + (335 \times 1 \times 5.5) + (496 \times 1 \times 6.5) + (666 \times 1 \times 7.5)$$
$$- [(842 \times 1 \times 8.5) + (1011 \times 1 \times 9.5)] = -6002 \text{ kNm.}$$

If the centre of rotation is lowered to 4.5 m, then

$$\sum M = 10759 + (798 \times 0.5 \times 8.25) - [(885 \times 0.5 \times 8.75) + 9604)]$$
$$= 14051 - 13476 + 575 \text{ kN,}$$

which is sufficiently close to zero for the purpose of this example. Then taking moments about the centre of rotation:

$$H_u \times 8.5 = (155 \times 4.0) + (335 \times 3.0) + (496 \times 2.0) + (660 \times 1.0)$$
$$+ (798 \times 0.5 \times 0.25) + (885 \times 0.3 \times 0.25) + (1011 \times 1 \times 1).$$

Thus $H_u = 530$ kN per metre width. Therefore the lowest value of the ultimate load results from drained shearing strength conditions. For a 900 mm pile, the ultimate horizontal load $= 0.9 \times 530 = 477$ kN.

Calculating the allowable horizontal load which limits the lateral deflection at ground level to 25 mm, from equation 6.50:

Depth to point of fixity = $1.4R = 1.4 \times 3.3 = 4.62$ m

From equation 6.20 with e = 0, $H = 3 \times 0.025 \times 837.38 \times 10^3 / 4.62^3 = 637$ kN

Therefore the allowable load is governed by the resistance of the pile to overturning. A factor of safety of 1.5 on the ultimate load of 477 kN will be appropriate giving an allowable load of 318 kN.

Example 6.5

A group of 36 steel box piles are spaced at 1.25 m centres in both directions to form six rows of six piles surmounted at ground level by a rigid cap. The piles are driven to a depth of 9 m into a medium-dense water-bearing sand and carry a working load of 240 kN on each pile. Calculate the bending moments, deflections and soil-resistance values at various points below the ground surface at the working load. Calculate the horizontal deflection of the pile cap if the horizontal load is applied in the direction resisted by the maximum resistance moment of the piles. Moment of inertia of the pile in direction of maximum resistance moment = 58064 cm^4 and elastic modulus of steel = 20 MN/cm^2.

From Figure 6.20, Terzaghi's value of n_h for a medium dense sand is 5 MN/m^3. Then from equation 6.12:

$$T = \sqrt[5]{\frac{20 \times 58064}{5 \times 10^{-6}}} = 188 \text{ cm.}$$

Because the embedded length of 9 m is more than $4T$ the pile behaves as a long elastic fixed-headed element. The ultimate moment of resistance of pile (modulus of section = 2950 cm^3) = $2950 \times 0.0247 = 72.86$ MNcm for a working stress on BS 4360 Grade 40A steel of 247 MN/m^2.

From equation 6.51, depth to point of fixity = $1.8 \times 188 = 338.4$ cm.

From equation 6.19, ultimate horizontal load = $H_u = 2 \times 72.86 \times 10^3/338.4 = 431$ kN.

Factor of safety on applied load = $431/240 = 1.8$, which is satisfactory if the pile head deflections and the pile group behaviour are within acceptable limits.

The deflections, bending moments and soil-resistance values for the single pile at the working load can be calculated from the curves in Figure 6.28. From equation 6.27:

$$\text{deflection } y_F = \frac{240 \times 188^3}{20 \times 10^3 \times 58064} F_y = 1.373F_y \text{ cm} = 13.73 \, F_y \text{ mm.}$$

From equation 6.28:

$$\text{bending moment } M_F = 240 \times 188 \, F_m = 45\,120F_m \text{ kNcm} = 451.2F_m \text{ kNm.}$$

From equation 6.29:

$$\text{soil reaction } P_F = 240 \, F_p/188 = 1.28 \, F_p \text{ kN per cm depth} = 128 \, F_p \text{ kN per m depth.}$$

$Z_{max} = L/T = 9.0/1.88 = 4.8$. Tabulated values of y_F, M_F, and P_F are as follows:

x(m)	0	0.5	1.0	1.5	2.0	2.5	3.0	4.0	5.0
$z = \dfrac{x}{T}$	0	0.27	0.53	0.80	1.06	1.33	1.60	2.13	2.66
F_y	+0.92	+0.90	+0.82	+0.71	+0.61	+0.50	+0.37	+0.18	+0.04
y_F (mm)	+12.6	+12.4	+11.3	+9.7	+8.4	+6.9	+5.1	+2.5	—
F_m	−0.91	−0.65	−0.40	−0.18	−0.03	+0.10	+0.19	+0.25	+0.21
M_F (kN/m)	−411	−293	−180	−81	−14	+4.5	+86	+113	+95
F_p	0	+0.25	+0.45	+0.57	+0.62	+0.62	+0.57	+0.38	+0.13
P_F (kN/m)	0	+32.0	+57.6	+73.0	+79.4	+79.4	+73.0	+48.6	+16.6

From the above table the pile head deflection is satisfactory and the moment of resistance of the pile is not exceeded. Because the piles are spaced at $125/46.7 = 2.67$ diameters the group will act as a single unit equivalent to a block foundation having a width of 5×1.25 m $= 6.25$ m and a depth below the ground surface of 9 m. The ultimate passive resistance to the horizontal thrust from a block foundation is

$$P_u = \tfrac{1}{2}B\gamma D^2 K_p = \tfrac{1}{2} \times 6.25 \times 1.3 \times 9.81 \times 9^2 \times 3.69 = 11\,912 \text{ kN}.$$

For a safety factor of 1.5 on the passive resistance, the total horizontal load on the pile group must be limited to $11\,912/1.5 = 7941$ kN. Thus the load on each pile must be limited to $7941/36 = 220$ kN, which provides a safety factor of 2 on the value estimated for the single pile.

It is also necessary to calculate the horizontal deflection of the pile group under the working load of 220 kN per pile. The values of P_F above show that the horizontal load is effectively distributed over a depth of 4 m below the ground surface. Thus the load on the group can be simulated by a block foundation having a 'width' B of 4 m, a 'length' of 6.25 m and a depth of 6.25 m. The elastic modulus of a medium-dense sand can be taken as 20 MN/m². From equation 5.22, with $H/B = 1000$, $L/B = 6.25/4 = 1.55$, $D/B = 6.25/4 = 1.55$, $\mu_1 = 0.85$, and $\mu_0 = 0.91$ as Figure 5.20:

$$\text{elastic settlement } \rho_i = \dfrac{0.85 \times 0.91 \times \dfrac{220 \times 36}{4 \times 6.25} \times 4 \times 1000}{20 \times 1000} = 49 \text{ mm}$$

This is within safe limits and, as would be expected, it is more than 10 times greater than the deflection of the single pile.

Example 6.6

In the previous example the horizontal wind load of $36 \times 220 = 7920$ kN is applied at the level of the centroid of a 7.25 m square by 2.0 m deep pile cap. If a centrally applied vertical load of 8000 kN also acts at the centroid, find the resultant vertical component of the load on each pile of the group due to the combined horizontal and vertical loading.

Weight of pile cap $= 7.25^2 \times 2.0 \times 9.81 \times 2.4 = 2475$ kN. Thus the total vertical load $= 8000 + 2475 = 10475$ kN. From Figure 6.45 the resultant of the horizontal and vertical loads cuts the underside of the pile cap at a point 0.76 m from the centroid. From equation 6.51.

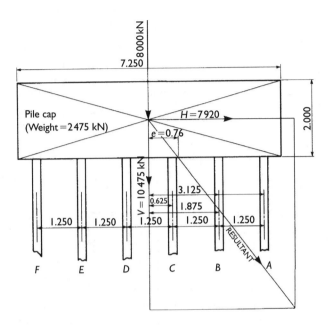

Figure 6.45

Vertical component on each pile in Row A $\quad = \dfrac{10\,475}{36} + \dfrac{10\,475 \times 0.76 \times 3.125}{12(3.125^2 + 1.875^2 + 0.625^2)}$

$$= 291.0 + \dfrac{24\,878}{164.1}$$

$$= 291.0 + 151.6 = 442.6 \text{ kN}$$

Vertical component on each pile in Row B $\quad = 291.0 + \dfrac{10\,475 \times 0.76 \times 1.875}{164.1}$

$$= 291.0 + 91.0 = 382.0 \text{ kN}$$

Vertical component on each pile in Row C $\quad = 291.0 + \dfrac{10\,475 \times 0.76 \times 0.625}{164.1}$

$$= 291.0 + 30.3 = 321.3 \text{ kN}$$

Vertical component on each pile in Row D $= 291.0 - 30.3 = 260.7 \text{ kN}$

Vertical component on each pile in Row E $= 291.0 - 91.0 = 200.0 \text{ kN}$

Vertical component on each pile in Row F $= 291.0 - 151.7 = 139.3 \text{ kN}$

Checking: $6(442.6 + 382.0 + 321.3 + 260.7 + 200.0 + 139.3) = 10\,475$ kN, and the calculation is thus correct.

Example 6.7

A tower is to be constructed on a site where 6 m of very soft clay overlie a very stiff glacial clay (undrained shearing strength $= 190$ kN/m^2). The tower and its base slab weigh 30 000 kN, and the tower is subject to a maximum horizontal wind force of 1 500 kN with a centre of pressure 35 m above ground level. The base of the tower is 12 m in diameter. Design the foundations and estimate the settlements under the dead load and wind loading.

Because of the presence of the soft clay layer, piled foundations are required and the heavy vertical load favours the use of large bored and cast in-situ piles. A suitable arrangement of piles to withstand the eccentric loading caused by the wind force is 22 piles in the staggered pattern shown in Figure 6.46.

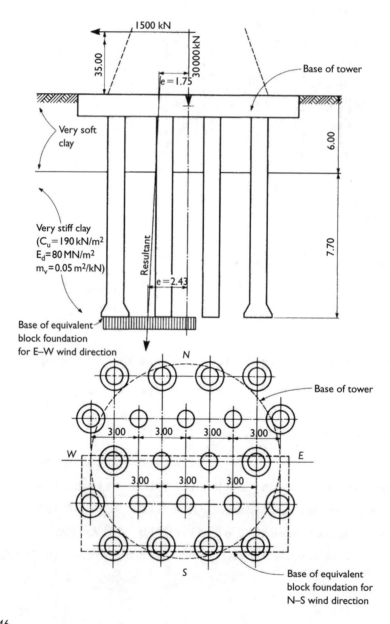

Figure 6.46

The resultant of the vertical and horizontal forces has an eccentricity of $1500 \times 35/30\,000 = 1.75$ m at ground level. From equation 6.54, the load on each of the outer four piles due to wind loading from an east–west direction is given by

$$V = \frac{30\,000}{22} + \frac{30\,000 \times 1.75 \times 6}{(4 \times 6^2) + (6 \times 4.5^2) + (4 \times 3^2) + (6 \times 1.5^2)}$$

$$= 1364 \pm 1000 \text{ kN}$$

Therefore uplift does not occur on the windward side and the maximum pile load is 2364 kN. Checking the maximum pile load for wind in a north–south direction:

$$V = 1364 \pm \frac{30\,000 \times 1.75 \times 5.20}{(8 \times 5.20^2) + (10 \times 2.60^2)} = 1364 \pm 962 \text{ kN}$$

Therefore, maximum pile load $= 2326$ kN.
For piles with a shaft diameter of 1 m:

$$\text{working stress on concrete} = \frac{2\,364 \times 1\,000}{\frac{1}{4}\pi \times 1\,000^2} = 3 \text{ N/mm}^2$$

which is within safe limits.
Adopting an under-reamed base to a diameter of 1.8 m, from equation 4.4,

$$\text{ultimate base resistance } Q_b = 9 \times 190 \times \tfrac{1}{4}\pi \times 1.8^2 = 4351 \text{ kN}$$

For a safety factor of two on the combined base resistance and skin friction, the required ultimate skin friction $= (2 \times 2364) - 4351 = 377$ kN.

If the required depth of penetration into the glacial clay to mobilize the required ultimate resistance is Lm, ignoring the small skin friction in the very soft clay and adopting an adhesion factor of 0.3 (to allow for delays in under-reaming), then from equation 4.7,

$$Q_s = 0.3 \times 190 \times \pi \times 1 \times L$$

and if $Q_s = 377$, $L = 2.10$ m. Thus the allowable pile load for a factor of safety of 3 in base resistance and unity in skin friction is $\frac{1}{3}Q_b + Q_s = (\frac{1}{3} \times 4351) + 377 = 1\,827$ kN, which is insufficient. Taking L as 4.9 m,

$$Q_s = 0.3 \times 190 \times \pi \times 1 \times 4.9 = 877 \text{ kN}$$

and the allowable pile load is $\frac{1}{3} \times 4351 + 877 = 2\,328$ kN, which is satisfactory.
It is necessary to add two shaft diameters and the depth of the under-ream to arrive at the total penetration of the piles below ground level. Thus

$$D = 6 \text{ m (soft clay)} + 4.9 + 2.0 + 0.8 = 13.7 \text{ m}$$

An adhesion factor of 0.5 is used for straight-shafted piles in a glacial clay (Figure 4.8). Therefore the allowable load on a straight-shafted pile drilled to the same depth as the under-reamed piles and adopting a safety factor of 2 on combined end bearing and shaft friction is given by

$$Q_a = \frac{(9 \times 190 \times \frac{1}{4}\pi \times 1^2) + (0.5 \times 190 \times \pi \times 1 \times 7.7)}{2} = 1820 \text{ kN}$$

Therefore straight-shafted piles can be used for the eight inner piles as shown in Figure 6.46. The maximum working load on these is one-half or less than one-half of the outer piles.

The overall depth to the base of the pile group of 13.7 m is only a little greater than the overall width of the group, i.e 13 m to the outsides of the pile shafts. Therefore it is necessary to check that block failure will not occur due to eccentric loading.

$$\text{Eccentricity of loading with respect to base of pile group} = \frac{1500 \times (35 + 13.7)}{30\,000}$$

$$= 2.43 \text{ m}$$

From equation 5.3a, the width of an equivalent block foundation for winds in a north-south direction $= 10.40 - (2 \times 2.43) = 5.54$ m. The overall dimensions of this block foundation are thus 13×5.54 m. Tangent of the angle of inclination of the resultant force $= \tan \alpha = 1\,500/30\,000 = 0.05$, and thus $\alpha = 2.87°$.

From Figure 5.6, for $\phi = 0°$, $N_c = 5.2$; from Figure 5.7 for $B'/L' = 5.54/13.0 = 0.43$, $s_c = 1.1$; from Figure 5.9 for $D/B = 7.7/5.54 = 1.4$, $d_c = 1.2$. The horizontal force of 1500 kN in Figure 6.46 is less than $c_u B'L' = 190 \times 5.54 \times 13.0 = 13\,684$ kN. Therefore equation 5.11 can be used to obtain the inclination factor $i_c = 1 - 1500/2 \times 190 \times 5.54 \times 13.0 = 0.95$. From Figure 5.6, $N_\gamma = 1.0$. From Figure 5.7, $s_\gamma = 0.95$; $d_\gamma = 1.0$. From equation 5.12, $i_q = 1 - 1500/30\,000 = 0.92$, therefore from equation 5.13, $i_\gamma = 0.92^2 = 0.85$.

The second term in equation 5.1 is zero, therefore

$$q_{ub} = 190 \times 5.2 \times 1.1 \times 1.2 \times 0.95 \times 1.0 + 0.5 \times 9.81 \times 1.8 \times 5.54 \times 1.0 \times 1.0$$
$$\times 0.95 \times 0.85$$
$$= 1238 + 39 = 1277 \text{ kN/m}^2$$
$$Q_{ab} = 1277 \times 5.54 \times 13.0 = 92\,005 \text{ kN}$$

Factor of safety against base failure $= 92\,005/30\,000 = 3.1$ which is satisfactory.

Checking for compliance with the EC7 recommendations, for the ultimate limit-state GEO and using the three design approaches (Section 4.1.4), for Design Approach DA1 combination (sets A1+M1+R1), the vertical load on the pile group in respect of overturning is a permanent stabilising action. Table 4.1 gives $\gamma_G = 1.0$, V'_d is $1.0 \times 30\,000 = 30\,000$ kN. The horizontal wind load is a variable unfavourable action, thus $H'_d = 1.5 \times 1500 = 2\,250$ kN

Eccentricity of loading in respect of base of pile group $= 2250(35 + 13.7)/30\,000 = 3.65$

For winds in a north–south direction, width of equivalent block foundation $= 10.4 - (2 \times 3.65) = 3.10$ m.

The overall dimensions of the transformed block foundation are $13.0 \times 10.4 \times 7.7$ m deep. The material factor, γ_{cu}, for set M1 in Table 4.1 is 1.0, giving characteristic $c_u = 190$ kN/m^2.

The Brinch Hansen bearing capacity factors for $\phi = 0°$ are obtained in the same way as shown above, giving $N_c = 5.2$, $s_c = 1.05$, $d_c = 1.3$ and $i_c = 0.96$.

It is evident from the preceding calculation to obtain q_{ub} that the third term in equation 5.1 is small and can be neglected.

Characteristic unit base resistance $= 190 \times 5.2 \times 1.05 \times 1.3 \times 0.96 = 1294.7$ kN/m². Applying the R1 resistance factor of 1.0, design resistance $= R_{cd} = 1294.7 \times 13.0 \times 3.1 = 52176$ kN which is greater than the design action V'_d of 30 000 kN.

For Design Approach DA1, combination 2 (sets A2+M2+R1)

Design actions, $V'_d = 1.0 \times 30000 = 30000$ kN

and $H'_d = 1.3 \times 1500 = 1950$ kN

Eccentricity of loading $= 1950(35.0 + 13.7)/30000 = 3.2$ m

Width of equivalent block foundation $= 10.4 - 2 \times 3.2 = 4.0$ m

Dimensions of equivalent block foundation are $13.0 \times 4.0 \times 7.7$ m deep

The material factor, γ_{cu}, for set M2 is 1.4 and R1 is 1.0

Design unit resistance to bearing failure $136 \times 5.2 \times 0.9 \times 1.3 \times 0.86 = 711.6$ kN/m²

Design resistance of foundation $= R_{cd} = 711.6 \times 13.0 \times 4.0 = 37000$ kN which is greater than V'_d of 30 000 kN

DA1 is used to check compliance with EC7 with respect to sliding. In the following calculations the passive resistance of the soil to horizontal movement of the piles has been ignored.

For DA1, combination 1, the base area of the equivalent block using the factored values of V' and H' is $13.0 \times 3.1 = 40.3$ m². From Table 4.2, M1 is 1.0 and R1 for spread foundations in respect of sliding is 1.0. Therefore design resistance to sliding $= 190 \times 1.0 \times 1.0 \times 40.3 = 7657$ kN which is greater than $H'_d = 1.5 \times 1500 = 2250$ kN.

For DA1, combination 2, base area $= 13.0 \times 4.0 = 52.0$ m², $R_{cd} = 1.0 \times 136 \times 52.0 = 7072$ kN which is greater than $H'_d = 1950$ kN.

It is also necessary to confirm that the total settlements and tilting of the structure are within safe limits. The following calculations are carried out using unfactored loadings to verify the serviceability limit-state.

Calculating first the immediate and consolidation settlements under dead and imposed load, but excluding the wind load. Because the piles have under-reamed bases which carry the major proportion of the load, the base of the equivalent raft will be close to pile base level. The approximate overall dimensions of the equivalent raft outside the toes of the pile bases are 13.8×11.2 m. Therefore

$$\text{overall base pressure beneath raft} = \frac{30\,000}{13.8 \times 11.2} = 194 \text{ kN/m}^2$$

Assume a value of E_u for the glacial clay of 80 mN/m² and a value of m_v of 0.05 m²/kN. From Figure 5.20 for $L/B = 13.8/11.2 = 1.2$, $H/B = \infty$ and $D/B = 7.7/11.2 = 0.69$ (ignoring the soft clay), $\mu_1 = 0.75$ and $\mu_0 = 0.92$. Therefore

$$\text{immediate settlement} = \frac{0.75 \times 0.92 \times 194 \times 11.2 \times 1000}{80 \times 1000} = 19 \text{ mm}$$

From Figure 5.13 the average vertical stress at the centre of a layer of thickness 2B is $0.3 \times 194 = 58 \text{ kN/m}^2$.

The depth factor μ_d for $D/\sqrt{LB} = 0.62$ is 0.81 and the geological factor μ_g is 0.5. Therefore, from equations 5.23 and 5.24

$$\rho_c = \frac{0.5 \times 0.81 \times 0.05 \times 58 \times 2 \times 11.2 \times 1000}{1000} = 26 \text{ mm}$$

Part of the imposed loading will not be sustained and will not contribute to the long-term settlement. Thus the total settlement under the vertical load of 30 000 kN will probably not exceed 30 mm. It is necessary to estimate the amount of tilting which would occur under sustained wind pressure, i.e. the immediate settlement induced by the horizontal wind force of 1500 kN producing a pressure under the combined vertical and horizontal loading of $30000/(13 \times 5.54) = 416 \text{ kN/m}^2$ on the equivalent raft caused by the eccentric loading. For $L/B = 13/5.54 = 2.3$, $H/B = \infty$ and $D/B = 7.7/5.54 = 1.4$, $\mu_1 = 1.0$ and $\mu_0 = 0.9$ giving

$$\text{immediate settlement} = \frac{1.0 \times 0.9 \times 416 \times 5.54 \times 1000}{80 \times 1000} = 26 \text{ mm}$$

Of this amount 16 mm is due to vertical loading only, giving a tilt of 10 mm due to wind loading. A movement of this order would have a negligible effect on the stability of the tower.

The horizontal force on each pile if no wind load is carried by the pile cap is $1500/22 = 68 \text{ kN}$. A pile 1 metre in diameter can carry this load without excessive deflection (see Example 6.4).

Example 6.8

Pressuremeter tests made at intervals of depth in a highly weathered weak broken siltstone gave the following parameters:

Pressuremeter modulus $= E_m = 30 \text{ MN/m}^2$
Limit pressure $= p_l = 1.8 \text{ MN/m}^2$
Creep pressure $= p_f = 0.8 \text{ MN/m}^2$

The above values were reasonably constant with depth. Draw the deflection curve for a horizontal load applied to the head of a 750 mm pile at the ground surface up to the ultimate load, and obtain the deflection for a horizontal load of half the ultimate.

$$\text{Moment of inertia of uncracked pile} = \frac{\pi \times 0.75^4}{64} = 0.0155 \text{ m}^4$$

Modulus of elasticity of pile $= 26 \times 10^3$ MN/m^2

Take a rheological factor of 0.8, then from equation 6.43a:

$$\frac{1}{k_m} = \frac{2 \times 0.6}{9 \times 30}\left(\frac{0.75}{0.6} \times 2.65\right)^{0.8} + \frac{0.8 \times 0.75}{6 \times 30}$$

$$k_m = 67 \text{ MN/m}^2$$

Over elastic range from $p = 0$ to $p = p_f$, then from equation 6.11:

$$\text{Stiffness factor} = R = \sqrt[4]{\frac{26 \times 10^3 \times 0.0155}{67 \times 0.75}} = 1.68 \text{ m}$$

To allow for surface heave assume no soil reaction from ground surface to assumed surface at $0.5 \times 0.75 = 0.375$ m

At creep pressure of 0.8 MN/m^2, corresponding deflection $= \dfrac{0.8 \times 0.75}{67} = 0.0090$ m

From equation 6.46 corresponding lateral load applied at assumed ground surface:

$$H = \frac{0.0090 \times 1.68 \times 67 \times 0.75}{2} = 0.380 \text{ MN}$$

From equation 6.47, slope at assumed ground surface

$$= -\frac{2 \times 0.380}{1.68^2 \times 67 \times 0.75} = -0.0054 \text{ radians}$$

Deflection at real ground surface

$$= 0.009 - \frac{0.380 \times 0.375^2}{6 \times 26 \times 10^3 \times 0.0155} - (-0.0054 \times 0.375)$$

$$= 0.0110 \text{ m}$$

$$= 11 \text{ mm}$$

Between $p = p_f$ and $p = p_l$ the upper curve in Figure 6.34b gives $k_m = \dfrac{67}{2} = 33.5$ MN/m^2 and

$$R = \sqrt[4]{\frac{26 \times 10^3 \times 0.0155}{33.5 \times 0.75}} = 2.00 \text{ m}$$

From upper curve in Figure 6.34b

At limit pressure of 1.8 MN/m^2 corresponding deflection $= \dfrac{1.8 \times 0.75}{33.5} = 0.0403$ m

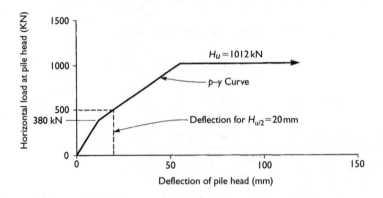

Figure 6.47

Corresponding lateral load at assumed ground surface $= \dfrac{0.0403 \times 2.00 \times 33.5 \times 0.75}{2}$

$$= 1.012 \text{ MN}$$

Slope at assumed ground surface $= -\dfrac{2 \times 1.012}{2^2 \times 33.5 \times 0.75} = -0.0201 \text{ radians}$

Total deflection at real ground surface

$= 0.0110 + 0.0403 - \dfrac{1.012 \times 0.375^2}{6 \times 26 \times 10^3 \times 0.0155} - (-0.0201 \times 0.375)$

$= 0.0588$

$= 59 \text{ mm}$

The load/deflection curve is shown in Figure 6.47. The deflection corresponding to an applied load of half the ultimate load of 1012 kN is 20 mm.

Chapter 7

Some aspects of the structural design of piles and pile groups

7.1 General design requirements

Piles must be designed to withstand stresses caused during their installation, and subsequently when they function as supporting members in a foundation structure. Stresses due to installation occur only in the case of piles driven as preformed elements. Such piles must be capable of withstanding bending stresses when they are lifted from their fabrication bed and pitched in the piling rig. They are then subjected to compressive, and sometimes to tensile, stresses as they are being driven into the ground, and may also suffer bending stresses if they deviate from their true alignment. Piles of all types may be subjected to bending stresses caused by eccentric loading, either as a designed loading condition or as a result of the pile heads deviating from their intended positions. Differential settlement between adjacent piles or pile groups can induce bending moments near the pile heads as a result of distortion of the pile caps or connecting beams.

The working stresses adopted for piles should take into account the effects of unseen breakage caused during driving, possible imperfections in concrete cast in-situ, and the long-term effects of corrosion or biological decay.

Pile caps, capping beams, and ground beams are designed to transfer loading from the superstructure to the heads of the piles, and to withstand pressures from the soil beneath and on the sides of the capping members. These soil pressures can be caused by settlement of the piles, by swelling of the soil, and by the passive resistances resulting from lateral loads transmitted to the pile caps from the superstructure.

In addition to guidance on structural design and detailing, matters of relevance to the design of piled foundations in EC2 (BS EN 1992-1-1: 2004) include the following:

(1) Dimensional tolerances of cast-in-place piles (see Table 4.9)
(2) Partial factors for the ultimate limit state of materials
(3) The influence of soil–structure interaction caused by differential settlement
(4) Strength classes of concrete and reinforcement cover for various exposure conditions
(5) Slenderness and effective lengths of isolated members
(6) Punching shear and reinforcement in pile caps
(7) Limits for crack widths and
(8) Minimum reinforcement for bored piles.

Many of the above items have been dealt with in the previous chapters. Structural analysis, design and detailing of reinforced concrete and prestressed concrete members will not, in general, be covered in this chapter, but a particular point to be noted is that whereas BS 8004 allows an increase in working stresses for temporary works, such increases are not permitted by EC2. BS EN 12699 allows for an increase in compressive stress generated during driving (see Sections 2.2 and 2.3).

7.2   Designing reinforced concrete piles for lifting after fabrication

The reinforcement of piles to withstand bending stresses caused by lifting has to be considered only in the case of precast reinforced (including prestressed) concrete piles. Bending takes place when the piles are lifted from their horizontal position on the casting bed for transportation to the stacking area. The most severe stresses thus occur at the time when the concrete is immature. Timber piles in commercially available lengths which have a cross-sectional area sufficiently large to withstand driving stresses will not be overstressed if they are lifted at the normal picking-up points. Splitting could occur if attempts were to be made to lift very long piles fabricated by splicing together lengths of timber, but there is no difficulty in designing spliced joints so that the units can be assembled and bolted together while the pile is standing vertically in the leaders of the piling frame. Again, steel piles with a cross-sectional area capable of withstanding driving stresses and of sufficient thickness to allow for corrosion losses will not fracture when lifted in long lengths from the horizontal position in the fabrication yard.

Reinforced concrete and prestressed concrete piles have a comparatively low resistance to bending, and the stresses caused during lifting may govern the amount of longitudinal reinforcing steel needed. The static bending moments induced by lifting and pitching piles at various points on their length are shown in Table 7.1. These considerations are principally concerned with piles cast on the project site using the techniques described in Section 2.2.2. In the UK, driven precast concrete piles usually consist of the proprietary jointed types described in Section 2.2.3. These are factory-made with specially designed facilities for handling and transport.

The design charts in Figure 7.2a to d show the bending moments due to self-weight which are induced when square piles of various cross-sectional dimensions are lifted from the head or centre (pick-up point A as in Figure 7.1g and h), from a point one-third of the length from the head (pick-up point B as in Figure 7.1d), and from points one-fifth of the length from the head and toe (pick-up point C as Figure 7.1a).

Also shown on the charts are horizontal lines representing the ultimate moment of resistance of each pile section using four main reinforcing bars of various sizes. Concrete with a

Table 7.1 Bending moments induced by lifting and pitching piles

Condition	Maximum static bending moment
Lifting by two points at $L/5$ from each end	$WL/40$ (Figure 7.1a)
Lifting by two points at $L/4$ from each end	$WL/32$ (Figure 7.1b)
Pitching by one point $3L/10$ from head	$WL/22$ (Figure 7.1c)
Pitching by one point $L/3$ from head	$WL/18$ (Figure 7.1d)
Pitching by one point $L/4$ from head	$WL/18$ (Figure 7.1e)
Pitching by one point $L/5$ from head	$WL/14$ (Figure 7.1f)
Pitching from head	$WL/8$ (Figure 7.1g)
Lifting from centre	$WL/8$ (Figure 7.1h)

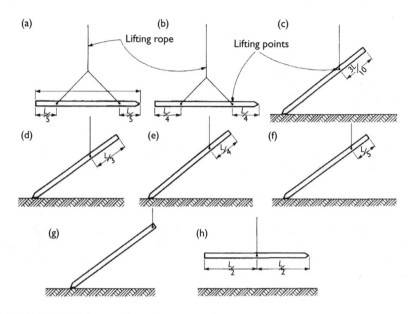

Figure 7.1 Methods of lifting reinforced concrete piles.

characteristic strength of 40 N/mm² has been used with a characteristic strength for the reinforcement of 250 N/mm², and 40 mm cover to the link steel.

It is desirable to limit the steel stresses to 250 N/mm² when determining ultimate resistance moments. This will ensure that a 4 bar reinforcement arrangement complies with Table 3.30 of BS 8110: Part 1 regarding maximum spacing requirements, that is, 300 mm assuming no redistribution. If a characteristic strength of 460 N/mm² were to be used, the maximum allowable bar spacing becomes 160 mm, thus requiring an 8 bar arrangement.

However, the situation can arise in which a moment is induced in a pile due to the pile being driven just within the permitted tolerance of 1 in 75 to the vertical. In such a case the pile can be checked as a column with an axial load and moment, by use of the design charts in BS 8110: Part 3. These charts relate only to high-yield steel with a characteristic strength of 460 N/mm². It is therefore advisable to use high-yield steel for pile reinforcement, but to limit the characteristic strength to 250 N/mm² when designing for lifting from the casting bed.

The charts can be used for a concrete having a higher characteristic strength but the lifting lengths will be somewhat conservative. The concrete strength is usually governed by the strength required to resist driving stresses (see Section 2.2.2), or to give durability in an aggressive environment (see Section 10.3 and Clauses 4 and 7 of EC2).

The length and cross-sectional dimensions of the pile are obtained from considerations of the resistance of the soil or rock to axial or lateral loading (Chapters 4 and 6). Then for the given length and cross-section the pick-up point is selected, having regard to the type of piling plant and craneage to be employed, and the economies which may be achieved by lifting the piles from points other than at the ends or the centre. The bending moment due to the *ultimate* dead load corresponding to the selected pick-up point is then read off from the appropriate curve. Next the ultimate dead load bending moment is compared with the ultimate moments of resistance for various sizes of reinforcement as shown on the charts and the appropriate reinforcement is selected. A partial safety factor of 1.4 on the bending moment due to dead load is suitable for normal designs, and this value has been used in the derivation of Figure 7.2 and Table 7.2.

It is evident that there is a maximum lifting length for a pile of given cross-sectional dimensions and size of main reinforcement. The design curves shown in Figure 7.2 have been used to obtain the maximum lifting lengths for the three pick-up conditions for square piles in Table 7.2. Also shown in these tables are the requirements for the link steel over a length equal to three times the pile width at the head and toe, and in the body of the pile, in accordance with BS 8004.

The reinforcement stress and bar spacing previously referred to ensure that, from considerations of exposure and durability, a surface crack width of 0.3 mm is not exceeded. This is permitted by EC2 (Clause 7.3.1) for concrete in foundations in non-aggressive soil, in wet or rarely dry conditions, and for concrete exposed to sea water including splash zone. Requirements for cover to reinforcement in aggressive environmental conditions, as recommended by EC2, are discussed in Chapter 10.

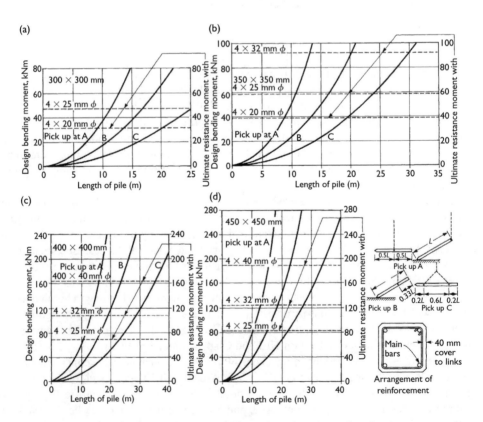

Figure 7.2 Diagrams showing required lifting points for reinforced concrete piles of various cross-sections (see also page 337).

Table 7.2 Maximum lengths of square section precast concrete piles for given reinforcement

Pile size (mm)	Main reinforcement (mm)	Maximum length in metres for pick up at			Transverse reinforcement	
		Head and toe	0.33 × length from head	0.2 × length from head and toe	Head and toe	Body of pile
300 × 300	4 × 20	9.0	13.5	20.5	6 mm at	6 mm at
	4 × 25	11.0	16.5	25.0	40 mm crs	130 mm crs
350 × 350	4 × 20	8.5	13.0	19.5	8 mm at	8 mm at
	4 × 25	10.5	16.0	24.0	70 mm crs	175 mm crs
	4 × 32	13.0	20.0	30.0		
400 × 400	4 × 25	10.0	15.0	22.5	8 mm at	8 mm at
	4 × 32	12.5	19.0	28.0	60 mm crs	200 mm crs
	4 × 40	15.5	23.0	34.5	or	
					10 mm at	
					100 mm crs	
450 × 450	4 × 25[a]	9.5	14.5	22.0	8 mm at	8 mm at
	4 × 32[a]	12.0	18.0	27.0	60 mm crs	180 mm crs
	4 × 40[a]	15.0	22.5	33.5	or	or
					10 mm at	10 mm at
					90 mm crs	225 mm crs

Notes
Piles designed in accordance with BS 8110 and BS 8004.
Characteristic strength of reinforcement limited to 250 N/mm^2.
Cover to link steel = 40 mm.
Characteristic strength of concrete = 40 N/mm^2.
a Alternatively, use a larger number of smaller diameter bars.

7.3   Designing piles to resist driving stresses

It is necessary to check the adequacy of the designed strength of a pile to resist the stresses caused by the impact of the piling hammer. Much useful data to aid the estimation of driving stresses came from the research of Glanville *et al.*[7.1] They embedded stress recorders in piles to measure the magnitude and velocity of the stress wave induced in the pile by blows from the hammer. The tests showed that the stress wave travels from the head to the toe of the pile and is partly reflected from there to return to the head. If the pile is driven onto a hard rock, the sharp reflection of the wave at the toe can cause a compressive stress at the point which is twice that at the head, but when long piles are driven into soil of low resistance, the tensile stress wave is reflected, causing tension to develop in the pile. It can be shown from simple impact theory that the *magnitude* of the stress wave depends mainly on the height of the drop. This is true for a perfectly elastic pile rebounding from an elastic material at the toe. In practice there is plastic yielding of the soil beneath the toe, and the pile penetrates the soil by the amount described as the 'permanent set'. The weight of the hammer is then important in governing the length of the stress wave and hence the efficiency of the blow in maintaining the downward movement of the pile.

   The simplest approach to ensuring that driving stresses are within safe limits is to adopt working stresses under static loading such that heavy driving is not required to achieve the depth of penetration required for the calculated ultimate bearing capacity. The usual practice is to assume that the dynamic resistance of a pile to its penetration into the soil is equal to its ultimate static load-bearing capacity, and then to calculate the 'permanent set' in terms of blows per unit penetration distance to develop this resistance, using a hammer of given rated energy or weight and height of drop. The driving stress is assumed to be the ultimate driving resistance divided by the cross-sectional area of the pile, and this must not exceed the safe working stress on the pile material. As already stated in Section 1.4, the dynamic resistance is not necessarily equal to the static load-bearing capacity. However, if soil mechanics calculations as described in Chapter 4 have been made to determine the required size and penetration depth necessary to develop the ultimate bearing capacity, then either a simple dynamic pile driving formula or, preferably, stress-wave theories can be used to check that a hammer of a given weight and drop (or rated energy) will not overstress the pile in driving it to the required penetration depth. If at any stage of penetration the stresses are excessive a heavier hammer must be used, but if greater hammer weights and lesser drops still cause overstressing then other measures, such as drilling below the pile toe or using an insert pile having a smaller diameter, must be adopted.

It is important to note that in many instances the soil resistance to driving will be higher than the value of ultimate bearing capacity as calculated for the purpose of determining the allowable pile load. This is because calculations for ultimate bearing capacity are normally based on average soil parameters. Where soil strength data are scanty it is necessary to assume conservative parameters. However, when considering resistance to driving, the possible presence of soil layers stronger than the assumed average must be taken into account. Hence, when assessing driving stresses it is advisable to make a separate calculation of ultimate bearing capacity based on soil strength values higher than the average. Also, in cases where negative skin friction is added to the working load, the soil strata within which the drag-down is developed will provide resistance to driving at the installation stage.

Methods of recording hammer blows and measurements of temporary compression and set as described in Section 11.3.1 are useful as a means of site control of driving operations, but they are not helpful for determining stresses caused in the pile body by hammer impact.

A widely used method of calculating driving stresses is based on the stress-wave theory developed by Smith[7.2]. The pile is divided into a number of elements in the form of rigid masses. Each mass is represented by a weight joined to the adjacent element by a spring as shown in the case of modelling a pile carrying an axial compression load in Figure 4.29. The hammer, helmet, and packing are also represented by separate masses joined to each other and to the pile by springs. Shaft friction is represented by springs and dashpots attached to the sides of the masses (Figure 4.29) which can exert upward or downward forces on them. The end-bearing spring can act only in compression. The resistance of the ground at toe is assumed to act as a resisting force to the downward motion of the pile when struck by the hammer. Friction on the pile shaft acts as a damping force to the stress wave as determined from the side springs and dashpots. For each blow of the hammer and each element in the hammer–pile system, calculations are made to determine the displacement of the element, the spring compression of the element, the force exerted by the spring, the accelerating force and the velocity of the element in a given interval of time. This time interval is selected in relation to the velocity of the stress wave and a computer is used to calculate the successive action of the weights and springs as the stress wave progresses from the head to the toe of the pile. The output of the computer is the compressive or tensile force in the pile at any required point between the head and toe.

The input to the computer comprises the following data:

(1) Length of pile
(2) Material of pile

(3) Weight per unit length of pile
(4) Weight and fall of hammer (or rated energy per hammer blow)
(5) Efficiency of hammer
(6) Weight of helmet, packing, and any dolly or follower used
(7) Elastic modulus of packing and of any dolly used
(8) Elastic modulus of pile
(9) Coefficient of restitution of packing (and dolly if used)
(10) Elastic compression (quake) of soil
(11) Damping properties of soil
(12) Required ultimate driving resistance.

Of the above input data, items 1, 2, 3, 4, 6, and 11 are factual. The efficiency of the hammer is obtained from the manufacturer's rating, but this can decrease as the working parts become worn. The elastic modulus and coefficient of restitution of the packing may also change from the commencement to the end of driving. The elastic compression of the ground is usually taken as the elastic modulus under static loading, and this again will change as the soil is compacted or is displaced by the pile. Thus the wave equation can never give exact values throughout all stages of driving, and its continuing usefulness depends on amassing data on correlations between calculated stress values and observations of driving stresses in instrumented piles. Further refinements of the calculation procedure may also be made to allow for the changing dynamic characteristics of the hammer–pile–soil system during driving. Smith[7.2] states that the commonly accepted values for the soil compression (quake), and the damping constants for the toe and sides of the pile are not particularly 'sensitive' in the calculations, i.e. small changes in these values produce a smaller change in the final calculated results.

The basic Smith idealization represents a pile being driven by a drop hammer or a single-acting hammer. Diesel hammers have to be considered in a different manner because the energy transmitted to the pile varies with the resistance of the pile as it is being driven down. At low resistances, there are low energies per blow at a high rate of striking. As the pile resistance increases the energy per blow increases and the striking rate decreases. Manufacturers of diesel hammers provide charts of energy versus rate of striking. When predictions are being made of the ability of a particular diesel hammer to drive a pile to a given resistance consideration should be given to the range of energy over which the hammer may operate. Goble et al.[7.3] have published details of the GRLWEAP computer program which models diesel and other hammer behaviour realistically. The program proceeds by iterations until compatibility is obtained between the pile–soil system and the energy/blows per minute performance of the hammer. Because of the problems of interpreting data from the pile driving analyser when operated with diesel hammers, the present tendency is to use hydraulic hammers which give a reasonably constant energy of blow. However, the resilience of the cushioning material can change with use causing variations in the energy transmitted to the pile head.

Pile driving resistance can be computed from field measurements of acceleration and strain at the time of driving by using the dynamic pile driving analyser in conjunction with the CAPWAP program[7.4]. Pairs of accelerometers and strain transducers are mounted near the pile head and the output of these instruments is processed to give plots of force and velocity versus time for selected hammer blows as shown in Figure 7.3. The second stage of the method is to run a wave equation analysis with the pile only modelled from the

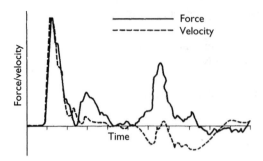

Figure 7.3 Force and velocity versus time for hammer blow on pile.

instrument location downwards. Values of soil resistance, quake, and damping are assigned and the measured time varying velocity is applied as the boundary condition at the top of the pile model. The analysis generates a force versus time plot for the instrument location and this is compared with the measured force versus time plot. Adjustments are made to the values of resistance, quake and damping until an acceptable agreement is reached between computed and measured values. At this stage the total soil resistance assigned in the analysis is taken as the resistance at the time of driving. The latter is a reliable assessment of the static resistance in soils and rocks where time effects are negligible.

The instrumentation described above including the field processing equipment which produces the force/time plot is a useful method of field control of pile driving in difficult ground conditions. The shape of the plot (Figure 7.3) can give an indication of a broken pile and a check can be made of the stresses in the pile induced by driving.

When assessing the results of wave equation analyses made at the project planning stage for the purpose of predicting the capability of a particular hammer to achieve the required penetration depth, due account should be taken of the effects of time on pile resistance as discussed in Section 4.3.8. Sufficient reserve of hammer energy should be provided to over-come the effects of set-up (increase of driving resistance) when redriving a partly driven pile after a delay period of a few hours or days. If pile driving tests are made at the planning stage it is helpful to make re-strike tests in conjunction with wave equation analyses at various time intervals after the initial drive.

Too much reliance should not be placed on immediate readings of the output from the field processing unit. Rigorous analysis of the data by experienced engineers is required in conjunction with the appropriate computer programs. Wheeler[7.5] described experiences of a field trial competition in the Netherlands when a number of firms specializing in dynamic pile testing were invited to predict the ultimate bearing capacity of four instrumented pre-cast concrete piles driven through sands and silts to penetrations between 11.5 and 19 m. A wide range of predicted capacities was obtained. In the case of one pile the range was 90 to 510 kN compared with a failure load of 340 kN obtained by static testing. Reliable estimates of ultimate bearing capacity may not be possible if the available hammer has insufficient energy to overcome the resistance mobilized by the soil against penetration of the pile. A downward movement of the toe of piles up to about 1 m in diameter of 2.5 mm or preferably more is required to mobilize sufficient soil resistance to obtain reliable results.

A heavier hammer can be used to achieve the necessary penetration but this may involve a risk of overstressing the pile.

7.4   The effects on bending of piles below ground level

Slender steel tubular piles and H-piles may deviate appreciably off line during driving. These effects were described in Section 2.2.4 where it was noted that, whereas the ill-effects of bending or buckling of tubular piles below ground level could be overcome by inserting a reinforcing cage and filling the pile with concrete, such a procedure could not be adopted with H-piles. Therefore, where long H-piles are to be driven in ground conditions giving rise to bending or buckling, a limiting value must be placed on their curvature.

It is not usual to take any special precautions against the deviation of reinforced concrete piles other than to ensure that the joints between elements of jointed pile systems (see Section 2.2.3) are capable of developing the same bending strength as the adjacent un-jointed sections. Reinforced concrete piles without joints cannot in any case be driven to very long lengths in soil conditions which give rise to excessive curvature. It is, of course, possible to inspect hollow prestressed concrete piles internally and to adopt the necessary strengthening by placing in-situ concrete if they are buckled.

It is impossible to drive a pile with a sufficient control of the alignment such that the pile is truly vertical (or at the intended rake) and that the head finishes exactly at the designed position. Tolerances specified in various codes of practice are given in Section 3.4.12. If the specified deviations are exceeded, to an extent detrimental to the performance of the piles under working conditions, the misaligned piles must be pulled out for redriving or additional piles driven. Calculations may show that minor excesses from the specified tolerances do not cause excessive bending stresses as a result of the eccentric loading. In the case of driven and cast-in-place or bored and cast-in-place piles it may be possible to provide extra rein-forcement in the upper part of the pile to withstand these bending stresses. For this reason Fleming and Lane[7.6] recommend that checks on the positional accuracy of in-situ forms of piling should be made before the concrete is placed. The methods described in Section 6.3.9 can be used to calculate the bending stresses caused by eccentric loading. The effect of the deviation is expressed as a bending moment Pe, where the load P deviates by a distance e from the vertical axis of the pile.

7.5   The design of axially loaded piles as columns

Buckling of axially loaded piles terminating at ground level in a pile cap or ground beam cannot occur if the piles are loaded to within the permissible working stresses on the pile material. Thus such piles need not be considered as long columns for the purpose of structural design. However, it is necessary to consider the column strength of piles projecting above the soil line, as in jetties or piled trestles.

BS 8004 recommends that the depth below ground surface to the point of contraflexure should be taken as 1 m in firm ground and as much as one-half of the penetration depth but not necessarily more than 3 m in a weak ground such as soft clay or silt. A stratum of liquid mud should be treated as if it were water. The column strength of the pile is then calculated as for a short column and a reduction factor is applied to the calculated ultimate load to allow for the slenderness of the column, where the slenderness is defined as the ratio of the effective length to the breadth or radius of gyration.

Effective lengths for reinforced concrete piles when regarded as columns are defined by EC2 (Clause 5.8.3.2):

Restrained at both ends in position and direction: 0.5 *L*
Restrained at both ends in position and one end in direction: 0.7 *L*
Restrained at both ends in position but not in direction: 1.0 *L*
Restrained at one end in position and direction and at the other end in
direction but not in position: 1.0 *L*
Restrained at one end in position and direction and free at the other end: 2.0 *L*

EC3 (BS EN 1993-1-1: 2005) is currently silent on effective lengths for steel piles when regarded as columns, requiring the application of the guidance given in BS 5950 part 1, which is slightly different from the above factors.

A pile embedded in the soil can be regarded as properly restrained in position and direction at the point of virtual fixity in the soil. The restraint at the upper end depends on the design of the pile cap and the extent to which the pile cap is restrained against movement by its connection with adjacent pile caps or structures. Some typical cases of the restraint of piles are shown in Figure 7.4a to e. The condition shown in Figure 7.4e

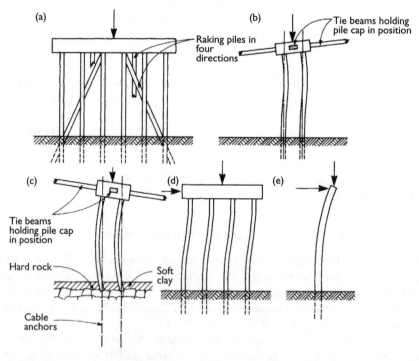

Figure 7.4 Conditions of restraint for vertical piles (a) Restrained at top and bottom in position and direction (b) Restrained at bottom in position and direction; restrained at top in position but not in direction (c) Restrained at top and bottom in position but not in direction (d) Restrained at bottom in position and direction; restrained at top in direction but not in position (e) Restrained at bottom in position and direction; unrestrained at top in position or direction.

should not arise if the piles are driven a short distance below rockhead to achieve a degree of fixity.

The appropriate codes of practice should be consulted to obtain the reduction factors on the working stresses to allow for the slenderness of the piles.

7.6   Lengthening piles

Precast (including prestressed) concrete piles can be lengthened by cutting away the concrete to expose the main reinforcement or by splicing bars for a distance of 40 bar diameters. The reinforcement of the new length is then spliced to the projecting steel, formwork is set up, and the extension is concreted. It is usual to lengthen a prestressed concrete pile by this technique in ordinary reinforced concrete. The disadvantage of using the method is the time required for the new length to gain sufficient strength to allow further driving.

A rapid method of lengthening which can be used where the piles carry compressive loads or only small bending moments is to place a mild steel sleeve with a length of four times the pile width over the head of the pile to be extended. The sleeve is made from 10 mm plates and incorporates a central diaphragm which is bedded down on a 10 to 15 mm layer of earth dry sand–cement mortar trowelled onto the pile head. After setting the sleeve a similar layer of mortar is placed on the upper surface of the diaphragm and rammed down by a square timber. The extension pile with a square end is then dropped down into the sleeve and driving commences without waiting for the mortar to set. An epoxy-resin–sand mortar can be used instead of sand–cement mortar. An epoxy-resin joint can take considerable tensile or bending forces, but the length of time over which the adhesion of the resin to the concrete is effective is indeterminate. The bond may be of rather short duration in warm damp conditions.

Another method of lengthening piles is to drill holes into the pile head. Then bars projecting from the extension piece are grouted into these holes using a cement grout or an epoxy-resin mortar.

Timber piles are lengthened by splicing as shown in Figures 2.3 and 2.4, and steel piles are butt-welded to lengthen them (Figure 7.5a and b). Backing plates or rings are provided

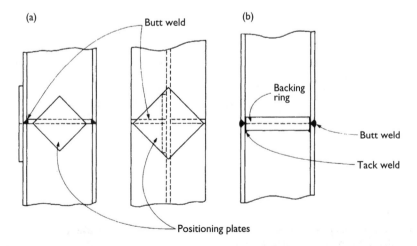

Figure 7.5 Splicing steel piles (a) Positioning plates for H-pile (b) Backing ring for tubular piles.

to position the two parts of the pile while the butt weld is made but the backing plates for the H-piles (Figure 7.5a) may not be needed if both sides of the pile are accessible to the welder. The backing ring for the tubular pile shown in Figure 7.5b is deliberately made thin so that it can be 'sprung' against the inside face of the pile. When lengthening piles in marine structures, the position of the weld should be predetermined so that, if possible, it will be situated below sea-bed level, and thus be less suceptible to corrosion than it would if located at a higher elevation.

The specification adopted for making welded splices in steel piles should take into account the conditions of loading and driving. For example, piles carrying only compressive loading and driven in easy to moderate conditions would not require a stringent specification with non-destructive testing for welding below the soil line. However, piles carrying substantial bending moments in marine structures would require a specification similar to that used for welding boilers or pressure vessels. Advice on specifications suitable for given conditions of loading and driving should be sought from the manufacturers of the piles.

7.7 Bonding piles with caps and ground beams

Where simple compressive loads without bending or without alternate compressive and uplift loading are carried by precast or cast in-situ concrete piles it is satisfactory to trim off the pile square so that the head without any projecting reinforcement is set some 75 to 100 mm into the cap (Figure 7.6a). Some uplift (but not bending) can be carried if the sides of the pile are roughened over a distance of about 300 mm and cast into the cap (Figure 7.6b). Where bending moments are to be transferred from the cap to the piles (or vice versa) the concrete must be cut away to expose the reinforcing steel or prestressing tendons, which are then bonded into the cap (Figure 7.6c). It is sometimes the practice to provide special mild steel splicing bars in the heads of prestressed concrete piles, which are exposed by cutting away the concrete after driving is complete (Figure 2.6). Alternatively, couplers can be set flush with the pile head to which further tendons or bars are attached for bonding into the cap. Splicing bars or couplers are satisfactory if the depth of penetration of

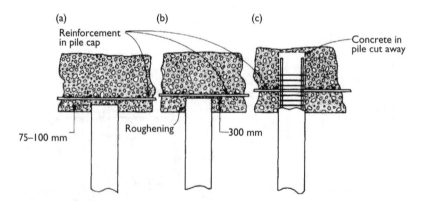

Figure 7.6 Bonding reinforced concrete piles into pile caps (a) Compressive loading only on piles (b) Compressive loading alternating with light to moderate uplift loading on piles (c) Bending moments or heavy uplift loads on piles.

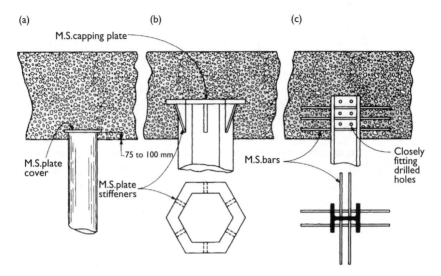

(a)   (b)   (c)

M.S.capping plate

M.S.plate
cover

└75 to 100 mm

M.S.plate
stiffeners

M.S.bars

Closely
fitting
drilled
holes

Figure 7.7 Bonding steel piles into pile caps (a) Compressive loads only on steel tubular piles
(b) Hexagonal box pile carrying heavy compressive loads or uplift loads (c) H-pile carrying
uplift loading or bending moments.

the pile can be predicted accurately. If the upper part of the pile has to be cut away, they no
longer have any useful function, but they can serve as a means of lengthening a pile should
this be necessary.

Proprietary hydraulic pile croppers, suspended from the dipper arm of an excavator, can
either break off the excess length of a concrete pile at the required level or nibble the concrete leaving the reinforcement exposed.

Steel box, tubular, or H-section piles carrying only compressive loads can be terminated
at about 100 to 150 mm into the pile cap without requiring any special modifications to
the pile to provide for bonding (Figure 7.7a). There must, however, be a sufficient thickness of concrete in the pile cap over the head of the pile to prevent failure in punching
shear. Research by the Ohio Department of Highways[7.7] has shown that if the concrete
forming the pile cap is of adequate thickness and if the reinforcement is correctly
disposed to withstand shearing and bending forces there is no need to provide a bearing
plate or other device for transferring load at the head of an H-pile. However, where steel
piles are carrying the maximum working load permitted by the material in cross-section,
the thickness of concrete in the pile cap to resist punching shear may be uneconomically
large. In such cases the head of the pile should be enlarged by welding on a capping plate
(Figure 7.7b) or by threading steel bars through close-fitting holes drilled in the pile
(Figure 7.7c). The capping arrangements shown in the latter two figures can be used to
bond the pile to the cap when uplift loads or bending moments are carried by the pile, or
alternatively bonding bars can be welded to the pile. Load transfer from large diameter
tubular piles to pile caps can be achieved by welding rectangular plates around the
periphery of the pile at its head.

7.8　The design of pile caps

A pile cap has the function of spreading the load from a compression or tension member onto a group of piles so that, as far as possible, the load is shared equally between the piles. The pile cap also accommodates deviations from the intended positions of piles, and by rigidly connecting all the piles in one group by a massive block of concrete, the ill-effects of one or more defective piles are overcome by redistributing the loads. The minimum number of small diameter piles which is permitted in an *isolated* pile cap is three. Caps for single piles must be interconnected by ground beams in two directions, and for twin piles by ground beams in a line transverse to the common axis of the pair.

A single large-diameter pile carrying a column does not necessarily require a cap. Any weak concrete or laitance at the pile head can be cut away and the projecting reinforcing bars bonded to the starter bars of the column reinforcement. Where a steel column is carried by a single large-diameter pile, the concrete is cut down and roughened to key to the pedestal beneath the column base. The heads of large-diameter piles are cast into the ground floor or basement floor concrete in order to distribute the horizontal wind forces on the superstructure to all the supporting piles.

To facilitate construction, ground beams should be arranged, where possible, to pass across the tops of the pile caps and not to frame into the sides of the caps, the connection between the cap and the ground beam being provided by column starter bars and by the friction and bond between cap and beam. The concrete forming the caps may then be placed in one operation and without the inconvenience and potential weakness that result from the formation of pockets to receive the ground beams. If the beams must frame into the cap sides, an alternative to providing pockets is to place the concrete in the caps in two operations, a horizontal construction joint being formed in each cap at the level of the underside of the ground beams.

Site setting out is also simplified by locating the ground beams on top of the pile caps. The caps (and column starter bars, if required) can be constructed with reference to the minimum number of drawings, and fixed points on the site are then available for setting out the formwork for the ground beams. Provision often has to be made for services to pass through a foundation. If the ground beams are all situated on top of the pile caps, the routes of the services are not obstructed by any pile caps, since the services may pass over the cap through holes or sleeves left in the ground beams. The apparent economy in materials and excavation gained by framing ground beams into the sides of pile caps can easily be lost by the inconvenience it causes to other operations.

A deep cap is suitable for four piles, as shown in Figure 7.8. By adopting this arrangement the column load is transferred directly into the pile heads in compression. The bending and shearing forces are negligible, requiring only the minimum proportion of steel in two directions at the bottom of the cap. EC2 (Clause 9.8.1) requires the distance from the outer edge of the pile to the edge of the pile cap to be sufficient to allow the tie forces in the cap to be properly anchored. The form of the distribution of compressive force from the top of the pile to the body of the cap is shown in Figure 7.9. The extent of the compressive zone can be allowed for when determining the anchorage length of the main reinforcement. This is most efficiently concentrated in the stressed zone between the tops of the piles (Figure 7.10). The minimum diameter of this reinforcement is required to be 8 mm. If the area of main reinforcement distributed over the pile heads is equal to the area required by considerations of control of cracking (Clause 7.3 of EC2) then evenly distributed bars along the

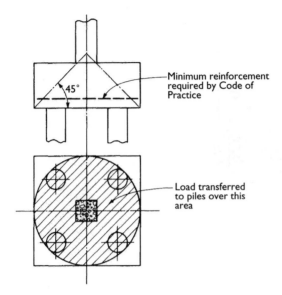

Figure 7.8 Load transfer from column to deep four-pile cap.

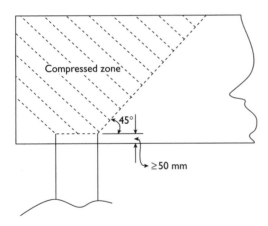

Figure 7.9 Distribution of compressive stress from pile head to pile cap.

bottom surface of the cap can be omitted. Also the sides and top surfaces of the cap may be unreinforced if there is no risk of tension developing in these zones (see Figure 7.9). BS 8110 requires minimum percentages of reinforcing steel of 0.24 and 0.13 for mild and high tensile steel respectively.

Pile caps constructed over large groups of piles as in Figures 7.10 and 7.11 can be designed as solid slabs. The design requirements for these members are set out in Clause 9.3 of EC2.

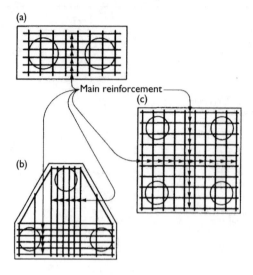

Figure 7.10 Arrangement of reinforcement in pile caps.

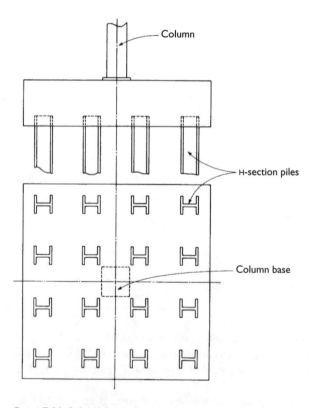

Figure 7.11 Solid slab cap for 16-pile group.

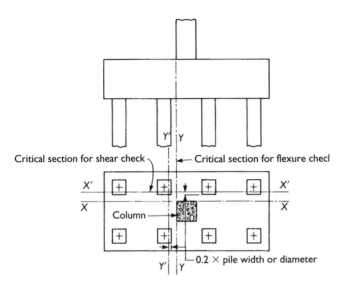

Figure 7.12 Calculation of bending moments and shearing forces on rectangular pile cap.

The bending moments on the pile cap are assumed to act from the centre of the pile to the face of the nearest column or column stem (Figure 7.12). When calculating bending moments an allowance should be made for deviations in the positions of the pile heads, up to the specified maximum tolerance (see Section 3.4.12).

In all cases the aim should be to preassemble reinforcement cages for pile caps and ground beams in order to avoid difficulties for steel fixers working in confined conditions in pits and trenches.

The cover to all reinforcement depends on the exposure condition, and the grade of concrete being used in the pile cap, and reference should be made to Clause 4 of EC2 and Clause 3.3 of BS 8110: Part 1. In particular, where concrete is cast directly against the earth, the cover should not be less than 75 mm.

Where columns carry a compressive load combined with a unidirectional bending moment, the line of action of the column load should be made to coincide with the centroid of the pile group in order to obtain a uniform distribution of load on the piles.

The dimensions of a number of standardized types of cap for use in design using computer programs are shown in Figure 7.13.

Deep pile caps are desirable for providing the stiffness necessary to distribute heavy concentrated column loads on to a large pile cluster. However, this can sometimes cause construction difficulties in unstable soils where the groundwater level is at a shallow depth below the ground surface. It is desirable, on the grounds of cost, to avoid construction expedients such as a wellpoint groundwater lowering system to enable the pile cap to be constructed in dry conditions. Consideration should therefore be given to raising the level of the pile cap to bring it above groundwater level or to such a level that sump pumping from an open excavation will not cause instability by upward seepage.

The design and construction of pile caps at over-water locations is discussed in Section 9.6.3.

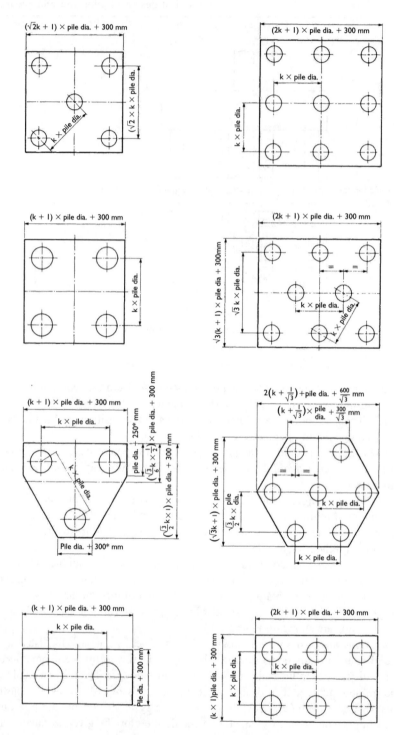

Figure 7.13 Standard pile caps (after Whittle and Beattie[7.8]).

7.9   The design of pile capping beams and connecting ground beams

Pile capping beams have the function of distributing the load from walls or closely spaced columns onto rows of piles. For heavy wall loading in conjunction with transverse bending moments the piles are placed in transverse rows surmounted by a wide capping beam (Figure 7.14a). The piles may be placed in a staggered row for walls carrying a compressive loading with little or no transverse bending moments (Figure 7.14b). A lightly loaded wall can be supported by a single row of piles beneath the centre-line provided that the beam capping the piles is restrained by tying it to transverse capping beams carrying cross walls in the structure. Attention should be given to providing adequate restraint to transverse movement and bending where ground beams are supported by minipiles.

A design method was proposed by the Building Research Establishment[7.9] which allows the bending moments and shearing forces in a beam to be reduced if the beam can be shown to be acting compositely with the brick wall built upon it. The method may be applied to the design of pile capping beams for house foundations, and is applicable to walls having a height of not less than 0.6 of the span of the beam. The walls must not have door or window openings near to the supports, as this would interfere with the arching action of the brickwork. The bending moment produced by a uniformly distributed load on a freely supported beam is $WL/8$. With full composite action between beam and wall, this moment may be reduced to a minimum of $WL/100$ for light loading, where W is the total distributed load on the brickwork (including self weight) and L is the span of the beam. A property of composite action is that the compression in the arch within the brickwork is directed radially towards the nearest firm supports; therefore, shearing reinforcement of the beam is theoretically unnecessary if the loaded lengths at the supports can be shown to be not greater than the depth of the beam. However, the BRE recommended that all beams designed for bending moments of greater than $WL/60$ should be designed to resist the shearing forces produced by the full dead and imposed loading.

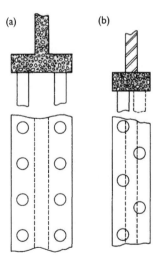

Figure 7.14 Arrangement of piles in capping beams (a) Heavy wall loading with transverse bending moments (b) Light wall loading with little or no transverse bending.

It is recommended that the depth of the beam should be between $L/15$ and $L/20$. However, with a heavily loaded wall only a small degree of composite action is allowed and it may be necessary to use a beam deeper than $L/15$. When considerable composite action is present, i.e. when the bending moment is less than $WL/40$, the reinforcement should be calculated for a beam having a depth equal to $L/15$ even if a deeper beam is to be provided for other reasons. This is to ensure that there is sufficient steel in the beam to act as a tie across the springing of the arch within the brickwork.

When designing pile capping beams by limit-state principles it is seldom necessary to consider the serviceability limit-state. However, an examination of the limit-state of cracking is necessary if the beam is to be exposed to soil or groundwater which can be expected to be corrosive. The limit-state of deflection should be checked if the beam is to support a wall faced with a material such as mosaic tiles, which are particularly susceptible to cracking due to small movements.

The BRE design method assumed that the ground floor slab is carried by the soil and is not connected to the capping beam. *If a suspended floor slab is provided the capping beam must be designed by conventional methods.* A disadvantage of the BRE design is that it can inhibit house owners from removing large areas of load-bearing walls in the course of making improvements such as the addition of rear conservatories. There is a danger that the house owner may not be aware of the basis of the design of the ground beams when proceeding with these and similar structural alterations.

It is a good practice to provide a suspended ground floor slab in cases where piles are provided to restrain a structure from lifting due to a *swelling* soil. A ground floor slab cast directly onto a swelling soil will lift and will, in turn, cause the lifting of internal partitions, with the consequent distortion of any floors carried by them, and the cracking of plaster finishes. Uplift pressures due to soil swelling against the underside of a pile capping beam must be considered. In clay soils where mature trees or hedges have been removed the clay may swell up to 100 mm over a long period of years. Swelling of pyritic mudstones and shales can occur due to the growth of gypsum crystals within the laminations of these rocks. Gypsum growth can be caused by chemical and microbiological changes consequent on changed environmental conditions[7.10]. Swelling pressures, if the upward movement of the soils is resisted by a reinforced concrete capping beam, can be of a magnitude which will cause the piles to fail as tension members, or which will lift the piles out of the soil.

It is essential to insert a layer of compressible material such as Clayboard or special low density polystyrene or to provide a void between the soil and underside of the capping beam to reduce the uplift forces transferred to the piles (Figure 7.15). Load/deflection tests should be made on specimens of the compressible material to ensure that the amount of compression required by the predicted degree of soil swelling does not generate a pressure on the ground beam that is sufficient to cause structural failure of the beams or piles, or lifting of the building. There have been a number of cases of failure and cracking of piles, ground beams, and superstructures to low-rise buildings constructed on swelling clays in recent years. These have been caused mainly by deficiencies in design such as inadequate tension reinforcement and lack of proper provision for uplift on ground beams. The latter should be of generous depth to provide for differential uplift forces caused by local tension failure in piles in unpredictable conditions of soil swelling.

Horizontal swelling forces can also impose loads on pile capping beams due to the restraint provided by the beam to the expansion of the mass of the soil. To avoid excessive

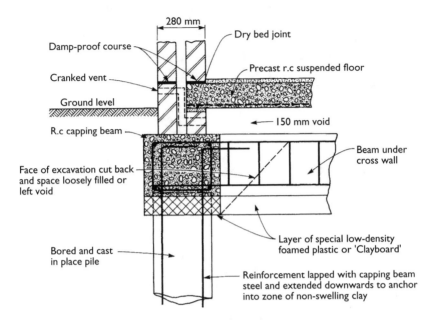

280 mm

Dry bed joint

Damp-proof course

Precast r.c suspended floor

Cranked vent

Ground level

150 mm void

R.c capping beam

Beam under
cross wall

Face of excavation cut back
and space loosely filled or
left void

Layer of special low-density
foamed plastic or 'Clayboard'

Bored and cast
in place pile

Reinforcement lapped with capping beam
steel and extended downwards to anchor
into zone of non-swelling clay

Figure 7.15 Design of pile capping beam for swelling clay soils.

swelling forces on the inner sides of beams they should not be left in contact with the clay. After casting the concrete has been completed, the clay should be cut back as shown in Figure 7.15. The space between the side of the beam and the cut-back of the excavation is left empty, or is only loosely backfilled.

Ground beams are provided to act as ties or compression members between adjacent pile caps, so providing the required restraint against sidesway or buckling of the piles under lateral or eccentric loading (see Section 7.5). Ground beams and pile capping beams may have to withstand horizontal loading from the soil due to the tendency to movement of vertical piles under lateral loading. They may also be subjected to bending in a vertical direction due to differential settlement between adjacent groups of piles.

It may be permissible to allow the passive resistance of the soil against the sides of pile caps and ground beams to supplement the resistance of the piles to lateral loading. However, in clay soils the ground will shrink away from the sides of shallow members in dry weather conditions. Trenching for building services alongside pile caps must also be considered a possibility. In any case quite appreciable yielding of the soil must take place before its passive resistance is fully mobilized. This movement may be sufficient to cause the failure in bending of vertical piles.

The superimposed loading on the ground beams or pile capping beams is transferred to the piles by bonding the longitudinal reinforcing steel to the beams into the pile caps. However, it is not good practice to carry the longitudinal steel through holes burned in the projecting parts of steel piles. It is quite likely that the pile head will have deviated from the correct position and it may be impossible to bend the beam reinforcing bars over a sufficient horizontal distance to pass through the holes in the steel pile without causing complications

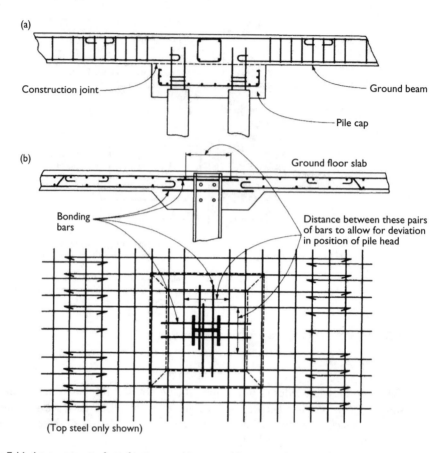

Figure 7.16 Arrangement of reinforcing steel in ground beams and ground floor slabs.

to the stirrups in the beams. If the main steel in ground beams or ground floor slabs must extend across the piles it should be carried above the pile heads as shown in Figure 7.16a, or at the sides of the projecting piles with ample spaces to allow for misalignment (Figure 7.16b).

7.10   References

7.1   GLANVILLE, W. G., GRIME, G., and DAVIES, W. W. The behaviour of reinforced concrete piles during driving, *Journal of the Institution of Civil Engineers*, Vol. 1, 1935, p. 150.

7.2   SMITH, E. A. L. Pile driving analysis by the wave equation, *Journal of the Soil Mechanics Division*, American Society of Civil Engineers, Vol. 86, No. SM4, 1960, pp. 35–61.

7.3   GOBLE, G. G., RAUSCHE, F., and LIKINS, G. E. Jr, The analysis of pile driving – A state of art, *Proceedings of the International Seminar on the Appplication of Stress Wave Theory on Piles*, Stockhoem, 1980, A. A. Balkema, Rotterdam, 1981.

7.4   GOBLE, G. G. and RAUSCHE, F. Pile driveability calculations by CAPWAP, *Proceedings of the Conference on Numerical Methods in Offshore Piling*. Institution of Civil Engineers, London, 1979. pp. 29–36.

7.5   WHEELER, P. Stress wave competition, and making waves, *Ground Engineering*, Vol. 25, No. 9, 1992, pp. 25–8.

7.6   FLEMING, W. G. K. and LANE, P. F. Tolerance requirements and construction problems in piling, *Proceedings of the Conference on the Behaviour of Piles*, Institution of Civil Engineers, London, 1971, pp. 115–18.

7.7   Investigation of the strength of the connexion between a concrete cap and the embedded end of a steel H-pile, *Research Report No. 1*, State of Ohio, Department of Highways, December 1947.

7.8   WHITTLE, R. T. and BEATTIE, D. Standard pile caps, *Concrete*, Vol. 6, No. 1, 1972, pp. 34–6, and Vol. 6, No. 2, 1972, pp. 29–31.

7.9   WOOD, R. H. and SIMMS, L. G. A tentative design method for the composite action of heavily-loaded brick panel walls supported on reinforced concrete beams, *Building Research Establishment Current Paper*, 26/69 July 1969.

7.10 HAWKINS, A. B. and PINCHES, G. M. Cause and significance of heave at Llandough Hospital, Cardiff – a case history of ground floor heave due to gypsum growth, *Quarterly Journal of Engineering Geology*, Vol. 20, 1987, pp. 41–57.

Chapter 8

Piling for marine structures

8.1   Berthing structures and jetties

Cargo jetties consist of a berthing head at which the ships are moored to receive or discharge their cargo and an approach structure connecting the berthing head to the shore and carrying the road or rail vehicles used to transport the cargo. Where minerals are handled in bulk the approach structure may carry a belt conveyor or an aerial ropeway. In addition to its function in providing a secure mooring for ships, the berthing head carries cargo-handling cranes or special equipment for loading and unloading dry bulk cargo and containers.

Berthing structures or jetties used exclusively for handling crude petroleum and its products are different in layout and equipment from cargo jetties. The tankers using the berths can be very much larger than the cargo vessels. However, the hose-handling equipment and its associated pipework are likely to be much lighter than the craneage or dry bulk-loading equipment installed on cargo jetties serving large vessels. The approach from the shore to a petroleum loading jetty consists only of a trestle for pipework and an access roadway. Where the deep water required by large tankers commences at a considerable distance from the shore-line, it is the usual practice to provide an island berthing structure connected to the shore by pipelines laid on the sea bed.

In spite of the considerable differences between the two types of structure, piling is an economical form of construction for cargo jetties as well as for berthing structures and pipe trestles for oil tankers. The berthing head of a cargo jetty is likely to consist of a heavy deck slab designed to carry fixed or travelling cranes and the imposed loading from vehicles and stored cargo. The berthing forces from the ships using the berths can be absorbed by fenders sited in front of and unconnected to the deck structure (Figure 8.1a), but it is more usual for the fenders to transfer the berthing impact force to the deck and in turn to the rows of supporting piles. The impact forces may be large and because the resistance of a vertical pile to lateral loading is small the deck is supported by a combination of vertical and raking piles (Figure 8.1b). These combinations can also be used in structures of the open trestle type such as a jetty head carrying a coal conveyor (Figure 8.2).

The piles in the berthing head of a cargo jetty are required to carry the following loadings:

(1) Lateral loads from berthing forces transmitted through fendering
(2) Lateral loads from the pull of mooring ropes
(3) Lateral loads from wave forces on the piles
(4) Current drag on the piles and moored ships

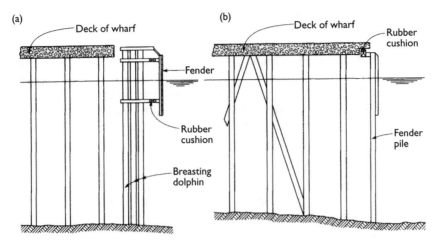

Figure 8.1 Fender piles for cargo jetties (a) In independent breasting dolphin (b) Attached to main deck structure.

Figure 8.2 Raking and vertical piles used to restrain berthing forces in bulk handling jetty.

(5) Lateral loads from wind forces on the berthing head, moored ships, stacked cargo, and cargo handling facilities
(6) Compressive loads from the dead weight of the structure, cargo handling equipment, and from imposed loading on the deck slab

(7) Compressive and uplift forces induced by overturning movements due to loads 1 to 5 above
(8) In some parts of the world piles may also have to carry vertical and lateral loads from floating ice, and loading from earthquakes.

The above forces are not necessarily cumulative. Whereas wind, wave, and current forces can occur simultaneously and in the same direction, the forces due to berthing impact and mooring rope pull occur in opposite directions. Berthing would not take place at times of maximum wave height, nor would the thrust from ice sheets coincide with the most severe wave action. Where containers are stored on the deck slab the possibility of stacking them in tiers above a nominal permitted height must be considered.

8.1.1   Loading on piles from berthing impact forces

The basic equation used in calculating the force on a jetty or independent berthing structure due to the impact of a ship as it is brought to rest by the structure is

$$\text{kinetic energy } E_k = \frac{m_s V^2}{2g} \tag{8.1}$$

where m_s is the displacement of the ship and the mass of water moving with the ship, and V is the velocity of approach to the structure.

The whole of the energy as represented by equation 8.1 is not imparted directly to the jetty piles. Kinetic energy is also absorbed by the deformation of the hull of the ship and by the compression of the fenders and of the cushioning between the fenders and their supporting structure. Ships normally approach the jetty at a narrow angle to the berthing line and the kinetic energy in the direction parallel to this line is generally retained in kinetic form but a part may be lost in overcoming the resistance of the water ahead of the ship's bows, in friction against the fenders, and in the pull on the mooring ropes if these are used to restrain longitudinal movement. A full consideration of the complexities involved in calculating the magnitude and direction of berthing forces cannot be dealt with adequately in this book, and the reader is referred to Part 4 of the British Standards *Code of Practice* (BS 6349-1: 2000) for maritime structures[8.1] for guidance on these problems.

On the assumption that the kinetic energy of the ship transverse and parallel to the berthing line has been correctly calculated the problem is then to assess the manner in which the energy is absorbed by the fenders and their supporting piles. Taking the case of a vertical pile acting as a simple cantilever from the point of virtual fixity below the sea bed, and receiving a blow from the ship with a force H applied at a point A (Figure 8.3a), the distance moved by the point A can then be calculated by the simple method shown in equation 6.20 and repeated here for convenience, namely:

$$\text{distance moved } y = \frac{H(e + z_f)^3}{3EI} \tag{8.2}$$

If the ship is brought to rest by the vertical pile as it moves the pile head over the distance y, then the work done by the force H over this distance is given by

$$\text{work done} = \frac{1}{2}Hy = \frac{H^2(e + z_f)^3}{6EI} \tag{8.3}$$

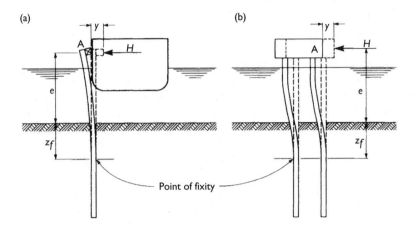

Figure 8.3 Lateral movement of fender piles due to impact force from berthing ship (a) Single free-headed pile (b) Group of fixed-headed piles.

The bending moment M on the pile is equal to $H(e + z_f)$ therefore

$$\text{work done} = \frac{M^2(e + z_f)}{6EI} \tag{8.4}$$

If required the more-rigorous methods described in Sections 6.3.3 and 6.3.4 can be used to calculate the deflection of the pile head and hence the work done in bringing the ship to rest.

The bending moment which can be applied to a pile is limited by the allowable working stress on the extreme fibres of the material forming the pile for normal berthing impacts, or by the yield stress with abnormal berthing velocities. Thus if the allowable resistance moment M_a is used in equation 8.4 the capacity of the pile to absorb kinetic energy can be calculated and compared to the kinetic energy of the moving ship which must be brought to rest. If the capacity of the pile is inadequate the blow from the ship must be absorbed by more than a single pile. In practice, vertical piles are grouped together and linked at the head and at some intermediate point (Figure 8.1a) to form a single berthing dolphin, or are spaced in rows or 'bents' in the berthing head of a jetty structure. In the latter case the kinetic energy of the ship may be absorbed by a large number of piles. In the case of a pile fixed against rotation by the deck slab of a structure (Figure 8.3b) it was shown in equation 6.21 that

$$\text{distance } y \text{ moved point A} = \frac{H(e + z_f)^3}{12EI} \tag{8.5}$$

The bending moment caused by a load at the fixed head of a pile is equal to $\frac{1}{2}H(e + z_f)$, and thus the work done is the same as shown in equation 8.4.

BS 6349 points out that in the case of a piled wharf erected parallel to a sloping shore line, the piles supporting the rear of the deck, being more deeply embedded than those at the front will resist a much higher proportion of the horizontal forces imposed on the fendering. It may be necessary to consider sleeving the rearward piles to equalize the flexural resistance.

If the rear of the deck is abutting a retaining wall such as a sheet pile wall, virtually the whole of the horizontal forces on the deck will be transmitted to the wall.

Where medium to large vessels are accommodated the berthing impact is not absorbed directly by a pile or by a deck structure supported by piles. Means are provided to cushion the blow, thus reducing the risk of damaging the ship, and limiting the horizontal movement of the jetty. It is also more economical during design to provide cushioning devices than to absorb forces directly on the structure. It must be noted that whereas independent berthing dolphins can be allowed to deflect over a considerable distance (and large deflections are the most efficient means of absorbing kinetic energy), the deck slab of a cargo jetty cannot be permitted to move to an extent which would cause instability in travelling cranes, stacked containers or mechanical elevators. This limitation restricts the allowable movement of such cargo jetties to a very small distance.

Where energy absorbing fenders are provided, the work equation 8.4 is modified. Taking the simplified case shown in Figure 8.4 of a fender pile backed by a cushion block transmitting the impact to a bent of piles transverse to the berthing line, the work equation becomes

kinetic energy of moving ship absorbed by system in Figure 8.4

$$= \tfrac{1}{2} \times H \times \Delta = \tfrac{1}{2} \times H(\Delta_1 + \Delta_2) \tag{8.6}$$

where H is the impact force of the first blow on the fender, Δ is the distance moved in bringing the ship to rest after the first impact, Δ_1 is the distance moved by the compression of the cushion block, and Δ_2 is the distance moved by the pile bent.

In a practical design case a limit is placed on Δ_2 by the operating conditions on the jetty. Then if the cushion block is to be fully compressed by the ship moving at the maximum design approach velocity, Δ_1 is known and Δ is the sum of Δ_1 and Δ_2. Hence, knowing the kinetic energy of the moving ship, the impact force H can be calculated. This force is the sum of the force in the cushion block and the shearing force at the head of the pile. The bending moment induced in the fender pile by the action of force H over distance Δ is compared with the moment of resistance of the selected pile, and the energy absorbing capacity of the cushion block is checked to ensure that the force required for full compression is not exceeded by the force H. The condition shown in Figure 8.4, of a single fender pile transmitting the full force of a moving ship to a single pile bent, does not occur in practice. In a cargo jetty the fender piles are spaced at equal distances along the berthing face and the impact is absorbed by a number of piles, depending on the closeness of their spacing, and

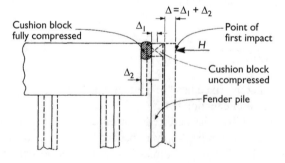

Figure 8.4 Energy absorption of fender pile cushioned at head.

the extent to which they are tied together by intercostal beams or by a longitudinal berthing beam. An approximate rule is to assume that the blow is absorbed over a length of berthing face equal to twice the width of the jetty. BS 6349 recommends a minimum distance between ships moored along a jetty of 15 m.

The design process is one of trial and adjustment to determine the most economical combination of vertical fender piles with rubber or spring cushion blocks that will limit the movement of the protected jetty structure to the desired value. If the impact is delivered at a point below the head (Figure 8.5) some of the energy is absorbed by the soil, some by the deflection of the pile considered as a beam fixed at the lower end and with a yielding prop at the upper end, and some by the yielding at the prop position (i.e. the yielding of the cushion block).

As alternatives to the system of fender piles, each backed by a cushion block as shown in Figure 8.4, a group of piles can carry a rubber fender (Figure 8.6a) or a link-suspended clump fender (Figure 8.6b). For these designs the energy transmitted to the supporting piles is equal to the kinetic energy of the moving ship, less than the energy expended in compressing, displacing, and raising the fender from its neutral position.

Forces act in a direction parallel to as well as normal to the berthing line. Assuming that there are no objects projecting beyond the side of the ship, the force acting parallel to the berthing line is equal to the coefficient of friction between ship and fender times the reaction normal to the berthing line. The longitudinal force tends to cause the twisting of fender piles and of pile bents set transversely to the berthing line. The rotational force on the pile bents is a maximum when the ship makes contact near the end of the jetty, and it is desirable to provide piles raking in a longitudinal direction at the two ends of the structure. The end piles in a jetty head are vulnerable to impact below the water-line from the bulbous bows of vessels provided with bow-thrust propellers.

Damage to fender piles or their connections to the main structure by longitudinal forces can be avoided by spiking timber rubbing strips onto the faces of the fenders. These will be torn off by a severe impact but the pile will remain relatively undamaged.

Rubber fenders are designed to deflect in a longitudinal as well as a transverse direction and are thus capable of absorbing impact energy from both directions. Suspended fenders are given a degree of freedom to swing in a longitudinal direction and they fall clear as the ship sheers off after the first impact. Fenders can also be provided with rollers mounted on vertical axles to reduce the longitudinal frictional force on the structure.

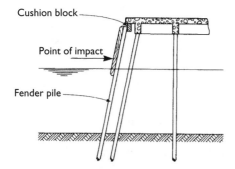

Figure 8.5 Impact force below head of raking fender pile.

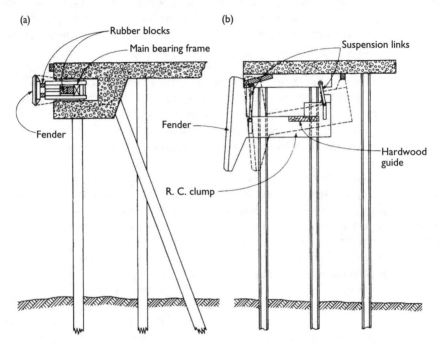

Figure 8.6 Pile-supported fendering systems (a) Rubber cushion fender (b) Link-suspended clump fender.

As already noted, the facilities provided at the berthing head of an oil jetty or island berthing structure are limited to hose-handling gear and pipework. A relatively small deck area is required and the berthing structure can take the form of two main fenders spaced at a distance equal to about 0.3 times the length of the largest tanker using the berth, with two or more secondary fenders having a lower energy-absorbing capacity sited between them to accommodate smaller vessels (Figure 8.7). Frequently, the main and secondary fenders are sited in front of the hose-handling platform and pile trestles to allow them to take the full impact of the tanker without transmitting any thrust to these structures. The independent breasting dolphins, as shown in Figure 8.7, are designed so that their collapse load is not exceeded by the thrust due to the maximum berthing velocity expected.

The type of piling required for independent breasting dolphins depends on the soil conditions. Where rock, stiff clay or granular soils offering a good resistance to lateral loads are present at or at a short distance below the sea bed, the dolphin can consist of a group of large-diameter circular or box-section vertical steel piles, linked together by horizontal diaphragms (Figure 8.8) and carrying a timber fender with rubber cushion blocks on the front face of the group. The face area of the fender should be large enough to prevent concentrated loading from damaging the hull of the ship. The horizontal bracing members are not rigidly connected to the pile group. This is to allow the piles to deflect freely to the maximum possible extent while performing their function of bringing the ship to rest.

The layout shown in Figure 8.7 can sometimes restrict the size and numbers of vessels using the berth. It can be more economical to adopt a berthing structure of the type used for cargo handling (Figure 8.1b). The berthing forces are transmitted directly to the deck so

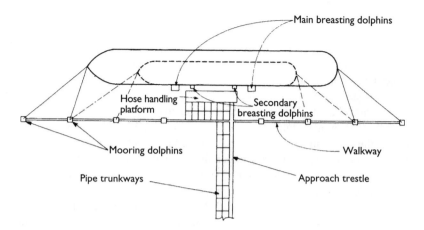

Figure 8.7 Layout at berthing head of oil jetty.

Figure 8.8 Steel tubular breasting dolphin.

permitting vessels to berth in any position along the face. Pairs of rakers resisting the ship impact are spaced at intervals along the deck or are grouped to form 'strong points' with the deck slab acting as a horizontal beam.

Breasting dolphins for the oil loading terminal of Abu Dhabi Marine Areas Ltd., at Das Island, were designed by the British Petroleum Company to consist of groups of vertical steel tubular piles. The main outer dolphins were formed from a group of seven piles, and the inner secondary dolphins were in three-pile groups. The conditions at sea-bed level, which consisted of a layer of shelly limestone cap-rock underlain by a stiff calcareous marl and then a dense detrital limestone, favoured the adoption of vertical piles to absorb the berthing forces. The 36.6 m piles varied in outside diameter from 800 mm at the top to 1300 mm at the bottom, the latter being closed by a full plate on which 15 roller cutters were mounted. The piles were pitched through a reinforced concrete template placed on the sea bed and then drilled down by rotating them by means of a hydraulically powered rotary table operated from a jack-up platform. The cuttings were washed up the annular space between the outside of the pile and the rock and this space was afterwards grouted with a sand–cement mix.

Broadhead[8.2] described a pulling test made on a mooring dolphin pile to confirm that the lateral resistance of the weak rocks below the sea bed would not be exceeded at the working load. The test pile had a bottom diameter of 1300 mm and the pull was applied at a point 24 m above the sea bed. The load/deflection curve obtained at a measuring point 22.86 m above the sea bed is shown in Figure 8.9 and is compared with the theoretical deflection curve assuming fixity at sea-bed level or support from an uncemented shell sand below sea bed, using the elastic analysis of Reese and Matlock (see Section 6.3.4).

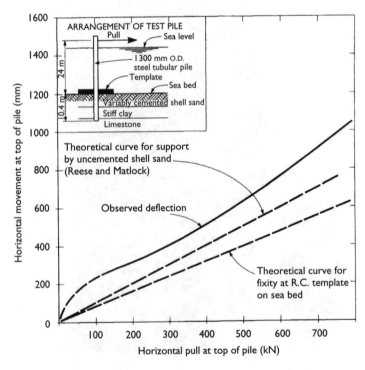

Figure 8.9 Load/deflection curve for 1300 mm O.D. steel tubular pile due to horizontal load at head of pile (after Broadhead[8.2]).

8.1.2   Mooring forces on piles

Mooring structures are not required to carry any pull from ropes during the operation of berthing ships other than a restraining longitudinal movement at the final stages of the berthing operation.

When the ship is fully moored four ropes are attached to bollards or bitts fixed to the jetty structure or mounted on independent mooring dolphins in positions such as those shown in Figure 8.7. Using this type of layout the ship is restrained from excessive ranging against the fenders and also from moving away from the berth under the influence of offshore waves or currents. The load on any individual rope due to winds or currents acting on the ship or to checking the way of a ship during berthing cannot be calculated with any accuracy. It depends on the tensioning of the rope and its angle to the berthing line.

The wind and current forces on the ship can be calculated using the equations given below for calculating the current force on a pile (equation 8.10) or the wind force on a pile (equation 8.14).

Mooring dolphins should be designed to be as rigid as possible. This is to restrict the ranging of ships which is exaggerated by the lifting and sagging of the mooring ropes. Independent mooring dolphins can take the form of pile groups set back from the berthing line as shown in Figure 8.7, or placed beyond the ends of the berthing head. Piles in mooring dolphins can be raked in two directions to resist longitudinal, transverse and torsional pulls (Figure 8.10). Where rock is present at or at a short distance below the sea bed, anchorages are required to withstand the uplift on tension piles as described in Section 6.2.4.

Guidance on the design of mooring structures is given in Part 4 of BS 6349.

8.1.3   Wave forces on piles

Jetties are normally sited in sheltered waters or in locations selected as not being subject to severe storm waves. Consequently, the forces on piles due to wave action are considerably less severe than those caused by the impact from berthing or the pull from mooring ropes. Also, berthing operations are not expected to take place when heavy wave action is occurring. Therefore, it is the usual practice to disregard wave forces on piles forming the berthing head of a jetty and any associated independent dolphin structures where these are sited in

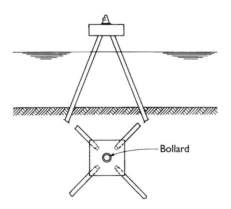

Figure 8.10 Mooring dolphin with piles raked in two directions.

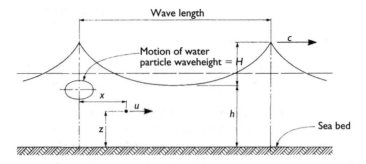

Figure 8.11 Shape of breaking wave.

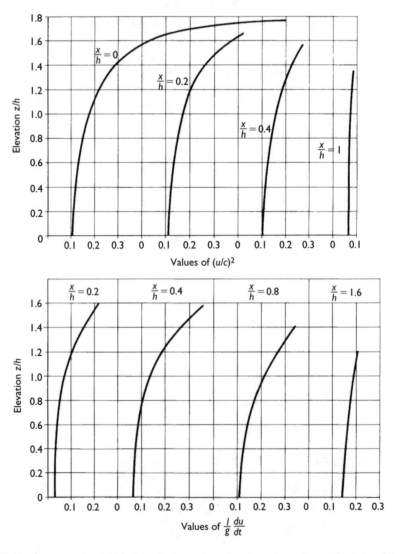

Figure 8.12 Design curves for calculating velocity and acceleration of water particles in breaking wave.

sheltered waters. However, in the case of island berthing structures for large vessels, which are sited in deep and relatively unsheltered waters, the wave forces may represent a significant proportion of the total force required to be calculated. Also, piles supporting the approach trestle to a jetty are not required to withstand berthing impact forces. Thus wave forces, even in fairly sheltered waters, when combined with wind pressures on the superstructure and current drag on the piles, may produce substantial loading transverse to the axis of the trestle.

A simple approach to the calculation of wave forces on fixed structures is to assume that the maximum wave force can be expressed as the equivalent static force caused by a solitary wave of the shape shown in Figure 8.11. This shape is representative of a breaking wave. An oscillatory wave has a different shape but the factors given in Figure 8.12 and Table 8.1 for use with equations 8.8 and 8.9 are applicable only to breaking wave conditions. Drag and inertial forces are exerted on the structure by the water particles which move in an elliptical path as shown. From the work of Wiegel et al.[8.3], Reid and Bretschneider[8.4], Dailey and Stephen[8.5], and Bretschneider[8.6], it is possible to calculate the water particle velocity u at any point having co-ordinates x horizontally from the wave crest and z vertically above the sea bed. The water particle velocity can be related to the velocity of advance of the wave crest (the wave celerity c) and expressed in terms of $(u/c)^2$ and $1/g \times du/dt$ for various ratios of x and z to the height h of the trough of the wave above the sea bed.

The solitary-wave theory is limited in its application to a range of conditions defined by the ratio of the wave period to the water depth. Because the equations given below are applicable only to breaking wave conditions they represent the maximum force which can be applied to a structure. Breaking wave conditions are unlikely to occur in deep water berths for large tankers, and these conditions are likely to be found only in fairly shallow water on exposed jetty sites, for example along the line of the approach structure from the

Table 8.1 Surface elevations, velocities, and accelerations for solitary breaking wave

Distance from crest x/h	Surface elevation z_s/h	Values of $(u/c)^2$					Values of $1/g \cdot (du/dt)$				
		At surface	At $z=h$	At bottom	Average value	Height to centroid	At surface	At $z=h$	At bottom	Average value	Height to centroid
0	1.78	1.000	0.176	0.109	0.226	1.19	0	0	0	0	
0.2	1.67	0.430	0.170	0.106	0.181	1.03	0.242	0.073	0.031	0.081	1.14
0.4	1.57	0.276	0.156	0.099	0.150	0.92	0.347	0.137	0.060	0.133	1.02
0.6	1.48	0.201	0.133	0.092	0.123	0.83	0.380	0.184	0.087	0.164	0.93
0.8	1.41	0.138	0.106	0.078	0.097	0.80	0.357	0.214	0.110	0.180	0.88
1.0	1.35	0.092	0.082	0.070	0.077	0.70	0.321	0.225	0.127	0.186	0.78
1.2	1.29	0.062	0.063	0.058	0.061	0.65	0.280	0.225	0.140	0.187	0.73
1.4	1.25	0.041	0.046	0.048	0.047	0.61	0.243	0.209	0.146	0.182	0.68
1.6	1.21	0.029	0.032	0.038	0.035	0.59	0.209	0.192	0.148	0.173	0.65
1.8	1.18	0.020	0.023	0.029	0.027	0.56	0.174	0.171	0.145	0.159	0.62
2.2	1.13	0.009	0.011	0.018	0.014	0.50	0.122	0.128	0.130	0.130	0.57
2.6	1.08	0.004	0.005	0.009	0.007	0.50	0.088	0.091	0.109	0.102	0.53
3.0	1.05	0.002	0.002	0.004	0.003	0.50	0.065	0.067	0.084	0.078	0.51
3.4	1.03	0.001	0.001	0.002	0.002	0.50	0.049	0.049	0.062	0.058	0.50
5.0	1.01	0.000	0.000	0.000	0.000	0.50	0.012	0.012	0.017	0.016	0.50

shore to a deep-water berth. However, as noted by Newmark[8.7] the solitary-wave theory is often applied to situations beyond its strict range of validity for want of a better theory. For deep-water structures the solitary-wave theory gives *over-conservative* values of wave force. However, equations 8.7 to 8.9 based on this theory together with the dimensionless graphs are simple and easy to use. It is suggested that the equations are used for all parts of a deep-water berthing-head structure and for the shallow-water approach whenever it is necessary to calculate wave forces. If these forces together with current drag, wind forces, and berthing impact forces do not produce excessive bending stresses on the piles then the calculations need not be further refined. It must be kept in mind that the cross-sectional area of a pile may be governed by considerations of corrosion and driving stress rather than the stress resulting from environmental forces. Where the wave forces calculated by the solitary-wave theory are a significant factor in the design of the piles more detailed calculations should be made taking into account the relationship between wave height, water depth, and wave period. Methods of general application can be found in the publications of the US Army Coastal Engineering Research Centre[8.8].

In general wave theories, the wave force on a fixed structure is taken as the sum of the drag and inertial forces exerted by the wave. These are expressed by the commonly used Morison equation[8.9]:

$$f = f_D + f_I = C_D \frac{wu^2}{2g} + C_M \frac{w\pi D}{g} \frac{4}{4} \cdot \frac{du}{dt} \tag{8.7}$$

where f, f_D, and f_I are the wave force, drag force, and inertial force, respectively, per unit area of object in the path of the wave, C_D is a drag coefficient, w is the density of water, g is the gravitational acceleration, u is the horizontal particle velocity of water, C_M is a coefficient of inertia force, D is the diameter of the cylindrical object, and du/dt is the horizontal acceleration of a water particle.

BS 6349-1, Clause 39.44, expresses the Morison equation in a somewhat different form and includes guidance on its limitations, together with equations for calculating the velocity of the water particles. The values for C_D shown in Table 8.2, Section 8.1.4 below, can be used in the version of the Morison equation given in equations 8.7 to 8.9. The values of C_1 in Table 8.2 can also be used for C_M in equations 8.7 to 8.9.

Newmark[8.7] reduced equation 8.7 to a simple expression given in lb-ft-sec units. By taking the weight of sea water as 64 lb/ft^3 and the gravitational acceleration as 32.2 ft/sec the equation becomes

$$f = f_D + f_I = \left[50C_D h \left(\frac{u}{c} \right)^2 + 50C_M D \cdot \frac{1}{g} \cdot \frac{du}{dt} \right] \text{lb/ft}^2 \tag{8.8}$$

Table 8.2 Drag force and inertia coefficients for square section piles

Flow direction	Figure no.	C_D	C_L
Perpendicular to face	8.13a	2.0	2.5
Against corner, in direction of diagonal	8.13b	1.6	2.2
Perpendicular to face, rounded corner, $r/y_s = 0.17$	8.13c	0.6	2.5
Perpendicular to face, rounded corner, $r/y_s = 0.33$	8.13c	0.5	2.5

In SI units, equation 8.8 becomes

$$f = \left[7.8C_D h\left(\frac{u}{c}\right)^2 + 8C_M D \cdot \frac{1}{g} \cdot \frac{du}{dt} \right] \text{kN/m}^2 \tag{8.9}$$

Values of $(u/c)^2$ and $1/g \cdot (du/dt)$ for different positions relative to the location of the wave crest are shown in Figure 8.12 and Table 8.1. This table also lists the average values of $(u/c)^2$ and $1/g \cdot (du/dt)$ together with the heights to the centroid of the two components. The wave forces and moments applied to each increment of height of pile projecting above the scoured sea bed up to wave crest level, and on any underwater bracing or jacket members, are integrated to obtain the total horizontal force on the pile or group of piles and also the overturning moment about the point of fixity below the sea bed.

For use with equations 8.8 and 8.9 Newmark[8.7] recommends a value for C_D of 0.5 to 0.6 for cylindrical members and 1.5 to 2.0 for the inertia coefficient C_M. For rectangular, H and I sections C_D can be taken as up to 2.0. Theoretically C_D is related to the Reynolds number R_e as discussed in the following section. Newmark also recommends that shielding effects produced by closely spaced piles or bracing members should be disregarded when calculating wave forces.

BS 6349 (Part 1) draws attention to the effect of impact forces (wave slam) on horizontal members exposed to the crests of advancing waves.

Barnacle growth on piles and bracings should be taken into account by allowing an appropriate increase in diameter. It has been reported[8.10] that marine growths more than 200 mm in thickness have occurred around steel piles of the North Sea gas production platforms after about eight years of exposure. The growths extend down to sea bed where the water depths were about 25 m. If drag forces due to marine growths are excessive, provision can be made for the members to be cleaned periodically by divers.

8.1.4   Current forces on piles

The velocities and directions of currents (or tidal streams) affecting the structure are obtained by on-site measurements which should include the determination of the variation in current velocity between the water surface and the sea bed. A curve is plotted relating the velocity to the depth and the current drag force is calculated for each increment of height of the pile above the sea bed. Any scour below the sea bed should be provided for.

Current forces are calculated from the equation:

$$F_D = 0.5C_D \rho V^2 A_n \tag{8.10}$$

The components of the above equation are defined in BS 6349 as

F_D = steady drag force (kN)
C_D = dimensionless time-averaged drag force coefficients
ρ = water density (tonne/m^3)
V = incident current velocity (m/sec)
A_n = Area normal to flow (m^2)

C_D is related to the Reynolds number, which for cylindrical members and normal water temperatures is given by the equation:

$$R_e = 9.3VD \times 10^5 \text{ in sec/m}^{-2} \text{ units} \tag{8.11}$$

Section 5 of BS 6349 includes graphs relating C_D for cylindrical members to their surface roughness and Reynolds number. They show that C_D for rough members is in the range of 0.4 to 0.6 for Reynolds numbers between 10^5 and 10^6. The code gives values for C_D and C_1 (C_M in equation 8.7) for square section piles as shown in Figure 8.13 and Table 8.2.

If piles or other submerged members are placed in closely spaced groups, shielding of current forces in the lee of the leading member will occur. Shielding can be allowed for by modifying the drag coefficient. Values of the shielding coefficient have been established by Chappelaar[8.11].

Where currents are associated with waves it may be necessary to add the current velocity vectorially to the water-particle velocity u to arrive at the total force on a member. Also, the possibility of an increase in the effective diameter and roughness of a submerged member due to barnacle growth must be considered.

Having calculated the current force on a pile it is necessary to check that oscillation will not take place as a result of vortex shedding induced by the current flow. This oscillation occurs transversely to the direction of current flow when the frequency of shedding pairs of vortices coincides with the natural frequency of the pile.

Determination of the critical velocity for the various forms of flow-induced oscillation of cylindrical members is given in BS 6349-1, Clause 38.3, by the equation:

$$V_{\text{crit}} = K f_N W_s \tag{8.12}$$

where K is a constant equal to

  1.2   for onset of in-line motion
  2.0   for maximum amplitude of in-line motion
  3.5   for onset of cross-flow motion
  5.5   for maximum amplitude of cross-flow motion

  f_N = natural frequency of the cylinder
  W_s = diameter of the cylinder.

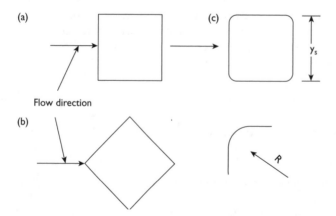

Figure 8.13 Flow conditions for determining drag conditions.

The natural frequency of the member is given by the equation:

$$f_N = \frac{K'}{L^2}\sqrt{\frac{EI}{M}}$$ (8.13)

where K' is a constant, L is the pile length, E is the elastic modulus, I is the moment of inertia, and M is the effective mass per unit length of pile. W_s should take into account the possibility of barnacle growth. K' is equal to 0.56, 2.45 and 3.56 respectively for cantilevered, propped, and fully fixed piles. The elastic modulus is expressed in units of force. In the case of a cylindrical pile the effective mass M is equal to the mass of the pile material plus the mass of water displaced by the pile. Where hollow tubular piles are filled with water the mass of the enclosed water must be added to the mass of the material. In the case of a tubular steel pile with a relatively thin wall the effective mass is approximately equal to the mass of the steel plus twice the mass of the displaced water.

BS 6349 provides graphs relating V_{crit} in equation 8.12 to L'/W_s where L' is the overall pile length from deck level, where the pile is assumed to be pin-jointed, to the level of apparent fixity below sea bed.

Very severe oscillations were experienced during the construction of the Immingham Oil Terminal. At this site in the Humber Estuary, piles were driven through water with a mean depth of 23 m and where ebb currents reach a mean velocity of 2.6 m/s (5 knots). The piles were helically welded steel tubes with outside diameters of 610 mm and 762 mm and a wall thickness of 12.7 mm. Before the piles could be braced together they developed a cross-flow motion which at times had an amplitude of ± 1.2 m. Many of the piles broke off at or above the sea bed. A completed dolphin consisting of a cap block with a mass of 700 tonnes supported by 17 piles swayed with a frequency of 90 cycles per minute and an amplitude of ± 6 mm.

Moored ships can transmit forces due to current drag onto the piles supporting the mooring bollards. The current drag on the ship is calculated from equation 8.10.

8.1.5   Wind forces on piles

Wind forces exerted directly on piles in a jetty structure are likely to be small in relation to the quite substantial wind forces transmitted to the piles from deck beams, cranes, conveyors, stacked, containers, sheds and pipe trunkways. In a jetty approach the combined wind and wave forces which usually act perpendicularly to the axis of the approach can cause large overturning moments on the pile bents, particularly when the wind forces are acting on pipe trunkways or conveyor structures placed at a high elevation, say at a location with a high tidal range. Wind forces on moored ships also require consideration, and allowance should be made where necessary for the accretion of ice on structures.

Wind forces can be calculated from equation 8.10 by taking the mass of air as 1.29 g/l or this equation can be conveniently expressed in Imperial units as

$$F = 0.00256V^2C_DA$$ (8.14)

where F is the wind force in pounds, V is the sustained wind velocity in m.p.h. at the elevation of the portion of the structure under consideration, C_D is a drag coefficient, and A is the projected area of the object in square feet (including an allowance for ice accretion).

The values of the drag coefficient for use with equations 8.10 and 8.14 are as listed in Section 8.1.3 and shielding coefficients[8.14] can be applied for closely spaced members. Wind velocities can be corrected for height by means of the equation:

$$V_2 = V_1 \left(\frac{H_2}{H_1} \right)^{\frac{1}{7}} \tag{8.15}$$

where H_2 and H_1 are the two elevations concerned. It should be noted that wind velocities based on short-duration gusts may be overconservative when considering wind forces on large ships.

8.1.6 Forces on piles from floating ice

Forces on piles caused by floating ice have characteristics somewhat similar to those from berthing ships, the principal difference being the length of time over which the ice forces are sustained. Ice floes are driven by currents and wind drag on the surface of the floe. Typically a floe consists of a consolidated layer, which may be up to 3 m thick in sub-arctic waters, underlain by a mass of 'rubble' in the form of loose blocks, and wholly or partly covered by loose debris and snow. When designing a structure to resist ice forces it is necessary to determine the dominant action, i.e. whether it is the pressure of the wind and current driven floe against the structure, or the resistance offered by the structure in splitting the advancing consolidated layer. In an extensive review of the subject Croasdale[8.14] stated that only on relatively small bodies of water will the wind-induced forces govern the design load.

Wind forces can be calculated from equation 8.10. Croasdale advises omitting the factor 0.5 when using this equation and gives values for C_D as 0.0022 for rough ice cover, $0.00335 < C_D < 0.00439$ for unridged ice, and 0.005 for ridged Arctic sea ice. In equation 8.10 the values for C_D are appropriate to m/sec units of the wind velocity at the 10 m level. Croasdale gives a typical force on a 4 m diameter cylindrical pier as 10 MN caused by an ice sheet 4.15×4.15 km in area, driven by a wind velocity of 15 m/sec.

On striking a vertical pile which is restrained from significant yielding, the consolidated ice layer is crushed at the point of impact. With further movement of the floe radial cracks are propagated in the ice sheet followed by buckling. The buckling dissipates the energy of the moving mass which is brought to rest locally against the pile. The surrounding cracked ice sheet and the underlying loose rubble are diverted to flow past the pile and in doing so they generate frictional forces on the contact surfaces. The force is likely to be at a maximum at the time of initial cracking of the ice sheet followed by lesser peaks due to jamming of the packed ice and adfreezing of the ice on to the structure (Section 9.4).

Croasdale gives the basic equation for the ice force on a narrow rigid structure as

$$F = p/tb \tag{8.16}$$

where

  p = effective ice stress
  t = ice thickness
  b = width of pier

The empirical equation of Korzhavin[8.12] for calculating p is

$$p = Imk\sigma_c \hspace{4cm} (8.17)$$

where

I = indentation factor
m = shape factor
k = contact factor
σ_c = uniaxial compression strength of the ice

I is stated to be equal to unity for a wide pier and 2.5 for a narrow pier ($t/b = 1$). The shape factor is approximately unity for a circular pier, k is also unity for perfect contact between the ice and the structure. The compression strength is difficult to determine by laboratory testing. It depends on the crystal structure, strain rate, temperature and sample size.

Croasdale gives an alternative calculation method based on plasticity theory. The penetration of the pier into the ice is analogous to the failure of a soil surface under the imposed loading of a strip foundation, when the ice sheet is displaced around the pier in the form of wedges, similar in shape to the soil heave around the foundation.

For wedges splitting at an angle of 45° to the edge of the ice sheet the equation for calculating the effective ice stress is

$$p = \sigma_c\,(1 + 0.304t/b) \hspace{3cm} (8.18)$$

It appears from Croasdale's paper that the contact factor k should be applied to the value of p calculated from equation 8.18. A factor of 0.5 is given for continuously moving ice, and 1.0 or more for ice frozen around a structure.

The equation of Tryde[8.13] based on wedge theory is

$$p = 0.8\sigma_c\left(1 + \frac{2.1}{(0.4 + b/t)}\right) \hspace{2cm} (8.19)$$

The forces on the pile from the rubble have been mentioned above. Frictional forces from loose blocks can be assumed to act as a granular material. Where the blocks are frozen together the stresses on the pile will be lower than that of the consolidated ice sheet because the bonds between the blocks will fracture at low strain levels.

It is evident that a single large pile or cylinder will be more effective in resisting ice forces than a cluster of smaller piles. A more efficient structure has a conical shape as shown in Figure 8.14. The impact force from the ice sheet is distributed in directions normal and tangential to the sloping face. Energy is dissipated as the ice sheet is levered up and cracked circumferentially. Further energy is dissipated as the broken blocks are pushed up the slope. Methods of calculating ice forces on conical structures are discussed by Croasdale[8.14] and more recently by Brown[8.15].

The structure shown in Figure 8.14 is designed for weak ground conditions needing support by a piled raft to resist horizontal and vertical forces. The shape is unsuitable for berthing large ships, but it is suitable as a single point mooring, or as a foundation for a wind generator.

8.1.7   Materials for piles in jetties and dolphins

For jetties serving vessels of light to moderate displacement tonnage and of shallow draught, timber is the ideal material for fender piles. It is light and resilient and easy to replace.

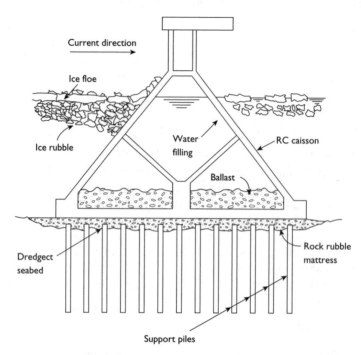

Figure 8.14 Conical structure for resisting ice forces.

As already noted the face of a timber fender pile can be protected by a renewable timber rubbing strip. The type of timber used for fender piles is governed by considerations of the attack by organisms present in the sea water. Suitable types of timber are described in Chapter 10.

For jetties and berthing structures in deep water serving large vessels, either steel or prestressed concrete tubular piles can be used. Steel piles have the advantage that they can withstand rough handling while being loaded onto barges and lifted into the leaders of the floating piling frame or jack-up platform (Figure 3.7). They can withstand hard driving to attain the penetration depths necessary to achieve the required uplift and lateral resistance. However, they require expensive cleaning and coating treatment above the soil line, supplemented by cathodic protection to enable them to resist corrosion in sea water. Losses in thickness of the pile section over the design life of the structure caused by corrosion need to be considered in relation to the working stresses under operating conditions. The types of steel suitable for piling in marine structure are discussed in Section 2.2.6.

Prestressed concrete piles also possess considerable resilience, but repair is a difficult problem if they are subjected to accidental heavy impact damage. Prestressed concrete piles are suitable for approach structures and for jetty heads protected by independent berthing structures. Problems of sea-water attack on steel and concrete structures are discussed in Chapter 10.

8.2   Fixed offshore platforms

Because of their location, frequently in deep water exposed to severe wave action, the forces acting on fixed platform structures are different in character from those on jetties in relatively shallow and sheltered waters. Whereas in berthing structures the dominant forces are those caused by the berthing of ships, the offshore platform is served only by small vessels and the environmental forces resulting from waves, winds, and currents have a dominating influence on design. In very deep water, the environmental forces can account for three-quarters of the total load on a main supporting member.

The economics in the design and construction of offshore platforms for petroleum and gas production and wind farms are viewed from a standpoint very different from that applied to jetty design. In the case of jetties and wind turbine installations the main requirements are low capital cost, ease of maintenance, and a long life. The time required for construction is not usually a critical factor in the design of shear-shore structures. However, only a limited life is required from oil and gas production platforms but assurance of stability in the most severe exposure conditions is of vital importance, and rapidity of installation at sea is essential. This is because of the limited periods during which the state of the sea will permit the operation of large floating cranes and other constructional plant.

The principal activity for the offshore construction industry in the 1970s and the 1980s was in the fabrication and installation of platforms for oil and gas production and laying sea-bed pipelines. The platforms constructed in the UK and Western Europe were mainly in the relatively shallow waters of the North Sea where multi-pile foundations of the type shown in Figure 8.15 were an economical form of construction. Gravity base platforms where the foundation consisted of a large caisson floated into place and sunk on to the prepared sea bed were also constructed.

Figure 8.15 Fabrication of platform for Forties Field (North Sea) of British Petroleum Co., showing guides for clusters of piles around each leg.

In recent years many of the offshore oil and gas fields were becoming or have become exhausted requiring the platforms and base structures to be dismantled and removed to restrict future pollution of the marine environment. The search for sources of oil and gas has been extended to deep-water areas where the piled jacket-type structure are not economically feasible because of the limitations of available construction equipment to operate in deep water and the associated sea conditions.

In the 1990s and continuing the present day a new outlet for the offshore industry has arisen in the installation of wind turbines. Offshore wind farms are required, from consideration of visual intrusion to be located at least 5 km from the shore line, but it has been possible to find sea areas having water depths sufficiently shallow to permit the construction of piled foundations for the turbines. The design and construction of wind farms present severe problems for the engineer which have been reviewed by Bonnett[8.16] and by Ffrench et al.[8.17] They give examples of wind turbines with rotor diameters up to 90 m, weighing with the associated machinery some 250 tonne mounted at a height of 70 m above sea level. At peak wind force conditions the dynamic forces generated by the turbines can act concurrently with peak wave action on the supporting structure to cause cyclic overturning moments on the foundations. A dominant design problem is in providing sufficient stiffness in the combined machinery and foundation system so that its natural frequency exceeds that of the excitation forces.

It has been possible, with the present generation of wind turbines, to erect them on a single large diameter pile (monopile) foundation. Tubular steel piles 5.4 m in diameter have been driven in water depths up to 20 m using equipment of the type shown in Figure 3.7.

Penetration depths of piles are determined from considerations of resistance of the soil to dynamically applied horizontal and vertical forces taking into account the possibility of sea-bed scour increasing the overturning moments. The risks of degradation of the soil around the shaft and beneath the base of the piles need to be assessed. The present limitations on the size of piles which can be handled and driven by available equipment may require the use of three piles driven through a sea-bed template carrying a tripod substructure for the foundations of the next generation of heavier turbines. Ffrench et al.[8.17] describe rotor diameters of 126 m for turbines of 4.5 to 5 MW capacity. Bonnett[8.16] refers to the unsuitability of codes of practice for the design of building structures to deal with the problems involved with wind turbines. He refers to a code used for structures erected in France[8.18].

Certifying authorities for oil and gas production usually demand a specific safety factor for a 100-year wave combined with the corresponding wind force and maximum current velocity, referred to as the *design* environmental conditions. The maximum forces due to operations on the platform such as drilling are combined with specified wind and sea conditions, and are known as the *operating* environmental conditions. The American Petroleum Institute[8.19] requires the safety factors on the ultimate bearing capacity of piled foundations not to be less than the minima given in Table 8.3.

The reader is referred to design and construction recommendations in the current publications of the American Petroleum Institute[8.19,8.20] and the UK Department of Energy[6.3]. Construction methods have been described in detail by Gerwick[8.21].

8.3   Pile installations for marine structures

Where marine structures are connected to the shore, as in the case of a jetty head with a trestle approach, the piles may be driven either as an 'end-on' operation with the piling

Table 8.3 Minimum safety factors for various loading conditions

Loading condition	Minimum safety factor
1. Design environmental conditions with appropriate drilling loads	1.5
2. Operating environmental conditions during drilling operations	2.0
3. Design environmental conditions with appropriate producing loads	1.5
4. Operating environmental conditions during producing operations	2.0
5. Design environmental conditions with minimum loads	1.5

equipment mounted on girders cantilevering from the completed pile bents, or as an operation from a floating or jack-up barge. In tidal waters there is usually sufficient water depth to float a barge with a draft of 1 to $1\frac{1}{2}$ m to a location close inshore. However, this can be inconvenient where tidal flats or saltings cover a long depth of the approach or where it is unsafe to ground the barge on the sea bed at low water.

Where the 'end-on' method is used the spacing between pile bents is limited by the ability of the girders to cantilever when carrying the weight of the piling frame, hammer, and suspended pile. Loading can be minimized by utilizing the buoyancy of tubular piles with permanently or temporarily closed ends, or by using trestle guides of the types shown in Figures 3.6 and 3.8 in conjunction with a pile-mounted hammer and a crane barge for lifting and pitching the piles.

Piling barges for deep-water locations range in length from about 60 to 120 m with a width of one-third to one-half of the length and an overall depth of $\frac{1}{12}$ to $\frac{1}{15}$ of the length. Adequate depth is necessary to provide sufficient strength for towing the barge to the site from a distant location, and to give sufficient freeboard for safe operation when moored at the work site. These barges are normally self-contained with accommodation for the barge and rig crew.

Jack-up barges operate most efficiently when provided with mechanically adjustable pile guides installed either by cantilevering from the side of the barge or spanning a 'moon-pool' inset in the barge hull.

If possible, piles should be driven to their full design penetration without the need to weld-on additional pile lengths, to drive insert piles, or to clean out the soil plug or drill below the initial refusal level of an open-ended tubular pile. Gerwick[8.21] gave an example of times required for welding add-on lengths of 1.37 m OD tubular piles; they varied from $3\frac{1}{4}$ hours for 25 mm wall thickness to $10\frac{1}{2}$ hours for 64 mm thickness. Such delays cause increased driving resistance due to 'take-up' (i.e. the increase of shaft friction). However, there are many situations where piles cannot be driven to their full penetration without the need for lengthening or for 'drilling-and-driving' techniques.

Cleaning out the soil plug is an effective way of reducing the driving resistance, thus obtaining deep penetration, because of the elimination of base resistance. It is particularly advantageous for obtaining deep penetration into coarse soils, say to develop uplift resistance, to avoid excessive settlement due to vibration effects, or to reach rockhead. This is because the base resistance in a coarse soil represents the major proportion of the total resistance to the driving of the pile. Removal of the soil plug is not particularly effective for piles penetrating deeply into clays where the base resistance is only a very small proportion of the total resistance. Drilling out the soil within the pile does not reduce the external shaft friction of the surrounding clay.

Equipment of the type described in Section 3.3 is used to drill-out the soil from within a tubular pile. Where rotary methods are used centralizers are required to keep the drilling pipes in line with the pile axis. A Calweld drill was used to clean out the soil from within the 2000 and 2200 mm outside diameter steel tubular piles used for the breasting and mooring dolphins of The British Petroleum Company's tanker terminal in the Firth of Forth. To install the vertical piles, a rotary table and guide frame were mounted on top of the pile as shown in Figure 8.16. A full-face drilling bit was used and the cuttings were removed by air–water reverse circulation up the 200 mm drilling pipe. The drill bit was maintained in correct alignment by centralizers and a heavy collar was provided to maintain pressure on

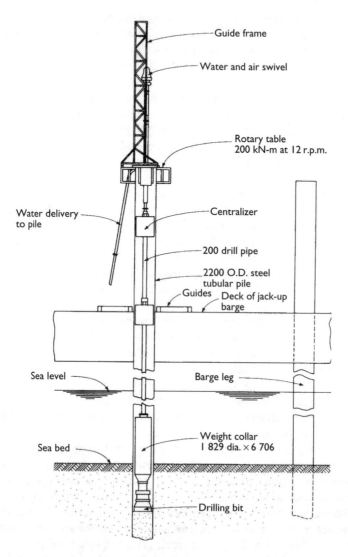

Figure 8.16 Drilling soil from within 2200 mm OD piles using reverse-circulation drill (Forth Tanker Terminal, The British Petroleum Co.).

the bit. Drilling was continued into rockhead, and followed by the separate operation of drilling-in a 560 mm steel tubular dead anchor to a depth of 15 m into the rock to provide an uplift resistance of 7.4 MN. It is evident from Figure 8.16 that the numbers of heavy components required to be assembled for drilling out the soil from within and below large diameter piles are considerably greater than those required for the simple operation of driving the pile by hammer. Space is limited on the deck of a construction barge, and where floating vessels are used the heavy equipment may need to be secured to the deck by bolting or chain tackle. Hence, the successive operations of driving the pile to refusal, removing the hammer, assembling the drilling gear, then drilling, and removing the equipment can be very protracted. Therefore, if 'drill and drive' operations are required the aim should be to restrict the drilling phase to only one operation.

Insert piles can be used where piles driven to their full design penetration fail to attain a satisfactory resistance, or where 'drilling and driving' techniques are unable to achieve the required penetration.

Where insert piles are used, or where single piles are driven within the tubular guides of a jacket, the transfer of load from the insert pile to the main pile, and from the main pile to the leg, is made by welded joints at the pile heads or by grouting the annular space between the members. Both methods can be used together. The grout is prevented from flowing out from between the bottom of the jacket leg and the pile by means of inflatable packers or wiper sleeves built into the bottom of the jacket legs. The design of the grout bond from the pile to the jacket or between piles is described in Section 6.2.5. The need for terminating an insert pile at the head of the exterior pile or jacket guide can be avoided, in the case of large-diameter sections by driving the insert pile with a 'slim-line' underwater hydraulic hammer.

Instead of relying on the bond stress between pile and grout, mechanical keying devices of the type described in Section 6.2.5 can be used. They may be essential to transfer the load from the legs of deep-water platforms to large-diameter piles where the large thickness of the annulus is of some significance concerning the development of sufficient bond strength between grout and steel. The shrinkage of a grout rich in cement can be quite significant within an annulus that is, say, 75 to 100 mm thick, and it has a weakening effect on the grout bond. In these conditions the development of an allowable bond stress even in the lower range recommended by the American Petroleum Institute may be impossible to achieve, and shear keys on both pile and sleeve are necessary to provide the means of transferring the load through a grouted annulus. Shear keys on the inner surface of a raking sleeve may prevent the pile from being lowered through the sleeve but they are unlikely to cause an obstruction when used in a vertical sleeve and pile.

The alternative to adopting insert piles or 'drilling and driving' techniques to mobilize compressive or uplift resistance in stiff to hard clays, is to provide an enlarged base to the piles. This can be achieved by using a rotary under-reaming tool operating below the toe of an open-ended steel tubular pile. The enlarged base provides both increased resistance to compressive loads and a positive anchorage against uplift. The uncertainty concerning the ability of available hammers to drive straight-sided piles to a deep penetration is avoided. However, there can be difficult problems when the arms of the expanding cutter fail to retract.

When open-end piles are driven into deep granular soil deposits the driving resistance may be very low for the reasons described in Section 4.3.3. As a result, calculations of resistance to axial compression loads based on dynamic testing are correspondingly low, indicating very deep penetration of the pile to achieve the required resistance. These penetrations are often much greater than those required for fixity against lateral loading. Although base resistance to axial loading can be achieved by grouting beneath the pile toe as described in Section 3.3.9, the operations of cleaning-out the pile and grouting are slow and relatively costly. An alternative method of developing base resistance of open-end piles which has been used on a number of marine projects is to weld a steel plate diaphragm across the interior of the pile. The minimum depth above the pile toe for locating the diaphragm is the penetration below sea bed required for fixity against lateral loading. However a further penetration is necessary to compact the soil within the plug and to develop the necessary base resistance. It is not possible to achieve a resistance equivalent to a solid-end pile but the penetration depths are much shorter than those required for an open-end pile.

The diaphragm method was used for the piling at the Hadera coal unloading terminal near Haifa[8.22]. Open-end piles 1424 and 1524 mm OD were proposed but initial trial driving showed that very deep penetrations, as much as 70 m below sea bed in calcareous sands, would be needed to develop the required axial resistance. The blow count diagram in Figure 8.17

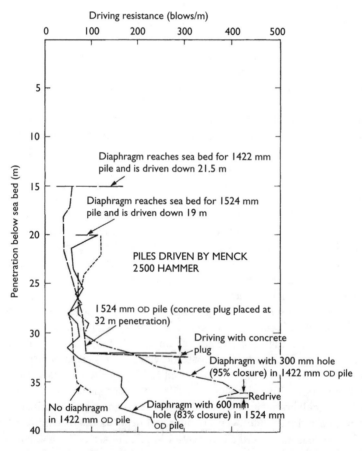

Figure 8.17 The effects of different methods of plugging steel tubular piles driven with open ends, Hadera coal unloading terminal.

showed quite low resistance at 36 m below sea bed. Another trial pile was driven to 32 m, cleaned-out, and plugged at the toe with concrete. An acceptable driving resistance of about 300 blows per metre was obtained by driving the plugged pile but it was appreciated that the plugging operations would be costly and would seriously delay completion of the project. Trials were then made of the diaphragm method. A diaphragm with a 600 mm hole giving 83% closure of the cross-section was inserted 20 mm above the toe. This increased the driving resistance at 39 m below sea bed and another trial with a 300 mm hole (95% closure) gave a higher resistance at 37 m (Figure 8.17).

The diaphragm method is ineffective if a very deep penetration is required because the long plug cannot compress sufficiently to mobilize the end-bearing resistance of the

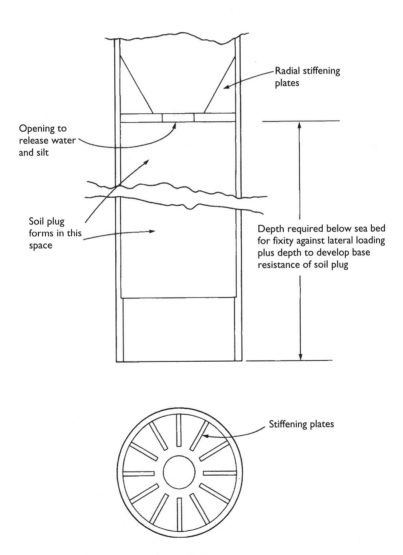

Figure 8.18 Internal diaphragm for tubular steel pile.

diaphragm and settlements at the working load would be excessive. It is also ineffective in clays or where clays are overlying the coarse soil bearing stratum. A hole is necessary in the diaphragm for release of water pressure in the soil plug and to allow expulsion of silt. Stresses on the underside of the diaphragm are high during driving and radial stiffeners are needed (Figure 8.18). The pile wall below the diaphragm must be sufficiently thick to prevent bursting by circumferential stresses induced by compression of the soil in the plug.

8.4   References

8.1   BRITISH STANDARDS INSTITUTION. BS 6349-1: 2000 *Maritime structures. Code of practice for general criteria*, BSI, London, 2000 (in revision 2007).

8.2   BROADHEAD, A. A marine foundation problem in the Arabian Gulf, *Quarterly Journal of Engineering Geology*, Vol. 3, No. 2, December 1970, pp. 73–84.

8.3   WIEGEL, L., BEEBE, K. E., and MOON, J. Ocean wave forces on circular cylindrical piles, *Proceedings of the American Society of Civil Engineers*, Vol. 83, No. HY2, 1957, pp. 1199/1–36.

8.4   REID, G. O. and BRETSCHNEIDER, C. L. Surface waves and offshore structures. The design wave in deep or shallow water. Storm tide and force on vertical piling and large submerged objects. *The A and M College of Texas Department of Oceanography*, October 1953 (unpublished).

8.5   DAILEY, J. W. and STEPHEN, S. C. Characteristics of the solitary wave, *Transactions of the American Society of Civil Engineers*, Vol. 118, 1953, pp. 575–81.

8.6   BRETSCHNEIDER, C. L. A theory for waves of finite height, *Proceedings of the 7th Conference on Coastal Engineering*, Vol. 9, 1961, pp. 146–83.

8.7   NEWMARK, N. The effect of dynamic loads on offshore structures, *Proceedings of the 8th Texas Conference on Offshore Technology*, Houston, Texas, September 1956, Paper No. 6.

8.8   *Shore Protection Planning and Design*, Technical Report No. 4, 1966, US Army Coastal Engineering Research Centre, Part 2, Chapter 4, pp. 278–96.

8.9   MORISON, J. R., O'BRIEN, M. P., JOHNSON, J. W., and SCHAAF, S. A. The force exerted by surface waves on piles, *Petroleum Transactions*, American Institute of Mining and Metallurgical Engineers, Vol. 189, TP 2846, 1950, pp. 149–54.

8.10   Steady pounding of North Sea gas platforms, *Ocean Industry*, Vol. 10, No. 8, 1975, pp. 64–71.

8.11   CHAPPELAAR, J. G. Wave forces on groups of vertical piles, *Journal of Geophysical Research*, American Geophysical Union, Vol. 64, 1959.

8.12   KORZHAVIN, K. N. *Action of ice on engineering structures*, USSR Academy of Science, Siberian Branch, 1962. (Translated by US Army Cold Regions Research and Engineering Laboratory, 1971).

8.13   TRYDE, P. Ice forces, *Journal of Glaciology*, Vol. 19, No. 2, 1977.

8.14   CROASDALE, K. R. *Ice forces on fixed rigid structures*, US Army Cold Regions Research and Engineering Laboratory, Special Report No. 80–26, 1980.

8.15   BROWN, T. G. Ice loads on the piers of the Confederation Bridge, Canada, *The Structural Engineer*, Vol. 78, No. 5, 2000, pp. 18–23.

8.16   BONNETT, D. Wind turbine foundations – loading, dynamics and design, *The Structural Engineer*, Vol. 83, No. 3, 2005, pp. 41–45.

8.17   FFRENCH, R., BONNETT, D., and SANDON, J. Wind power – a major opportunity for the UK, *Proceedings of the Institution of Civil Engineers*, Vol. 158 (Special Issue 2), 2005, pp. 20–27.

8.18   Regles BAEL 91 (Béton Armé aux Etats Limites), Modifée 99, publ Eyrolles, May 2000.

8.19   Recommended practice for planning, designing and constructing fixed offshore platforms, American Petroleum Institute, Publication RP2A, Washington DC, 1987.

8.20   Recommended practice for planning, designing and constructing tension leg platforms, American Petroleum Institute, Publication RP2T, Washington DC, 1993.

8.21 GERWICK, B. C. *Construction of Offshore Structures*, Wiley, 1986.

8.22 YARON, S. L. and SHIMONI, J. The Hadera offshore coal unloading terminal, *Proceedings of the Offshore Technology Conference*, Paper OTC 4396, Houston, 1982.

8.5 Worked examples

Example 8.1

A breasting dolphin is constructed by linking at the head four 350×350 mm reinforced concrete piles which are driven through 2.5 m of soft clay into a stiff clay, to a total penetration below sea bed of 9.0 m. Find the kinetic energy which can be absorbed by the pile group for an impact at a point 8 m above the sea bed. The maximum energy absorption value is to be taken as the figure which stresses the piles to their yield point.

The piles can be considered as fixed at the surface of the stiff clay stratum, and the ultimate resistance moment of each pile at the yield point is 125 kN m. Therefore from equation 8.4:

work done in deflecting piles to yield point

$$= \frac{4 \times 125^2(8 + 2.5)}{6 \times 26 \times 10^6 \times 0.0833 \times 0.35^4} = 3.37 \text{ kJ}$$

Example 8.2

A steel tubular pile having an outside diameter of 1300 mm and a wall thickness of 30 mm forms part of a pile group in a breasting dolphin. The pile is fabricated from high-tensile alloy steel to BS 4360 Grade 55C. The piles are driven into a stiff over-consolidated clay ($c_u = 150$ kN/m^2). Calculate the maximum cyclic force which can be applied to the pile at a point 26 m above the sea bed at the stage when the failure in the soil occurs at sea-bed level, the deflection of the pile head at this point, and the corresponding energy absorption value of the pile.

Steel to Grade 55C should have a minimum yield strength of 417 N/mm^2 and an elastic modulus of 2×10^5 MN/m^2.

Moment of inertia of pile $= \pi(1.30^4 - 1.24^4)/64 = 0.024$ m^4

Moment of resistance of pile at yield point $= \dfrac{417 \times 0.024}{0.65} = 15.4$ MN m

The first step is to establish the p–y curves. In equations 6.36 and 6.37, the submerged density of the soil is 1.2 Mg/m^3, and a value of 0.25 can be taken for the factor J.

At sea-bed level

Critical depth $x_r = \dfrac{6 \times 1.3}{\dfrac{1.2 \times 9.81 \times 1.3}{150} + 0.25} = 22.1$ m

$N_c = 3 + 0 + 0 = 3$

$p_u = 3 \times 150 \times 1.3 = 585$ kN per m depth

For cyclically applied loading take

$$p_b = 0.72p_u = 0.72 \times 585 = 421 \text{ kN per m depth}$$

In the absence of laboratory compression tests, the appropriate value of ε_c in equation 6.39 can be taken as 0.01, and the p–y curves will be derived in the same manner as for a normally consolidated clay.

Therefore

$$y_c = 2.5 \times 0.01 \times 1.3 = 0.0325 \text{ m} = 32.5 \text{ mm}$$

The deflection corresponding to p_b is $3y_c = 3 \times 32.5 = 97$ mm.

Other points of the p–y curve are calculated from equation 6.38. Thus for y = 15 mm:

$$p = 0.5 \times 585 \times \sqrt[3]{\frac{15}{32.5}} = 226 \text{ kN per m depth.}$$

Similarly for

$y = 25$ mm, $p = 268$ kN per m depth
$y = 50$ mm, $p = 338$ kN per m depth
$y = 75$ mm, $p = 386$ kN per m depth.

Beyond the critical point at $3y_c$, the p–y curve decreases linearly from $p_b = 0.72p_u$ to zero at $y = 15y_c = 487$ mm for $x/x_r = 0$.

The p–y curve at sea-bed level for the six points established above is shown in Figure 8.19a.

At 0.5 m below sea bed

$$N_c = 3 + \frac{1.2 \times 9.81 \times 0.5}{150} + \frac{0.25 \times 0.5}{1.3} = 3.13$$

$$p_u = 3.13 \times 150 \times 1.3 = 610 \text{ kN per m depth.}$$

$$p_b = 0.72 \times 610 = 439 \text{ kN/m at } y = 97 \text{ mm.}$$

For $y = 15$ mm, $p = 610 \times 0.5\sqrt[3]{15/32.5} = 236$ kN per m depth. Similarly

$y = 25$ mm, $p = 280$ kN per m depth
$y = 50$ mm, $p = 352$ kN per m depth
$y = 75$ mm, $p = 403$ kN per m depth.

The p–y curve falls linearly at $15y_c = 487$ mm to a value of $p = 0.72 \times 610 \times 0.5/22.1 = 10\,\text{kN/m}$.

The p–y curve for $x = 0.5$ m is also plotted in Figure 8.19a and the curves for values of x of 1.0, 1.5, 2.0, and 2.5 m below sea bed, established in a similar manner, are also shown on this figure.

The value of $p = 421$ kN/m represents the pressure at which yielding of the soil at the sea bed occurs. Therefore

$$\text{bending moment at sea-bed level } M_t = 26 \times 0.421 = 10.9 \text{ MN m}$$

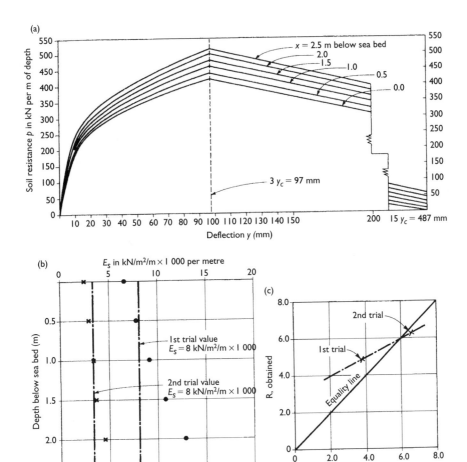

Figure 8.19

The deflections at various points below the sea bed are obtained from Figures 6.29a and 6.29b, taking as a first trial $R = 3.78$, corresponding to a k value from equation 6.11 of about 24 MN/m^2. Required penetration depth is $3.5 \times 3.78 = 13.2$, say 14 m. Then

$$Z_{max} = \frac{14}{3.78} = 3.7$$

From equation 6.30, $M_A = 10.9\, M_m$ MNm

From equation 6.32, $M_B = 0.421 \times 3.78 \times M_h = 1.6 M_h$ MNm

From equation 6.31, $y_A = \dfrac{10.9 \times 3.78^2 \times 1\,000}{2 \times 10^5 \times 0.024} y_m = 32.4 y_m$ mm

From equation 6.33, $y_B = \dfrac{0.421 \times 3.78^3 \times 1000}{2 \times 10^5 \times 0.024} y_h = 4.7 y_h$ mm

x (m)	$Z = \dfrac{x}{R}$	y_m (mm)	$y_A = 32.4 y_m$ (mm)	y_h	$y_B = 4.7 y_h$ (mm)	$y_A + y_B = y$ (mm)
0	0	+1.0	+32.4	+1.40	+6.6	+39.0
0.5	0.13	+0.78	+25.3	+1.32	+6.2	+31.5
1.0	0.26	+0.63	+20.4	+1.15	+5.4	+25.8
1.5	0.40	+0.50	+16.2	+1.00	+4.7	+20.9
2.0	0.53	+0.40	+13.0	+0.90	+4.2	+17.2
2.5	0.66	+0.32	+10.4	+0.80	+3.8	+14.2

The above values of y are referred to the p–y curves to obtain the corresponding values of p and hence to obtain E_s from the linear relationship $E_s = -p/y$, as tabulated below.

x (m)	y (mm)	p (kN/m)	$p' = p/1.3$ (kN/m²)	$E_s = -p'/y$ (kN/m²/m)
0	39.0	320	246	6.3
0.5	31.5	310	238	7.6
1.0	25.8	295	227	8.8
1.5	20.9	290	223	10.7
2.0	17.2	285	219	12.7
2.5	14.2	280	215	15.1

The values of E_s are plotted against depth in Figure 8.19b, from which an average constant value of E_s of 8×10^3 kN/m²/m is obtained. From equation 6.11:

$$R \text{ (obtained)} = \sqrt[4]{\frac{2 \times 10^5 \times 0.024}{8}} = 4.9$$

This value of R (obtained) is plotted against R (tried) in Figure 8.19c, from which a second trial value of R of 6.5 is taken. This higher value requires a deeper penetration of the pile, i.e. $L > 3.5 \times 6.5 = 22.75$; say 23 m. Thus $Z_{max} = 23/6.5 = 3.5$, and from equation 6.31:

$$y_A = \frac{10.9 \times 6.5^2 \times 1000}{2 \times 10^5 \times 0.024} y_m = 95.9 y_m \text{ mm}$$

From equation 6.33:

$$y_B = \frac{0.421 \times 6.5^3 \times 1000}{2 \times 10^5 \times 0.024} y_h = 24.1 y_h \text{ mm}$$

From Figures 6.29a and b the computed deflections are as tabulated below.

x (m)	Z = x/R	y_m	$y_A = 95.9 y_m$ (mm)	y_h	$y_B = 24.1 y_h$ (mm)	$y_A = y_B$ (mm)
0	0	+1.00	+95.9	+1.45	+34.9	130.8
0.5	0.08	+0.85	+81.5	+1.37	+33.0	114.5
1.0	1.15	+0.75	+71.9	+1.30	+31.3	103.2
1.5	0.23	+0.65	+62.3	+1.20	+28.9	91.2
2.0	0.31	+0.57	+54.7	+1.11	+26.7	81.4
2.5	0.38	+0.52	+49.9	+1.05	+25.3	75.2

From the p–y curve

x (m)	y (mm)	p (kN/m)	$p' = p/1.3$ (kN/m²)	$E_s = -p'/y$ (kN/m²/m)
0	130.8	385	296	2.3
0.5	114.5	420	323	2.8
1.0	103.2	455	350	3.4
1.5	91.2	470	362	4.0
2.0	81.4	470	362	4.4
2.5	75.2	470	362	4.8

From Figure 8.19b, the second trial value of $E_s = 3.3 \times 10^3$ kN/m², and

$$R \text{ (obtained)} = \sqrt[4]{\frac{2 \times 10^5 \times 0.024}{3.3}} = 6.2$$

This is sufficiently close to the equality line for 6.5 to be accepted as the final value of R (see Figure 8.19c).

The deflection of the pile head at the loading for the critical value of $H = 421$ kN for soil rupture is the sum of the following deflections (a) to (c).

(a) Deflection of pile considered as cantilever fixed at sea bed

$$= \frac{0.421 \times 26^3 \times 1000}{3 \times 2 \times 10^5 \times 0.024} = 514 \text{ mm}$$

(b) Deflection of pile at sea bed due to soil compression (from table above) = 130.8 mm.
(c) Deflection of pile head due to slope of pile below sea bed.

This can be obtained from the difference of the deflections at the sea bed and 1.0m below the sea bed. From the above table the deflection at 1 m below sea bed = 103.2 mm. Therefore, slope below sea bed = 130.8 − 103.2 = 27.6 mm in 1 m. Thus, deflection at pile head = 26 × 27.6 = 718 mm.

Total deflection at pile head = 514 + 131 + 718 = 1363 mm.

It is necessary to check the bending moments at and below the sea bed to ensure that the resistance moment of the pile section is not exceeded. From Figures 6.29a and b, for

$Z_{max} = 23/6.5 = 3.5,$

x (m)	Z = x/R	M_m	$M_A = 10.9M_m$ (MNm)	M_h	$M_B = 0.421 \times 6.5M_h$ $= 2.74M_h$ (MNm)	$M = M_A + M_B$ (MNm)
0	0	+1.00	+10.9	0	0	+10.9
0.5	0.08	+0.98	+10.7	+0.10	+0.3	+11.0
1.0	1.15	+0.97	+10.6	+0.15	+0.4	+11.0
1.5	0.23	+0.95	+10.4	+0.20	+0.5	+10.9
2.0	0.31	+0.94	+10.2	+0.27	+0.7	+10.9
4.0	0.62	+0.85	+9.3	+0.40	+1.1	+10.4
8.0	1.23	+0.55	+6.0	+0.45	+1.2	+7.2

The maximum bending moment of 11.0 MNm provides a safety factor of 15.4/11.0 = 1.4 against yielding of the steel.

From equation 8.3, the kinetic energy absorption value of the pile for horizontal movement at the stage of soil rupture at sea-bed level:

$$= \tfrac{1}{2} \times 421 \times 1363/1\,000 = 287 \text{ kJ}$$

In a similar manner to that set out above, it is possible to obtain pile head deflections and bending moments for various stages of horizontal loading up to the stage of yielding of the steel and hence to draw curves of deflection and energy absorption against horizontal load.

The deflection of the pile at sea-bed level caused by a lateral force of 421 kN applied at the sea bed can be calculated using Randolph's curves (Section 6.3.8).

Effective Young's modulus of equivalent solid section pile:

$$= E'_p = \frac{4 \times 2 \times 10^5 \times 0.024}{\pi \times 0.65^4} = 34.2 \times 10^3 \text{ MN/m}^2$$

An average constant soil modulus of 3.3 MN/m^2 from Figure 8.19b was used to calculate pile deflections and bending moments. For undrained loading take Poisson's ratio $v_u = 0.5$.

$$\text{Shear modulus} = G_c = \frac{3.3}{2(1 + 0.5)} = 1.1 \text{ MN/m}^2$$

$$G^* = 1.1(1 + 0.75 \times 0.5) = 1.5 \text{ MN/m}^2$$

$$\text{Critical length} = l_c = 2 \times 0.65 \left(\frac{34.2 \times 10^3}{1.5} \right)^{2/7} = 22.9 \text{ m}$$

Homogeneity factor = 1

In Figure 6.36a,

$$\frac{yr_0 G_c}{H_0} \left(\frac{E'_p}{G_c} \right)^{1/7} = \frac{y \times 0.65 \times 1.5}{0.421} \left(\frac{34.2 \times 10^3}{1.5} \right)^{1/7} = 9.7y$$

At 0.5 m below sea bed

$$z/l_c = 0.5/22.9 = 0.02 \text{ m}$$

$$\text{giving } 9.7y = 0.26, \quad y = \frac{0.26 \times 10^3}{9.7} = 27 \text{ mm}$$

Example 8.3

A cross-section of an approach trestle giving roadway access to a cargo jetty is shown in Figure 8.20. The trestle is sited at right-angles to the direction of maximum current velocity and travel of storm waves. The distribution of current velocity with depth is shown on the cross-section. The deck slab and other components of the superstructure impose a total horizontal wind force of 25 kN on each pile bent. Storm waves have a maximum height from crest to trough of 3 m. Determine the distribution of current and wave forces on the pile bent and calculate the bending moments on the piles produced by these forces.

The maximum horizontal force on the piles will be due to the combined current and wave action at HWST (+6.0 m). At this stage of the tide the storm wave crest will be at +7.5 m. The underside of the transom beam is at +8.0 m and therefore the wind force on the exposed length of pile from +7.5 to +8.0 m will be relatively small and can be neglected. It is convenient to divide the length of the pile into 2 m elements. Allowance is made for barnacle growth on the piles. Thus,

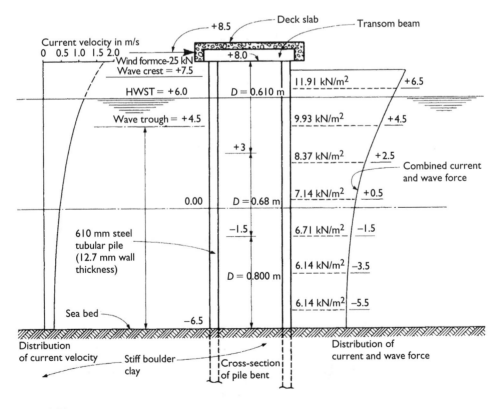

Figure 8.20

Table 8.4 Calculations for current and wave forces in Example 8.3

Elevation (m) (CD)	Average height above sea bed (m)	Pile diameter (m)	Current		$\frac{z}{h}$	Wave drag		Wave inertia		Total force (kN/m²)	Horizontal load on element (kN)	Bending moment (kNm)
			Velocity (m/s)	Force (kN/m²)		$(u/C)^2$	Force (kN/m²)	$\frac{1}{g}\cdot\frac{du}{dt}$	Force (kN/m²)			
+7.5 − +5.5	13	0.610	1.2	0.36	1.18	0.21	9.01	0.26	2.54	11.91	14.53	×14.5 = 210.69
+5.5 − +3.5	11	0.610	0.8	0.16	1.00	0.18	7.72	0.21	2.05	9.93	12.15	×12.5 = 151.87
+3.5 − +1.5	9	0.680	0.6	0.09	0.82	0.15	6.43	0.17	1.85	8.37	11.38	×10.5 = 119.49
+1.5 − −0.5	7	0.680	0.4	0.04	0.64	0.13	5.58	0.14	1.52	7.14	9.71	×8.5 = 82.53
−0.5 − −2.5	5	0.680	0.3	0.02	0.45	0.12	5.15	0.12	1.31	6.48	8.81	×6.5 = 57.25
−2.5 − −4.5	3	0.800	0.25	0.01	0.27	0.11	4.72	0.11	1.41	6.14	9.82	×4.5 = 44.19
−4.5 − −6.5	1	0.800	0.25	0.01	0.09	0.11	4.72	0.11	1.41	6.14	9.82	×2.5 = 24.55

Maximum bending moment (at −6.5 m) in kN m = 690.57

From +7.5 to +3.0 m: no increase of diameter   (i.e. $D = 0.61$ m)
From +3.0 to -1.5 m: increase of 70 mm   ($D = 0.68$ m)
From -1.5 m to sea bed: increase of 190 mm   ($D = 0.80$ m)

Taking Newmark's values a drag force coefficient of 0.5 is used to calculate the current and wave drag forces, and an inertia coefficient of 2.0 is used to calculate the wave inertia forces. Thus in equation 8.10:

$$F_D = 0.5 \times 0.5 \times \rho \times V^2 \times A_n = 0.25 V^2 A_n \text{ kN} \quad \text{(for } \rho = 1 \text{ mg/m}^3\text{)}$$

In equation 8.9

$$f = 7.8 \times 0.5 \times 11(u/c)^2 + 8 \times 2 \times D\left(\frac{1}{g} \cdot \frac{du}{dt}\right) = 42.9(u/c)^2 + 16D\left(\frac{1}{g} \cdot \frac{du}{dt}\right).$$

The calculated wave and current forces are shown in Table 8.4 and Figure 8.20. The bending moments shown in Table 8.4 have been calculated on the assumption of virtual fixity of the pile at a point 1.5 m below the sea bed in the stiff boulder clay. Scour would not be expected around the piles in this type of soil. From Table 8.4, the combined wave and current forces produce a maximum bending moment at the point of fixity of 690.57 kN m.

Bending moment due to wind force on deck slab:

$$= \tfrac{1}{2} \times 25 \times (15.0 + 1.5) = 206.25 \text{ kN m}$$

Total bending moment $= 896.82$ kNm/pile.

Moment of inertia of pile section $= \pi(0.6100^4 - 0.5846^4)/64 = 1.063 \times 10^{-3} \text{m}^4$.

Extreme fibre stress of pile $= \dfrac{896.82 \times 0.305}{1.063 \times 10^{-3} \times 10^3} = 257 \text{ MN/m}^2$.

The direct stress resulting from the dead load of the deck slab and self weight of the pile is added to the bending stress calculated above. It is also necessary to calculate the susceptibility of the pile to current-induced oscillations.

Assuming the pile to be filled with fresh water, the effective mass is approximately equal to the mass of metal plus twice the mass of the displaced water. Therefore

$$M = 187 + (2 \times \tfrac{1}{4}\pi \times 0.61^2 \times 1\,000) = 771.5 \text{ kg/m}$$

When the pile is in an unsupported condition cantilevering from the sea bed, from equation 8.13:

$$f_N = \frac{0.56}{14^2} \sqrt{\frac{200 \times 10^9 \times 1.063 \times 10^{-3}}{771.5}} = 1.50 \ H_z$$

From equation 8.12 critical velocity for onset of cross-flow oscillation $= 5.5 \times 1.5 \times 0.61 =$ 5 m/sec.

Therefore cross-flow or in-line oscillations should not take place for the flow velocities shown in Figure 8.20.

Chapter 9

Miscellaneous piling problems

9.1   Piling for machinery foundations

9.1.1   General principles

The foundations of machinery installations have the combined function of transmitting the dead loading from the machinery to the supporting soil and of absorbing or transmitting to the soil in an attenuated form the vibrations caused by impacting, reciprocating, or rotating machinery. In the case of impacting machinery or equipment such as forging hammers or presses, and reciprocating machines, piston compressors and diesel engines, the dynamic loads transmitted to the soil take the form of thrusts in a vertical, horizontal or inclined direction. Rotating machinery, such as gas and steam turbines, creates a torque on the shaft, resulting in lateral loads or moments applied to the foundation block. Rock crushers and metal shredders produce random dynamic loads as a result of rotating imbalances depending on the particular operation. Dynamic loading from hammers or presses, or from low-speed reciprocating engines has a comparatively low frequency of application, but the vibrations resulting from out-of-balance components in high-speed rotating machinery can have a high frequency.

The higher the frequency of dynamic loading, the less is the amplitude which can be permitted before damage to the machinery occurs, or before damage to nearby structures, and noise and discomfort to people in the vicinity becomes intolerable. When the frequency of vibration of a machine and its foundations approaches the natural frequency of the supporting soil, resonance occurs and the resulting increased amplitude may result in damage to the plant and excessive settlement of the soil. The latter is particularly liable to happen when the vibrations are transmitted to loose or medium-dense coarse soils.

When the mass of the machine and its foundations and vibration characteristics of the soils are known, it is possible to calculate the resonant frequency of the combined machine–foundation–soil system. In order to avoid resonance, the frequency of the applied dynamic loading should ideally not exceed 50% of the resonant frequency for most impact hammers or reciprocating machinery. In the case of high-speed rotating machinery it is probable that the applied frequency will be higher than the resonant frequency of the machine–foundation–soil system. For this condition the aim should be to ensure that the applied frequency is at least 1.5 times the resonant frequency. The need for the wide divergency is to allow for the starting-up and shutting-down periods when the frequency of the machine passes through the resonant stage. If the applied frequency is too close to the resonant frequency the stage of resonance at the acceleration or slowing down of the machine might be too protracted.

When designing shallow foundations for machinery, vibrations which might cause damage or nuisance to the surroundings can be absorbed or attenuated by increasing the mass of the foundation block. There are old 'rules-of-thumb' which require the ratio of the mass of the foundation to the mass of the machine to be in the range of 1:1 to 4:1 depending on the type of machine. The resulting required mass of the foundation may be excessive for loose or weak soils leading to excessive settlement, even under static loading conditions, and necessitating the provision of a piled foundation. Also, it may be necessary to employ piles on sites where the water table is at a depth of less than one-half of the width of the block below the underside of the base or even within a depth of twice the width of the block. This is because water transmits amplitudes of vibration almost undamped over long distances which might result in damaging effects over a wide area surrounding the installation. Similarly, piles may be desirable if a rigid stratum of rock or strongly cemented soil exists within a depth of $1\frac{1}{2}$ times the block width. Such a stratum reflects energy waves and magnifies their amplitude of vibration.

Generally, the effect of providing a piled foundation to a reciprocating or rotating machine is to increase the natural frequency of the installation in the vertical, rocking, pitching, and also possibly longitudinal, modes. This is because of the behaviour of the mass of soil enclosed by the pile group acting with the pile cap and the piles themselves. The soil mass may be relatively small where the piles act in end bearing, or large in the case of friction piles. The natural frequency may be decreased in the lateral and yawing modes of vibration because of the low resistance of piles to lateral loads at shallow depths.

To ensure that the ratio of the frequency of the disturbing moment or disturbing force applied by the machinery to the natural frequency of the machine–foundation–soil system is either greater or less than the required value, it is necessary to calculate the natural frequency of the system. This is a complex matter, particularly for piled foundations, and is beyond the scope of the present book. The reader is referred to the publications of Barkan[9.1], Hsieh[9.2], Whitman and Richart[9.3], and Richart et al.[9.4] for general guidance. Irish and Walker[9.5] have established design curves relating the natural frequency of piles of various types to their effective length, both for the vertical and the rocking modes of vibration. The American Concrete Institute[9.6] presents various design criteria and methods of analysis, design and construction as currently applied to dynamic equipment foundations.

9.1.2   Pile design for static machinery loading

Piles and pile groups carrying static loads from machinery should be designed by the methods described in Chapters 4 and 5. Particular attention should be paid to the avoidance of excessive differential settlement of the pile cap; the differential movement should not exceed 8 mm. The centre of gravity of the machine combined with the pile cap and supporting piles should be located as nearly as possible on a vertical line through the centroid of the pile group, and the eccentricity of the combined masses should not be greater than 5% of the length of the side of the pile group. If possible the centre of gravity of the machine and soil mass should be below the top of the pile cap.

9.1.3   Pile design for dynamic loading from machinery

Generally, it can be stated that the effect of applying dynamic loads to piles in fine-grained soils is to reduce their shaft friction and end-bearing value, i.e. to reduce their ultimate carrying capacity, and the effect in coarse-grained soils is to reduce their shaft friction but to increase their end-bearing resistance at the expense of increased settlement under working load.

The reduction in the shaft friction and end-bearing resistance of piles in fine-grained soils is the result of a reduction in the shearing strength of these soils under cyclic loading. The amount of reduction for an infinite number of load repetitions depends on the ratio of the applied stress to the ultimate stress of the soil. It is the usual practice to double the safety factor on the combined shaft friction and end bearing to allow for the dynamic application of load (see Section 6.2.2).

The torque of rotating machinery can cause lateral loading on the supporting piles. The deflection under lateral loading can be calculated by the methods described in Chapter 6. To allow for dynamic loading the deflections calculated for the equivalent static load should be doubled.

The type of pile, whether driven, driven and cast-in-place, or bored and cast-in-place, is unlikely to have any significant effect on the behaviour of piles installed wholly in fine-grained soils. It is possible that the lateral movements of piles with driven pre-formed shafts (e.g. precast concrete or steel H-piles) will be greater than those of cast-in-place piles, because of the formation of an enlarged hole around the upper part of the shaft (see Figure 4.5).

The frictional resistance of a pile to static compressive loading in a coarse soil is relatively low. This resistance is reduced still further when the pile is subjected to vibratory loading, and it is advisable to ignore all frictional resistance on piles carrying high-frequency vibrating loads. If such piles are terminated in loose to medium-dense soils there will be continuing settlement to a degree which is unacceptable for most machinery installations. It is therefore necessary to drive piles to a dense or very dense coarse soil stratum and even then the settlements may be significant, particularly when high end-bearing pressures are adopted. This is due to the progressive attrition of the soil grains at their points of contact. The continuing degradation of the soil particles results in the slow but continuous settlement of the piles. If possible, piles carrying vibrating machinery should be driven completely through a coarse soil stratum for termination on bedrock or within a stiff clay. The ACI Report[9.6] considers the complex interaction of piles in a group under dynamic loading when piles are closer than 20 diameters and recommends suitable computer programs to consider group dynamic stiffness and damping effects in such cases.

A problem of piled foundations for machinery sensitive to small differential settlements was experienced at John Brown Company's shipyard at Clydebank. At this site 18.6 m of loose to medium-dense silty sand were overlying stiff glacial till. Gear-cutting machinery comprising large hobbing, shaving and grinding machines had to operate to an accuracy of 0.009 mm and each machine was installed in a separate enclosure under conditions of constant temperature and humidity. It was essential to avoid any appreciable settlement of the machines due to vibrations caused by their own motion, or transmitted from elsewhere on the shipyard and the adjacent main road. It was expected that settlements of raft or piled foundations terminated in the medium-dense sand (with a standard penetration test N-value of 20 to 30 blows per 300 mm) would be excessive and a type of pile had to be selected which could be driven through the deep sand layers to reach the glacial till. The possibility of the compaction of the sand due to driving a number of piles in a closely spaced group was considered and this led to the choice of a small displacement pile in the form of a Larssen BP2 box-section driven with an open end. It would have been possible to use water jetting to assist the penetration of these piles, but all the piles were driven by a double-acting hammer into the glacial till without recourse to jetting. The building surrounding the plant was carried by driven and cast-in-place piles terminated at a penetration of about 4.6 m into the sand stratum.

9.2   Piling for underpinning

9.2.1   Requirements for underpinning

Underpinning of existing foundations may be required for the following purposes:

(1) As a remedial measure to arrest the settlement of a structure
(2) As a precautionary measure carried out in advance to prevent the excessive settlement of a structure when deep excavations are to be undertaken close to its foundations, and
(3) As a strengthening measure to enable existing foundations to carry increased loading, or to replace the deteriorating fabric of a foundation.

An example of the use of piling as a remedial measure is shown in Figure 9.1a. The column has settled exclusively due to the consolidation of the soft clay beneath its base. Piles are installed on each side of the base and the load transferred to the pile heads by needle beams inserted below the base.

A typical use of piles as a precautionary underpinning measure is shown in Figure 9.1b. A deep basement is to be constructed close to an existing building on shallow strip foundations. Underpinning of the foundation adjacent to the basement is required since yielding of the ground surface as a result of the relief of lateral pressure due to the excavations would cause excessive settlement. Rows of piles are installed close to the wall foundation inside and outside the building and the loads are transferred to them by a system of longitudinal and transverse beams. The external row of piles also serves as a support for the horizontal sheeting members used to retain the face of the excavation.

Piling as a strengthening measure is shown in Figure 9.1c. Pits are excavated beneath the existing foundation and piles are jacked down to a bearing on a hard incompressible stratum. Underpinning of the foundations may be required where the existing piles have deteriorated due to attack by aggressive substances in the soil or groundwater. New piles can be installed in holes drilled through the cap or raft (Figure 9.1d). The new pile heads are bonded to the reinforcement of the existing substructure.

Piling has a somewhat limited application to underpinning work. This is because it is usually necessary to excavate pits below the existing substructure to place supporting beams or pads. In a high proportion of the cases where remedial or strengthening works are required a suitable bearing stratum exists at no great depth and it is cheaper to take the pits down to this stratum and to backfill the void with mass concrete rather than to install piles in conditions with a low headroom and a restricted working area. Also a considerable force may be required to jack down an underpinning pile even though the soil is removed from the piling tube at each stage of jacking. There may be insufficient mass in the existing structure to provide the required reaction to this jacking force. It is not usually feasible to employ jacked piles beneath two-storey or three-storey buildings of load-bearing wall construction, since the usual mass concrete strip foundation and brick footing walls have insufficient bending strength to withstand the loading that results from jacking down a pile, even though a spreader beam is used between the ram of the jack and the foundation.

9.2.2   Piling methods in underpinning work

Before underpinning by piling is considered, it is essential to determine the cause of structure–foundation instability and confirm the ground conditions at depth. Means of checking pile capacity and integrity once installed should be available.

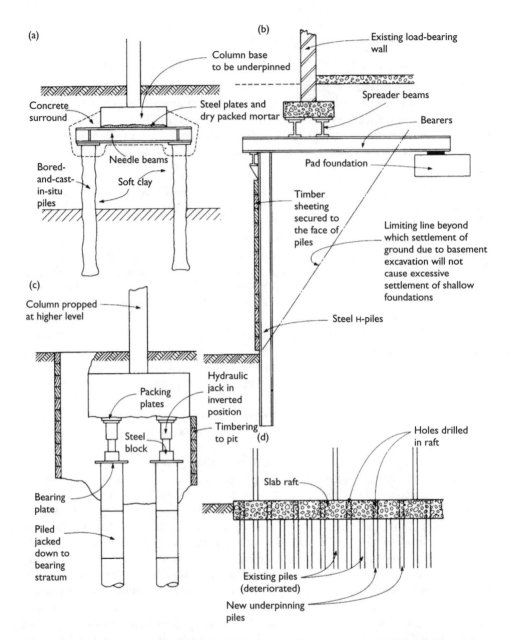

(a)

Column base
to be underpinned

(b)

Existing load-bearing
wall

Concrete
surround

Steel plates and
dry packed mortar

Spreader beams

Bearers

Bored-
and-cast-
in-situ
piles

Needle beams

Soft clay

Pad foundation

Timber
sheeting
secured to
the face of
piles

Limiting line beyond
which settlement of
ground due to basement
excavation will not
cause excessive
settlement of shallow
foundations

(c)

Column propped
at higher level

Packing
plates

Hydraulic
jack in
inverted
position

Steel H-piles

Steel
block

Timbering
to pit

Holes drilled
in raft

Bearing
plate

(d)

Slab raft

Piled
jacked
down to
bearing
stratum

Existing piles
(deteriorated)

New underpinning
piles

Figure 9.1 Use of piles in underpinning (a) Remedial measures to support column base
(b) Precautionary measures in underpinning strip foundation adjacent to deep excavation
(c) Jacked piles to strengthen column base (d) Drilled piles to replace existing piles beneath
raft slab.

Bored piles are suitable for underpinning light structures and they are installed outside the periphery of the existing foundations as shown in Figure 9.1a. In addition to light tripod percussion rigs many short-masted rotary auger rigs are available for installing the piles inside buildings in conditions of low headroom. However, it is desirable, for reasons of economy, to install piles outside a building as far as possible. This can be done at the corners of buildings as shown in Figure 9.2. Precast reinforced concrete sections or steel H-piles can be concreted or grouted into the pile boreholes in cases where it is desired to transfer the loading to underpinning piles as quickly as possible after installing them. When using bored piles inside basements spoil removal can be a drawback.

Light structures can be underpinned from a single row of bored piles located outside the building. After concreting the piles, cantilever brackets are cast onto their heads as shown in Figure 9.3a. The bending resistance of a small-diameter pile is relatively low, and therefore the form of construction is limited to strip foundations of light buildings or to lightly

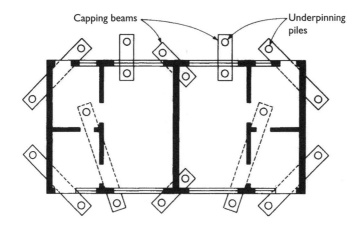

Figure 9.2 Layout of piles for light structures.

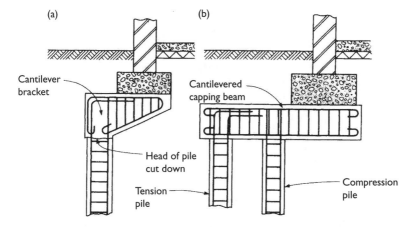

Figure 9.3 Cantilevered brackets for supporting light structures (a) From single piles (b) From pairs of piles.

loaded columns. Heavier structures can be underpinned by pairs of piles located outside the building but carrying a cantilevered bracket as shown in Figure 9.3b. This system can cause difficulties in pile design. The compression pile is required to carry heavy loading and there may be problems in achieving the required resistance to uplift in shaft friction on the tension pile.

The Fondedile piling system employing the 'Pali Radice' (root pile) is suitable as a means of underpinning structures undergoing settlement or for strengthening existing foundations to enable them to carry heavier loads. Small-diameter holes lined temporarily with casing tubes are drilled through the existing foundations or through both a load-bearing wall and its foundation (Figure 9.4). A rich sand–cement mortar is then pumped down a tremie pipe to fill the borehole and any cavities in the existing structure. Reinforcement is provided in the form of a single bar for small-diameter (100 mm) piles or a cage or tube for the larger-diameter (250–300 mm) piles. The casing is extracted with the assistance of compressed air to push the grout into the cavities and against the walls of the drilled holes.

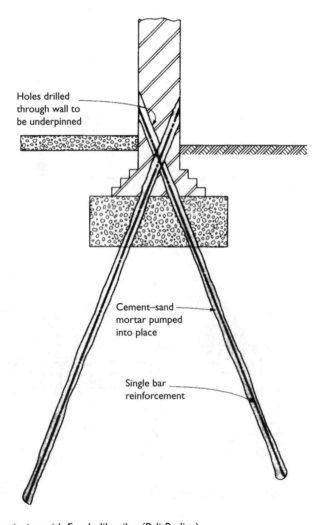

Holes drilled through wall to be underpinned

Cement–sand mortar pumped into place

Single bar reinforcement

Figure 9.4 Underpinning with Fondedile piles (Pali Radice).

The usual method of application is to install the piles in relatively large numbers at differing angles of rake so as to provide resistance to lateral loads and overturning moments as well as to support loads in a vertical direction. The rotary drilling machines used by Fondedile require a working space of only 2×1.5 m and a headroom of 1.8 m, and the drilling is performed with relatively little vibration. Thus the piles can be installed in confined surroundings and very close to the foundations to be underpinned (Figure 9.5). Other systems employ minipiles of the type described in Section 2.6 such as small-diameter steel tubes which are drilled down or set in drilled holes. Grout is pumped down the tubes which remain in place as permanent reinforcement. A detailed account of the Fondedile system and underpinning with other types of small-diameter piles with numerous case histories is given by Lizzi[9.7].

Bored piles installed by continuous flight auger as described in Section 2.4.2 are also suitable for strengthening and underpinning work since the rotary drilling methods and continuous support given to the soil result in little vibration and a negligible loss of ground in properly controlled operations.

Bottom driven piles using thin wall steel tubes up to 300 mm diameter may be used in difficult conditions such as brownfield sites, peaty soils and soft clay, and founded on a competent stratum. Such piles will have limited tension capacity and should not be used in heaving ground or where vibration may cause problems. A pneumatic piling hammer may be used for driving tubes up to 100 mm diameter.

Heavily loaded foundations can be underpinned by jacking piles down to the bearing stratum using the dead load of the existing foundations and superstructure as the reaction to the jacking operation. The *Presscore* precast jacked-in pile with a central hole is described

Figure 9.5 Fondedile piling rig working close to foundations to be underpinned.

in Section 2.2.3 and Figure 2.15. Typical jacked piles require a pit excavation beneath the foundation, and a hole in the floor of the pit to receive the bottom-pointed unit of the pile. A hydraulic jack is placed on top of this unit and is surmounted by short lengths of steel plate or beam sections to spread the load onto the underside of the existing foundation. The bottom unit is then jacked down until it is flush with the bottom of the pit. The jacks and packers are removed and another precast concrete section added, which is in turn jacked down. Additional units are now added and jacked down until the bearing stratum is reached or until the resistance of the pile as measured by the pressure gauge on the jack indicates that the desired 'preload' has been attained. The elements are next bonded together by inserting short steel bars into the longitudinal central hole and grouting them with cement. On the completion of the jacking operation short lengths of steel beam are driven hard into the space between the pile head and the foundation, or between the pile head and the spreader beams. The jack is then removed and the head of the pile, packers, and spreaders encased solidly in concrete. Grout bags which are inflated and pressurized with cement grout are a convenient alternative to the steel packing.

It is essential to maintain the load on the jack until the packing is completed. This is to avoid any rebound of the pile head, and subsequent settlement when the load from the structure is transferred to the piles.

An alternative method which 'pre-tests' the jacked-in pile once it reaches the bearing stratum, or the desired value of preload has been attained, requires a pair of hydraulic jacks to be inserted between the head of the pile and a bearing plate packed up to the underside of the existing foundation. The thrust on the rams of these jacks is adjusted to apply a load of 1.5 times the working load onto the pile. When downward movement of the pile has ceased a short length of steel H-section with end-bearing plates is wedged tightly into the space between the jacks (Figure 9.6). The latter can then be removed and used for the same

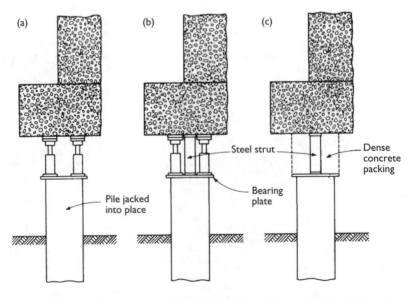

Figure 9.6 Underpinning with pre-test load (a) Jacking down underpinning pile (b) Insertion of steel strut (c) Steel strut wedged into place before encasement in concrete.

procedure on the adjoining piles. Where piles are installed in rows or closely spaced groups by preloading or 'pre-testing' methods, the operation of jacking an individual pile relieves some of the load on the adjacent piles which have already been installed and wedged-up. It then becomes necessary to replace the jacks and re-load these piles, after which the inserted struts are re-wedged. Alternatively, all the pre-testing jacks can remain in position until the last pile in the group or row is jacked down. Then all the loads on the jacks are balanced, the struts installed, and the jacks removed. The final operation is to encase the struts and pile heads in concrete well rammed-up to the underside of the existing foundation.

Steel tube or box section 'jacked piers' such as those manufactured by Atlas Systems and A B Chance in the USA and Rautoruukki Metform in Finland support the foundation being underpinned on a steel bracket fixed to the top of the pile section. Diameters range from 60 to 320 mm, provided in short lengths appropriate to the jack stroke, with friction or welded joints for the thin wall sections and threaded joints for 10 mm wall. The bearing capacity of slender piles will be governed by the buckling resistance in weak soils. Corrosion protection should be provided.

Where steel sections have to be added by welding during pile jacking to reach the required stratum or resistance, the alignment and welding quality can be difficult to control when working in constricted spaces excavated under existing foundations.

Whichever system of jacked piles is used, safeguards are needed to avoid a sudden drop in the ram due to loss of oil pressure. Also care must be taken to restrain the existing foundation, or the rows of jacks and struts, from moving horizontally due to lateral or eccentric thrusts. Raking shores to the superstructure, strutting of the existing foundation to the walls of the underpinning pit, or bracings between jacks and pile heads can be used to restrain lateral movement.

The existing columns or walls of the structure to be underpinned can be used to provide the reaction to jacking if they are sufficiently massive. Niches are cut into the faces of the structure and concrete corbels or brackets are cast into these pockets to form the bearing members for the jacks, as shown in Figure 9.7. The pairs of jacks must, of course, be operated simultaneously to avoid applying eccentric loading to the existing structure.

Where H-section piles are used to provide underpinning combined with lateral support to a deep excavation, as shown in Figure 9.1b, they can be installed by placing them in holes previously drilled by mechanical auger or, in stable ground, by tripod rigs. If the drill holes are given continuous support by a bentonite slurry or by casing there should be a negligible loss of ground around the borehole and the installation of the piles is effected with little noise or damaging vibration. Where it is necessary to use tripod rigs in unstable coarse soils, there is a risk of loss of ground around the boreholes as described in Section 3.3.7. If there is such a risk it will be necessary to shore the building temporarily with supports bearing on the ground outside the zone of subsidence. Below the level planned for the base of the excavation the space between the pile and the borehole is filled with a weak sand–cement mortar. This provides the required passive resistance to lateral loads on the piles and allows the latter to be removed if permitted by the planned sequence of underpinning and construction of the permanent work.

Bored piles used in the combined role of underpinning and lateral support to the sides of the excavation can be arranged in a single row (Figure 9.8a) or a double row of abutting piles, or in a single row of interlocking piles (Figure 9.8b). When they are abutting the system is known as 'contiguous piling' and when interlocking as 'secant piling'. Contiguous piles are cheaper to install, but because it is impossible to drill the holes in a truly vertical

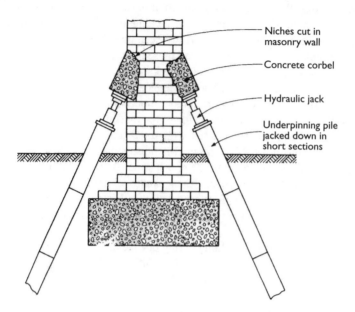

Figure 9.7 Underpinning load-bearing wall by jacking piles from corbels.

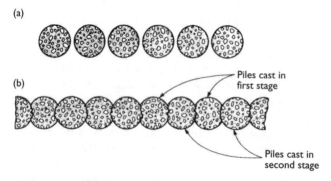

Figure 9.8 Bored piles used for combined underpinning and lateral support (a) Contiguous piles (b) Secant piles.

direction or to maintain a constant shaft diameter, there are always gaps present between adjacent piles. Below the water table, sand and silt can bleed through these gaps causing a considerable loss of ground, and the installation of the piles in a double staggered row cannot prevent this happening. Contiguous piles are best suited to underpinning and to support excavations in firm to stiff clays or damp silts and sands above the water table.

Where the excavation is to be performed in water-bearing coarse soils, any loss of ground can be avoided by adopting secant piling (Figure 9.8b). Alternate piles are first installed by conventional drilling and casting relatively weak concrete in-situ. The soil in the space between the pile shafts is then drilled out and a 'secant' is cut into the wall of the 'soft' pile on each side, using appropriate drilling tools, including CFA techniques. Concrete is next placed to fill the drill hole, thus forming the interlocked and virtually watertight wall.

Longitudinal reinforcement is provided in the piles to the extent necessary to carry vertical loading, eccentric loads from the underpinning bearers, and lateral loading from earth and hydrostatic pressure.

Screw or helical piles comprising solid square shafts up to 60 mm and tubular shafts up to 90 mm with helical plates between 200 and 350 mm diameter attached at intervals along the shaft, such as the AB Chance systems and ScrewFast Foundations in the UK, are effective as underpinning support. They are screwed into the ground adjacent to the foundation using rotary drives, either attached to hydraulic excavators with torque capacity up to 68 kNm or hand-held units with around 3 kNm torque – resisted by a torque bar. The bearing capacity is related to the area of the helical plates, four plates being typical and spaced so that overlap of bearing zones does not occur. Shaft friction is not usually applied unless the shaft diameter is greater than 90 mm. As for the slender steel-jacked piers above, buckling has to be considered and corrosion protection provided (note this form of slender pile should not be confused with the large-diameter displacement pile formed by screwing a mandrel into the ground – see Section 2.3.5).

9.3 Piling in mining subsidence areas

The form in which subsidence takes place after extracting minerals by underground mining depends on the particular technique used in the mining operations. In Great Britain the problems of subsidence mainly occur in coal-mining areas where the practice in the remaining working collieries is to extract the coal by 'longwall' methods. Using this technique the entire coal seam is removed from a continuously advancing face. The roof of the workings behind the face is supported by multiple rows of hydraulically operated props. As the face moves forward the props in the rear are systematically lowered to allow the roof of the workings to sink down onto heaps of mine dirt or 'stowage'. The overlying rock strata and over-burden soil follow the downward movement of the roof and the consequent subsidence of the ground surface is in the form of a wave which advances parallel to and at approximately the same rate as the advancing coal face. The subsidence is accompanied by very substantial horizontal strains of the ground surface, these strains being tensile at the crest of the wave and compressive at the trough and thus taking the form shown in Figure 9.9.

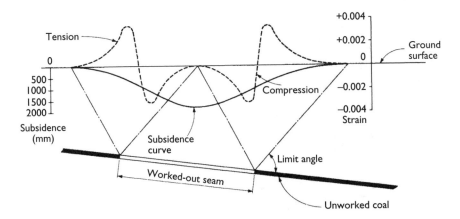

Figure 9.9 Profile of ground subsidence over longwall mine workings.

The magnitude of the strain can range from as much as plus or minus 0.8% of the overburden thickness above shallow workings to 0.2% over deep seams.

The horizontal ground movements make it virtually impossible to use piled foundations in areas where longwall mining is proposed, or is currently being practised. The horizontal shearing forces accompanying the strains are so high that it is quite uneconomical to attempt to resist them by heavily strengthening the pile shafts. However, in areas where subsidence following longwall mining has virtually ceased it is possible to use piled foundations if it is recognized that some residual movement will take place as the collapsed strata slowly reach final equilibrium. In these cases it is desirable to terminate piles in a soil layer overlying rockhead, as shown on the left-hand side of Figure 9.10. The soil acts as a cushion, preventing any concentration of load on the broken rock strata. Long-term movements may be substantial near the boundary of the worked-out seam. If the workings are shallow, piles may be taken down through the collapsed overburden to intact rock layers below the coal seam as shown on the right-hand side of Figure 9.10. Bored and cast-in-place piles are used for this purpose, but it is essential to isolate the shaft of the pile from the overburden above the coal seam in order to avoid heavy compressive loading caused by downdrag from the collapsing strata. This isolation is achieved by placing the concrete within a shell formed from stiffened light-gauge steel sheeting, the sheeting terminating at the base of the coal seam. Below this level the concrete can be cast against the surface of the stable strata to form a 'rock socket', as shown in Figure 9.11. The space between the shell and the wall of the drill hole through the overburden can be filled with bentonite slurry, soft bitumen or loosely placed rock fragments. A minimum clearance of 150 mm should be provided to accommodate minor

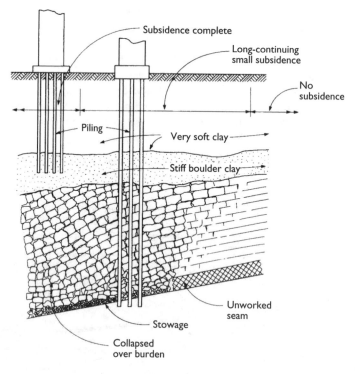

Figure 9.10 Piling through collapsed ground over longwall mine workings.

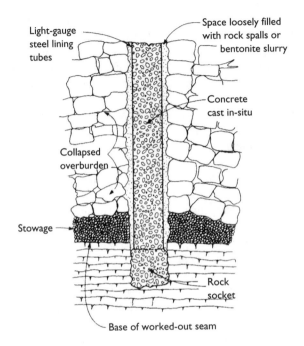

Figure 9.11 Isolating shafts of bored piles from surrounding collapsed ground.

lateral movements as the rock strata adjust themselves to their equilibrium position. A bentonite or bitumen infill to seal the annulus completely would be used if there is a risk of emission of natural gas from the coal seam.

During the nineteenth century and the early years of the present century, coal and other minerals were extracted using mining techniques variously known as 'pillar and stall', 'board and pillar', and 'stoop and room'. A main heading or road was driven from the shaft to follow the coal seam to the planned boundary of the workings. Transverse galleries were then driven from the main roadway to form a rectangular, triangular, or lozenge-shaped pattern of galleries separated by pillars of unworked coal. These pillars served to support the overlying rock strata until the general area had been mined. The pillars were then either left intact or were wholly or partially removed as the coal extraction operations retreated towards the shaft. Where the pillars were wholly removed the pattern of subsidence followed that of longwall mining (Figure 9.9). Chalk was mined in south-east England for flints and agricultural purposes from pre-historic times until comparatively recently. The mining was usually in the form of a rather haphazard pillar and stall method. Where pillars were left in place they remained, and still remain, in an unpredictable state of stability which has resulted in complex problems concerning building over abandoned mineworkings, problems which are still encountered in the built-up areas of Britain to the present day[9.8].

The instability of coal pillars may be due to the slow decay of the coal, to changes in the groundwater regime in flooded workings, to increased loading on the ground surface, to an increase in the load transferred to pillars due to the collapse of neighbouring areas, or to longwall mining in deeper coal seams. If massive rock strata such as the thick sandstones of the Coal Measures are overlying the partly worked seam they may form a bridge over the cavities such that the collapse of the weak strata forming the roof of the working will not extend above the base of the massive rock stratum (Figure 9.12a). Provided that the coal pillars themselves do not decay, the workings may remain in a stable condition for centuries and it will be quite satisfactory to construct piled foundations overlying them. Again it is desirable to terminate the piles in a soil layer to avoid any concentration of loading at the rockhead.

Where massive rock strata are not present and the overburden consists only of weak and thinly bedded shales, mudstones and sandstone bands overlain by soil, a collapse of the roof will eventually work its way up to the ground surface to form a chimney-like cavity known as a 'crownhole' (Figure 9.12b).

Piling should be avoided above these unstable, or potentially unstable, areas, but if the workings lie at a fairly shallow depth, it is possible to install bored and cast-in-place piles completely through the overburden, terminating them in a stable stratum below the coal seam as shown in Figure 9.12b. The pile shaft must be isolated from the soils and rocks of the overburden in the manner illustrated in Figure 9.11. Any collapse of the strata over pillar and stall workings usually takes place in a vertical direction with little lateral movement, but nevertheless a generous space (a minimum of 150 mm) should be allowed between the pile shaft and the walls of the drill hole. As noted above, it is necessary to seal the annulus around the piles to prevent gas seepage to the surface. Large-diameter piles are preferable to small sections because of their higher resistance to lateral loading that may be due to local distortions of the rock strata. The large-diameter drill holes also serve as a means of access for inspecting any cavities and to enable geologists to judge the stability conditions of the overburden.

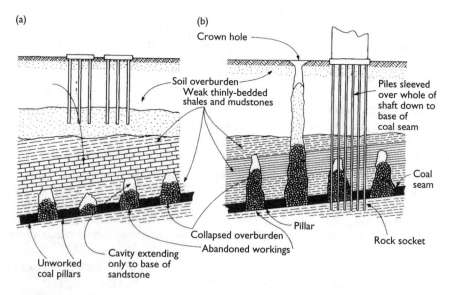

Figure 9.12 Piling in areas of abandoned 'pillar and stall' mine workings (a) Where massive rock forms stable roof over workings (b) Where roof over workings is weak and unstable.

When inspecting the geological conditions in the shafts full precautions should be taken against the collapse of the walls or base of the drill hole and against the presence of explosive or asphyxiating gases (see Section 11.3.3). The advice of an experienced mining engineer should be sought before exploring abandoned workings from the pile borehole.

9.4   Piling in frozen ground

9.4.1   General effects

In most parts of the UK the depth of penetration of frost into the ground does not exceed 600 mm, and consequently frozen soil conditions are not detrimental to piled foundations. However, in countries lying in the northern latitudes with continental-type climates the penetration of frost below the surface gives rise to considerable problems in piling work. In the southern regions of Canada and in Norway the frost penetrates to depths of 1.2 to 2.1 m. In far-northern latitudes the ground is underlain by great depths of permanently frozen soil known as 'permafrost'. About 49% of the land mass of the former USSR is a permafrost region, which generally lies north of latitude 50°. The depth of permafrost extends to 1.5 km in some areas. Permafrost regions are also widespread in Northern Canada, Alaska, and Greenland.

In areas where frost penetration is limited to a deep surface layer overlying non-frozen soil, the effect on pile foundations is to cause uplift forces on the pile shaft and on the pile caps and ground beams. These effects occur in frost-susceptible soils, i.e. soils which exhibit marked swelling when they become frozen. Frost-susceptible soils include silts, clays, and sand–silt–clay mixtures. Swelling of these soils occurs when water in the pores migrates into layers or lenses and becomes frozen. The increase in volume when the ice lenses form results in a heave of the ground surface. When these soils are frozen onto the shafts of piles, or onto the sides of pile caps and ground beams, the uplift forces tend to lift the foundations. The soil heave also causes uplift forces to develop on the undersides of the pile caps and ground beams. The uplift forces on the sides of the sub-structure are referred to as 'adfreezing' forces and measures to prevent the upward movement of piled foundations are described in the next section of this chapter.

The foundation problems presented by permafrost are much more severe, because of the extreme conditions of instability of this material within the depths affected by piling work. The permanently frozen ground is overlain by an 'active layer' that is subject to seasonal freezing and thawing. In the winter adfreezing occurs on foundations sited within frost-susceptible soils in the active layer. In summer there is rapid and massive collapse of thawing ice lenses in the active zone. Severe freeze-thaw conditions in highly frost-susceptible soils can result in the formation of dome-shaped ice-caverns as much as 6 m high above the permafrost. The thickness of the active layer is not constant, but varies with cyclic changes in the climate of the region, with changes in the cover of vegetation such as mosses and lichens, and with the effects of buildings and roads constructed over the permafrost. The laws governing the physical, chemical and mechanical properties of frozen soil have been reviewed by Anderson and Morgenstern[9.9], and Andersland and Ladanyi[9.10] provide extensive soil mechanics data for frozen ground conditions with worked examples of a variety of foundation support systems.

Tsytovich[9.11] has described three modes of formation of permafrost: these are when water-bearing soils are frozen through, when ice and snow are buried, and when ice is formed in layers in the soil. The latter formation, referred to by Tsytovich as 'recurrent vein ice', frequently occupies some 50% of the top 20 m of the soil in the northern parts of Russia and is responsible for the severe foundation problems in the region. The recurrent vein ice can contain layers of unfrozen water within the permanently frozen soil and foundation pressure applied to such ground can result in substantial settlement. Because of the variation in thickness of the active layer the upper zone of the permafrost can undergo considerable changes such as major heaving, the collapse of ice caverns, and the migration of unfrozen water.

9.4.2   The effects of adfreezing on piled foundations

Penner and Irwin[9.12] measured the uplift forces caused by adfreezing on 89 mm steel pipe anchored into unfrozen soil. The measurements were made in the Leda clay of Ontario in a region where a deep penetration of frost occurs below the ground surface. The Leda clay consists of a 70% clay fraction and a 30% silt fraction. The formation of ice lenses in the soil caused a surface heave of 75 to 100 mm where the frost penetrated to a depth of 1.2 m. The adfreezing force on the steel pipe was 96 kN/m^2.

Further measurements in the Leda clay have been reported by Penner and Gold[9.13]. When the frost penetrated to a depth of 1.09 m, causing a surface heave of 100 mm, the measured peak adfreezing forces on anchored columns were as follows:

Steel: 113 kN/m^2
Concrete: 134 kN/m^2
Timber: 86 kN/m^2

Penner and Irwin[9.12] quote similar measurements by Kinoshita and Ono[9.14] as follows:

Iron pipe: 204 kN/m^2
Vinyl pipe: 193 kN/m^2
Concrete pipe: 134 kN/m^2
Epoxy-resin coated concrete pipe: 600 kN/m^2

Grain size can affect adfreeze strength, with clays and coarse soils, particularly those with low moisture content, being lower than sandy soils.

Dalmatov[9.15] (as quoted by Andersland and Ladanyi[9.10]) expressed adfreezing forces by the equation:

$$F = Lh_a(c - 0.5b\, T_m)$$
(9.1)

where F is the total upward force due to frost heave (kgf), L is the perimeter of the foundation in contact with the soil in centimetres, h_a is the thickness of the frozen zone in centimetres, T_m is the surface temperature (°C), and b and c are constants determined experimentally.

Dalmatov's values of b and c for timber piles in a silty sandy clay were 0.1 and 0.4 respectively. The measured forces on the steel pipe in Leda clay when expressed by Dalmatov's equation gave the same values of b and c but Penner and Irwin[9.12] regard this as somewhat coincidental.

Measurements of the magnitude of frost heave forces on steel plates restrained from uplift in Leda clay were made by Penner[9.16], the plates being placed at various depths in the frozen layer. When the surface heave was 55 mm the maximum swelling pressures of 54 to 91 kN/m^2 occurred at the base of the frozen soil layer.

Andersland and Ladanyi[9.10] offer a more comprehensive theoretical solution to the heave rate and mobilized adfreeze stress taking account of climate data, soil thermal properties, frost penetration and creep which compares well with the observations by Penner and Gold and the other field researchers. Design for frost heave must ensure that uplift forces are not sufficient to cause movement of the structure and that the adfreeze bond is not ruptured causing increased rate of uplift in the permafrost zone.

Adfreezing forces on the shafts of piles and on the sides of pile caps and ground beams can be eliminated or greatly reduced by removing frost-susceptible soils from around these substructures to a depth equal to the maximum penetration of frost predicted. These soils are replaced by suitable non-susceptible material such as clean sandy gravel or crushed and graded rock. Open gravels should not be used since groundwater movements at periods of thawing might wash fine soil particles into the voids in the gravel, leading to the formation of a silty gravel susceptible to frost heave. Bond breakers such as grease or polyethylene wrap on pile surfaces or an oil–wax mix in the annulus in the active zone can effectively reduce uplift forces.

Uplift forces on the undersides of pile caps and ground beams can be reduced by interposing a layer of compressible material between the substructure and the soil. Cellular cardboard or low-density expanded polystyrene can be used for this purpose as shown in Figure 7.15.

Instantaneous deformation and time-dependent deformation (creep) will occur in frozen soil under load due to the breaking of ice bonds and the melting of a film of water around the soil particles under compression. The degree of deformation is complex and will depend on the stress applied, soil type and temperature. Piles in ice-rich frozen soil can be expected to creep at a steady rate at stresses below the adfreeze strength.

9.4.3   Piling in permafrost regions

Piled foundations are generally employed where structures in permafrost regions are sited in areas of frost-susceptible soils. Shallow foundations cannot normally be used because of the massive volume changes which take place in the active layer under the influence of seasonal freezing and thawing.

The general principle to be adopted when designing piled foundations is to anchor the piles securely into a zone of stable permafrost (which can be difficult to locate) or into non-susceptible material such as well-drained sandy gravel or relatively intact bedrock. Where the piles are anchored into the permafrost layers their stability must be maintained by conserving as far as possible the natural regime which existed before construction was commenced in the area. Thus buildings must be supported well clear of the ground (Figure 9.13) to allow winds at sub-zero temperatures to remove the heat from beneath the buildings, and so prevent thawing of the active layer in the winter season.

The depth to which piles should be taken into the permafrost depends on the state of stability of this zone. Consideration must be given to the recurrence of cyclic changes in the upper layers, to the presence of layers of unfrozen water, and to the pre-treatment which can be given to the permafrost by thawing, compaction of the soil, and re-freezing.

Compressive loads on the piles are carried almost entirely by adfreezing forces on the pile shaft in the permanently frozen zone. Little end-bearing resistance is offered by the frozen

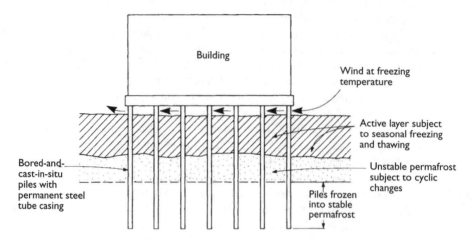

Figure 9.13 Piling into permafrost.

ground due to the re-packing and re-crystallization of ice under pressure and the migration of unfrozen water. Uplift forces on the piles which occur as a result of adfreezing in the active layer in the winter season must be allowed for. Results of tests on laterally loaded piles in permafrost and computer simulations of the displacements are described by Foriero *et al.*[9.17]

Generally, it is not recommended to drive piles into permafrost at temperatures less than −5°C since this will cause splitting of the frozen ground, allowing thawing waters to penetrate deeply into the cracks, and so upsetting the stable regime. However, reinforced H-piles and tubular steel piles with wall thickness greater than 12.5 mm can be driven into relatively warm permafrost (−1°C to −5°C) using vibratory hammers without pilot holes. In a research project, Canadian Petroleum Engineering Inc has driven 500 mm diameter steel piles 53 m into permafrost at −7°C using a high frequency pile driver. Generally, adfreeze occurs earlier in driven piles, but driving resistance should not be used to calculate long-term capacity of piles in permafrost. Driving can be easier in saline permafrost in fine and coarse soils, because of the greater quantities of unfrozen moisture around the pile; however, the bond is reduced due to the saline porewater pressure, reducing the pile capacity by as much as 50% with salt concentrations of 10 g/litre. Enlarged base piles in ice-rich permafrost formed using jet cutting are considered by Sego *et al.*[9.18] While improvements in end-bearing capacity of 30% to 40% are claimed as a result of the belling, this is from a low base value and when the costs of the high-alumina grout used to fill the bell are considered, the benefits of the belling are marginal.

Drilled and cast-in-place piles are suitable but the concrete must not be placed in direct contact with the frozen ground. North American practice is to use powered rotary augers to drill into the permafrost to the required depth, but wear on bits will be high in silts and sands. A permanent steel casing is then placed in the drill hole and filled with concrete. The heat of hydration thaws the surrounding ground and as the concrete cools the freezing of the melt water bonds the pile permanently to the permafrost. Water or a sand slurry can be poured into the annulus between the casing and the permafrost to ensure full bonding, but this may result in high creep. The annulus should be less than 100 mm to ensure adequate natural 'freeze-back' around the pile. Korhonen and Orchino[9.19] point out that freezing of the concrete need not be the sole reason for rejecting the pile: concrete in cast-in-place piles that has frozen after attaining a certain minimum strength and remains frozen will continue to gain strength while frozen. Biggar and Sego[9.20] comment on the use of high-alumina cement-based grout for infilling the annulus at temperatures of −10°C.

Timber piles installed in pre-drilled holes or driven in conjunction with steam jetting have been used for many years in northern Canada. Timber piles will generally remain well preserved in permafrost, but must be protected against deterioration in the active zone.

In Russia a form of reverse-circulation drilling with steam or gas burners is used to thaw the soil and flush the cuttings to the surface before freezing-in the permanent steel casing. Holes should be provided in the casing to drain off the water before the concrete is placed. If this is not done freezing of the water inside the casing may cause bursting of the steel. Precast concrete piles driven into pre-formed holes are extensively used in Siberia, but low concrete tensile strength may result in cracking during frost heave. Prestressed concrete piles perform better.

'Thermal piles' are piles on which natural convection or forced circulation cooling systems have been installed to retain the adfreeze during warmer weather and lower the pile surface temperature during the autumn. Johnson[9.21] describes two types of thermal piles, including piles used to support the above-ground section of the Trans-Alaska oil pipeline.

9.5   Piled foundations for bridges on land

9.5.1   Selection of pile type

Bridge construction for developed countries is subject to many constraints concerned with access to sites and environmental conditions. These have an important influence on the selection of a suitable pile type and equipment for installation. In undeveloped territories the constraints are fewer and selection of suitable pile types is influenced mainly by the ground conditions.

When constructing new main highways it is desirable to complete under- and over-bridges at an early stage in the overall construction programme in order to facilitate the operation of earthmoving and paving equipment along the length of the highway without the need for detours or the use of existing public highways by construction equipment. Hence, access to bridges will be difficult at this early stage and it may be impossible to route the piling equipment and material deliveries along the cleared highway alignment without interfering with the early earthmoving operations.

In the case of small bridges, such as those carrying minor roads over or beneath the main highway, it is desirable to use light and easily transportable equipment to install a number of small- or medium-diameter piles rather than a few large-diameter piles requiring heavy equipment. Suitable types are precast concrete or steel sections which have the advantage over bored piles of the facility to drive them on the rake thus providing efficient resistance to lateral forces, which are an important consideration in most bridge structures. Only small angles of rake are feasible with bored piles (see Section 3.4.11) and it is usually preferable to provide only vertical bored piles suitably reinforced to resist horizontal loads and bending moments.

Some of the most difficult access problems are involved with bridges in deep cuttings where the bridge is constructed in an isolated excavation in advance of the main earthmoving operations. It is possible to install the piling for piers of bridges with spill-through abutments at the toe of the cutting, and in the median strip from plant operating from ground level before bulk excavation is commenced for the bridge. This operation is similar to that adopted for 'top-down' basement construction in building work (see Section 5.9). It may also be possible to excavate the cutting to a temporary steep slope to enable piles to be driven at the toe of a cutting using trestle guides (Figure 3.6), the piles being pitched by a crane standing at the crest of the cutting. However, such operations involve a risk of instability of the slope due to surcharge load, and, in the case of clay slopes, to excess pore pressures caused by soil displacement.

Bridge construction, or reconstruction, in urban areas involves piling in severely restricted sites with the likely imposition of noise abatement regulations. Driven types of pile have the advantages of speed and simplicity. Compliance with noise regulations may be possible by adopting a bottom-driven type (see Sections 2.3.2 and 3.2) in conjunction with sound absorbant screens surrounding the piling equipment. If possible pile caps should be located above ground-water level in order to avoid pumping from excavations which could cause loss of ground or settlement of adjacent buildings due to general drawdown of the groundwater table.

Piling over or beneath railways involves special difficulties. The presence of overhead electrification cables will probably rule out any form of bored or driven pile requiring the use of equipment with a tall mast or leaders. The railway authority will insist on piling operations being limited to restricted periods of track possession by the contractor if there is any risk of equipment or materials falling on to the track. Soil disturbance by large-displacement-driven piles may cause heave or misalignment of the rails. If it is at all possible the design of the bridge should avoid the need for piling the foundations.

As noted above, many of the constraints described in the preceding paragraphs do not apply to bridges in undeveloped territories. However, conditions of access to remote bridge sites should be investigated. Equipment should be capable of being transported over poor roads and across weak bridges of limited width.

9.5.2   Imposed loads on bridge piling

The various types of loading imposed on bridge foundations have been reviewed by Hambly[9.22] in a wide-ranging report published by the Building Research Establishment:

- Dead and live loads on superstructure
- Dead load of superstructure
- Earth pressure (including surcharge pressure) on abutments
- Creep and shrinkage of superstructure
- Temperature variations in superstructure
- Traffic impact and braking forces on bridge deck (longitudinal and transverse)
- Wind and earthquake forces on superstructure
- Impact from vehicle collisions, locomotives and rail wagons
- Construction loads including falsework.

In UK practice the loading requirements are specified in British Standard 5400, and the relevant parts of Actions on Structures as given in Eurocode 1 (BS EN 1991: 2003). The Highways Agency's Design Manual for Roads and Bridges gives guidance on the use of BS 5400.

Dead and live load combinations should be considered in relation to permissible differential settlements between piers or between piers and abutments in longitudinal and transverse directions. Permissible settlements are often poorly defined or not defined at all by bridge designers. Hambly[9.22] states that foundations for simply supported deck bridges are frequently designed for differential settlements of up to 1 in 800 relative rotation (25 mm in a 20 m span). In reasonably homogeneous soils, differential settlements between adjacent foundations are often assumed to be half of the total settlement, thus a total settlement of 50 mm would be permissible under this criterion. Differential settlements of the order of 1 in 800 in a continuous deck bridge are required to be treated as a load producing bending moments in the superstructure. This can add to the cost of the bridge, but it should also be noted that limitation of total settlement to 5 to 10 mm is difficult to achieve with spread foundations on soils of moderate to low compressibility. Some designers expect the rotation to be limited to 1 in 4000, which is equivalent to a differential settlement of only 5 mm in a 20 m span bridge. This would be difficult to ensure for bridges with longer spans even when supported by piles taken down to a competent bearing stratum. Larger rotations have to be anticipated in special conditions such as bridges in mining subsidence areas.

The distribution of live load when assessing total and differential settlement is usually a matter of judgement. Full live load on the whole or part of the spans should be allowed for calculating immediate settlements but the contribution of live load to consolidation settlement may be small in relation to that from the dead loading. Figure 9.14 shows the loading on a typical pier foundation for the 4 km-long elevated section of the Jeddah–Mecca Expressway designed by Dar al-Handasah, consulting engineers. The piers support the 36 m continuous spans of the three-lane carriageway. It will be noted that the predominant horizontal force on the piers was in a longitudinal direction, the resulting bending moments increasing the loads on the outer piles of the eight-pile group by about 25% above the combined vertical dead and live loads. It was possible to carry the horizontal forces and bending moments by 770 mm diameter bored and cast-in-place base-grouted piles of the type described in 3.3.9 using the 'flat-jack' process.

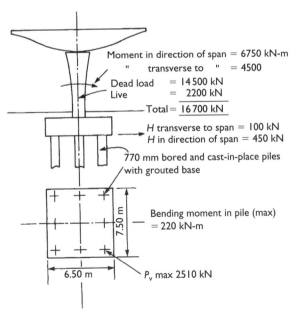

Moment in direction of span = 6750 kN-m
"   transverse to   "   = 4500
Dead load   = 14500 kN
Live   = 2200 kN
Total = 16700 kN

H transverse to span = 100 kN
H in direction of span = 450 kN

770 mm bored and cast-in-place piles with grouted base

Bending moment in pile (max) = 220 kN-m

7.50 m

6.50 m

P_v max 2510 kN

Figure 9.14 Vertical and horizontal loads on viaduct piers of Jeddah–Mecca Expressway.

Horizontal earth and surcharge pressures on bridge abutments and wing walls are resisted much more efficiently by raking piles than by vertical piles. However, rakers provide a high degree of rigidity to the foundation in a horizontal direction which may require designing for at-rest earth pressures rather than the lower active pressures which depend on permitted yielding of the retaining structure. At-rest pressures are likely to be operating when the top of the abutment is strutted by the bridge deck as well as being restrained at the toe by the rows of raking piles.

The simple and computer-based methods of determining individual pile loads in groups of vertical and raking piles carrying a combination of vertical and lateral loads were described in Section 6.5. Hambly[9.22] has pointed out the desirability of varying the angle of rake in order to avoid concentration of load on the bearing stratum, although some designers consider that because of 'buildability' considerations, raking piles should only be used when absolutely necessary. The choice of bored or driven piles for combined vertical and raked pile groups should take account of the need to install casing and reinforcement and place concrete in a raking bored pile compared to achieving the designed set with a reduced efficiency when driving on a rake.

In the case of bridges with spill-through abutments and embanked approaches the piles supporting the bank seats are best installed from the surface of the completed embankment (Figure 9.15a). In this way the drag-down forces from the settling embankment and any underlying compressible soils are carried preferably by vertical piles. The drag-down force can be minimized by using slender sections in high strength materials. If the piles are constructed at ground level with the bank seat supported on columns erected on a pile cap, the latter will act as a 'hard-spot' attracting load from the embankment fill (Figure 9.15b). Unless precautions are taken the higher loading on the piles supporting the low level pile cap will result in greater tendency for them to settle relatively to the piles supporting the adjacent bridge pier with consequent differential movement in the bridge deck.

Vertical piles as shown in Figure 9.15b are preferable to rakers for supporting bridge abutments constructed on ground underlain by a soft deformable layer, whether or not the abutments are of the spill-through type or in the form of vertical retaining walls and inclined

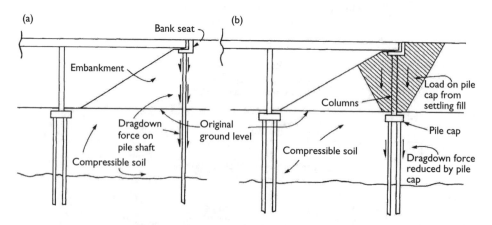

Figure 9.15 Piling for bridges with spill-through abutments (a) Bank seat carried by piles driven from completed embankment (b) Bank seat carried by columns with pile cap at original ground level.

wing walls. The abutment is only partially restrained from moving forward under the influence of the retained soil. A small degree of restraint is provided at the top of the wall by friction or rotation in the bearings supporting the deck structure. At pile cap level higher restraint is provided by the stiffness of the supporting piles, but the amount of forward movement should, theoretically, result in earth pressure on the back of the abutment corresponding to the 'active' state. However, heavy compaction of the embankment filling is required to prevent settlement of the road surface, such that the earth pressure, particularly near the top of the wall, can be higher than the 'at rest' (K_0) condition. The UK Department of Transport[9.23] specifies that the abutment retaining walls should be designed to resist a pressure of $1.5K_0$ at the Ultimate Limit State.

Raking piles to support abutments should be avoided if at all possible because rigidity at pile cap level could result in earth pressures at low levels approaching 'passive' conditions. If, because of ground or loading conditions, the use of rakers is unavoidable, the angle of rake should be varied as shown in Figure 9.16 to prevent a high proportion of load being carried by a single row of rakers when used in combination with vertical piles on the rearward side of the foundation.

Bending moments and deflections in rows of vertical piles caused by earth pressure on the abutment can be calculated by the methods described in Sections 6.3 to 6.5. Where the abutment is underlain by a weak deformable layer such as soft clay, horizontal and vertical movements take place in the soft layer under the loading of the embankment. The vertical movements are restrained if there is a stiff underlying layer, but the only restraint to

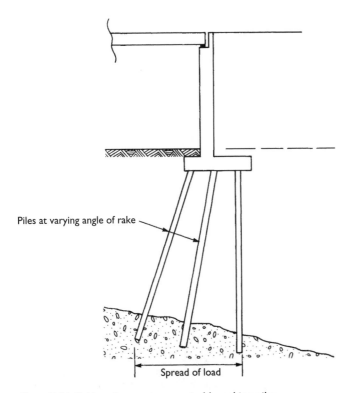

Figure 9.16 Bridge abutment supported by raking piles.

horizontal movement is shear resistance between the soft clay and the underside of the pile cap and at the interface between the soft clay and the stiff layer. As the embankment loading increases plastic deformation occurs in the soft clay which flows horizontally away from the abutment. In effect, the clay layer is extruded between the piles accompanied by horizontal pressure on the 'upstream' face of the piles and an upward pressure on the underside of the pile cap. The horizontal pressure is low at pile cap level because the pile and soil are moving together. It is also low at the interface with the stiff layer because the pile movement at this level is relatively small and the stiff layer is also moving forward as a result of shear stress on it from the soft clay.

Springman and Bolton[9.24] undertook research on behalf of the Department of Transport into the behaviour of a single vertical free-head pile subjected to one-sided surcharge pressure caused by placing fill on a weak deformable layer underlain by a stiffer but yielding stratum. They used finite element analyses confirmed by centrifuge modelling to reproduce the pressure distribution on the pile and the resulting deflections. These are shown diagrammatically in Figure 9.17. The research was subsequently extended by Springman et al.[9.25] to deal with the case of a full-height bridge abutment supported by two rows of vertical piles driven through a soft clay layer into a dense sand stratum (Figure 9.18). Two loading cases were considered: (a) rapid placing of the embankment fill at a rate equivalent to a one-day construction period in the prototype, and (b) slow placing of the embankment over a 21-day period. Vertical drains beneath the embankment were also incorporated in the model. These extended from a drainage layer beneath the embankment into the sand stratum.

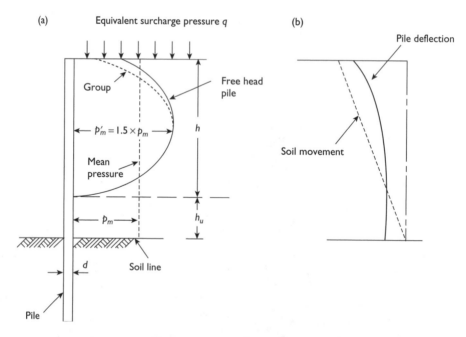

Figure 9.17 The effect of asymmetrical surcharge loading on a vertical pile with a free-head pile driven through soft deformable soil on to a stiff layer (a) Lateral pressure distribution (b) Relative soil/pile movement (after Springman and Bolton[9.24]). Crown copyright 1995. Reproduced by permission of HM Stationery Office.

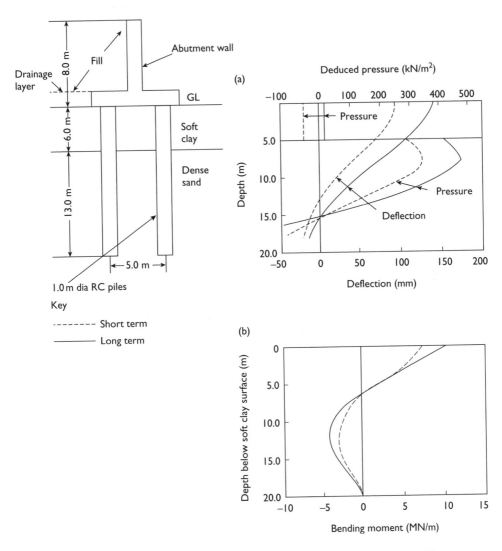

Figure 9.18 Lateral pressure distribution on full height bridge abutment supported by two rows of piles driven through soft clay into dense sand (a) deflection (b) bending moments (after Springman et al.[(9.25)]). Crown copyright 1995. Reproduced by permission of HM Stationery Office.

Measurements made in the centrifuge model of pile head deflections, bending moments, and deduced horizontal soil pressures on the surfaces of the central pile of the three-pile row furthest from the embankment are shown in Figure 9.18. They show a marked difference in the magnitude of deflection and pressure between the short-term (end of construction) and long-term (125 weeks after end of construction) loading periods. The pressure on the pile surface within the soft clay was negative at the end of construction as a result of the large deflections causing the pile to pull away from the soil. With time the clay closed-up against the pile causing a small positive pressure to develop. Generally, the measurements on the model piles showed increases in maximum bending moments over the 125 week (prototype) loading period of 30% for the rear (furthest from the embankment) and 15% for the front row of piles.

The research studies of Springman and Bolton, and Springman *et al.* were based on considerations of the undrained shear strength and undrained shear modulus of the soft clay. These undrained soil parameters formed the basis, in the earlier work, of a computer program given the name SIMPLE[9.26] from which the distribution of lateral pressures, bending moments and deflections of over the full length of piles subjected to similar loading conditions can be determined. The general pattern of the effects on piles as predicted by the SIMPLE program was to some extent confirmed by full-scale measurements on piles supporting the south abutment of the Wiggenhall Overbridge near Kings Lynn in Norfolk[9.27]. Four rows of 11 piles were driven through 10.4 m of weak Marine Alluvium to a depth of about 5 m into a very stiff Kimmeridge Clay. Horizontal soil movements were recorded by a tube close to the instrumented pile. Embankment filling was undertaken over a period of about 5 months after completing pile driving.

The instrumented ground tube showed that flow of the soft clayey alluvium had occurred between the piles over the period of the embankment placing with lateral movements at pile cap level of 29 mm in the soil and 19 mm in the pile. The SIMPLE program had over-predicted both the pile deflections and bending moments. The instrumented pile head had deflected about 19 mm compared with the prediction of 60 mm. However, it should be pointed out that the design of the bridge as a spill-through structure with fill on both sides of the abutment and the multiple rows of supporting piles did not correspond to the arrangements studied in the laboratory research.

Generally, before commencing detailed design studies into the behaviour of piles subjected to the loading conditions described above it is desirable to consider the maximum lateral pressure which can be applied to the piles within the soft layer. This was shown by Randolph and Houlsby[9.28] to correspond to $9.14c_u$ for a perfectly smooth pile and $11.94c_u$ for a perfectly rough pile. When the pressure on the leading face of a pile reaches these values the clay flows past them and cannot exert any higher pressure. Hence, if it can be shown that the pile section, as designed to resist vertical and horizontal forces on the abutment or bank seat, has an adequate safety factor against failure or excessive deflection when subjected to additional forces caused by soil movements applied directly to the supporting pile shafts, then further detailed design work may be judged unnecessary.

Springman and Bolton[9.24] recommended that the embankment–pile–soil system should be designed to ensure that the ratio of the mean horizontal soil pressure (p_m) to the undrained shear strength (c_u) should lie within the pseudo-elastic zone shown in the interaction diagram (Figure 9.19). In this diagram the ratio p_m/c_u is plotted as the ordinate with an upper limit of 10.5. At this stage the clay flows plastically around the pile (it was noted above that Randolph and Houlsby put this limit between 9.14 and 11.94 depending on the surface roughness of the pile). The ratio of the embankment surcharge pressure (q) to c_u is plotted as the abscissa, for which the limit is given by $q = (2 + \pi)c_u$ which represents the stage of plastic yielding of soil beneath the embankment. To avoid excessive deformation of the embankment causing soil to flow between the piles supporting the abutment, there should be a safety factor of at least 1.5 against base failure. Elastic behaviour of the system is defined by the limits of h/d in Figure 9.19 between 4 and 10. The height h, the pile diameter d,

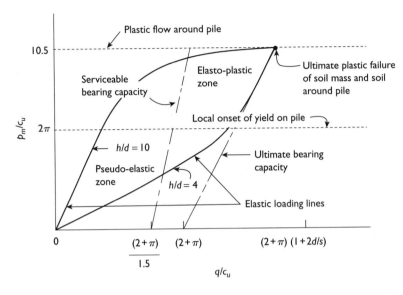

Figure 9.19 Interaction diagram for horizontal soil pressure on vertical pile driven through soft clay into an underlying stiff stratum (after Springman and Bolton[9.24]). Crown copyright 1995. Reproduced by permission of HM Stationery Office.

and the mean horizontal pressure p_m are shown in Figure 9.17. Methods of calculating p_m and the height h_u are described by Springman and Bolton.

When calculating lateral forces on the piles for a range of values of c_u the higher values should be used to obtain the bending moments and pile deflections, and the lower values for assessing the stability of the embankment. It is also important to ensure that the side slopes of the embankment have an adequate safety factor against rotational shear failure.

De Beer and Wallays [9.29] have established an empirical method of calculating the lateral pressure on vertical piles due to unsymmetrical surcharge loading. The surcharge is represented by a fictitious fill of height H_f with a sloping front face, as shown for three arrangements of piles and embankment loading in Figure 9.20a to c. The height H_f is given by

$$H_f = H \frac{\gamma}{1.8} \tag{9.2}$$

where γ is the density of the fill in tonne/m³.

The fictitious fill is assumed to slope at an angle α which is drawn by one of the methods shown in Figure 9.20a to c, depending on the location of the surcharge loading in relation to the piles. The lateral pressure on the piles is then given by

$$p_z = fp \tag{9.3}$$

where f is a reduction factor given by

$$f = \frac{\alpha - 0.5\phi'}{90° - 0.5\phi'} \tag{9.4}$$

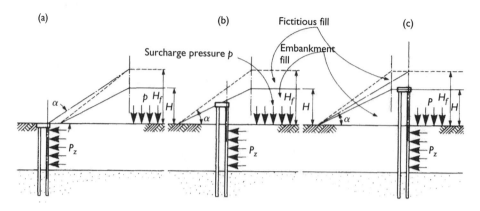

Figure 9.20 Calculation of lateral pressure on vertical piles due to unsymmetrical surcharge loading (after De Beer and Wallays[9.29]).

p is the surcharge pressure, and ϕ' is the effective angle of shearing resistance of the soil applying pressure to the pile.

It should be noted that when $\alpha \leq 0.5\phi'$ the lateral pressure becomes negligible. De Beer and Wallays point out that the method is very approximate. It should not be used to obtain the variation in bending moments along the pile shaft but only to obtain the maximum moment. They also make the important point that the calculation method cannot be used if the safety factor for conditions of overall stability of the surcharge load is less than 1.6. It has the advantage of being based on drained soil conditions.

Driving piles within or close to the toe of clay slopes can result in the development of excess pore pressure which may cause slipping of the slope. Massarsch and Broms[9.30] have developed a method of predicting the excess pore pressures induced by the soil displacement.

It is very difficult to avoid relative settlement between a piled bridge abutment and the fill material forming an embanked approach behind the abutment. Settlement of the fill often occurs even when well-compacted granular material is used. Relative settlement can be large where the embankment is placed on a compressible clay. The concept of allowing piles to yield under load was adopted by Reid and Buchanan[9.31] for the purpose of reducing the relative settlement of a piled bridge abutment and the approach embankment which was founded on soft compressible clay. The arrangement of piles is shown in Figure 9.21. The piles beneath the embankment close to the abutment were at close-spacing and were designed to carry the whole of the embankment load with a safety factor of 2. After the first four rows the spacing was increased to a 3 to 4 m grid and the piles were made successively shorter so that they would yield under a progressively increasing proportion of the embankment load. The piles had circular caps 1.1 to 1.5 m diameter. Loading from the embankment was distributed to the pile heads by a flexible membrane consisting of two layers of Terram plastics fabric reinforced with Paraweb strapping. If piles are used to support a bridge approach slab, the embankment design and construction and the subsoil conditions will affect the drag-down load on the piles.

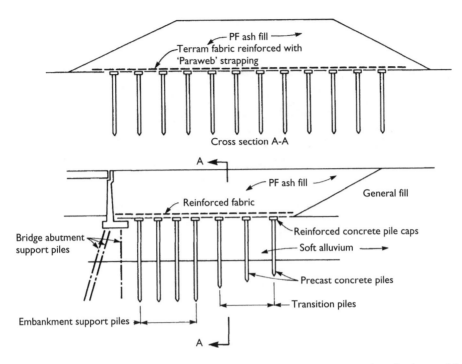

Figure 9.21 Arrangement of settlement reducing piles beneath bridge approach embankment (after Reid and Buchanan [9.31]).

9.6   Piled foundations for over-water bridges

9.6.1   Selection of pile type

Because of the desirability of avoiding different types of piling on the same bridge project the piling used for piers constructed in over-water locations will usually dictate the type to be used for the abutments. Driven piles are the favoured type for over-water piers. The installation of bored piles is limited to work carried out either in a pumped-out cofferdam, or in a permanent casing driven below river bed. In fast-flowing rivers the casing will have to be taken down to a sufficient depth below the river bed to obtain fixity against overturning particularly in conditions of bed scour. Tubular steel piles or precast concrete piles of cylindrical section are preferred to H-sections in order to minimize current drag and eddies causing bed scour. The need for raked piles for efficient resistance of lateral forces again favours a driven type of pile. Where precast prestressed cylindrical piles are used in deep-water locations or for deep penetrations below bed level there can be problems with handling long heavy piles. Also, forming joints to extend partly driven piles can cause difficulties and delays.

Attrition by soil particles of the exterior surface of piles at the sea or river bed can be a factor influencing the material of the pile and its wall thickness. This is more likely to be a problem where the bed level is constant or changing over a limited range rather than rivers where seasonal floods cause wide variations in bed contours.

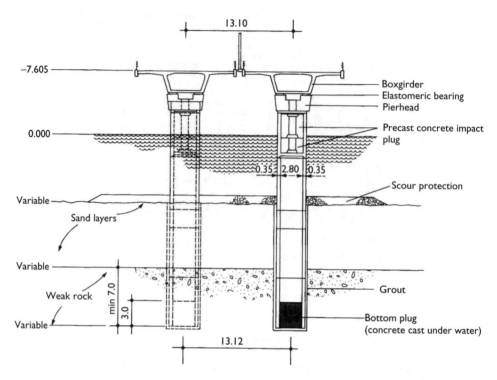

Figure 9.22 Cylinder pile foundations for the Saudi Arabia–Bahrain Causeway (after Beestra et al.[9.32] Courtesy of Ballast Nedam).

A notable example of precast concrete piling for bridge works is the over-water sections of the 25 km causeway between Saudi Arabia and Bahrain Island.[9.32] The bridge sections of the causeway form a total length of 12.5 km and were constructed in water depths ranging from 5 to 12 m. A single 3.50 m OD × 0.35 m wall thickness precast concrete cylinder supports the 50 m span box girder carrying the two-lane carriageway of the dual carriageway bridge (Figure 9.22). The cylinders were cast vertically in short sections at the shore-based casting yard. The sections were then formed into complete piles by longitudinal prestressing and transported to the bridge locations by a 1000 tonne crane barge. The crane lifted the piles into the guides of the twin drilling rigs mounted on a jack-up platform (Figure 9.23). Reverse-circulation drilling using a Wirth rotary table was used to take the piles down to the required penetration below sea bed where the base was plugged with concrete.

The foundations for the cable-stayed Sutong Bridge[9.33] over the lower Yangtze River had to deal with water depths of 30 m with maximum flow rates of 3 m/s and layers of silty sands and silty clays extending up to 270 m below river level to bedrock. Drilled shafts 131No 2.8/2.5 m diameter, with ultimate capacity of 92 MN, support the two main pylon piers constructed on a 13 m deep pile cap. Construction of the shafts was carried out from a steel platform fixed over the pier 3 m above high water, and the 2.8 m casings driven by vibratory hammers at the north pier and diesel hammer at the south pier to depths of around 60 m. Eight rotary drills, using a variety of soft formation drill tools 2.5 m diameter, were used on each

Figure 9.23 Drilling equipment for sinking 3.5 m OD precast concrete cylinder piles, Saudi Arabia–Bahrain Causeway.

platform to extend the shafts to depths of 114 to 117 m using bentonite to maintain hole stability. Reinforcement cages were inserted and a batching plant rated at 100 m³/hour moored downstream of the platform supplied concrete. Post-grouting of the pile tip was carried out using methods similar to those shown in Figure 3.39, increasing pile capacity by 20% as indicated by before and after tests.

9.6.2   Imposed loads on piers of over-water bridges

In addition to the loadings listed in Section 9.5.2 the piles of over-water bridges are required to withstand lateral forces from current drag and wave action, pressure from floating flood debris or ice, and impact from vessels straying from the designated navigation channels.

Current drag and *wave forces* can be calculated using the methods described in Sections 8.1.3 and 8.1.4. The profile of the current velocity with depth varying from a maximum at the water surface to a minimum at bed level must be considered in relation to the bending moments on piles in deep fast-flowing rivers. Current-induced oscillation can also be a problem in these conditions. It is also necessary to calculate the lateral deflections in the direction of the river flow at pile head level because these can induce bending of the bridge superstructure in the horizontal plane.

The depth of *scour* below river bed around piles at times of peak flood must be estimated for the purpose of calculating bending moments due to current drag forces and wave action on piles. The scour consists of three components: (a) general scour from changes in bed levels across the width of the channel, (b) formation of troughs in 'sand waves' which move downstream with the passage of the flood and (c) local scour around the piles. Rip-rap, armouring, cable-tied concrete block mats and grout bag mats are used to protect piers and abutment foundations. Care has to be taken to prevent failure due to 'winnowing' of sediments between the mats and blocks, causing uplift and rolling up of the leading edge of the mat if not anchored. May *et al.*[9.34] review the causes and effects of, and remedies for, scour around bridge piers.

An extreme example of the influence of bed scour on bridge foundations is given by the design of the foundations of the multi-purpose bridge over the Jamuna river near Sirajgang in Bangladesh[4.33,4.34]. The bridge provides a dual two-lane roadway, a metre gauge railway, pylons carrying a power connector and a high-pressure gas pipeline. At the bridge location the river was 15 km wide. The waterway had a braided configuration with numerous deep scour channels and shifting sandbanks. In order to limit the overall length of the bridge the waterway was narrowed by constructing massive armoured training bunds on each bank which reduced the width to 4.8 km. It was calculated that the result of constriction of flow would cause the river bed to scour to a depth of 40 to 45 m below bank level at the time of a 1 in 100 year flood discharging 63 000 m³/sec. An additional 1 m of scour was estimated to occur around the foundation piles.

The bridge structure consists of 52 segmental box girder spans carried on piers, each pier being supported by a pair of raking piles (Figure 9.24). The 3.15 m OD × 40/60 mm wall thickness piles were driven with open ends and have outside diameters of 2.50 and 3.15 m depending on their location relative to the training bunds. The piles were driven to a depth of about 70 m below bank level into a loose becoming medium-dense to dense silty medium to fine sand containing upto 5% of micaceous particles. Support to the piles is provided partly by shaft friction and partly by base resistance. The maximum load in compression on a 3.15 m pile was estimated to be 57.1 MN resulting from the bridge loading combined with current drag forces caused by the 1 in 100 year flood and by earthquake forces. The maximum lateral load on each pile was calculated to be 1.5 MN.

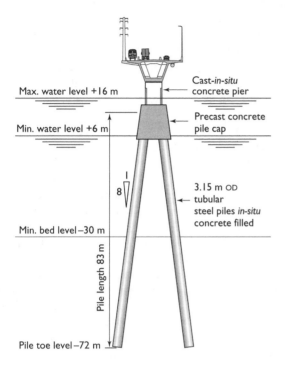

Max. water level +16 m

Cast-*in-situ*
concrete pier

Min. water level +6 m

Precast concrete
pile cap

8

3.15 m OD
tubular
steel piles *in-situ*
concrete filled

Min. bed level –30 m

Pile length 83 m

Pile toe level –72 m

Figure 9.24 Two-pile bent supporting intermediate piers of Jamuna River Bridge, Bangladesh (after Tappin et al.[4.33, 4.34]).

At the time of a major flood more than one-half of the shaft friction available from the soil below river bed level under dry conditions could be lost due to scour. Furthermore, the frictional resistance in the upper part of the piles could be reduced as a result of relief of overburden pressure (see Section 4.3.6). These conditions could not be reproduced at the site of the pre-construction trial piling, nor could loading tests to failure be contemplated on piles with such large diameters. Accordingly, the tests were made on 762 mm tubular piles instrumented to measure the distribution of shaft resistance during driving and test loading. The driving test measurements were analysed by the CAPWAP method (see Section 7.3) to confirm that the hammer selected to drive the piles was adequate for the purpose. This was a Menck 1700T hydraulic hammer with a 102 tonne ram delivering 1700 kJ of energy per blow. The damping constants and other characteristics obtained from the driving tests were used to correlate the dynamic measurements made at the time of driving the permanent piles. The results of the measurements of shaft friction resistance on the trial piles are discussed in Section 4.3.7.

On completion of driving the permanent piles the sand within the shafts was cleaned-out by reverse-circulation drilling to within 3 m of the toe. A grid of tubes-à-manchette was placed on the levelled sand surface, and the pile was filled with concrete followed by grouting with cement through the tubes at a pressure of 50 bar.

Scour protection at the main piers is a major feature of the Sutong Bridge[9.33] where the steel casings for the piles are exposed above river bed level. The initial inner protection zone, extending 20 m around the piles, comprises sand-filled geotextile bags (1.6 × 1.6 × 0.6 m) dumped on the river bed, through which the pile casings were driven. On completion of piling, protection was provided by layers of quarry-run filter and 1 m of rock armour with a density of 2.65 t/m^3. The outer zone, 20 m around the inner, consists of a layer of sandbags topped by a filter layer and 1 m rock armour. A 'falling apron', in which the material in the apron is intended to fall down a scoured slope to produce a stable profile, forms the next variable width zone, set at 1.5 times the expected scour depth, and comprises quarry-run stone overlain by armour with a D_{50} of 0.4 to 0.6 m (Figure 9.25). Dumping of the materials was monitored by echo sounders.

Impact by ships can be a severe problem in the design of bridge support piles in situations where impact cannot be absorbed by massive structures such as caissons or piers constructed inside cofferdams. It is difficult to achieve an economical solution to the problem particularly at deep-water locations. The incidence of random collisions between ships straying from the navigable channel and bridge piers has not decreased since the introduction of shipborne radar. In fact, it may have increased because of the false sense of security given by such equipment.

Three possible methods of protecting piled foundations may be considered. In shallow water not subject to major bed changes and with a small range between high and low water the pile group can be surrounded by an *artificial island* protected against erosion by rockfill. Figure 9.26 shows a cross-section of one of four islands protecting the piers of the Penang Island Bridge[9.35]. The Muroran Bay Bridge in Hokkaido features a 67 m diameter man-made island formed by placing self-setting fly-ash slurry under water on the soft sea bed within a cofferdam. These forms of protection have the added advantage of preventing local scour around the foundations. The island must be large enough to prevent impact between

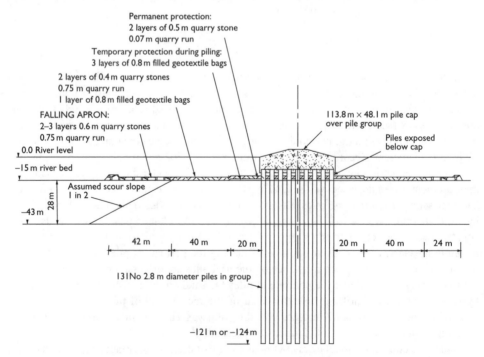

Figure 9.25 Sutong Bridge, scour protection at main pylons (after Bittern et al.[9.33]). Copyright Deep Foundations Institute 2005.

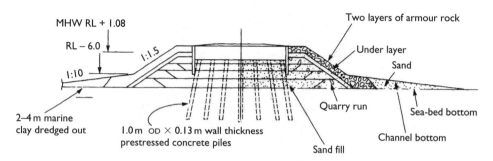

Figure 9.26 Artificial Islands protecting the piers of the Penang Island Bridge Malaysia (after Chin Fung Kee and McCabe[9.35]).

the overhanging bows of a ship and the bridge pier or pile if the vessel should ride up the slope of the island when drifting out of control in a fast-flowing river.

Piles can be strengthened against buckling under direct impact by increasing the wall thickness and a group of piles can be given lateral restraint by a diaphragm connecting them at some point between the cap and bed level. The cylinder piles of the Bahrain Causeway bridge were strengthened by the insertion of precast concrete elements to increase the thickness over the zone of possible impact (Figure 9.22).

Fender piles installed independently of the piers can be installed in deep-water locations. Piles are required to protect the sides of the piers as well as the ends in case of impact at an angle to the axis of the pier. The arrangement of fender piles capped by a massive reinforced concrete ring beam to protect the piers of the Sungei Perak Bridge[9.36] in Malaysia is shown in Figure 9.27. The ring beam was constructed by placing precast concrete trough sections on the piles, sealing the joints between the sections, and placing the reinforcement and concrete infill in dry conditions. The loading on fender piles is calculated in the same way as fender piles for berthing structures (see Section 8.1.1).

9.6.3 Pile caps for over-water bridges

It can be advantageous to locate pile caps at or below low river or low tide level. It avoids floating debris building-up between piles, and ensures that if collision by vessel does occur the impact will be on a massive part of the substructure instead of directly on a pile. Also a vessel is likely to sheer off at the first impact with a pile cap whereas it might become trapped when colliding with a group of piles. A pile cap at or below water level is preferable, aesthetically, to one exposed at low water. However, high-level pile caps are economical for a bridge requiring a high navigation clearance, but such an arrangement would have to be restricted to approach spans in water too shallow to be navigable by vessels which could demolish piles supporting a high-level deck bridging the navigation channel.

Pile caps partly submerged or wholly below water level can be constructed within sheet pile cofferdams (Figure 9.28a). The sheet piles can be cut-off at low water to give protection against scour. Alternatively, if a heavy lifting barge is available a precast concrete cap in the form of an open-topped box can be lowered on to collars welded to the heads of the piles and prevented from floating by clamps. The annulus between the pile wall and the opening in the box can be sealed by quick-setting concrete or by rubber rings. The box is then pumped out and reinforcement and concrete is placed in dry conditions. The concrete seal is used in tidal

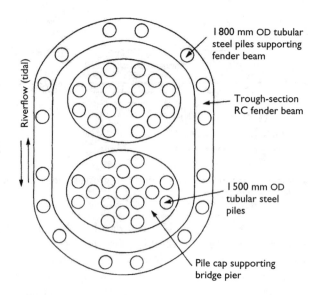

Figure 9.27 Fender beam and piles protecting the river piers of the Sungei Perak Bridge, Malaysia (after Stanley[9.36]).

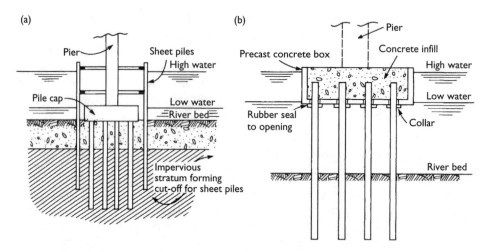

Figure 9.28 Construction of submerged pile caps (a) In cofferdam (b) In open-topped box.

conditions where a sufficient period of time is available for the concrete to set before the bottom of the box is submerged. Arrangements should be made to flood the box to equalize pressures above and below the seal until the concrete has hardened (Figure 9.28b).

Where piers are located in deep water and there is a risk of ship collision it is desirable to construct the pile cap at bed level in order to eliminate any unsupported length of piling. This arrangement is also desirable if lateral forces from earthquakes are transmitted from the bridge superstructure and piers on to the piles. The pier and pile cap can be constructed on shore as a single buoyant unit lowered on to the sea bed followed by driving piles through peripheral skirts in a manner similar to the piled foundations of offshore drilling platforms. Alternatively, the piles can be driven in the form of a raft with their heads projecting above a rock blanket or geotextile mattress. A prefabricated pier unit is then lowered over the pile group and the connection between the two formed by underwater concrete. The availability of heavy-lift cranes on barges or jack-up platforms favours this type of design.

Prefabricated piers were used for 15 of the piers carrying the 3.3 km-long bridges between the islands of Sjaelland and Falster in Denmark.[9.37] A group of forty-nine 700 mm tubular steel piles in two concentric rings supported the deep-water piers in the navigable channel. The first operation was to form a level bed by dredging with protection against scour by rockfill. Then the piles were driven leaving their heads projecting about 6 m above the prepared bed (Figure 9.29). During these operations a precast reinforced concrete conical base unit weighing 440 tonne was being fabricated on shore. It was taken out to the pier site and lowered over the projecting pile heads on to three pinning piles. Concrete was then pumped into place under water to about mid-height of the base unit. The next operation was to lower a temporary circular steel cofferdam on to the top of the base unit to which it was locked by stressed rods. The cofferdam consisted of an assembly of steel plate rings 10 to 11 m in diameter and 3 m deep. The joints between the rings were sealed by rubber sheeting. This was followed by dewatering the cofferdam and constructing the pier, after which the cofferdam was flooded and lifted off the base unit by floating crane for transport to the next pier location.

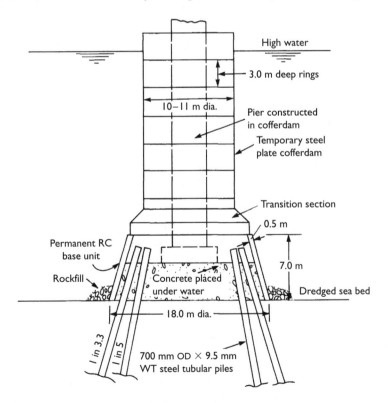

Figure 9.29 Construction of deep-water piers for the Sjaelland–Faro–Falster Bridge, Denmark (after Levesque[9.37]).

9.7   Piled foundations in karst

The design and construction of piles for structures on land underlain by limestone formations which exhibit karst conditions such as wide fissures and solution cavities present several unique challenges. Because variations in rockhead and cavitation can occur over short distances it is difficult to produce an overall geological model of the site to determine if shallow foundations can be used or whether piles can be founded on 'competent rock'. The first requirements are therefore to assess the depth and strength of the overburden, the extent of cavities and the degree of infilling under each foundation by drilling a series of closely spaced probe holes using a combination of rotary percussive rigs capable of installing casing and rotary coring drills. Waltham and Fookes[9.38] give an engineering classification of karst as a means of identifying foundation difficulties but they point out that there is no simple answer to the number of probes which may be required. The probes are usually taken to a depth of at least 3 m below rockhead and any void encountered, or to a similar depth below the anticipated depth of rock socket of each pile. Because of the possibility of vertical faces in the rockhead and cavities it is advisable to include a percentage of raked probe holes in the investigation.

The selection of the pile installation method is critical, as it may be necessary to overcome random boulders in the overburden, remove and replace weak material in cavities through which the pile has to pass, and finally found on competent rock or form a socket in rock, ensuring that sound rock also exists within the bearing zone. Large driven piles are not usually feasible and the most effective method is the drilled and cast-in-place pile, usually with permanent steel casing sealed into the rock at the top of the socket.

The removal of cavity and fissure infill debris and replacement with cement grout to allow uncased holes to be drilled for piles is expensive and rarely achieves the desired results. Flushing/grout holes are required at less than 1 m centres under and around the pile group and flushing water is necessary in quantities greater than 150 l per min and pressure greater than 10 bar – potentially causing pollution of surrounding water courses. If sufficient grout can then be injected it may be possible to place concrete in the open pile hole, or as temporary casing is withdrawn, without loss of fluid concrete. However, for pile diameters up to 1 200 mm rotary-percussive rigs which can simultaneously install permanent casing (duplex drilling) are generally considered the most cost-effective installation method. For larger-diameter piles, the use of a powerful casing oscillator and a drilling method to clean out the pile and form the rock socket are recommended (see Section 3.3.2); above this diameter, shaft or caisson construction techniques may be necessary. Whichever method is used it is essential to probe below the base of the pile to check for cavities. Pile drilling may have to be continued until a sound bearing for the rock socket is located. Jet grouting could be used to consolidate any cavity infill within the bearing zone below the sound rock socket – again high grout pressure and volume (450 bar and 350 l per min) will be required with adequate venting to the surface and pollution control.

Grouting to form a mattress of sound bearing material at rockhead for driven H-piles can be effective in karstic limestone[9.39]. In order to limit the flow of grout away from closely centred grout holes it was necessary in this case to vary the slump depending on injection rate and pressure and to use secondary and tertiary injection holes where grout loss occurred. Verifying the treatment is difficult and reliance has to be placed on detailed grouting records. Specially serrated driving shoes were needed to minimize the slipping of the pile tip along uneven bedrock surfaces to achieve full bearing capacity.

'Micropiles', with working capacity of 890 and 1 160 kN, were used for bridge piers by inserting 245 mm diameter thick wall steel tubes into grout-filled pre-drilled holes as cost-effective alternatives for 1371 mm diameter, steel 'caisson' piles up to 30 m deep, in karstic dolomite[9.40]. 160No micropiles replaced forty caissons at each of three piers. Where the conditions were highly variable with pinnacles, voids and clay-filled solution cavities, the use of 'Tubex' duplex drilling was necessary to install 327 mm temporary casing to depths up to 59 m with a 1 m deep rock socket to insert the permanent pile tube. At other locations where the karstic conditions were less variable, a down-the-hole rotary percussive drill was used to drill 305 mm diameter holes up to 23 m deep without casing to insert the specified 245 mm steel tube – with the assistance of a D5 pile hammer (Figure 9.30). The pile holes in each case were grouted using a tremie pipe, ensuring that grout level was stable at the top of the hole prior to inserting the permanent tube.

Drilling 'slim' holes, with or without simultaneous casing, or driving long H-piles in karstic conditions can cause significant problems due to deviations compromising the axial capacity of the piles. Concreting or grouting open holes or as temporary casing is withdrawn runs a risk of loss of material into weak cavity infill or undetected voids requiring pre-grouting using a low slump mix injected in several stages and re-drilling. Micropile test piles, installed with open hole drilling methods, have failed in karst geology due to contamination of the bond zone in the time between withdrawal of the drill tools and installation of the permanent pile tube and concreting.

Natural overburden and decomposed debris overlying the karst formation can be treated by various ground improvement techniques prior to piling – such as vibroflotation, compaction grouting, and jet grouting. Fischer[9.41] describes the foundations for a nuclear power plant on karst terrain which comprised tubular steel piles driven into relatively flat limestone bedrock,

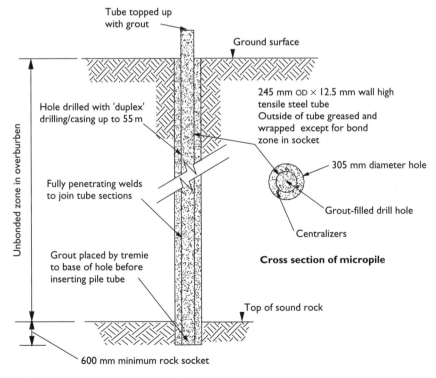

Tube topped up with grout

Ground surface

245 mm OD × 12.5 mm wall high tensile steel tube
Outside of tube greased and wrapped except for bond zone in socket

Hole drilled with 'duplex' drilling/casing up to 55 m

305 mm diameter hole

Unbonded zone in overburden

Fully penetrating welds to join tube sections

Grout-filled drill hole

Centralizers

Grout placed by tremie to base of hole before inserting pile tube

Cross section of micropile

Top of sound rock

600 mm minimum rock socket

Figure 9.30 Micropile in Karst (Uranowski et al.[9.40]).

with a 5 m deep probe at each pile tip to locate cavities. The pile hole was extended by under-reaming where the probes located cavities, the tube re-driven as necessary to sound rock and filled with concrete. The overburden sand, up to 20 m deep, was treated by vibroflotation to improve the relative density to 85% to 90% in order to reduce liquefaction potential.

9.8   Energy piles

Ground temperatures in much of Europe are reasonably constant at 10°C to 15°C (and in the tropics as high as 20°C to 25°C), below a depth of 10 m. This near-surface geothermal energy potential is being exploited to provide a consistent low level, but cost-effective and environ-mentally friendly, source of heating for buildings, using the thermal properties of the building foundations. Concrete has a high thermal storage capacity and good thermal conductivity and heat from the ground taken up by the pile, diaphragm wall or other foundation can be trans-ferred from the concrete to a heat exchanger coil buried within the concrete and moved by a simple heat pump to heat the building. Conversely, in suitable soils the heat from the building can be transferred to the concrete and ground for cooling during the summer. Brandl[9.42] describes the heat transfer mechanisms in the ground and between the absorber fluid in the exchanger pipework and the structural concrete and provides recommendations for the design and operation of geothermal piles and other 'earth-contact' concrete elements.

The geothermal properties of the ground (thermal conductivity and capacity) and groundwater flow and direction have to be determined for the complex heat exchange calculations, but the pile diameter and length should be designed to resist the applied struc-tural loads and not increased to suit the geothermal requirements. The primary circuit within the pile comprises absorber pipes of high density polyethylene plastic, 25 mm diameter and 2–3 mm wall thickness, formed into several closed-end coils or loops and fixed evenly around the inside of a rigid, welded reinforcement cage for the full depth. Typically loops of eight vertical runs would be provided in a 600 mm diameter pile. The geothermal effective-ness of piles less than 300 mm diameter is much reduced due to lower surface area and the limited number of loops which can be fitted; the economically minimum depth of an energy pile is about 6 m. Each loop is filled with the heat transfer fluid, water with antifreeze or saline solution, and fitted with a locking valve and manometer at the top of the pile cage. This may necessitate off-site fabrication. The piling method must produce a stable hole for the careful insertion of the cage and absorber pipework. Bored piles, with or without drilling fluid support, or a cased or withdrawable tube method, are acceptable for most schemes. Before concreting, the absorber pipes are pressurized to around 8 bar for an integrity test and to prevent collapse due to the head of fluid concrete. The pressure has to be maintained until the concrete has hardened and then re-applied before the primary circuit is finally enclosed. Concreting should be by tremie pipe placed to the base of the pile to avoid damaging the pipework. Plunging the absorber pipes, either as separate tubes or attached to the reinforcement cage, into fluid concrete in a CFA pile is not currently recommended.

The primary circuits in each pile are connected via header pipes to manifold blocks which in turn are connected usually through a heat pump to the secondary circuit embedded in the floors and walls of the building. Using a heat pump with a coefficient of performance of 4 (the ratio of the energy downstream of the heat pump to the energy input of the pump), the ground tem-perature can be raised from 10°C–15°C to between 25°C and 35°C at the building. Depending on soil properties and installation depth of the absorbers, Brandl notes that 1 kW heating needs between 20 m^2 of saturated soil and 50 m^2 of dry sand in contact with the pile surface.

The ground temperature around the pile in a heat extraction system using brine will be lowered by around 5°C, and there is no evidence that the shaft resistance and bearing capacity of the pile are affected by the heat transfer process in this case. Also any temperature-induced settlement/heave is likely to be less than the displacements due to the applied loads on the foundations. If excessive heat is extracted using a lower temperature refrigerant as the absorber, temperature around the foundation can drop to near freezing.

The piling technique necessary for the installation of energy piles may not be the most cost-effective for the ground conditions, but depending on energy prices, climatic conditions and the type of building, pay-back for the higher initial cost in terms of energy saving is estimated to be between 2 and 10 years. Development of CFA and vibro-concrete piles to accommodate more robust heat transfer systems could provide more economical energy piles.

9.9 References

9.1 BARKAN, D. D. *Dynamics of bases and foundations* (Translated from the Russian), McCraw-Hill Book Co. 1962.

9.2 HSIEH, T. K. Foundation vibrations, *Proceedings of the Institution of Civil Engineers*, Vol. 22, June 1962, pp. 211–26.

9.3 WHITMAN, R. V. and RICHART, F. E. Design procedures for dynamically loaded foundations, *Proceedings of the American Society of Civil Engineers*, Vol. 93, No. SM6, November 1967, pp. 169–93.

9.4 RICHART, F. E., HALL, F. R., and WOODS, R. D. *Vibrations of Soils and Foundations*, Prentice Hall, New Jersey, 1970.

9.5 IRISH, K. and WALKER, W. P. Foundations for reciprocating machines, *Concrete*, October 1967, pp. 327–37.

9.6 Foundations for Dynamic Equipment. *American Concrete Institute*, Report ACI351.3R-04, 2004.

9.7 LIZZI, F. Pali radice and reticulated pali radice, Micropiling, *Underpinning*, Thorburn, S. and Hutchison, J. N. (eds), Surrey University Press, 1985, pp. 84–161.

9.8 PRICE, D. G., MALKIN, A. B., and KNILL, J. L. Foundations of multi-storey blocks on the Coal Measures with special reference to old mineworkings, *Quarterly Journal of Engineering Geology*, Vol. 1, No. 4, June 1969.

9.9 ANDERSON, D. M. and MORGENSTERN, N. R. Physics, chemistry and mechanics of frozen ground, *Proceedings of the 2nd International Conference on Permafrost*, National Academy of Sciences, Washington DC, 1973.

9.10 ANDERSLAND, O. B. and LADANYI, B. *Frozen Ground Engineering 2nd Edition*. ACSE Press and John Wiley & Sons, New York, 2004.

9.11 TSYTOVICH, N. A. Permafrost in the USSR as foundations for structures, *Proceedings of the 6th International Conference, ISSMFE*, Montreal, Vol. 3, 1965, pp. 155–67.

9.12 PENNER, E. and IRWIN, W. W. Adfreezing of Leda Clay to anchored footing columns, *Canadian Geotechnical Journal*, Vol. 6, No. 3, August 1969, pp. 327–37.

9.13 PENNER, E. and GOLD, L. W. Transfer of heavy forces by adfreezing to columns and foundation walls in frost-susceptible soils, *Canadian Geotechnical Journal*, Vol. 8, No. 4, November 1971, pp. 514–26.

9.14 KINOSHITA, S. and ONO, T. Heavy forces on frozen ground, *Low Temperature Science Laboratory*, Teron, Kagaka, Serial A, 21, 1963, pp. 117–39. (Translated by National Science Library, National Research Council, Ottawa, Translation No. TT 1246, 1966.)

9.15 DALMATOV, B. L. Effect of frost-heaving on foundation of structures, *Gosizid literatuy po streitch stuv i architektur*, Leningrad-Moscow, 1957.

9.16 PENNER, E. Frost heaving forces in Leda Clay, *Canadian Geotechnical Journal*, Vol. 7, No. 1, February 1970, pp. 8–16.

9.17 FORIERO, A., St-LAURENT, N., and LADANYI, B. Laterally loaded pile study in permafrost of Northern Quebec, Canada, American Society of Civil Engineers, *Journal of Cold Regions Engineering*, Vol. 19, No. 3, 2005, pp. 61–84.

9.18 SEGO, D. C., BIGGAR, K. W., and WONG, G. Enlarged base (belled) piles for use in ice or ice-rich permafrost, *ASCE Journal of Cold Regions Engineering*, Vol. 17, No. 2, 2003, pp. 68–88.

9.19 KORHONEN, C. and ORCHINO, S. Effects of low temperature on concrete strength, *ASCE Proceedings of 10th International Conference on Cold Regions Engineering*, 1999, pp. 677–83.

9.20 BIGGAR, K. W. and SEGO, D. C. The curing and strength characteristics of cold setting cement fondu grout, *Proceedings of 5th Canadian Permafrost Conference*, Quebec, 1990, pp. 349–56.

9.21 JOHNSON, G. H. (ed.) *Permafrost: Engineering Design and Construction*, Wiley, New York, 1981.

9.22 HAMBLY, E. C. *Bridge Foundations and Substructures*, Building Research Establishment, HMSO, 1979.

9.23 BD 30/87. *Backfilled retaining walls and bridge abutments, Highways and Traffic Departmental Standard*, Department of Transport, 1987.

9.24 SPRINGMAN, S. M. and BOLTON, M. D. *The effect of surcharge loading adjacent to piles*, Transport and Road Research Laboratory, Contractor Report, 1990.

9.25 SPRINGMAN, S. M., NG, C. W. W., and ELLIS, E. A. *Centrifuge and analytical studies of full height bridge abutment on piled foundation subject to lateral loading,* Transport Research Laboratory, Project Report 98, 1995.

9.26 SPRINGMAN, S. M. *Manual for the SIMPLE – for evaluating the effect of surcharge loading adjacent to piles*, Transport and Road Research Laboratory/Cambridge University, 1992.

9.27 DARLEY, P., CARDER, D. R., and RYLEY, M. D. *Lateral loading on piled foundations at Wiggenhall Overbridge (A47)*, Transport Research Laboratory, TRL Report 246, 1997.

9.28 RANDOLPH, M. F. and HOULSBY, G. T. The limiting pressure on a circular pile loaded laterally in cohesive soil, *Geotechnique*, Vol. 34, No. 4, 1984, pp. 613–23.

9.29 DE BEER, E. and WALLAYS, M. Forces induced by unsymmetrical surcharges on the soil around the pile, *Proceedings of the 5th European Conference on Soil Mechanics and Foundation Engineering*, Madrid, Vol. 1, 1972, pp. 325–32.

9.30 MASSARSCH, K. R. and BROMS, B. B. Pile driving in clay slopes, *Proceedings of the 10th International Conference, ISSMFE*, Stockholm, Vol. 3, 1981, pp. 469–74.

9.31 REID, W. M. and BUCHANAN, N. W. Bridge approach support piling, *Proceedings of the Conference on Piling and Ground Treatment*, Institution of Civil Engineers, London, 1983, pp. 267–74.

9.32 BEETSTRA, G. W., VAN DEN HOONARD, J., and VOGELAAR, L. J. J. Bridges and viaducts, in *The Netherlands Commemorative Volume*, E. H. de Leew (ed.), *Proceedings of the 11th International Conference, ISSMFE*, San Francisco, 1985.

9.33 BITTNER, R. B., ZHANG, X., and JENSEN, O. J. Design and construction of the Sutong bridge foundations, *Deep Foundations Institute, Marine Foundations Speciality Seminar*, New York, 2005, pp. 19–27.

9.34 MAY, R. W. P., ACKERS, J. C., and KIRBY, A. M. Manual on scour at bridges and other hydraulic structures, *CIRIA Report C551*, Construction Industry Research and Information Association, London, 2002.

9.35 CHIN FUNG KEE and McCABE, R. Penang Bridge Project: foundations and reclamation work, *Proceedings of the Institution of Civil Engineers*, Vol. 88, No. 1, 1990, pp. 531–49.

9.36 STANLEY, R. G. Design and construction of the Sungei Perak Bridge, Malaysia, *Proceedings of the Institution of Civil Engineers*, Vol. 88, No. 1, 1990, pp. 571–99.

9.37 LEVESQUE, M. Les fondations du pont de Färo en Danemark, *Travaux*, November 1983, pp. 33–6.

9.38 WALTHAM, A. C. and FOOKES, P. G. Engineering classification of karst ground conditions. *Journal of Engineering Geology and Hydrology*, Vol. 36, No. 2, Geological Society of London, 2003, pp. 101–18.

9.39 BYLE, M. J. and MACKEY, R. Grouting karst to support driven pile foundation for a truss railway bridge. In *Geotrans 2004 – Geotechnical engineering for transportation projects*, Yegian, M. K. and Kavazanjian, E. (eds), American Society of Civil Engineers Geotechnical Special Publication No. 126, 2004, pp. 1915–25.

9.40 URANOWSKI, D. D., DODDS, S., and STONECHECK, P. E. Micropiles in karstic dolomite; similarities and differences of two case histories. In *Geo-Support 2004, Drilled shafts, deep mixing, remedial methods and speciality foundation systems*, Turner, J. P. and Mayne, P. W. (eds), American Society of Civil Engineers Geotechnical Special Publication No. 124, 2004, pp. 674–81.

9.41 FISCHER, J. A. A nuclear power plant on karst terrane? In *Sinkholes and the Engineering and Environmental Impacts of Karst*, Beck B. F. (ed), American Society of Civil Engineers Geotechnical Special Publication No. 122, 2003, pp. 485–91.

9.42 BRANDL, H. Energy foundations and other thermo-active ground structures, *Geotechnique*, Vol. 56, No. 2, 2006, pp. 81–122.

9.10   Worked example

Example 9.1

An embankment 9 m high consisting of fill having a density of 2.1 Mg/m^3 is placed with its toe 1.5 m from a row of vertical piles supporting a bridge pier. The piles are driven through 8 m of soft to firm clay into a stratum of stiff clay. Calculate the lateral pressure on the piles within the stratum of soft firm clay which has drained shearing strength parameters of $c' = 0$ and $\phi' = 28°$.

From equation 9.2, $H_f = 9 \times 2.1/1.8 = 10.5$ m. From Figure 9.31, $\alpha = 23.3°$. Therefore from equation 9.4,

$$f = \frac{23.3° - (28° \times 0.5)}{90° - (28° \times 0.5)} = 0.12$$

Surcharge pressure from embankment $p = 2.1 \times 9 \times 9.81 = 185$ kN/m^2
From equation 9.3, lateral pressure on piles $= 0.12 \times 185 = 22$ kN/m^2

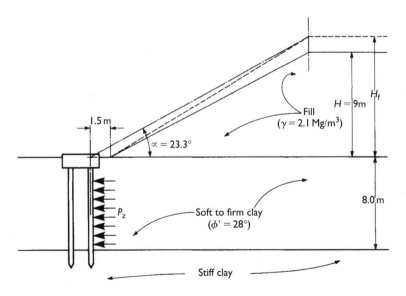

Figure 9.31

Chapter 10

The durability of piled foundations

10.1 General

In all situations consideration must be given to the possibility of the deterioration of piled foundations due to aggressive substances in soils, in rocks, in ground waters, in the sea and in river waters. Piles in river or marine structures are also exposed to potentially aggressive conditions in the atmosphere, and they may be subjected to abrasion from shifting sand or shingle, or damage from floating ice or driftwood.

In considering schemes for protecting piles against deterioration due to these influences, the main requirement is for detailed information at the site investigation stage on the environmental conditions. In particular, adequate information is required on the range of fluctuation of river or sea levels and of the groundwater table. In the latter case, the highest levels are required when considering the likely severity of sulphate attack on concrete piles or the corrosion of steel piles, and the lowest possible levels are of considerable importance in relation to the decay of timber piles. The possibility of major changes in groundwater levels due to, say, drainage schemes, irrigation, or the impoundment of water must be considered.

In normal soil conditions it is usually sufficient to limit chemical analyses of soil or groundwater samples to the determination of pH-values, water-soluble sulphate content (mg per litre) and chloride content. Where the sulphate content exceeds 0.24% in soils it is advisable to determine the water-soluble sulphate content, expressing this in mg of SO_4 per litre of water extracted. For brownfield sites, full chemical analyses are required to identify potentially aggressive substances[2.6]. Methods of investigating and assessing brownfield sites are given by Harris *et al.*[10.1], drawing attention to the health and safety precautions necessary, the need to employ specialist personnel, and care in selecting representative samples.

Bacterial action can be an influence in the corrosion of steel piles. Samples of soil and groundwater should be obtained in sterilized containers, which are then sealed for transportation to the bacteriological laboratory for later analyses. Where steel piles are used for foundations in disturbed soils or fill material on land, an electrical resistivity survey is helpful in assessing the risk of corrosion and in the design of schemes for cathodic protection (see Section 10.4.2).

Investigations for marine or river structures should include a survey of possible sources of pollution which might encourage bacteriological corrosion, such as contaminated tidal mud flats, discharges of untreated sewage or industrial effluents, dumping grounds for industrial or household refuse, and floating rubbish discharged from ships or harbour structures. The pattern of sea or river currents should be studied and water samples taken at various stages of spring and neap tides or at dry-weather and at flood and dry discharge stages in rivers. Particular attention should be paid to sampling water from currents originating at the areas of contamination previously identified. Chemical and bacteriological analyses should be made on the full range of samples to assess the daily or seasonal variation in potentially aggressive substances. Other items for study include the presence and activity of organisms such as weeds and barnacles, and molluscan or crustacean borers (see Section 10.2.2).

10.2   Durability and protection of timber piles

10.2.1   Timber piles in land structures

Timber piles permanently below groundwater level have an indefinite life. There are numerous examples of stumps of timber piles that are more than 2000 years old being found in excavations below the water table. While timber does not decay from fungal attack if the moisture content is kept below 20% it is impossible to maintain it in this dry condition when buried in the ground above water level. Hence damp timber which does not have natural durability is subject to decay by fungal attack, resulting in its complete disintegration. Figure 10.1 shows an example of the decay of timber piles above the water table. Figure 10.1a shows the cavities left by the complete decay of the timber. The timber capping beams have also decayed, allowing the stone lintels to sink down onto the ground surface. Figure 10.1b is a view down a cavity which is partly filled by soil debris and fragments of decayed timber. The piles were driven into clay fill in the early nineteenth century. Preservative treatment can, however, give a useful life to timber piles in the zone above groundwater level. If treatment is applied to properly air-seasoned wood at the correct moisture content for the impregnation of the preservative, a life of several decades may be achieved.

The Building Research Establishment (BRE)[10.2] has classified various grades of durability in terms of their approximate life when in contact with the ground (Table 10.1).

Types of softwoods and tropical hardwoods suitable for timber piles as BS 5268 are listed in Section 2.2.1 (Table 2.1). Their natural durability and 'treatability' is classified by BRE

Table 10.1 Natural durability classification of the heartwood of untreated timbers

Grade of durability	Approximate life in ground contact (years)	European Standards Class (Resistance to fungal attack)
Very durable	More than 25	I
Durable	15 to 25	2
Moderately durable	10 to 15	3
Non-durable	5 to 10	4
Perishable	Up to 5	5

(a)

(b)

Figure 10.1 Decay of timber piles above groundwater level (Crown copyright reserved) (a) Cavities left by complete decay of piles and timber capping sills (b) View down cavity left in clay after complete decay of timber pile.

and European Standards as shown in the following table:

Species	Durability	European Standards Classes			
		Durability	Treatability		Treatability
Douglas fir	Non-durable	4	Resistant		3
Pitch pine (Caribbean)	Durable	2	Moderately resistant		2
Larch	Moderately durable	3	Resistant		3
Western red cedar	Durable	2	Resistant		3
Greenheart	Very durable	I	Extremely resistant		4
Jarrah	Very durable	I	Extremely resistant		4
Opepe	Very durable	I	Moderately resistant		2
Teak	Very durable	I	Extremely resistant		4

Precautions against fungal attack must be commenced at the time that the timber is felled. It should be carted away from the forest as quickly as possible and then stacked on firm, well-drained, and elevated ground from which all surface soils which might harbour organisms have been stripped. The timber should be stacked clear of the ground with spaces between the baulks to encourage the circulation of air and the drying of the timber to the moisture content suitable for the application of the preservative treatment. Treatability, based on the heartwood, is an assessment of the take up of the preservative and depends on the structure of the wood and the method of treatment.

Suitable methods of preserving timber for piling work which have been on the market for many years involved pressure impregnation with creosote, waterborne preservatives such as copper–chrome–arsenic (CCA) (e.g. various Celcure formulations) and copper–chrome–boron compounds (CCB). Although these products have a good safety record, European Directives effective in 2005/6 proscribe many of these compounds. As a result, alternative, more benign preservatives are being developed such as copper azole compounds (e.g. tebuconazole), Celcure P60, based on copper and chromium with phosphoric acid and light organic solvents as a carrier for fungicides. British Standards BS 1282: 1999, BS 8417: 2003, and the publications of the Timber Research and Development Association specify the hazard classes, service factors for 15, 30, and 60 years service life, and 'loadings' of the preservative solution which are adequate for British and other temperate climates for timber piles in fresh- and salt-water. In the USA the specifications of the American Wood Preservers' Institute are followed. Biological deterioration including termite attack is much more severe in tropical countries and the loadings or the selection of resistance species for these conditions should be specified in consultation with a specialist authority in the country under consideration.

Softwoods such as Scots pine and southern pine can be impregnated completely with preservative solutions, but the harder, 'resistant' woods, including Douglas fir, can only be treated to a limited depth (3 to 6 mm) after the timber has first been incised and then subjected to a long-sustained application of pressure. The 'moderately resistant' woods are fairly easy to treat to depths up to 18 mm. There can be a useful advantage in using round timber for piles, in which the outer zone of sapwood can be thoroughly impregnated to a depth which will resist fungal decay over a long period of years. For example, the sapwood of Scots pine or Baltic redwood can be treated to a depth of 75 mm, whereas if squared timbers are used much of the sapwood is cut away to expose the less absorptive heartwood which cannot be impregnated properly. Because the interior of the harder timbers remains untreated, careful attention should be paid to bolt holes. When these are drilled after the main impregnation treatment, preservative should be poured into the holes. Incisions made by lifting hooks, dogs, or slings should be painted with the solution.

Similar attention should be given to the end grain after trimming the tip to receive the shoe or preparing the butt for the driving cap or ring (Figure 2.2). The exposed end grain should be given two heavy coats of the preservative.

Some hardwoods, for example, ekki, greenheart, jarrah, okan and opepe can be used without preservative treatment, but in these cases it is essential to specify that no sapwood is left on the prepared timber. It is difficult to distinguish between sapwood and heartwood in greenheart and either expert advice should be sought to ensure exclusion of the former or a preservative should be used to treat the sapwood as a precautionary measure. Timber used for piling is normally required to have large cross-sectional dimensions making it impracticable to remove the sapwood. BS 8004 strongly recommends using round logs when the preservative-treated sapwood provides a deep uniformly treated protective zone around the pile.

The adoption of preservative treatment by using creosote or some other solution does not give indefinite life to the timber above groundwater level, and it may be preferable to adopt a form of composite pile having a concrete upper section and timber below the water line, as shown in Figure 2.1a.

10.2.2 Timber piles in river and marine structures

The moisture and oxygen in the *atmospheric zone* of timber marine piles above the water line creates a favourable environment for fungal growth, which usually starts in the centre portion where preservatives have not penetrated. Fungal activity occurs in the *splash zone* but is limited due to poor oxygen supply. Marine borers do not attack wood in these zones. Brown rot decay is the most common type of fungal decay in coniferous wood species, and in the early stages of attack the wood will have lost weight and, while visually appearing sound, will have suffered considerable loss of elasticity. Fungal attack does not occur below a maintained water table and immersion in salt-water protects against fungal decay.

The most destructive agency which can occur in piles *fully immersed* in brackish or saline waters in estuaries or in the sea is attack by molluscan or crustacean borers. Conditions in the *tidal zone* are also likely to be favourable for attack by borers where adequate oxygen and salt-water are present, but crustacean borers can often attack near an exposed mud line. Below the mud line, adequate oxygen is not available for the survival of marine borers. These organisms burrow into the timber, forming networks of holes that eventually result in the complete destruction of the piles. Timber jetties in tropical waters have been destroyed in this way in a matter of months.

The main types of marine boring organisms are

Molluscan borers	*Teredo* ('shipworm')
	Bankia
	Martesia (in tropical waters only)
	Xylophaga dorsalis
Crustacean borers	*Limnoria* ('gribble' or 'sealouse')
	Cheluria
	Sphaeroma

The young molluscan borers enter the timber through minute holes in the surface or through incisions. They then grow to a considerable size (*bankia* can grow to a diameter of 25 mm and to nearly 2 m long) and destroy the wood as they grow (Figure 10.2a). The crustaceans work on the surface of the timber, forming a network of branching and interlacing holes (Figure 10.2b). Their activity depends on factors such as the salinity, temperature, pollution level, dissolved oxygen content and current velocity of the water. A salinity of more than 15 parts per 1000 (the normal salinity of seawater is between 30 and 35 parts per 1000) is necessary for the survival of most species of borer, but *sphaeroma* have been found in nearly fresh tropical waters in South America, South Africa, India, Srilanka (formerly Ceylon), New Zealand and Australia. Attack by *cheluria* is usually dependent on the presence of *limnoria*. *Limnoria* cannot survive in fresh water.

Chellis[10.3] states that *teredo* and *limnoria* do not attack in current velocities higher than 0.7 m/s (1.4 knots) and 0.9 m/s (1.8 knots) respectively. Although activity from some species may be marked in tropical waters, borers have been found above the Arctic Circle. They show cyclic activity rising to a peak in some years, and not infrequently dying away completely. Conversely, previously trouble-free areas can become infested with borers brought in by ships or driftwood.

It was stated in the twenty-first report of the Sea Action Committee of the Institution of Civil Engineers[10.4] that no species of timber is absolutely free from borer attack, but certain species are highly resistant and in many conditions of exposure they may be considered to have practical immunity. The report lists the more-resistant species (the 'very durable' class), greenheart, pynkadou, turpentine, totara and jarrah, as being suitable for conditions of heavy attack by *limnoria* and *teredo* in temperate and topical waters. 'Moderately durable' woods will resist moderate attack by *limnoria*. In commenting on the suitability of various types of preservative, the report concludes that ordinary coal–tar creosote is the most satisfactory, and states that in British waters any timber which is *efficiently* impregnated with creosote should be practically immune to borer attack.

A comparative study was made by the Sea Action Committee of the relative effectiveness as preservatives of creosote, Celcure (copper sulphate–potassium dichromate), and creosote with the addition of copper napthenate. The latter material was tried as there was evidence that some copper salts were poisonous to borers. The treated specimens consisted of Douglas fir and they were exposed at Singapore, where the borer attack was mainly by

(a)

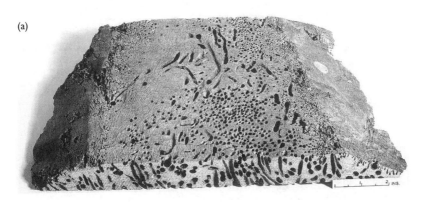

Figure 10.2 Attack on timber piles by marine borers (Crown copyright reproduced with permission of BRE) (a) Attack by *Teredo* (b) Attack by limnoria.

(b)

Figure 10.2 Continued.

martesia with some *teredo*, and at Colombo. The observations of the relative severity of attack after 10½ years of exposure were as shown in the following table:

Treatment	Singapore	Colombo
Coal–tar creosote to BS 144	Badly damaged	Slightly to badly damaged
5% Celcure	Very badly damaged	Very badly to badly damaged
Coal–tar creosote with 1% of copper napthenate	Very badly to badly damaged	Slightly to badly damaged
Coal–tar creosote with 5% of copper napthenate	Very badly damaged	Very badly to badly damaged

It was concluded from the experiments that creosote gave marginally the best treatment and that the addition of copper napthenate gave no advantage. *Limnoria tripunctata* are tolerant to creosote but the species can be effectively controlled by the addition of copper pentachlorophenate to the creosote. Although the range of preservatives available is much reduced under EU Marketing Directives, creosote applied by qualified personnel is still acceptable.

The Building Research Establishment (Farmer *et al.*[10.5]) lists the following timbers as having heartwood resistant to borer attack. Those marked with an asterisk are believed to be the best for marine work and their properties, durability, preservation and uses are described in detail in the BRE publication.

African padauk (*Pterocarpus soyauxii*)
Afrormosia (*Pericopsis elata*)
Andaman padauk (*Pterocarpus dalbergioides*)
*Basralocus (Angelique) (*Dicorynia guianensis*)
*Belian (*Eusideroxylyn zwageri*)
*Brush box (*Tristania conferta*)
*Ekki (*Lophira alata*)
*Greenheart (*Ocotea rodiaei*)
Iroko (*Chlorophora excelsa*)
Ironbark (*Eucalyptus siderophloia*)
Jarrah (*Eucalyptus marginata*)
Kapur (*Dryobalanops spp*)
*Manbarklak (*Eschweilera longipes*)
Muninga (*Pterocarpus angolensis*)
*Okan (*Cyclicadiscus gabunensis*)
*Opepe (*Nauclea diderrichii*)
*Pyinkado (*Xylia xylocarpa*)
*Red louro (*Ocotea rubra*)
*Southern blue gum (*Eucalyptus globulus*)
Teak (*Tectona grandis*)
*Turpentine (*Syncarpia laurifolia*).

The sapwood of these timbers is liable to be attacked by borers, and if it is impossible to ensure the removal of all sapwood the timber should be treated with creosote as a precautionary measure. Greenheart fenders in Milford Haven were attacked in the sapwood by *teredo*, causing about 10 mm of damage in five years.

The methods of preparing, air-seasoning and preserving timber against borer attack are the same as those described for fungal decay in Section 10.2.1 above. However, great care is necessary to avoid making incisions through which borers can enter the untreated wood in the interior of the pile. The timber should be handled by slings rather than hooks or dogs after creosoting, and purpose-made devices should be used to give pressure impregnation of the bolt holes after drilling.

Chellis[10.3] describes the following other methods of protecting timber piles against attack by borers:

(1) Tipping stone around the piles (this protects only the length covered by the stones)
(2) Sleeving the timber with galvanized iron, copper, or aluminium sheeting
(3) Encasing the piles
(4) Jacketing the piles with precast concrete tubes and filling the space between the timber and the tubes with cement grout and
(5) Coating the piles with cement–sand mortar, applied with a spray gun (e.g. the 'Gunite' process).

Reliable methods of repairing decayed marine timber piles to provide substantial recovery of original strength are not available, not least because of the difficulty in gaining access to the critical zones. Experimental techniques which first remove the decayed material, treat the remaining wood with preservative, and infill the void with epoxy resin mortar followed by wrapping with glass fibre have shown some small-scale success. Voids left by rotting timber piles below the Royal Scottish Academy in Edinburgh were successfully treated by Keller Ground Engineering using their 'Soilfrac' process. This entailed installing horizontal tubes-à-manchette 2 m below the pile cap stonework from a trench around the building so that each 41 m long tube intersected an average of 15 piles. This allowed about 40% of the piles to be directly injected using a low viscosity grout, with the remaining piles filled by overflow and pressure grouting. The stability of the building was extensively monitored during the process.

In tropical and sub-tropical countries timber piles can be destroyed by termites above the waterline unless a resistance species is used, or they are given the usual preservative treatment. Also the end grain at the heads of piles is particularly susceptible to attack by fungi or beetles when in a damp condition. The pile heads can be protected by heavy coats of hot-applied creosote followed by capping with metal sheeting, bituminous felt or glass fibre set in coal-tar pitch.

Some species of wood corrode iron fastenings by the secretion of organic acids. Either non-ferrous fastenings should be used or steel components should be heavily coated with tar or sheathed in plastics. Stainless steel fastenings can be used if the type of steel is resistant to corrosion by seawater.

The abrasion of timber piles by shingle on the sea bed has been mentioned. While protection by metal sleeving can be adopted, non-ferrous metal is expensive and it may be preferable to use sacrificial timber strapped around the main bearing piles, or to accept the cost of periodical renewal.

10.3   Durability and protection of concrete piles

10.3.1   Concrete piles in land structures

Properly mixed concrete compacted to a dense impermeable mass is one of the most permanent of all constructional materials, and gives little cause for concern about its long-term durability in a non-aggressive environment. However, concrete can be attacked by sulphates and sulphuric acid occurring naturally in soils, by corrosive chemicals which may be present in industrial waste in fill materials, and by organic acids and carbon dioxide present in groundwater as a result of decaying vegetable matter[10.6]. Attack by sulphates is a disruptive process whereas the action of organic acids or dissolved carbon dioxide is one of leaching. Attack by sulphuric acid combines features of both processes.

The naturally occurring sulphates in soils are those of calcium, magnesium, sodium and potassium. The basic mechanism of attack by sulphates in the ground is a reaction with hydrated calcium aluminate in the cement paste to form calcium sulphoaluminate. The reaction is accompanied by an increase in molecular volume of the minerals, resulting in the expansion and finally the disintegration of the hardened concrete. Other reactions can also occur, and in the case of magnesium sulphate, which is one of the most aggressive of the naturally occurring sulphates, the magnesium ions attack the silicate minerals in the cement in addition to the sulphate reaction. Ammonium sulphate, which attacks Portland cement very severely, does not occur naturally. However, it is used as a fertilizer and may enter the ground in quite significant concentrations, particularly in storage areas on farms or in the factories producing the fertilizer. Ammonium sulphate is also a by-product of coal-gas production and it can be found on sites of abandoned gasworks. Because calcium sulphate is relatively insoluble in water, it cannot be present in sufficiently high concentrations to cause severe attack. However, other soluble sulphates can exist in concentrations that are much higher than that possible with calcium sulphate. This is particularly the case where there is a fluctuating water table or flow of groundwater across a sloping site. The flow of groundwater brings fresh sulphates to continue and accelerate the chemical reaction. High concentrations of sulphates can occur in some peats and within the root mass of well-grown trees and hedgerows due to the movement and subsequent evaporation of sulphate-bearing groundwater drawn from the surrounding ground by root-action. The severity of attack by soluble sulphates must be assessed by determining the soluble sulphate content and the proportions of the various cations present in an aqueous extract of the soil. These determinations must be made in all cases where the concentration of sulphate in a soil sample exceeds 0.5%.

The thaumasite form of sulphate attack which consumes the binding calcium silicate hydrates in Portland cement, thereby weakening the concrete, has been investigated extensively in recent years[10.7]. The reaction requires the presence of sulphates, calcium silicate, carbonate, and water, and appears to be more vigorous at temperatures below 15°C. Carbonation of concrete due to atmospheric carbon dioxide acting on the calcium hydroxide in the concrete matrix causes a reduction in the pH rendering the concrete susceptible to sulphate reactions forming thaumasite.

Free sulphuric acid may be formed in natural soil or groundwater as a result of the oxidation of pyrites in some peats, or in ironstone or alum shales. Sulphuric acid can also be present in industrial waste materials which have been contaminated by leakages from copper and zinc smelting works, and from dyeing processes. The acid has an effect on the cement in hardened concrete that is similar to that of sulphate attack, but the degradation may not result in significant expansion. Figure 10.3 shows the disintegration of the concrete in the shaft of a bored and cast-in-place pile caused by the seepage of sulphuric acid into porous fill material.

In the UK, sulphates occurring naturally in soils are generally confined to the Keuper Marl (Mercia Mudstone), and to the Lias, London, Oxford, Kimmeridge and Weald Clays. They are also found in glacial drift associated with these formations. Sulphates may be present in the form of gypsum plaster in brick rubble fill.

Figure 10.3 Disintegration of concrete in bored and cast-in-place pile due to attack by sulphuric acid leaking into fill from industrial processes.

The sulphate content of the groundwater gives the best indication of the likely severity of sulphate attack, particularly that resulting from soluble sulphates. Where the water samples are taken from boreholes care should be taken to ensure that the sample is not diluted by water added to assist the drilling. If possible the groundwater should be sampled after a long period of dry weather. Groundwater flow across a sloping site through sulphate-bearing ground results in the highest concentration on the downhill side of the site and the flow may continue into permeable soil deposits which are not naturally sulphate-bearing. An account of the distribution of sulphates in various ground conditions in Great Britain is given by Bessey and Lea[10.8]. Methods of analysis to determine the sulphate content and pH-value of soils and ground waters are set out in BS 1377 and by Bowley[10.9] in BRE Report 279 and are critically reviewed by Eglinton[10.6].

A dense, well-compacted concrete provides the best protection against the attack by sulphates on concrete piles, pile caps and ground beams. The low permeability of dense concrete prevents or greatly restricts the entry of the sulphates into the pore spaces of the concrete. For this reason high-strength precast concrete piles are the most favourable type to use. However, for the reasons explained in Chapter 2, precast concrete piles are not suitable for all site conditions and the mixes used for the alternatives of bored and cast-in-place or driven and cast-in-place piles must be designed to achieve the required degree of impermeability and resistance to aggressive action.

In British practice recommendations for the types of cement and the mix proportions are given in BRE Special Digest 1: 2005[2.6]. There are several significant changes compared with the previous BRE recommendations, mainly designed to harmonize with the new British and European standards and to take account of research into combating the thaumasite form of sulphate attack on concrete. The five classes of severity of attack ('Design Sulphate' classes DS1 to 5) have been retained from which are derived the new 'Aggressive Chemical Environment for Concrete' (ACEC) classes (AC1 to 5) for natural ground and brownfield sites, subject to certain conditions (e.g. pH should be greater than 2.5). The AC classes provide for adjustment from one DS class to another depending on the conditions of exposure, the pH and mobility of groundwater, and other environmental conditions. For a given AC class a 'Design Chemical' (DC) class is derived for the intended working life, either 50 or 100 years, together with recommended 'additional protective measures' specific to highly aggressive ground types. Concrete mixes are then tabulated to suit the DC class giving a wide selection of free-water/cement ratios and aggregate sizes down to 10 mm and the appropriate cement and cement combinations in accordance with BS EN 197-1 and BS 8500-2. The use of sulphate-resisting Portland cement is covered in BS 4027.

The workability of the BRE suggested cast in-situ concrete mixes may in some cases be too low for placing for bored and driven small diameter cast-in-place piles. Slightly modi-fied mixes are given for certain precast products, including surface-carbonated precast con-crete, which would be suitable for precast piles. As it is not possible to cover in this text the various comments and qualifications to the recommendations given in Special Digest 1, it is important to follow the step by step approach to determine the appropriate concrete quality for a particular assessment of ground conditions. It should be noted that the Digest does not purport to assess and advise on the use of sand–cement grouts in minipiles and around the permanent sleeves to piles.

Mixes suitable for concrete in pile caps, ground beams and blinding concrete depend on the size shape and amount of reinforcement of the members which govern the workability requirements. Footnotes to the Special Digest 1, Table D1, provide for modifications to the DC class depending on the size of a structural member.

Generally, no additional protection measures (APMs) are necessary where the ground-water is considered 'static', but other conditions may over-ride this (e.g. thickness of concrete section). When in doubt the 'mobile' groundwater condition should be used. For example, it would be unwise to assume a static groundwater table at a shallow depth for cast-in-place concrete piles where the concrete may be weaker than in the body of the pile due to accumulation of laitance. Weak concrete used as a binding layer beneath pile caps is also vulnerable to sulphate attack when the resulting expansion of the blinding concrete could lift the cap; hence the quality of blinding concrete should match the structural quality.

Pile caps and ground beams can be protected on the underside by a layer of heavy gauge polyethylene sheeting (designated APM3) laid on a sand carpet or on blinding concrete. The vertical sides can be protected after removing the formwork by applying hot bitumen spray coats, bituminous paint, trowelled-on mastic asphalt, or adhesive plastics sheeting. The recommendation for placing a membrane between floors and fill, or hardcore containing sulphates, should be considered for the undersides of slender pile capping beams, or shallow pile caps.

Coatings of tar or bitumen on the surface of precast concrete piles do not give adequate protection against sulphate attack since they are readily stripped off by abrasion as the piles are driven down in all but the softer soils. Protection can be given to the pile surface by metal sheathing or glass fibre wrapping impregnated with bitumen, but the latter is likely to be torn when piles are driven into gravelly or stony soils. If a sacrificial layer of concrete (APM4) is added to friction piles, consideration must be given to the effects of sulphate and thaumasite attack causing expansion and reducing frictional resistance.

The use of high-alumina cement (BS 915 under revision as prEN 14647) or supersulphated cement (BS 4248) for high sulphate concentrations is referred to in Special Digest 1. The latter cement is attacked by ammonium sulphate to which high-alumina cement alone is resistant. Also, there is some experience to indicate that supersulphated cement has less resistant properties to attack by magnesium sulphate than those of sulphate-resisting cement.

Neither high-alumina cement nor supersulphated cement is favoured for piling work. In any case, approval of the use in structural concrete of the former type has at present been withdrawn from codes of practice in Britain and in some other countries. Structural concrete is deemed to include all concrete in foundations. The withdrawal of approval has been due to the property of the cement to 'chemical conversion'[10.6] which results in a serious loss of strength. While this reduction of strength may not be critical in the case of foundations subjected to relatively low levels of stress, the conversion is accompanied by a marked reduction in the sulphate-resisting properties of the cement. Conversion is particularly liable to take place in warm and damp conditions. These may occur in piles above water level in marine structures, and in large-diameter bored and cast-in-place piles where the heat of hydration of the cement is dissipated only slowly. The use of this cement also causes serious practical difficulties in placing the concrete in pile shafts due to its rapid setting.

Supersulphated cement is costly and difficult or impossible to obtain in many countries including the UK. It has a low heat of hydration and is therefore rather slow to harden. This makes it unfavourable for use in precast concrete piles because of the long period required between casting and driving. Special care is required when using this cement in cold weather. Table D3 in Special Digest 1 provides for the use of Portland cements incorporating ground granular blastfurnace slag (ggbs) or pulverized fuel ash (pfa – now designated 'flyash' in BS 8500) and for a variety of Portland cement–pozzolanic combinations mixed on site to give enhanced sulphate-resisting properties. Concrete containing ggbs is now recommended in place of sulphate-resisting cement to combat thaumasite attack in the UK.

The leaching of concrete exposed to flowing river or groundwater containing organic acids or dissolved carbon dioxide was mentioned at the beginning of this section. Organic acids are present in run-off water from moorlands, and in groundwater in peaty and lignitic soils. The recommendations for concrete exposed to acid attack as determined by the pH value of the soil or groundwater are covered by the ACEC Tables in Special Digest 1. Good quality concrete, made with any of the tabulated cements and non-degradable aggregates, is essential.

On a site where there was severe contamination with acid industrial waste the authors advised a protective scheme for the foundation piling consisting of precast concrete shell piles coated externally with bitumen over the portion of the shaft within the fill. As a safeguard against partial stripping of the bitumen by bricks and concrete in the fill, the concrete shells were regarded as sacrificial. The main load-bearing element consisted of a PVC sleeve (weight 800 gm/m^3) lowered down the shells after completion of driving onto a concrete plug in the lower part of the pile. The PVC sleeve was then filled with concrete. A flexible PVC membrane (Bituthene sheeting) was provided beneath the pile caps. This was lapped and bonded to the PVC pile sleeve.

10.3.2 Concrete piles in marine structures

Precautions against the aggressive action by seawater on concrete need only be considered in respect of precast concrete piles. Cast in-situ concrete is used only as a hearting to steel tubes or cylindrical precast concrete shell piles, where the tube or shell acts as the protective element. A rich concrete, well-compacted to form a dense impermeable mass, is highly resistant to aggressive action and, provided a cover of at least 50 mm is given to all reinforcing steel, precast concrete piles should have satisfactory durability over the normal service life of the structures they support.

When the disintegration of reinforced concrete in seawater does occur it is usually most severe in the 'splash zone' and is the result of porous or cracked concrete caused by faulty design or poor construction. Evaporation of the seawater in the porous or cracked zone is followed by the crystallization of the salts and the resulting expansive action causes spalling of the concrete and the consequent exposure of the reinforcing steel to corrosion by air and water. The expansive reaction that occurs when corrosion products are formed on the steel accelerates the disintegration of the concrete. Freezing of seawater in porous or cracked concrete can cause similar spalling. However, where concrete piles are wholly immersed in seawater there is no degradation of properly made and well-compacted concrete.

In an extensive review of literature and the inspection of structures which had been in the sea for 70 years, Browne and Domone[10.10] found no disintegration in permanently immersed reinforced concrete structures even though severe damage had occurred in the splash zone. They concluded that corrosion of the steel cannot occur with permanent immersion because the chloride present is restricted to a uniform low level and the availability of oxygen is low.

Although seawater typically has a sulphate content of about 230 parts per 100 000, the presence of sodium chloride has an inhibiting or retarding effect on the expansion caused by its reaction with ordinary Portland cement. The latter material is, therefore, quite satisfactory for the manufacture of precast concrete piles for marine conditions but to avoid disintegration in the splash zone the concrete should have a minimum cement content of 360 kg/m^3 and a maximum water/cement ratio of 0.45 by weight. Special Digest 1 does not provide recommendations for concrete exposed to seawater, but reference should be made to BS 6349-1: 2000 Marine Structures. Air entrainment of concrete as a safeguard against frost attack on piles above the water line is unnecessary if the water/cement ratio is less than 0.45.

The concrete in precast piles should be moist-cured for 7 days after the removal of the form-work (with a further 10 days exposure to air in order to be classified as 'surface carbonated'). Great care should be taken in handling the piles to avoid the formation of transverse cracks which would expose the steel to corrosion in the splash zone. Coatings on precast concrete piles to protect them against deterioration in the splash zone are of little value since they are soon removed by the erosive action of waves, and by abrasion from floating debris or ice.

10.4   Durability and protection of steel piles

10.4.1   Steel piles for land structures

Corrosion of iron or steel in the electrolyte provided by water or moist soil is an electro-chemical phenomenon in which some areas of the metal surface act as anodes and other areas act as cathodes. Pitting occurs in anodic areas, with rust as the corrosion product in cathodic areas. Air and water are normally essential to sustain corrosion but bacterial corrosion can take place in the absence of oxygen, i.e. in anaerobic conditions. Anaerobic corrosion is caused by the action of sulphate-reducing bacteria which thrive below the sea or river bed in polluted waters, particularly in relatively impermeable silts and clays.

An exhaustive investigation of the corrosion rates of steel sheet piles and bearing piles in soils was made by Romanoff[10.11] on behalf of the US National Bureau of Standards. Steel piles which had been in the ground for periods of between 7 and 40 years were examined. The soil types ranged from permeable sands to relatively impervious clays. Soil resistivities ranged between 300 and 50200 ohm-cm and pH values between 2.3 and 8.6. Romanoff concluded from observations of the condition of the piles that where they were driven into *undisturbed* natural soil, the type and amount of corrosion was so small that it would not significantly affect the strength or useful life of the piling to support structures. Some localized pitting corrosion and loss of mill-scale were seen on steel surfaces but the loss of metal was considered to have a negligible effect on the serviceability of the piles. Corrosion had occurred in some instances where piles had been driven through fill above the water table, or in the zone extending 0.6 m above and below the water table.

Romanoff pointed out that undisturbed natural soils are so deficient in oxygen that they will not sustain the process of corrosion. Romanoff also found that determinations of soil resistivity and pH-value had no relevance to the incidence of corrosion in the undisturbed soil conditions covered by the Bureau of Standards research. He did not encounter any cases of anaerobic corrosion by sulphate-reducing bacteria but the possibility of their occurrence should not be overlooked at the site investigation stage. Undisturbed samples of the soil should be sealed in their containers and submitted for bacteriological examination.

In a later study, Romanoff[10.12] examined steel sheet piles which had been driven through fill material. Inspections were made at 13 locations where piles had been installed for periods of between 11 and 30 years. With only one exception the piles showed only shallow attack on the metal with some localized pitting corrosion. The single exception was at a site where sheet piles had been driven through 6 m of clinker filling. Severe attack on the metal and pitting up to 6 mm deep had occurred over large areas. However, it was pointed out that these piles were continuing to give useful service 23 years after they had been driven. Romanoff concluded that the relatively small amount of corrosion over the portion of the pile in fill or in undisturbed soil above the water table is the result of the formation of a galvanic corrosion cell between the upper part of the pile above the water table and the lower permanently immersed part. The upper portion is small in volume compared with the lower portion and it acts as a cathode, while the lower part in soil deficient in oxygen is the anode. Because of the much greater mass of steel in the anodic portion only a small proportion is sacrificed in protecting the cathodic part.

Similar corrosion rates for piling in land structures have been recorded by Morley[10.13]. British Steel Corporation investigations of piles extracted from UK sites in the 1970s showed corrosion losses below the soil line varying from nothing to 0.03 mm per year with a mean of 0.01 mm per year. No precautions are required for such low rates of loss of thickness. Where piles in land structures are extended above ground, mild steel thickness losses of 0.2 mm per year were measured over a 10-year period in a marine environment. Morley considered that a more usual figure for the UK would be less than 0.1 mm per year. For steel bearing piles in natural soils, BS 8002 and BS 8004 advised a maximum corrosion allowance of 0.015 mm per year per side where no other corrosion protection is required; this is consistent with corrosion rates derived from Eurocode EC3-5 (Piling). The long-term corrosion rate of piles in normal atmospheres in urban conditions given in EC3–5 is 0.01 mm per year per side and for coastal areas 0.02 mm per side per year. In areas where localized conditions give rise to more aggressive microclimates the greater allowances in BS 8004 may be needed. Paint treatment[10.14] would be a suitable precautionary measure for the exposed steel provided that it is accessible for maintenance. If the aesthetic appearance of the steel is important, Arcelor[10.15] suggest application of coating systems using zinc silicate epoxy primer and aliphatic polyurethane topcoat. Where the water table is shallow the pile cap can be extended down to a depth of 0.6 m below water level to protect the steel of the piles.

Morley[10.13] reported a corrosion rate of 0.05 mm per year for steel piling immersed in fresh water except at the waterline in canals where the rate was as high as 0.34 mm per year. This locally higher corrosion zone may be due to abrasion by floating debris or to cell action between parts of the structure in different conditions of oxygen availability. The pH range of fresh water has little effect on corrosion, but to reflect the variability due to potential pollution, the corrosion rate allowances derived from EC3-5 are approximately 0.02 to 0.05 mm per year per side. Corus suggest[2.4] that glass flake epoxy coating with nominal dry film thickness of 400 μm be used for piers and jetties to extend the time to the first maintenance period to beyond 20 years. An alternative for shorter maintenance periods, in both immersed and atmospheric exposures[10.15], is a polyamine-cured epoxy with dry film thickness of 300 μm. The coatings must be applied over blast-cleaned steel. Isocynate-cured pitch epoxy and cheap coal tar coatings are no longer recommended and are being phased out for health and safety reasons.

Paint coatings are not generally satisfactory for protection against bacterial corrosion. Any pinholes in the coating or areas removed by abrasion serve as points of attack by the organisms. Cathodic protection (see Section 10.4.2) is effective but higher current densities are required than those needed to combat normal corrosion in aerobic conditions.

Where steel piles are buried in fill or disturbed natural soil, the thickness of metal in a bearing pile should be such that the steel section will not be overstressed due to wastage of the metal by corrosion over the period of useful life of the structure. Taking a figure of 0.08 mm per year as a maximum in the range established by the US Bureau of Standards for disturbed ground, a steel H-pile with web and flange thicknesses of 15.5 mm exposed to the soil on both sides will lose 50% of its thickness in a period of 48 years, although there may be localized areas of deeper pitting. Long-term corrosion allowances for service periods up to 100 years provided in EC3-5 for non-aggressive and aggressive non-compacted fills are approximately 0.02 mm per year per side and 0.06 mm per year per side respectively. In compacted fills these figures may be halved. Marsh and Chao[10.16] have refined the contamination guidelines so that more accurate long-term corrosion allowances can be made. Protection coating of piles in severely contaminated ground should resist abrasion, impact, and acidic attack using, for example,[10.15] a polyamide-cured epoxy system with increased chemical resistance and a nominal dry film thickness of 480 μm onto blast-cleaned surfaces. Protection should extend to around 0.6 m below water table.

Other protective measures in contaminated disturbed ground include jacketting the pile with concrete or filling the shafts of hollow piles with concrete capable of carrying the full load.

10.4.2   Steel piles for marine structures

Steel piles supporting jetties, offshore platforms and other river or marine structures must be considered for protection against corrosion in five separate zones. These are as follows:

(1) *Atmospheric zone*: exposed to the damp conditions of the atmosphere above highest water levels or to airborne spray.
(2) *Splash zone*: above mean high water level and exposed to waves and spray and wash from ships.
(3) *Intertidal zone*: between mean high and mean low water levels.
(4) *Continuous immersion zone*: below lowest water level.
(5) *Underground zone*: below the soil line.

Morley and Bruce[10.17] made an extensive survey of the extent of corrosion on steel piling in marine structures at various sites in the UK, Cyprus and the United Arab Emirates. Average and probable maximum corrosion loss rates for the five zones are shown diagrammatically in Figure 10.4. The EC3-5 guidance on corrosion rates, which apply to seawater and fresh water, is not as detailed as these survey values. Design thicknesses to allow for corrosion loss and methods of protection should take into account the variation in corrosion

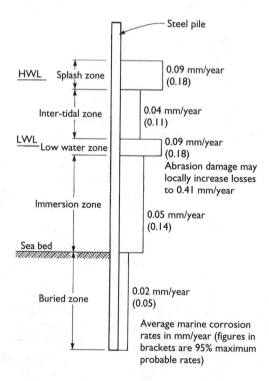

Figure 10.4 Loss of thickness by corrosion for steel piles in seawater (after Morley and Bruce [10.17]).

over these zones, particularly in the low-water and splash zones (see BS 6349[8.1]). The development of 'macro-cells' in the steel at the junction of the tidal and low water zones can limit the corrosion rate to that in the immersed zone. As in the case of canals, abrasion from fenders and floating debris also causes increased corrosion in the low-water zone.

The presence of marine growth also has a considerable influence on protective measures. There is no growth within the atmospheric and splash zones, but in the intertidal and continuously immersed zones heavy growths of barnacles and weeds can develop, which damage the paint treatment and prevent its renewal. However, the growth can shield the steel from exposure to oxygen and in this way reduce the rate of corrosion, counter-balanced by the removal of the growth by abrasion and wash from ships, particularly those with bow thrust propellers. Evidence for bacterial corrosion is limited and it is generally concluded that, although bacterial activity occurs to some extent on most marine structures, it does not cause a significant amount of corrosion damage.

In discussing protective measures Morley and Bruce noted that it is improbable that the life of paint coatings from application to first maintenance will exceed 12 to 15 years, although recent improvements in glass flake epoxy and polyester coatings can extend the maintenance period to over 25 years. However, the cost of painting should be balanced against the alternative of increasing steel thickness, or the use of high tensile steel at mild steel stresses (e.g. grade S355GP steel at S270GP steel stresses, see Section 2.2.6). This provides an additional corrosion loss of 30% without loss of load-bearing capacity at an additional steel cost of about 7%. It should also be noted that steel thicknesses may be determined by the stresses caused during driving (see Section 2.2.6) giving a reserve available for the lower stresses under service conditions. Also, maximum stresses for working conditions in marine structures may be at or near the soil line where corrosion losses are at the minimum rate.

Corus[2.4] recommend the following protective measures for marine structures:

Atmospheric zone and splash zone   Organic coatings or high-quality concrete encasement, well compacted with appropriate cover, extending 1m below mean high water level. Care is needed to ensure that the splash zone, and possibly the tidal zone, is fully encased, otherwise increased electro-chemical corrosion can occur at the steel-concrete junction. Coatings should have a 400 μm dry film thickness to give estimated 20-year life.

Intertidal zone   Bare steel to nominal or increased thickness to allow for corrosion loss (because of uncertainty in driving depths, it may be necessary to extend the coating from the splash zone into the intertidal zone).

Continuously immersed zone   Bare steel or cathodic protection.

Underground zone   No protection necessary.

Piles forming the main supporting structures in important jetties or in offshore platforms exposed to a marine environment require elaborate and relatively expensive treatment to ensure a long life. The steel in the *atmospheric zone* is protected by paint and the first essential is to obtain thorough cleaning of the metal. This is achieved by the application of sand or grit blasting to obtain a white metal or near white metal condition. Coating systems using zinc silicate epoxy primer and aliphatic polyurethane or epoxy topcoats at a total dry film thickness of 240 μm would be suitable for 10- to 15-year maintenance periods.

The most severe conditions of corrosion are experienced in the *splash zone* where Hedborg[10.18] quotes corrosion rates of 0.13 to 0.25 mm per year in the Canal Zone of the USA and the Hawaiian Islands, and a rate of 0.88 mm per year which has been observed on a platform at Cook Inlet, Alaska. Paint coatings used in the past comprised a zinc silicate primer followed by three or more coats of epoxy coal-tar paint to obtain an overall dry film thickness of 400 μm. However, such coatings had a life of only a few years, after which wastage on the exposed steel continued at the rates quoted above and are no longer recommended. Alternative high-build, shop-applied, organic coatings, such as glass flake epoxy[2.5], are more durable. These have been subjected to bioreactor tests and have shown good protection against localized corrosion due to bacteria at the low water level. The problem of pinholes and abrasion allowing potential bacterial attack remains. Cathodic protection is ineffective in the splash zone and the thickness of metal should be such as to ensure that wastage due to corrosion will not curtail the design life of the structure. This is achieved over the length in the splash zone, either by increasing the thickness of the steel, or by providing cover plates of steel to the same specification as the piles, bracings, or jacket members or of a corrosion-resistant material such as Monel metal. *Above the splash zone* continual maintenance by periodic cleaning and painting is needed on exposed steelwork.

Below the splash zone the bare or painted steel is protected by cathodic means or a compatible combination of coating and cathodic measures. Cathodic protection utilizes the characteristic electrochemical potential possessed by all metals (see BS 7361-1:1991 Cathodic Protection–except for offshore applications). The metals which are higher in the electromotive series act as anodes to the metals lower in the series which form the cathodes. Thus if a steel structure is connected electrically to a zinc anode the current escapes to the soil or water through the anode, and the structure thus forms the cathode, so preventing the escape of metallic ions from the structure. The two methods of cathodic protection used in marine structures are the sacrificial anode system and the power-supplied (or impressed-current) system. In the former, large masses of metal such as magnesium, aluminium, or zinc, which are higher in the electromotive series than steel, are used as the anodes. In the power-supplied system the anodes are non-wasting and consist of graphite, lead-silver, or other noble metals. They supply direct current from a generator or transformer rectifier to the structure acting as the cathode. The anodes are not welded directly to the piles as they would become detached when driving through guides or jacket members. Instead they are electrically bonded to the jacket or to bracing members. Appropriate coating can reduce the direct current requirements.

The sacrificial anode system is generally preferred for marine structures since it does not require the use of cables which are liable to be damaged by vessels or objects dropped or lowered into the water from the structures. However, Hedborg[10.18] points out that divers are required to replace sacrificial anodes which are designed to have, for example, a 10-year life. In depths of water of up to 60 m the anodes can be replaced by divers at a reasonable cost, but in deep-water platform structures, the diving costs increase steeply with increasing depths. Sacrificial systems can be designed to be replaced without using divers, and the life of the anodes can be extended by reducing the area of steel requiring protection, i.e. by painting the steel. While a long life cannot be achieved by painting, marine growth will replace the paint and hence maintain the protection. The choice between sacrificial anode systems, with or without a coated structure, and power-supplied systems is a matter of economics, taking into account the capital costs of installation, the current consumption, the costs of maintenance, and the intended life of the structure.

Where hollow steel piles are plugged at their base with concrete or impervious soil it is sometimes the practice to pump out the seawater and replace it with fresh water containing a corrosion-inhibiting compound. The addition of sodium nitrite and sodium carbonate to form a 2% solution can be used for this purpose. However, the need for this has been questioned because an empty or seawater-filled pile contains little oxygen which is quickly used up in the early corrosion process, leaving none to maintain the corrosion.

10.5   References

10.1   HARRIS, M. R., HERBERT, S. M., and SMITH, M. A. *Remedial treatment for contaminated land,* Vol. 111: Site investigation and assessment, CIRIA Report SP103, Construction Industry Research and Information Association, London, 1995.

10.2   British Research Establishment. Timbers: the natural durability and resistance to preservative treatment, BRE Digest 429, Construction Research Communications Ltd, Watford, 1998.

10.3   CHELLIS, R. D. Pile foundations, *Foundation Engineering,* G. A. Leonards (ed.), McGraw-Hill, New York, 1962, pp. 723–31.

10.4   Deterioration of structures of timber, metal, and concrete exposed to the action of seawater, *21st Report of the Sea Action Committee of the Institution of Civil Engineers,* London, 1965.

10.5   FARMER, R. H., MAUN, K. W., and CODAY, A. E. *Handbook of hardwoods. BR400,* Building Research Establishment, Construction Research Communications Ltd, Watford, 2000.

10.6   EGLINTON, M. S. *Concrete and its Chemical Behaviour,* Thomas Telford, London, 1987.

10.7   Thaumasite in Cementitious Materials. *Proceedings of First International Conference.* AP147, Building Research Establishment, Construction Research Communications Ltd, Watford, 2002.

10.8   BESSEY, G. E. and LEA, F. M. The distribution of sulphates in clay soils and groundwater, *Proceedings of the Institution of Civil Engineers,* Vol. 2, Part 1, No. 2, March 1953, pp. 159–81.

10.9   BOWLEY, M. J. *Sulphate and acid attack in the ground: recommended procedures for analysis, BRE Report BR279,* Building Research Establishment, Construction Research Communications Ltd, Watford, 1995.

10.10   BROWNE, R. D. and DOMONE, P. L. J. The long term performance of concrete in the marine environment, *Proceedings of the Conference on Offshore Structures,* Institution of Civil Engineers, London, 1974, pp. 31–41.

10.11   ROMANOFF, M. Corrosion of steel pilings in soils, National Bureau of Standards, NBS Monograph 58, US Department of Commerce, Washington, DC, October 1962.

10.12   ROMANOFF, M. Performance of steel pilings in soils, *Proceedings of the 25th Conference of National Association of Corrosion Engineers,* Houston, Texas, 1969, pp. 14–22.

10.13   MORLEY, J. *The Corrosion and Protection of Steel Piling,* British Steel Corporation, Teesside Laboratories, 1979.

10.14   A corrosion protection guide – For steelwork exposed to atmospheric environments. *Corus Construction and Industrial,* Scunthorpe, 2004.

10.15   Piling Handbook. *Arcelor RPS,* Luxembourg and Scunthorpe, 2005.

10.16   MARSH, E. and CHAO, W. T. The durability of steel in fill soils and contaminated land, *Corus Research, Development and Technology,* Swindon, 2004.

10.17   MORLEY, J. and BRUCE, D. W. *Survey of steel piling performance in marine environments,* final report, Commission of the European Communities, Document EUR 8492 EN, 1983.

10.18   HEDBORG, C. E. Corrosion in the offshore environment, *Proceedings of the Offshore Technology Conference,* Houston, Texas, 1974, Paper No. OTC 1949, pp. 155–68.

Chapter 11

Ground investigations, piling contracts, pile testing

The importance of a thorough ground investigation as an essential preliminary to piling operations cannot be over-emphasized. Accurate and detailed descriptions of soil and rock strata and an adequate programme of field and laboratory tests are necessary for the engineer to design the piling system in the most favourable conditions.

Detailed descriptions of the ground conditions are also essential if the piling contractor is to select the most appropriate equipment for pile installation, while giving prior warning of possible difficulties when driving or drilling through obstructions in the ground.

The engineer must have assurance that the piles have been correctly designed and installed in a sound manner without defects which might impair their bearing capacity. To this end piling contracts must define clearly the responsibilities of the various parties, and the installation of piles must be controlled at all stages of the operations. It will have become evident from the earlier chapters of this book that load testing cannot be dispensed with as a means of checking that the correct assumptions have been made in design and that the deflections under the working load conform, within tolerable limits, to those predicted. Load testing is also one of the most effective means of checking that the piles have been soundly constructed.

The various aspects of ground investigations, piling contracts and specifications, control of installation, load testing and other forms of test are discussed in the following sections of this chapter.

11.1 Ground investigations

11.1.1 Planning the investigation

At the time when a ground investigation is planned it is not always certain that piled foundations will be necessary. Therefore, the programme for the site work should follow the usual pattern for a foundation investigation with boreholes that are sufficient in number to give proper coverage of the site both laterally and in depth. If it becomes evident from the initial boreholes that piling is required, or is an economical alternative to the use of shallow spread foundations, then special attention should be given to ascertaining the level and characteristics of a suitable stratum in which the piles can take their bearing. Where loaded areas are large in extent, thus requiring piles to be arranged in large groups rather than in isolated small clusters, the borings should be drilled to a depth of 1.5 times the width of the group below the intended *base level* of the piles, or 1.5 times the width of the equivalent raft below the base of the raft (Figure 11.1). This depth of exploration is necessary to obtain

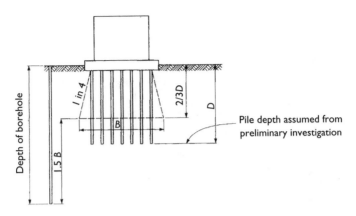

Figure 11.1 Required depth of boreholes for pile groups in compressible soils.

information on the compressibility of the soil or rock strata with depth, thus enabling calculations to be made of the settlement of the pile groups in the manner described in Sections 5.2 and 5.3. If the piles can be founded on a strong and relatively incompressible rock formation the drilling need not be taken deeper than a few metres below 'rockhead' (the buried interface between overburden or superficial sediments and rock), to check that there are no layers or lenses of weak weathered rock which might impair the base resistance of individual piles. However, before permission is given for the drilling depth to be curtailed in this manner there must be reliable geological evidence that the bearing stratum is not underlain by weak compressible rocks which might deform under pressures transmitted from heavily loaded pile groups, and that large boulders have not been mistaken for bedrock. Rockhead contours formed due to erosion prior to the deposition of the overburden may be unrelated to current topographical surface. It is sometimes the practice, when preparing borehole records, to define rockhead or bedrock as the level at which auger or cable per-cussion boring in weak rock is terminated and core drilling in the stronger rock commences. This practice is quite wrong. The decision to change to core drilling may have nothing to do with the perceived strength of the rock. It may be no more than a routine changeover of equipment at the end of a working shift.

Particular care is necessary in interpreting borehole information where the site is underlain by weathered rocks or by alternating strong and weak rock formations dipping across the site.

Without an adequate number of *cored* boreholes and their interpretation by a geologist, wrong assumptions may be made concerning the required penetration depth of end-bearing piles. Two typical cases of misinterpretation are shown in Figure 11.2.

Where piles are end bearing on a rock formation it may be desirable, for economic reasons, to obtain a detailed profile of the interface between the bearing stratum and the overburden, so enabling reliable predictions to be made of the required pile lengths over the site. Cased light cable percussive rig ('shell and auger') borings followed by rotary core drilling to prove the rock conditions can be costly when drilled in large numbers at the close spacing required to establish a detailed profile. Geophysical exploration by seismic refraction on land and by continuous seismic profiling at sea are economical methods of establishing bedrock profiles over large site areas. However, the success of these indirect methods depends on there being a sufficient contrast in seismic velocity between the rock stratum

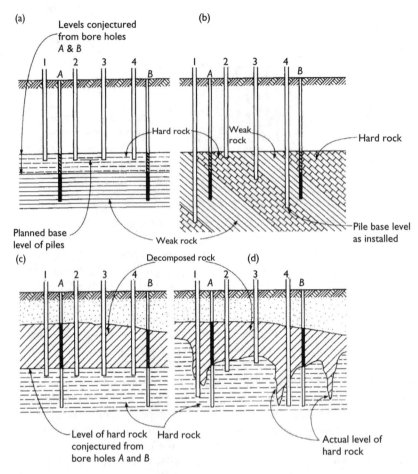

Figure 11.2 Misinterpretation of borehole information (a) Horizontal stratification interpreted by interpolation between boreholes A and B. Piles I to 4 planned to have uniform base level (b) Actual stratification revealed by drilling boreholes for piles I, 3, and 4, showing base level required by dipping strata (c) Uniform level of interface between decomposed rock and hard rock interpreted by interpolation between boreholes A and B. Piles I to 4 planned to have uniform base level (d) Actual profile of hard rock surface.

(or other strong relatively incompressible material) and the overburden. Also there must be adequate correlation with the control boreholes.

Geophysical methods are not usually economical for small site areas, but where the overburden is soft or loose, either uncased wash probings or continuous dynamic cone penetration tests (see Section 11.1.4) are cheap and reliable methods of interpolating ground conditions between widely spaced cable percussion boreholes.

Information on groundwater conditions is vital to the successful installation of driven and cast-in-place and bored and cast-in-place piles. The problems of installing these pile types in water-bearing soils and rocks are discussed in Sections 3.4.8 and 3.4.9. Standpipes or piezometers should be installed in selected boreholes for long-term observations of the fluctuation in groundwater levels. Simple forms of in-situ permeability tests are described in Section 11.1.4.

Trial pits are often a useful adjunct to borehole exploration for a piling project. Shallow trial pits are excavated in filled ground to locate obstructions to piling such as buried timber or blocks of concrete (note that by statute all excavations deeper than 1.2 m must be supported). Deep trial pits, properly shored, may be required for the direct inspection of a rock formation by a geologist, or to conduct plate-bearing tests to determine the modulus of deformation of the ground at the intended pile base level (see Sections 4.7 and 5.5). It may be more convenient and economical to make these tests at the preliminary test piling stage.

It is not the intention to describe ground investigation techniques in detail in the following section of this chapter. Details of drilling and sampling methods, geophysical surveying, and the various forms of in-situ tests are given in BS 5930. Detailed information on soil testing procedure is given in BS 1377 (methods of test for soils for civil engineering purposes). Eurocode EC7-Part 2: 2007, Ground investigation and testing, gives 'guidance for planning and interpretation of geotechnical laboratory and field tests'. However, it is not as comprehensive as code of practice BS 5930 and does not provide the detail in testing methods given in BS 1377; reference has to be made to this standard and various other procedural documents such as ISRM 'suggested methods'. The UK National Annex to EC7-2 has not yet been published.

11.1.2   Boring in soil

Cable percussion borings give the most reliable information for piling work. Operation of the boring tools from the winch rope gives a good indication of the state of compaction of the soil strata. If the casing is allowed to follow down with the boring and drilling, and water to aid drilling is used sparingly, reliable information can be obtained on groundwater conditions, but where groundwater fluctuates seasonally and tidally, standpipe readings over a period are essential. Such information cannot be obtained from wash borings or by drilling in uncased holes supported by bentonite slurry. Borings by continuous flight auger are satisfactory provided that there is a hollow drill stem down which sample tubes can be driven below the bottom of the boring and measurements of the groundwater level obtained.

Information on the size of boulders is essential for a proper assessment of the difficulties of driving piles past these obstructions or of drilling through boulder deposits for bored piles. Light cable percussion tools or flight auger drilling cannot penetrate large and hard boulders, and it is the usual practice to bring a rotary drill over the hole to core through the boulder, so obtaining information on its size and hardness. To avoid delays and standing time of the two types of drilling equipment it is more economical to continue rotary drilling past the boulder down through the remaining soil overburden to bedrock while the percussion rig is moved to an adjacent hole. It is usually possible to obtain information on the soil conditions below the level of the boulder from these adjacent boreholes.

Investigation of glacial tills for piled foundations presents particular problems in addition to the potential for random boulders, for example, identification of mixed sequence of strata, laminations of silty clay, perched water tables, infilled buried channels and samples for testing. Care is needed to ensure that compressible clays are identified below the anticipated pile tip so they are not overloaded, also soft clays which may be subjected to lateral loads from the pile shaft.

In UK practice 'undisturbed' samples of fine-grained soils are obtained from cable percussion boreholes by means of 100 mm open-drive thick-wall sample tubes. Recent developments in ground investigation techniques include the use of thin-wall sample tubes pushed into the soil and core drilling using triple core barrels to obtain continuous samples in over-consolidated clay and some coarser soils.[11.1] The samples from these techniques are of high quality with very little disturbance of the fabric of the material.

11.1.3   Drilling in rock

Weak rocks can be drilled by percussion equipment, but this technique is useful only to determine the level of the interface between the rock formation and the soil overburden. Little useful information is given on the characteristics and structure of the rock layers because they are reduced to a gritty slurry by the drilling tools and drilling should be stopped as soon as it is evident that a rock formation has been reached. Some indication of the strength of weak rocks can be obtained from standard penetration tests (see Section 11.1.4). Percussion boring can provide reliable information from rocks which have been weathered to a stiff or hard clayey consistency such as weathered chalk, marl or shale. Open drive tubes, or preferably the thin-wall pushed-in tubes, can be used in these weathered rocks to obtain undisturbed samples for laboratory testing. The improved core drilling equipment described by Binns[11.1] is now the preferred method of sampling weak and weathered rock. Hammering sample tubes into shattered rock will not produce useable samples and frequently leads to confusion and error in determination of rockhead.

The most reliable information on the strength and compressibility of rocks is obtained by rotary core drilling, supplemented as necessary by in-situ tests. The core diameter must be large enough to ensure complete or virtually complete recovery of weak or heavily jointed rocks to allow reliable assessment to be made of bearing capacity. The percentage core recovery achieved and the Rock Quality Designation (RQD) should be recorded. All cores should be stored in secure, correctly sized core boxes and selected cores should be coated in wax or wrapped in aluminium foil to preserve in-situ moisture content. Generally, the larger the core size the better will be the core recovery. Drilling to recover large diameter cores, say up to the ZF size (165 mm core diameter), can be expensive, but the costs are amply repaid if claims by contractors for the extra costs of installing piles in 'unforeseen' rock conditions can be avoided. Also, by a careful inspection and testing of the cores to assess the effects of the joint pattern on deformability, and to observe the thickness of any pockets or layers of weathered material, the required depth of the rock socket (see Section 4.7.3) can be reliably determined. It must be remembered that drilling for piles in rock by chiselling and baling or by the operation of a rotary rock bucket (Figure 3.28) will form a weak slurry at the base of the pile borehole which may make it impossible to ascertain the depth to a sound stratum for end-bearing piles. Whereas if there has been full recovery of the cores from an adequate number of boreholes together with sufficient testing of core specimens the required base level of the piles can be determined in advance of the piling operations.

Investigation of chalk for piled foundations requires attention to defining the 'marker beds' (marl and flints), variability of the chalk with depth, possible fissures and dissolution cavities, leading to determination of the 'grades' as given in the revised engineering classification of chalk[4.43] (also see Appendix). Exploration should continue for at least 5 m below the tip of the longest pile anticipated. Percussion boring can cause disturbance, and is best used in low and medium-density chalks. Rotary drilling in most grades will produce cores, but even with high-quality large-diameter cores identification of the fracture size is difficult.

11.1.4   In-situ and laboratory testing in soils and rocks

Vane tests to determine the undrained shearing strength of soft silts and clays have little application to piling operations. Shaft friction in these soils contributes only a small proportion of the total pile resistance and it is of no great significance if laboratory tests for shearing strength on conventional 'undisturbed' soil samples indicate shearing values that are somewhat lower than the true in-situ strengths. The lateral resistance of piles is particularly sensitive to the shearing strength of clays at shallow depths, and if the calculation methods can be refined to a greater degree of certainty than that exists at present then the vane test may have a useful application.

The most useful all-round test for piling investigations is the *standard penetration test* (SPT), which in clays, silts, and sands is performed with an open-ended tube and in gravels and weak rocks is made by plugging the standard tube with a cone end, when the test is sometimes known as the *dynamic cone penetration test* (CPT). The blow counts (blows per 300 mm of penetration desegnated the '*N*-value') for the SPT and CPT have been correlated with the angle of shearing resistance of coarse-grained soils (Figure 4.10) by Peck *et al.*[4.18]. Terzaghi and Peck[11.2] have given the following approximate correlation with the consistency of fine-grained soils:

N-value (blows/300 mm)	Consistency	Approx. unconfined compressive strength kN/m²
Below 2	Very soft	Below 25
2–4	Soft	25–50
4–8	Medium	50–100
8–15	Stiff	100–200
15–30	Very stiff	200–400
Over 30	Hard	Over 400

Stroud[5.7] has established a relationship between the standard penetration test and the undrained shear strength of stiff over-consolidated clays as shown in Figure 5.22. The cone-ended standard penetration test can also be made in weak rocks and hard clays. Useful correlations have been established between the blow-count values of stiff to hard clays and the modulus of volume compressibility (see Figure 5.22). The test should also be made if percussion borings are carried down below rockhead.

The standard penetration test is liable to give erroneous results if the drilling operations cause loosening of the soil below the base of the borehole. This can occur if the borehole is not kept filled with water up to ground level, or above ground level, to overcome the head of groundwater causing 'blowing' of a granular soil. Careful manipulation of the 'shell' or baler is also necessary to avoid loosening the soil by sucking or surging it through the clack valve on the baler. It is particularly necessary to avoid misinterpretation of SPT data on piling investigations since denser conditions than indicated by the test may make it impossible to drive piles to the required penetration level.

The standard penetration test cannot be performed satisfactorily at deep-sea locations, say, for example, in ground investigations for piled foundations for offshore oil production platforms. This is because the hammer is operated at the surface and the inertia of 100 m or more of drill rods from the hammer to the SPT sampler would make it impossible to achieve anything like the standard blow as performed at normal drilling depths on land investigations. Underwater hammers operating in air in sealed containers are available for deep-water ground investigations.

The application of the *static cone penetration test* to the design of individual piles is described in Section 4.3.6 and to the design of pile groups in Section 5.3. Because of the experience gained in these applications the test is particularly useful for piling investigations. A correlation has been established between the static cone resistance and the angle of shearing resistance of coarse-grained soils (see Figure 4.11). It also gives information on the resistance to the driving of piles over the full depth to the design penetration level. The standard mechanical ('Dutch') cone as developed by the Soil Mechanics Laboratory at Delft is shown in Figure 11.3a. This cone is generally used in conjunction with mechanically operated penetration devices which measure the thrust on the cone and on the sleeve separately by means of hydraulic cylinders mounted on the machines. The electric cone developed by Fugro shown in Figure 11.3b has electrical-resistance strain gauges mounted behind the cone and inside the sleeve, giving continuous readings of penetration resistance by means of electrical signals recorded on data loggers at the surface. Research by Meigh[5.20], and modified by Fugro Engineering Services Ltd for UK soils, shows that the ratio of the local side friction to cone-end resistance q_c (the 'friction ratio') can be useful in identifying the soil type. The cone shown in Figure 11.3b can be used in deep-sea ground investigations. In this case, the cone together with rods and a drive unit are lowered together to the bottom of the borehole and take the reaction from the drill string through latches bearing against the core barrel. The signals from the electrical strain gauges are transmitted by cable to the recording unit on the drilling vessel. In a comparison of the penetration resistance readings reported by Joustra[11.3] the cone resistance by the electric cone was 3.3% less than that of the mechanical cone in sands in Amsterdam.

The *continuous dynamic cone penetration* test is a useful and much neglected method of logging the stratification of layered soils such as interbedded sands, silts, and clays. The Borros penetrometer employs a 63 kg hammer impacting on a 50.5 mm cone at a rate of 20 blows per minute. The number of blows required for a penetration of 100 mm is denoted as n. The torque on the cone is measured to provide an additional means of interpreting the data. There is very little published information in the UK on correlation between n values and the SPT N-value or q_c values from static cone penetration tests. Cearns and McKenzie[11.4] have published relationships between n and the SPT N for sands, gravels, and chalk as shown in Figure 11.4.

Approximate determinations of the deformation modulus of soil and rocks can be made by expanding a cylindrical rubber membrane against the walls of the borehole and measuring

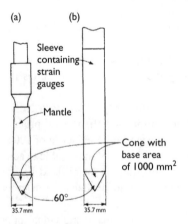

(a)　　　(b)

Sleeve
containing
strain
gauges

Mantle

Cone with
base area
of 1000 mm²

60°

35.7 mm　　35.7 mm

Figure 11.3 Types of cone for static cone penetration test (a) Mechanical cone (b) Electric cone.

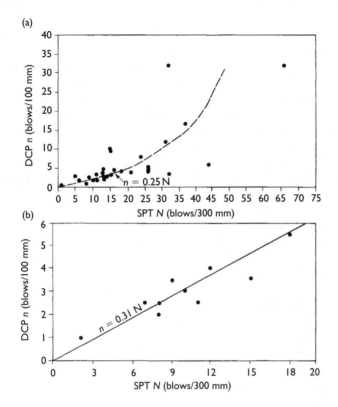

Figure 11.4 Relationship between Dynamic Cone Penetration Tests *n* and Standard Penetration Test *N* (after Cearns and McKenzie[11.4]) (a) in sands and gravels (b) in chalk.

the increase in diameter of the cylinder over an increasing range of cell pressures. Apparatus developed for this purpose includes the *Ménard pressuremeter*[6.22], the *Camkometer*[11.5], or *Cambridge Self-boring pressuremeter* (SBP), and the *High Pressure Dilatometer (HPD)*[11.6]. The Ménard pressuremeter is operated in boreholes ranging in diameter from 34 to 80 mm. Three independent rubber cells are mounted, one above the other, in the probe to apply a uniform pressure to the soil. The central cell is filled with water pressurized pneumatically, and the upper and lower cells are expanded by air. The measured co-ordinates of pressure and displacement can be inserted directly into design equations using empirical correlations based on a large data bank from standard tests. The Cambridge Insitu SBP, developed from the simpler Camkometer, is a self-boring pressuremeter with a rotating drilling bit mounted at the base of the probe so that a plug of soil is removed and replaced with the instrument in order to preserve the state of stress in the soil. The cell is expanded by compressed air and the radial deformation is measured by electrical gauges attached to feelers within the probe. The device is particularly useful in a very soft or loose soil where disturbance of the soil around a pre-drilled hole could be detrimental to accurate measurements. The Cambridge Insitu HPD, 73/75 or 93/95 mm diameter, for use in stiff soils or weak rocks is placed in a good quality pre-drilled cored borehole and inflated using either hydraulic oil or compressed gas to a maximum pressure of 300 bar. Displacement is measured on three axes by six full bridge strain gauged cantilever springs with a maximum capacity of 20 MN/m². Two hundred and forty pressuremeter tests using the HPD to determine in-situ deformation properties in Trias and Carboniferous strata were carried out to depths up to 50 m for the piling to the viaduct piers on the Second Severn Crossing[11.7]. The *Marchetti dilatometer*[11.8] is a spade-shaped device which is pushed or hammered into the soil. A load-cell is mounted on the vertical face of the spade and pushed against the soil or rock. The pressuremeter should be distinguished from a borehole jack which applies forces to the sides of boreholes by forcing apart circular plates, imposing different boundary conditions on the test.

The *cone pressuremeter*, developed by Fugro and Cambridge Insitu (Whittle[11.9]) initially for offshore site investigations, is a self-boring device which incorporates a 1500 mm² cone below the friction sleeve of the pressuremeter module and is pushed into the soil using standard cone rods. A piezo-cone may replace the cone to assist in identifying soil for pressuremeter testing.

As noted in Section 6.3.7, the pressuremeter has useful applications to determinations of the ultimate resistance to lateral loads on piles and the calculation of deflections for a given load. Because the pressuremeter only shears a soil or rock (there is no compression of the elastic soil or rock) the slope of the pressure/volume change curve in Figure 6.34 gives the shear modulus G. This can be converted to the Young's modulus from equation 6.49. In recently developed instrumentation, the data points on the load/unload loop are now very frequent so that the change in strain can be accurately measured for each successive point from a selected zero – with the smallest increment being around 0.01% radial strain. G calculated in this way more accurately reflects actual strain produced in the ground by structures, and is greater than G obtained from slopes of lines through the loops. When using the pressuremeter to obtain E values for pile group settlements using the methods described in Chapter 5, it is necessary to take into account the drainage conditions in the period of loading.

Plate-bearing tests can be used to obtain both the ultimate resistance and deformation characteristics of soils and rocks. When used for piling investigations these tests are generally made at an appreciable depth below the ground surface, and rather than adopting costly methods of excavating and timbering pits down to the required level it is usually more economical to drill holes 1 to 1.5 m in diameter by power auger or grabbing rig. The holes are lined with casing and the soil at the base is carefully trimmed by hand and the plate accurately levelled on a bed of cement or plaster of Paris[11.10]. The deformation of the soil or rock below the test plate can be measured at various depths by lowering a probe down a tube inserted in a drill hole beneath the centre of the plate. This device[11.11] is helpful in obtaining the modulus of deformation of layered soils and rocks. The load is transmitted to the plate through a tubular or box-section strut and is applied by a hydraulic jack bearing against a reaction girder as described for pile loading tests (see Section 11.4.1). Loading tests on 500 mm diameter plates were carried out in 600 mm holes 20 m deep drilled offshore from a jack-up platform to determine deformation properties in Trias rocks for the main span foundations for the Second Severn River crossing[11.7].

Small-diameter plate loading tests can be made using a 143 mm plate in a 150 mm borehole, but it is, of course, impossible to trim the bottom of the hole or to ensure even bedding of the plate. However, these tests can be useful means of obtaining the ultimate resistance of stiff to hard stony soils[11.12] or weak rocks[11.13]. They do not give reliable values of the deformation modulus.

Simple forms of *in-situ permeability test* can give useful information for assessing problems of placing concrete in bored and cast-in-place piles in water-bearing ground.

The *falling-head test* consists of filling the borehole with water and measuring the time required for the level to drop over a prescribed distance. In the *constant-head test* water is poured or pumped into the borehole and the quantity required to maintain the head at a constant level above standing groundwater level is recorded. The procedure for obtaining the coefficient of permeability is described in BS 5930. Pumping-in tests made through packers in a borehole, or pumping-out tests with observations of the surrounding drawdown are too elaborate for most piling investigations. Sufficient information to evaluate groundwater problems can often be obtained by baling the borehole dry and observing the rate at which the water rises to its standing level. Such simple tests provide useful background information for contractors tendering for bored piling work and it is false economy to neglect them.

It was noted in Section 11.1.2 that high-quality undisturbed samples of soil can be obtained by means of pushed-in samples in thin-walled tubes. Equipment has also been developed for measuring very small strains in samples undergoing triaxial compression tests. Thus reliable Young's modulus values can be obtained from samples of soil or weak rock without the need for employing special in-situ testing equipment or making plate-bearing tests in deep boreholes. Small-strain Young's modulus values are useful for calculating consolidation settlements in the mass of ground beneath pile groups rather than the base settlement of individual piles. They are also applicable to the determination of the deflection of laterally loaded piles. The adoption of improved sampling and triaxial compression testing techniques will result in undisturbed shear strength values higher than those obtained in earlier practice, particularly in very stiff to hard clays. These higher values may require modification of correlations established between the shear strength of clays and shaft friction and end-bearing resistance of piles.

Laboratory tests on rock cores should include the determination of the unconfined compression strength of the material, either directly in the laboratory or indirectly in the field or laboratory by means of point load strength tests. Young's modulus values of rock cores can be obtained by triaxial compression testing using the transducer equipment for small strain measurements.

The *point load test*[11.14] is a quick and cheap method of obtaining an indirect measurement of the compression strength of a rock core specimen. It is particularly useful in closely jointed rocks where the core is not long enough to perform uniaxial compression tests in the laboratory. The equipment is easily portable and suitable for use in the field. The tests are made in the axial and diametrical directions on cores or block samples. The failure load to break the specimen is designated as the point load strength (I_s) which is then corrected to the value of point load strength which would have been measured by a diametral test on a 50 mm diameter core using a standard correction (Table 11.1) to obtain $I_{s(50)}$.

Table 11.1 Relationship between uniaxial compression strength (q_{comp}) and point load strength ($I_{s(50)}$) of some weak rocks

Rock description	Average q_{comp} MN/m^2	$q_{comp}/I_{s(50)}$
Jurassic limestone	58	22
Magnesium limestone	37	25
Upper Chalk (Humberside)	3–8	18
Carbonate siltstone/sandstone (UAE)	2–5	12
Mudstone/siltstone (Coal Measures)	11	23
Tuffaceous rhyolite (Korea)	15–90	8
Tuffaceous andesite (Korea)	40–160	10

11.1.5   Offshore investigations

Offshore investigations for deep-water structures and oil production platforms are highly specialized and, although the basic procedures contained in the British Standards and Eurocodes for geotechnical investigations should be adhered to for UK waters, there are many additional Statutes and Regulations which apply to such work and are outside the scope of this text. Draft Guidance Notes on Site Investigations[11.15] have been prepared by the Society for Underwater Technology to provide a basic framework for offshore investigations, particularly for renewable energy projects, citing practices and regulations for the UK, other European regulatory bodies and under API regulations. Foundation design considerations include driven piles, driven and drilled piles, suction piles, and gravity bases.

11.2   Piling contracts and specifications

11.2.1   Contract procedure

In Britain it is the usual practice for the piling works for foundations on land to be executed as a sub-contract to the main general contract. Foundation piling rarely forms a high proportion of the total cost of a project on land, and the administrative arrangements and preliminary works required for carrying out the piling as a separate advance contract are unjustified in most cases. Works which are required in advance of a piling contract include such items as demolition, fencing, levelling and grading, the construction of site roads, the erection of site offices, and the supply of electrical power and water. On some large contracts it is the practice to let a separate contract for these advance works, in which case it may be feasible also to let a separate contract for the piled foundations. Piling contractors prefer to do the work in this way rather than as sub-contractors. The responsibilities of each party are defined more clearly and, from the piling contractor's point of view, a directly employed contractor has the advantage that he is paid retention money after the usual period following the completion of his work. Where a piling contractor works as a sub-contractor it is usual for part of his retention money to be withheld until the main contractor's retention is released, which may be several years after the completion of the piling work.

Piling normally forms a high proportion of the cost of marine construction and it is usually undertaken directly by the main contractor. Any sub-contracting is limited to specialist services such as grouting or the construction of anchorages to tension piles, and to the supply of prefabricated components.

Piling carried out as sub-contract work may be done by a nominated contractor, i.e. a specialist piling contractor, who has submitted a tender to the Employer/Engineer in advance of the main contract and is then selected for nomination in the main contract documents. Alternatively, the main contractor can, after appointment, invite tenders from selected specialist contractors, or from an open tender list, to be employed as sub-contractor.

The Civil Engineering Contractors Association has prepared a form of sub-contract (the CECA 'Blue Form') which can be used with the addition of a suitable clause to cover piling work. On building contracts under the control of an architect, the contract conditions are set out in the Standard Forms of Building Contract prepared by the Joint Contracts Tribunal (the 'JCT' forms), with various provisions for 'domestic' and nominated sub-contractors. In a similar manner the National Federation of Building Trades Employers has prepared standard forms of contract for nominated or non-nominated sub-contractors. In addition, ad hoc and bespoke forms of sub-contract prepared by main contractors are increasingly used to change the liabilities and risk-sharing obligations given in standard forms. Collateral warranties, which are separate from, but operate alongside, the works contract, are now frequently requested by the Employer, developer or project funders. They provide for liability to a beneficiary (who may not be the same person as the principal works owner) in respect of defective performance over a period of time, possibly longer than the statutory defects period, and must be treated with caution by specialist contractors.

Model procedures for piling contracts and specifications are given in a publication of the Institution of Civil Engineers[2.5] for use with the Institution's standard forms of contract, but with provisions for drafting to conform to other forms. Account should also be taken of the BS EN Standards for execution of special geotechnical works (see Section 2.1).

Building and construction works on behalf of government departments are usually carried out under a form of contract designated GC/WORKS/1, similar to the ICE form. Here the Employer, advisers and designers are termed the 'Authority' and the supervisory duties of the Engineer are delegated to the 'Superintending Officer'. As with the other forms, responsibility for design of piles may be placed on an independent engineer, the contractor or piling sub-contractor.

There is a growing requirement, particularly for large public sector works, for 'best value' procurement, based on 'partnering' and co-operation using specially prepared documents or 'charters'. These documents may be based on the standard forms, particularly the new Engineering and Construction Contract[11.16] (part of the Institution of Civil Engineers' NEC3 suite of contracts) which sets out various standard options for the Employer and Contractor to select: for example, 'design and build', 'build-own-operate-transfer' schemes or 'management contract' forms. One of the main benefits claimed for partnering is that it provides for a limited list of proven or preferred suppliers and sub-contractors who 'negotiate' rather than tender competitively to produce the perceived best value for the project owner/funder. Under these arrangements the specialist sub-contractors are required to enter into back-to-back agreements with the Employer and the other parties which frequently place additional risk-taking burdens on the specialist. Complex partnering, funding and project management arrangements are features of the 'private finance initiative' for major public policy works – for example, hospitals, new rail and roadworks.

In the following comments on the responsibilities for piling work it must be understood that the Engineer is not a party to the contract but acts as the agent for an Employer and accepts certain responsibilities on behalf of the Employer under, say, an ICE form of contract. The Architect or Contracts Administrator acts in a similar way for the Employer in the JCT form of contract. The term 'Contractor' refers to the piling contractor whether as the main contractor or a nominated or non-nominated sub-contractor.

The Institution of Civil Engineers' publication provides for four basic types of traditional contractual arrangement under which piling may be undertaken. These are as follows:

(a) Civil engineering works with an Engineer responsible to an Employer for design and supervision
(b) Building works with an Architect responsible to Employer for design and supervision and advised by an Engineer
(c) Building or civil engineering works with a Contractor responsible to an Employer (who has in-house expertise) for design and construction and
(d) Building works with an Architect responsible to an Employer for design and supervision but having no engineering adviser.

The need for the Engineer to define the respective status and responsibilities is discussed below in relation to procedures (b) and (c), but it should be noted that an Architect may not have supervisory responsibilities (only co-ordination authority) under a JCT contract with the Employer.

Irrespective of whether piling work is executed as a main contract or by sub-contract, in traditional British practice tenders are invited by one of three principal methods. These are as follows:

Method 1:   The Engineer is responsible for deciding the type or alternative type of pile, the working loads, and the allowable settlement under test load. The Engineer specifies the material to be used, the working stresses, fabrication methods and penetration depths. Tenders are invited on the basis of a detailed specification and drawings, which should be accompanied by a ground investigation report, and a site plan for contract works showing existing structures and surface levels, proposed re-grading levels, and the operating levels for the piling rigs.

Method 2:   The Engineer invites tenders for one or an alternative system of piling from specialist contractors. The invitation to tender is accompanied by a pile layout showing individual pile loads or column and wall loadings, and by a detailed specification including such items as materials, working stresses, performance under load test, and other criteria of acceptability. The Contractor decides on the required type (or alternative type) of pile, the diameter and the penetration depth for the specified working loads, and bases a tender on in-house estimates of performance. Site information as described for Method 1 should also be supplied.

Method 3:   The Engineer supplies a drawing to the tendering Contractor showing the wall and column layout of the structure together with the loadings; the site information as described for Method 1 is also supplied. No specification is issued and the Contractor is expected to submit a specification with his tender, and to guarantee the successful performance of the piles. Because an increasing number of piling contractors are offering new types of piles, particularly displacement piles which are designed to optimize bearing capacity and minimize concrete usage and spoil disposal, this procurement method can give best value to the Employer in these circumstances.

Method 1 has the advantage that the responsibility of each party is clearly defined. The Contractor has the responsibility only of selecting the most efficient type of plant to do the job and to install the piles in a sound manner complying with the specification. The method has the disadvantage that the knowledge and experience of the Contractor may not be fully utilized, since the Engineer may not always select the most suitable pile for the job. In exercising responsibility for deciding on the pile diameters and penetration depths, the Engineer may instruct the Contractor to install preliminary test piles before making final decisions on the dimensions of the working piles. The liability for unforeseen adverse ground conditions generally falls to the Employer on the advice of the Engineer.

Method 2 provides the widest choice of piling systems and utilizes the experience of the Contractor to the fullest extent, but greater care is needed in defining responsibility. In particular, the Engineer must specify precisely the requirements for performance under loading tests, both on preliminary and working piles. While the Contractor is responsible for selecting the type, diameter and penetration depth of the piles, calculations to justify these selections are required to be submitted for the approval of the Engineer. The statement concerning working loads on columns, walls or individual piles should make it clear as to whether or not the loads have been factored in compliance with Eurocodes or other structural codes of practice.

Method 3 is satisfactory in most respects provided that tenders are invited only from those firms that have the necessary experience. It is usually stated in the tender invitation or it is implied that the Contractor assumes responsibility for all aspects of the piling work. This would include accepting liability for unforeseen variations in the ground conditions and the possibility of having to increase substantially the penetration depth, or increase the number of piles or even to abandon a particular system. The Contractor is usually responsible for deciding whether or not load testing is required and the criteria for deciding successful performance under test loading. However, the insurers providing the Contractor with the essential warranty for his work and the Employer with cover for the structure may need to be involved in the decision. As the design and construction rules in EC7 are adopted, the Employer and Contractor will have to demonstrate that the pile design is related to static load tests and, as noted in Section 11.4, testing will be mandatory in certain conditions. The intention of Method 3 contract arrangements is to prevent unforeseen ground conditions leading to claims by the Contractor, but it is incumbent on the Employer to ensure that all information available at tender stage is passed to the Contractor, including any expert interpretations of ground conditions obtained. It is in the interest of all parties that the Contractor be allowed access to the site to carry out any additional investigations deemed necessary.

The Engineering and Construction Contract[11.16] aims to be simpler and more flexible than other standard forms mentioned. For example, it gives more than 20 options for the Employer to build up a contract to suit individual requirements; it allows for design responsibility to be carried by either the Employer or Contractor depending on which party has the competency. The traditional Engineer's role has been removed and replaced with a 'Project Manager' having defined duties including acting as the Employer's agent, issuing instructions and certifying payments, and a separate 'Supervisor' with duties for testing and inspecting the works. Disputes are referred directly to a nominated Adjudicator.

The responsibility for providing information should be clearly defined in all the forms of contract mentioned. However, the provision by the Employer through the Project Manager of 'additional information', not in the contract Works Information but necessary for the Contractor to complete the works, is only an implied term under NEC3 forms of contract. It may therefore fall to the Contractor to fill in gaps in this case. Some items of information and responsibility which should be clarified in the contract are as follows:

All relevant details of *ground investigations* undertaken on behalf of the Employer before inviting tenders for the piling. Geological data and interpretations may be excluded by the Employer from the main contract Site Information under a NEC3 contract and only be available as 'reference' data – with implications for contractual Compensation Events. The piling contractor should be aware of the need for additional ground information either prior to tendering or following award of a contract in such circumstances.

The *facilities* provided by the main contractor, or those to be included in the piling contract, should be stated. These include such items as access roads, hardstandings for piling plant, storage areas, fencing, watching, lighting, and the supply of electrical power and water. Hardstandings (working platforms)[3.20] for large piling plant may need to be of substantial construction and the Engineer/Designer should state the form in which they will be provided including the level of the platform in relation to the pile commencing surface and cut-off level.

Underground services and *obstructions* can be a contentious item. Under traditional ICE contracts it is normally the Engineer's responsibility to locate all known buried services and other obstructions to pile installation. It is unfair to the Contractor for the Engineer to disclaim all responsibility for the accuracy of the location plan, and to expect the Contractor to accept the consequences of damage to services. However, the Engineer has the right to expect that the Contractor will not push on blindly with the piling work with complete disregard for the safety of the operatives or the consequences of damage. The Contractor is expected to keep a close watch on the conditions as the piling progresses and make enquiries as to the likely presence of further underground obstructions. The consequential damages can be very severe if, for example, a water main is broken which floods the running lines of an underground railway. Hence, the clause in the conditions of contract covering underground obstructions needs to be carefully worded to be fair to the interests of all parties. For example, it may be an express condition of the contract for piling in an urban environment for the Contractor to provide a consultant services engineer to carry out investigations prior to piling.

Compliance with the *Construction (Design and Management) Regulations 1994* (CDM), the *Health and Safety at Work Act 1974* and the *Environment Protection Acts* (EPA) involves all parties to the contract even if not expressly stated. The piling contractor may be the first and only party on the site initially, requiring him under the CDM regulations to undertake the statutory duties of the 'principal contractor', and possibly the 'planning supervisor' and 'designer' all as defined. The statutory regulations under the EPA control the disposal of arisings from bored piles and waste drilling fluids, and the Health and Safety regulations cover all aspects of construction from protective clothing, lifting and hoisting appliances to access into excavations and welfare facilities.

There is usually an obligation for the Contractor to operate a *Quality Management System* (conforming to BS 5750/ISO9001) within the company and to produce a Project Quality Plan as a means of assuring the Employer that the required standards for the works have been met through traceable documentation. The System and the Plan will be subject to certification and audit either by an independent third party or by the Employer, but self-certification by the Contractor to assure compliance with the specification may be acceptable – except for laboratory testing. Surveillance and intervention by the Engineer or the Supervisor will be in addition to the Contractor's demonstration of conformance under his plan.

Responsibility for *risk assessments* to identify the hazards (risk events), probability of a risk event occurring, ensuing injury, damage and loss, and any general uncertainties will initially fall to the Employer's team as part of the project feasibility study. The piling contractor will have to advise the Employer of potential hazards involved in the particular method proposed. In the NEC3 contract a 'risk register', which contains descriptions of the risks and the actions to be taken to avoid or reduce the risks, is included in the Contract Data.

11.2.2 Piling specifications

The Institution of Civil Engineers' Specification[2.5] details items which should be included in the 'Particular Specification' for a piling contract. These include stating responsibility for design, performance criteria to be applied, requirement for additional ground investigation as well as routine matters on site location, personnel, and so on. It also states that all materials and workmanship 'shall be in accordance with the appropriate British Standards, Codes of Practice and other specified standards' and specifications should quote the relevant standard under the various work classifications. BS 8004 can be quoted for working stresses on the various types of pile, but this is not provided for in limit-state design to Eurocode EC7. It is not the intention in this chapter to give model clauses for piling work. These can be drafted with the guidance of the ICE model specifications, BS 8004 or other codes, particularly EC7, BS EN 1536 for bored pile execution, BS EN 12699 for displacement piles and BS EN 14199 for minipiles. Some matters which require particular attention are listed below:

Setting out. The responsibility for setting out is clear if the piling contractor is the main contractor. The Engineer has no responsibility in the matter but should check the positions of the piles from time to time, since if these are inaccurately placed the remedial work can be very costly. Problems arise when a piling sub-contractor does the setting-out from a main contractor's grid-lines. If these are inaccurate or if, as sometimes happens, the numbering is obscured (or level pegs are confused with line pegs) then there can be major errors in pile positions and the main contractor may decline to accept responsibility for the cost of the replacement piling. If the specification does not define the responsibility for setting out, the piling sub-contractor must have a clear understanding with the main contractor on this matter.

Ground heave. In the case of the Method 1 type of contract the Engineer, in specifying the type and principal dimensions of the pile, must accept responsibility for the effects of ground heave, as described in Section 5.7. However, if the contract is of the Method 2 category the matter is not so clear, and piling contractors are reluctant to accept responsibility for ground heave, either for remedial work to risen piles, or for repairing damage to surrounding structures. In the case of both Method 2 and Method 3, the authors are of the opinion that as it is the Contractor who decides on the type and dimensions of the pile, and therefore should have experience of ground heave effects, the Contractor should accept full responsibility for the site operations. If pre-boring or other measures are insufficient to prevent ground heave, the Contractor would be well advised to decline to tender.

Loss of ground due to boring. The consequences of a loss of ground while boring for piles were described in Section 5.7. The responsibilities for these are similar to those for ground heave.

Noise and vibration. The Contractor is responsible for selecting the plant for installing piles and is therefore responsible for the effects of noise and vibration (see Section 3.1.7). The current statutory and local authority regulations limiting noise emissions should be stated in the specification (or conditions of contract).

Piling programme. If the Engineer wishes to install the piles for the various foundations in a particular sequence to suit the main construction programme the sequence should be stated in the specification, since it may not be the most economical one for the piling contractor to follow.

'Set' of driven piles. This should not be stated in precise terms in specifications for driven, or driven and cast-in-place piles. The 'set' for a particular site and working load cannot be established until preliminary piles have been driven and the driving records checked against the ground conditions assumed in design. The set may have to be modified as a result of loading tests.

Tolerances. Tolerances in plan position, vertical deviation from the required rake, and deviation in level of the pile head should be specified. Suitable values for tolerances are given in Section 3.4.12.

Monitoring of piling is mandatory under EC7-1 Clause 7.9 and the BS EN standards for execution of special geotechnical works, in accordance with a method statement and 'pile installation plan' which are consistent with the design.

Piling records. The Engineer and the Contractor should agree to the form in which records should be submitted (see Section 11.3).

Cutting down pile heads. The specification should define whether it is the main contractor's or the piling contractor's responsibility to remove excess lengths of pile projecting above the nominal cut-off level. The responsibility for cutting away concrete to expose reinforcement, and trimming and preparing the heads of steel piles should also be stated.

Method of measurement. The method of measuring pile lengths as installed should be based on an appropriate standard, for example, as given in Civil Engineering Standard Method of Measurement[11.17] or in the ICE Specification, modified as necessary to meet particular contract circumstances, structures and sites. Care is needed to define the length of pile to be measured (i.e. from cut-off level to pile toe or ground level to toe) and to state that it is the total length of a particular type and diameter of pile is to be measured. Credits for short piles installed and extra payment for additional lengths of individual piles are generally excluded from the standard methods.

Removal of spoil. The respective responsibilities for the removal of spoil from bored piles, the removal of cut-off lengths of pile, trimming off laitance and ground raised by ground heave, and the disposal of used bentonite slurry should be covered in the specification and conform with the statutory regulations for waste disposal currently in force (see Section 11.2.1).

11.3   Control of pile installation

11.3.1   Driven piles

Control of driven pile operations commences with the inspection and testing of the prefabricated piles before they are driven. Thus timber piles should be inspected for quality, straightness and the application of preservative. The operations of casting precast concrete piles on site or in the factory should be inspected regularly and cubes or cylinders of the concrete should be made daily for compression testing at the appropriate age. Materials used for concrete production should be tested for compliance with the relevant standards. In the case of steel piles, tests should be made for dimensional tolerances and full documentation of the quality of the steel in the form of manufacturers' test certificates should be supplied with each consignment. Welding tests should be made for piles fabricated in the factory or on site. Full radiographic inspection of welds may be necessary only for marine piles, where the exposure conditions are severe (see Section 10.4.2). The coating treatments should be checked for film thickness, continuity, and adhesion. Degaussing may be needed to counter magnetization of the pile heads caused by driving. This can be detrimental to the quality of welds made for pile extensions. BS EN 288 covers degaussing steel by generating a counter-active magnetic field when welding is in progress.

The ICE Specification lists the information which should be recorded for each type of pile; Table 11.2 is a typical compliant form. A separate record should be provided for each pile and records should be signed by the Contractor's and Employer's representatives and submitted daily. Records to comply with EC7, Clause 7.9, are similar to Table 11.2, but should be provided in two parts according to BS EN 12699 for each displacement pile driven. Part 1 should give general information on the contract and type of pile, methods and quality of materials; Part 2 'particular information' as tabulated in Clause 10 of this standard for each pile. 'As built' records of piles have to be submitted to the Employer. It is advisable to keep all records for a period of 5 years after completing the works.

Table 11.2 Daily pile record for driven pile

DAILY PILE RECORD FOR DRIVEN PILES												
PILE RECORDS TO BE SUBMITTED TO OFFICE DAILY
A SEPARATE SHEET TO BE USED FOR EACH PILE

BLOCK NUMBER					DRAWING NUMBER						
SUB CONTRACTOR					PILE TYPE						

1.GENERAL	PILE REF. NO.		PILE DIA.		LEVEL OF TOE	
	GROUND LEVEL		CUT OFF LEVEL		CONCRETED LEVEL	
2.DRIVING	DATE STARTED		DATE COMPLETED		AIR TEMPERATURE	
	ERROR IN POSITION ON PLAN		ERROR IN PLUMB		DEPTH DRIVEN	
3. OBSTRUCTIONS	TYPE		DEPTH ENCOUNTERED		PENETRATION TIME	
4.*STEEL MAIN LINKS OR HELIX	NO. OF BARS		DIAMETER		LENGTH	
	CENTRES OF BARS/PITCH		DIAMETER		COVER TO ALL STEEL	
5.CONCRETE CAST-IN-SITU WORK ONLY	DATE STARTED		TYPE OF CEMENT		CONCRETE TEMPERATURE	
	MIX		SLUMP		SUPPLIER	

6.*HAMMER	TYPE		MASS	CAST-IN-SITU CONCRETE PILES ONLY ESTIMATED QUANTITY OF CONCRETE IN BULB		
7.FINAL SET	NO. OF BLOWS			DROP OF HAMMER	MOVEMENT OF PILE	

8.PRECAST CONCRETE PILES ONLY	COMPONENTS OF PILE	A	B	C	D
	LENGTH OF COMPONENT				
	REF. NO./DATE CAST				
	O.D. LEVEL OF JOINT				

9.PILE DRIVING LOG	DEPTH DRIVEN	NO. OF BLOWS	HAMMER DROP	DEPTH DRIVEN	NO. OF BLOWS	HAMMER DROP	DEPTH DRIVEN	NO. OF BLOWS	HAMMER DROP	DEPTH DRIVEN	NO. OF BLOWS	HAMMER DROP
OTHER REMARKS MAY BE RECORDED ON REVERSE SIDE OF THIS SHEET.												

NOTE * IF THERE ARE NO CHANGES TO BE RECORDED ITEMS 4 & 6 NEED BE COMPLETED FOR THE FIRST PILE ONLY IN EACH BLOCK.	SIGNED	CONTRACT SITE ENGINEER

While it is essential for the toe level and final set of every pile to be recorded, BS EN 12699 does not mandate a full record of sets during driving. There are, however, advantages in providing a log of the blow count against penetration over the full depth for every pile driven. If, for example, piles are to be driven to end bearing on a hard stratum it may be sufficient to record the sets in blows for each 25 mm of penetration after the pile has reached the hard stratum. On the other hand, where piles are supported by shaft friction, say in a stratum of firm to stiff clay, or in a granular soil overlain by weak soils, it is essential to record for every pile the level at which the bearing stratum is encountered and hence to check that the required length of shaft to be supported is obtained. For this purpose, the blows required for each 500 mm or each 250 mm of penetration must be recorded over the full depth of driving of each pile, until the final metre or so when the sets are recorded in blows for each 25 mm. Sometimes final sets are recorded as penetration depths for 10 to 25 blows of the hammer. The advantage of recording the full driving log for piles of every category is that if troubles arise, such as pile breakage, the records of each pile can be scrutinized, and any which show peculiarities can be singled out for special examination or testing.

At the preliminary piling stage the driving records are compared with the ground investigation data, the static design criteria and with the results of loading tests, and suitable criteria regarding final sets for terminating driving are established. If the methods of Chapter 4 have been used for calculating the penetration depth of friction piles, the depth into the bearing stratum should, theoretically, be the only criterion, and final sets should be irrelevant. However, because of natural variations in soil properties piles with identical lengths in the bearing stratum will not necessarily have identical ultimate loads. By driving to a minimum depth into the bearing stratum and to a constant final set (or to within a specified range of set) the variations in the soil properties can be accommodated.

A minimum penetration is necessary because random compact layers in the soil may result in localized areas of high driving resistance. The driving records within these layers should be compared with the ground investigation data, so that suitable termination levels can be established. The establishment of criteria for controlling the termination of piles driven into layered soils is described in Section 4.5.

It is advisable to conduct re-driving tests on preliminary piles, and on random working piles. These tests are a check on the effects of heave and on possible weakening in resistance due to pore pressure changes. Re-driving can commence within a few hours in the case of granular soils, after 12 hours for silts, and after 24 hours or more for clays. If the re-driving shows a reduction in resistance after about 20 blows, driving should continue until the original final set is regained.

Diagrams of the driving and re-driving tests should be made for the preliminary piles, and compared with the borehole records and with in-situ and laboratory test data.

The temporary compression at various intervals of pile driving is irrelevant if working loads have been obtained by the methods described in Chapter 4. However, if dynamic formulae are adopted the temporary compression values must be taken at intervals after the pile enters the bearing stratum. The values are obtained by securing a sheet of graph paper to the pile by adhesive tape. A straight-edge is held horizontally close to the pile and using the straight-edge a pencil line is drawn across the paper during the impact of the hammer (Figure 11.5a). The pattern of the pencil line is shown in Figure 11.5b from which the temporary compression is measured.

Other items to be recorded include any obstructions to driving or damage to the pile and deviations in alignment which might indicate breakage below the ground surface. Methods of checking the alignment of steel tubular and H-piles are described in Section 2.2.4.

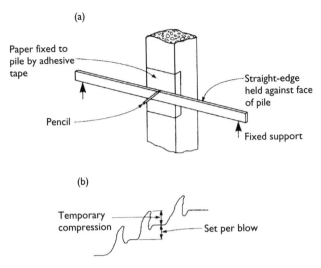

(a)

Paper fixed to
pile by adhesive
tape

Pencil

Straight-edge
held against face
of pile

Fixed support

(b)

Temporary
compression

Set per blow

Figure 11.5 Measuring set and temporary compression on driven pile (a) Arrangement of straight-edge
and paper card (b) Pencil trace showing set and temporary compression.

Hollow precast concrete piles can be checked for alignment in a similar way to steel tubes.
It would be advantageous if manufacturers of jointed precast concrete piles were to provide
a central hole in each unit, or in a proportion of the piles cast, down which an inclinometer
could be lowered on the completion of driving.

The use of accelerometers and strain transducers to measure stresses in piles at the time
of driving has been described in Section 7.3. This equipment gives a useful check on
workmanship at the construction stage.

11.3.2   Driven and cast-in-place piles

Table 11.2 is a suitable form of record. Generally the procedure for recording driving
resistances and sets is similar to that described in the preceding section, but in the case of
proprietary piles the piling contractor decides the criteria for the final set.

The concrete should be tested for compliance with the specification for materials and mixing and cubes or cylinders taken daily for compression tests. The quantity of concrete placed
in the shaft of each pile should be recorded as a check against the possible collapse of the soil
during the withdrawal of the tube. By checking the level of the concrete as each batch is
placed an indication is given of possible 'necking' of the shaft. The volume of concrete in an
enlarged base should be recorded as a check on the design assumptions for the diameter of
base. A check should be made on the level of the reinforcing cage after withdrawing the drive
tube on every pile driven. This is a safeguard against the concrete being lifted with the tube.

Thin shell piles should be inspected before placing the concrete by shining a light down
the hole. This reveals any torn or buckled shells.

11.3.3   Bored and cast-in-place piles

The Table 11.3 record gives information required in the ICE Specification and complies
generally with EC7, Clause 7.9, but as for displacement piles, records have to be provided

Table 11.3 Record for bored pile

DAILY PILE RECORD FOR LARGE-AND SMALL-DIAMETER BORED PILES
PILE RECORDS TO BE SUBMITTED TO OFFICE DAILY
A SEPARATE SHEET TO BE USED FOR EACH PILE

BLOCK NUMBER			DRAWING NUMBER		/	/
1. General	PILE REF. NO.		PILE DIA.		LEVEL OF BASE	
			UNDERREAM DIA.			
	GROUND LEVEL		CUT OFF LEVEL		CONCRETED LEVEL	
2. Drilling	DATE STARTED		DATE COMPLETED		AIR TEMP	
	ERROR IN POSITION ON PLAN		ERROR IN PLUMB		DEPTH BORED	
3. *Obstructions Natural	TYPE		DEPTH ENCOUNTERED		PENETRATION TIME	
Unnatural	TYPE		DEPTH ENCOUNTERED		PENETRATION TIME	
4. *Steel main steel	NO. OF BARS		DIAMETER		LENGTH	
links or helix	CENTRES OF BARS/PITCH		DIAMETER		COVER TO ALL STEEL	
5. Concrete	DATE STARTED		DATE COMPLETED		CONCRETE TEMP.	QUANTITY ACTUAL: THEORETICAL:
	MIX		SLUMP		SUPPLIER	
6. Borehole log and rock excavation	DEPTH OF SOIL	DESCRIPTION OF SOIL	DEPTH OF ROCK	DESCRIPTION OF ROCK	DEPTH OF ROCK AUGERED	DEPTH OF ROCK CHISELLED
7. *Casing	DEPTH OF TEMPORARY CASING		DEPTH OF PERMANENT CASING		REASON FOR USE OF PERMANENT CASING	
8. *Water	DEPTH ENCOUNTERED		DETAILS OF STRONG FLOW		DETAILS OF REMEDIAL MEASURES	
	DEPTH TO STRONG FLOW					

Note: * If there are no changes to be recorded, items 3, 4, 7 and 8 need be completed for the *first pile only* in each block.

Remarks

SIGNED   CONTRACT SITE ENGINEER in two parts according to BS EN 1536 Annex B for each bored pile. Records for CFA piles should include the pitch of the screw and the factors included on data loggers used to monitor construction, for example, the penetration per revolution, torque of drilling motor and pumping pressure of grout or concrete (Figure 2.32). Clause 9.2.5 of BS EN 1536 states that 'ground behaviour' during excavation shall be observed and any changes which may be important shall be communicated to the supervisor and designer. Reference should be made

Figure 11.6 Safety cage used for inspection of pile boreholes.

to the comprehensive set of tables which detail the information and frequencies required under this Standard.

If the boreholes are free of water, the conditions at the base of small-diameter piles in dry boreholes can be checked by shining a light down to the bottom before placing the concrete. In the case of large-diameter piles the base of all piles should be inspected from a safety cage of the type shown in Figure 11.6 or by lowering the inspector in an approved chair or harness (see BS EN 813). The safety precautions should follow the procedure described in BS 8008 (safety precautions and procedures for the construction and descent of machine-bored shafts for piling and other purposes). The procedures and problems in placing concrete in pile boreholes are described in Sections 3.4.5 and 3.4.6. An essential factor in controlling these operations is the maintenance of records of the quantity of concrete placed in each pile, and preferably in addition, the level of the concrete in the shaft as each batch is placed. Tests to control the quality of materials and mixing of concrete are, of course, required and must be specified.

11.4   Load testing of piles

EC7-1 Clause 7.5 defines when pile tests are to be considered mandatory, in summary:

- when using a type of pile or installation method for which there is no comparable experience
- where piles have not been tested under similar conditions
- where theory and practice are insufficient to give confidence in the design and
- when observations during installation indicate behaviour which deviates from anticipated behaviour.

General requirements are given for static and dynamic tests, trial piles, and tests on working piles.

11.4.1   Compression tests

Two principal types of test are used for compressive loading on piles. The first of these is the constant rate of penetration (CRP) test developed by the Building Research Establishment,[11.18] in which the compressive force is progressively increased to cause the pile to penetrate the soil at a constant rate until failure occurs. The second type of test is the maintained load (ML) test in which the load is increased in stages to some multiple, say 1.5 times or twice the working load with the time/settlement curve recorded at each stage of loading and unloading. The ML test may also be taken to failure by progressively increasing the load in stages.

The CRP method is essentially a test to determine the ultimate load on a pile and is therefore applied only to preliminary test piles or research-type investigations. The method has the advantage of speed in execution and because there is no time for consolidation or creep settlement of the ground the load/settlement curve is easy to interpret. BS 8004 states that penetration rates of 0.75 mm per minute are suitable for friction piles in clay and 1.55 mm per minute for piles end bearing in a granular soil. The CRP test is not suitable for checking the compliance with the specification requirements for the maximum settlement at given stages of loading. There is also the difficulty of pricing tenders for this form of load testing since the failure load on the pile is not known with any certainty until the test is made.

The ML test is best suited for contract work, particularly for proof loading tests on working piles. It is also suitable for use where empirical methods are employed to predict the ultimate load from measurement of residual deflections after returning the test load to zero at four or five stages up to the maximum (see below). The load at each stage is held until the rate of settlement has decreased to less than 0.25 mm/hour and is still decreasing.

EC7-1 Clause 7.5 deals with pile load tests in general and outlines procedures for static and dynamic load tests, trial piles and testing working piles. BS EN 1536 refers to EC7 requirements giving recommendations for ML, CRP, dynamic and integrity testing which are slightly different from the BS 8004 procedures. For example, in ML tests the loads should be constant at each of at least six stages for a specified duration and unless otherwise stated the displacement rate should be less than 0.1 mm/20 minutes at the end of each stage. For CRP tests the rate of displacement should be constant at approximately 1 mm per minute, unless otherwise agreed. BS EN 12699 is less prescriptive for displacement piles, but requires testing to be in accordance with the relevant parts of EC7 and the specifications.

The section of the ICE Specification dealing with static load tests defines the *Specified Working Load* (SWL) as 'the specified load on the head of a pile as stated in the relevant Particular Specification' or in provided schedules. This is differentiated from the *Design Verification Load* (DVL) which is defined as 'a load which will be substituted for the specified working load for the purpose of a test and which may be applied to an isolated or singly loaded pile at the time of testing the given conditions of the Site'. The DVL takes into account special conditions which may not apply to all piles on the site such as negative shaft friction, or variations in pile head casting level.

The ICE recommends that a proof load test should normally be the sum of the DVL plus 50% of the SWL applied in the sequence shown in Table 11.4 for multi-cyclic pile tests (reproduced with permission of Thomas Telford Limited).

Table 11.4 Loading sequence for proof load test to a maximum of 100% DVL plus 50% SWL

Load	Minimum time of holding load
25% DVL	30 minutes
50% DVL	30 minutes
75% DVL	30 minutes
100% DVL	6 hours
75% DVL	10 minutes
50% DVL	10 minutes
25% DVL	10 minutes
0	1 hour
100% DVL	1 hour
100% DVL + 25% SWL	1 hour
100% DVL + 50% SWL	6 hours
100% DVL + 25% SWL	10 minutes
100% DVL	10 minutes
75% DVL	10 minutes
50% DVL	10 minutes
25% DVL	10 minutes
0	1 hour

With regard to loading procedures, EC7 Clause 7.5.2.1 only refers in a note to the recommended procedure 'Axial Pile Loading Test, Suggested Method' as published in the ASTM *Geotechnical Testing Journal*, June 1985, pp. 79–90. This is generally followed in current European practice. However, ISSMGE has produced a recommendation document[11.19] for the execution and interpretation of axial static pile load tests which is consistent with EC7 philosophy and is likely to form the basis of a new European standard. A note in Clause 7.5.3 refers to the procedures for dynamic tests in ASTM Designation D4945, 'Standard Test Method for High-Strain Dynamic Testing of Piles'.

The ASTM axial load procedure can be used for a test to twice the working load. If it is desired to obtain the ultimate load on a preliminary test pile it is useful to adopt the ML method for up to twice the working load, and then to continue loading to failure at a constant rate of penetration. A further modification of the ML test consists of returning the load to zero after each increment. This form of test is necessary if the net settlement curve is used as the basis of defining the failure load (see Section 11.4.2). It is essential to maintain a constant rate of load application as specified in EC7. It has been found that the slower the rate the smaller is the ultimate failure load. Hence the need for standardization both for ML and CRP testing.

CRP and ML tests use the same type of loading arrangements and pile preparation. A square cap is cast onto the head of a concrete pile with its underside clear of the ground surface. Steel piles are trimmed square to their axis and a steel plate is welded to the head, stiffened as necessary by gussets. Suitable loading arrangements for applying the load to the pile by a hydraulic jack using as the reaction either kentledge, tension piles or cable anchors are shown in Figures 11.7, 11.8 and 11.9 respectively. The clearances between the pile and the reaction support systems are noted in each case. These are necessary to avoid the induced horizontal pressures from the supports having an appreciable effect on the shaft friction and base load of the test pile. It is uneconomical to space the supports so widely apart that all effects are eliminated, and if necessary the contribution of these surcharge effects should be calculated and allowed for in the interpretation of the test results.

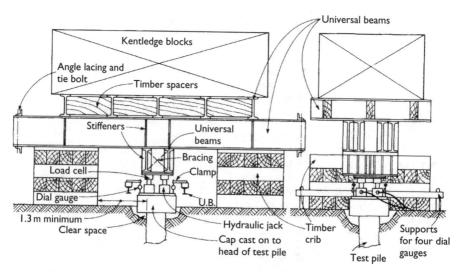

Figure 11.7 Testing rig for compressive test on pile using kentledge for reaction.

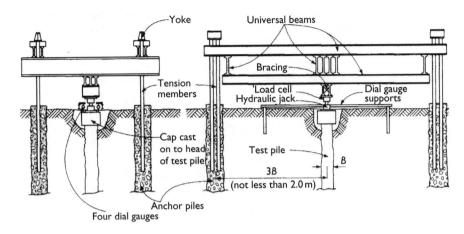

Figure 11.8 Testing rig for compressive test on pile using tension piles for reaction.

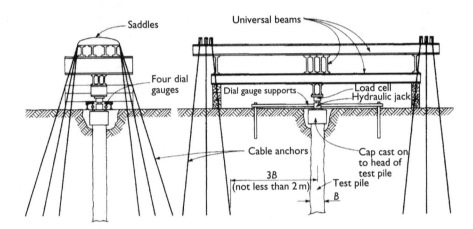

Figure 11.9 Testing rig for compressive test on pile using cable anchors for reaction.

Where piles are installed through fill or soft clay these materials give positive support in shaft friction to the test pile, whereas they may add to the working load in negative shaft friction on the permanent piles. It may therefore be desirable to sleeve the pile through these layers by using a double sleeve arrangement. Alternatively, the outer casing can be withdrawn, after filling the annular space between it and the steel tube encasing the test pile with a bentonite slurry.

It is inadvisable to test raking piles by a reaction from kentledge or tension piles since the horizontal component of the jacking force cannot be satisfactorily restrained by the jacking system. Cable anchors inclined in the same direction as the raking piles can be used but it is preferable to determine the ultimate or allowable loads on raking piles by installing special vertical piles for loading tests.

BS EN 1536 advises that the supports or anchorages of a reaction system for pile loading tests should observe minimum clearances from the test pile (diameter D) as follows:

(a) Kentledge supports: clearance $\geq 3 \times D$
(b) Bond lengths of vertical anchorages: clearance $\geq 3 \times D$ and 3 m
(c) Bond lengths of inclined anchorages: clearance $\geq 5 \times D$ and ≥ 5 m

BS EN 12699 for tests on displacement piles and EC7-1 are silent on support spacing.

The combined weight of the kentledge and reaction girders, or the calculated resistance capacity of tension piles or cables, must be greater than the maximum jacking force required. In the case of kentledge loading the combined weight should be about 20% greater than this force. Cable anchorages or tension piles should have an ample safety factor against uplift. The former can be tested by stressing the anchors after grouting them in. If there is any doubt about the uplift capacity of tension piles a test should be made to check the design assumptions. Increased capacity of tension piles in clays can be obtained by under-reaming them (see Section 6.2).

The reaction girders and load-spreading members should be so arranged that eccentric loads caused by any lateral movement of the pile head will not cause dangerous sidesway, or buckling of the girders. Connections should be bolted so that they will not become dislodged if there is a sudden rebound of load due to the failure of the pile shaft or of the jack. Similarly the kentledge stack should not be arranged in such a way that it may topple over.

Restraint by a pair of anchors from a single pile to each end of the reaction girder is not a good practice as it can cause dangerous sidesway of a deep girder. The piles or anchor cables should be placed in pairs at each end of the girders, as shown in Figures 11.8 and 11.9. Permanent piles can be used as anchorages for ML tests on working piles but it is unwise to use end-bearing piles for this purpose when the shaft friction will be low and the pile may be lifted off its seating. When using tension piles special threaded anchor bars extending above the pile head should be cast into the piles for attachment to the reaction girders. It is inadvisable to weld such bars to the projecting reinforcing bars because of the difficulty in forming satisfactory welds to resist the high tensile forces involved.

The hydraulic jack should have a nominal capacity which exceeds by 20% or more the maximum test load to be applied to the pile. This is necessary in order to avoid heavy manual pumping effort when nearing maximum load and to minimize the risks of any leakage of oil through the seals. The jacking force on the pile head should be measured by a load cell or pressure capsule since the pressure gauge fitted to most jacks is not sufficiently accurate, particularly when working towards the maximum capacity of the jack. However, the jack should have a pressure gauge mounted on the pumping unit which is calibrated to read in terms of the force on the ram. This gauge is necessary since the load cell mounted on the pile head may not be visible from the pumping position. For high-capacity piling tests, much heavy manual effort is saved by providing a mechanical pumping unit, and where CRP tests are being made a load pacer is a useful addition. The ram of the jack should have a long travel where piles are being loaded near to the failure condition. This avoids the necessity of releasing oil pressure and repacking with steel plates above the ram as the pile is pushed into the ground. Equipment is available for monitoring and restoring jacking loads at intervals of only a few seconds[11.20].

The reaction girders, anchorages and jacking arrangements for a 5800 tonne static load test in Taipei are shown in Figure 11.10.

Figure 11.10 Patented arrangement for a 5800 tonne static load test.

The settlement of the pile head under load can be measured optically by means of a surveyor's level reading onto graduated scales fixed to the pile in four positions or by laser beam producing an image on a photo-electric detector. Using a digital optical theodolite set up 3 to 10 m from the pile and a suitable target the movement of the pile can be measured to an accuracy of ±0.5 mm which is more than adequate for most piling tests. An alternative method is to set a dial gauge on each of four reference points on the pile head. The dial gauges are clamped to a datum frame securely mounted well clear of the ground around the pile or the reaction support system, and the gauges are calibrated to read to 0.1 mm. This order of accuracy is not realized in practice since wind, temperature effects, and ground vibrations can cause the datum frame to move by much more than 0.1 mm. However, it is helpful to be able to read to such an accuracy when making each increment of jacking force since the time/settlement curve can then be plotted accurately and the rate of decrease of movement is readily obtained. Levels should be taken on the datum frame before and after the loading test to check that the frame has not been displaced during the test. A linear potentiometer can be used to obtain the pile movements, which are read on a dial or print-out mechanism at an instrument station well clear of the pile. The kentledge support system must be carefully designed to give technicians safe access into the confined space under the kentledge to install and read the dial gauges.

As reported by Fleming[11.20], it is now usual practice to record pile head loads and settlements directly on a portable computer to plot load/settlement and time/settlement curves as the test is in progress. The data can be reproduced in the format of the test report and used to analyse the pile behaviour throughout the whole range of loading.

Where piles have been designed by the methods described in Chapter 4 it is very helpful to provide devices whereby the shaft and base loads can be evaluated separately. The load on the base of the pile can be measured by inserting load measuring devices in a cylindrical unit interposed between the pile base and the shaft. A typical installation consists of a ring of pillar-type load cells around the periphery of the unit recording to a data logger at the ground surface[11.21]. The distribution of shaft friction on the pile shaft can be measured by fixing electrical-resistance strain gauges onto the interior surface of a hollow steel pile, or to a steel pipe embedded in a precast or cast-in-place concrete pile. Gauges of this type can withstand the impact of pile driving and have given satisfactory service on piles which have remained in the ground for a year or more.

A simple method described by Hanna[11.22] for obtaining the distribution of shaft friction on the shaft of long hollow-section piles consists of installing metal rods down the interior of the pile. The rods are terminated at various levels as shown in Figure 11.11 and are free to move in guides as the pile settles under load. By means of dial gauges mounted on the heads of the rods the elastic shortening of each length of pile between the toes of the rods can be measured. Thus the load reaching the pile shaft at the toe of each rod is given by the

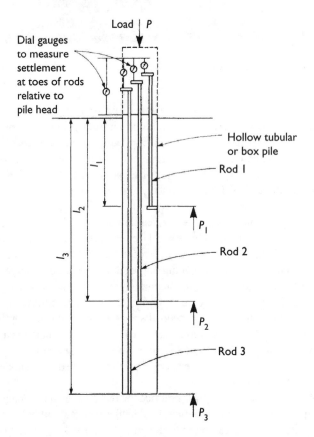

Figure 11.11 Use of rod strain gauges to measure load transfer from pile to soil at various levels down pile shaft.

following expressions:

$$\text{Load at pile } P_3 = \frac{2AE\Delta_3}{l_3} - P \tag{11.1}$$

$$\text{Load on pile at toe of rod 2 } P_2 = \frac{2AE\Delta_2}{l_2} - P \tag{11.2}$$

where P_3 and P_2 are the loads reaching the pile at the level of the toe of the rod considered, A is the cross-sectional area of the pile shaft, E is the elastic modulus of the material forming the pile, Δ_3 and Δ_2 are the elastic shortenings of the pile between the pile head and rod 3, and the pile head and rod 2, respectively, l_3 and l_2 are the lengths of the rods, and P is the load on the pile head.

The load at the toe of rod 1 is obtained in a similar manner.

Where the rod strain gauges are used in the interior of a steel tubular pile filled with concrete the elastic shortening between each length of pile is that due to the elastic modulus of the composite section. Thus

$$\Delta_l = \frac{P \times l}{A_s E_s \left(1 + \dfrac{A_c}{A_s} \cdot \dfrac{E_c}{E_s} \right)} \tag{11.3}$$

where Δ_l is the elastic shortening over length l, P is the load on length l, A_s is the area of steel, A_c is the area of concrete, E_s is the elastic modulus of the steel, and E_c is the elastic modulus of the concrete.

While these forms of instrumentation are used mainly for research-type investigations they can be adopted for the preliminary test piling to give useful design information at a relatively small additional cost.

Further guidance on the procedure for pile load testing is given by Weltman[11.23].

Dynamic load tests and high strain integrity testing have developed from the need to determine the static load capacity of driven piles (designed using empirical dynamic formulae as Section 7.3) at the time of driving. The analyses provide the soil resistance mobilized at the time of test and may therefore not show time-dependent effects of consolidation on settlement, particularly when working load is near to ultimate pile capacity. EC7 Clause 7.6.2 sets stringent criteria for the use of dynamic load tests for assessing the compressive resistance of piles:

- an adequate ground investigation has been carried out and
- the method has been calibrated against static load tests on the same type of pile, of similar length and cross-section and in comparable ground conditions.

If more than one type of dynamic test is performed then cross-checking of results is mandatory.

The *SIMBAT*[11.24] technique for bored cast-in-place piles applies a series of blows to the pile and measures the set at each blow as in the static tests. Extensions of the stress wave analytical methods are then used, which, under EC7 Clause 7.6.2.6, must also be calibrated against static load tests. BS EN 1536 requires any rapid loading tests to be correlated with maintained static load tests in similar ground. Clearly, cast-in-place piles should not be tested until they have gained sufficient strength.

The *Statnamic test* developed jointly by the Berminghammer Corporation in Canada and TNO-IBBC in the Netherlands[11.25] is now widely used. The load is applied to the pile head by the reaction of an explosive force designed to raise weights mounted on the pile head to a height of about 2.5 m at accelerations of up to 20 g. It is claimed that over 10 MN of force is generated by the explosion of fuel in a combustion chamber beneath the stack of cylindrical weights. It is also claimed that the duration of the explosive reaction of about 100 milliseconds can reproduce the effect of static loading up to the working load, with little divergence at twice the working load; thus simulating a static rather than a dynamic load test of the type described in Section 7.3. The force on the pile is measured by a load cell and the deflections of the pile by a laser beam and light-sensitive sensor.

11.4.2   Interpretation of compression test records

A typical load/settlement curve for the CRP test and a load/time/settlement curve for the ML test are shown in Figure 11.12. The ultimate or failure load condition can be interpreted in several different ways. While there is no doubt that failure in the soil mechanics sense occurs when the pile plunges down into the ground without any further increase in load, from the point of view of the structural designer the pile has failed when its settlement has reached the stage when unacceptable distortion and cracking is caused to the structure which it supports. The latter movement can be much less than that resulting from ultimate failure in shear of the supporting soil.

With reference to Figure 11.12, some of the recognized criteria for defining failure loads are listed as follows:

(1) The load at which settlement continues to increase without any further increase of load (Point A)
(2) The load causing a gross settlement of 10% of the least pile width (Point B)
(3) The load beyond which there is an increase in gross settlement disproportionate to the increase in load (Point C)
(4) The load beyond which there is an increase in net settlement disproportionate to the increase of load (Point D)
(5) The load that produces a plastic yielding or net settlement of 6 mm (Point E)
(6) The load indicated by the intersection of tangent lines drawn through the initial, flatter portion of the gross settlement curve and the steeper portion of the same curve (Point F) and
(7) The load at which the slope of the net settlement is equal to 0.25 mm per MN of test load.

EC7, Clause 7.6.2.2, prescribes a method for assessing design pile loads from the load/settlement curves obtained from a series of static load tests as described in Section 4.1.4.

With experience the load/settlement curve from a compression test can be used to interpret the mode of failure of a pile. A defective pile shaft is also indicated by the shape of the curve. Some typical load/settlement curves and their interpretation are shown in Figure 11.13.

A method of analysing the results of either CRP or ML tests to obtain an indication of the ultimate load is described by Chin[11.26]. The settlement Δ at each loading stage P is divided by the load P at that stage and plotted against Δ/P as shown in Figure 11.14. For an undamaged pile a straight line plot is produced. For an end-bearing pile the plot is a single line (Figure 11.14a). A combined friction and end-bearing pile produces two straight lines which

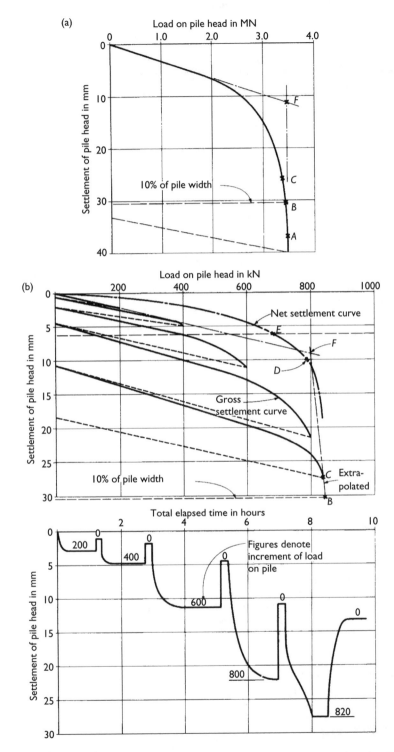

Figure 11.12 Compression load tests on 305 × 305 mm pile (a) Load/settlement curve for CRP test for pile on dense gravel (b) Load/settlement and time/settlement curves for pile on stiff clay.

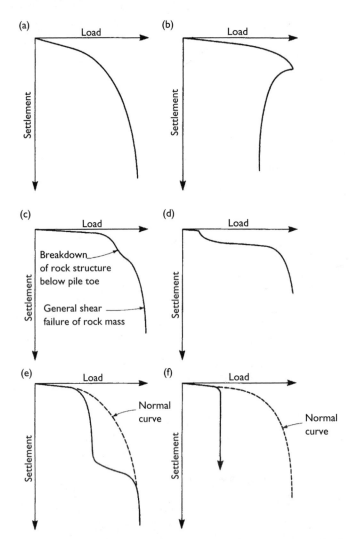

Figure 11.13 Typical load/settlement curves for compressive load tests (a) Friction pile in soft-firm clay or loose sand (b) Friction pile in stiff clay (c) Pile bearing on weak porous rock (d) Pile lifted off seating on hard rock due to soil heave and pushed down by test load to new bearing on rock (e) Gap in pile shaft closed up by test load (f) Weak concrete in pile shaft sheared completely through by test load.

intersect (Figure 11.14b). The inverse slope of the line gives the ultimate load in each case. Chin describes how a broken pile is detected by a curved plot (Figure 11.14c).

11.4.3   Uplift tests

Uplift or tension tests on piles can be made at a continuous rate of uplift (CRU), or an incremental loading basis (ML). Where uplift loads are intermittent or cyclic in character, as in wave loading on a marine structure, it is good practice to adopt repetitive loading on

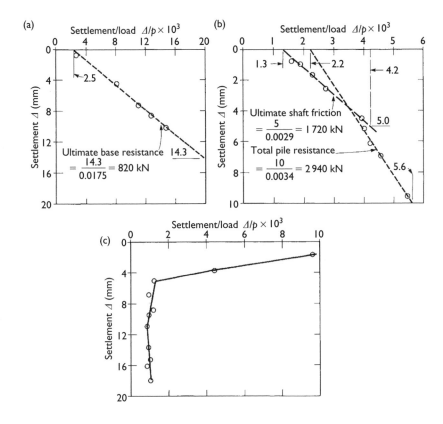

Figure 11.14 Analysis of load/settlement curves from pile loading test (after Chin[11.26]) (a) End-bearing pile (b) Friction and end-bearing pile (c) Broken pile.

the test pile. The desirable maximum load for repeated application cannot be readily determined in advance of the load testing programme since the relationship between the ultimate load for a single application and that for repeated application is not known. Ideally a single pile should be subjected to a CRU test to obtain the ultimate load for a single application. Then two further piles should be tested; one cycled at an uplift load of, say, 50% of the single-application ultimate load, and the second at 75% of this value. At least 25 load repetitions should be applied. If the uplift continues to increase at an increasing rate after each repetition, the cycling should be continued without increasing the load until failure in uplift occurs. Alternatively, an incremental uplift test can be made with say 10 repetitions of the load at each increment.

A typical load/time/uplift curve for an ML test is shown in Figure 11.15. The criteria for evaluating the failure load are similar to those described in Section 11.4.2.

EC7, Clause 7.6.3.2, prescribes a method of deriving the design tensile capacity, R_{td}, of a single pile from tension tests as described in Section 6.2.2.

A loading rig for an uplift test is shown by Figure 11.16. This utilizes the same components as the compressive load testing rig shown in Figure 11.8. The methods used for measuring the jacking force and the movement of the pile head are the same as those used for compressive tests. It is particularly important to space the ground beams or bearers at an

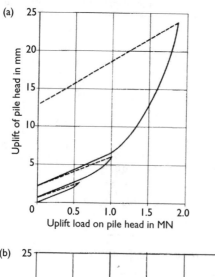

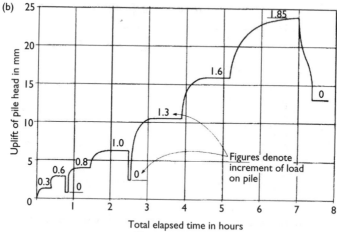

Figure 11.15 Uplift load on test pile (ML test) (a) Load/uplift curve (b) Time/uplift curve.

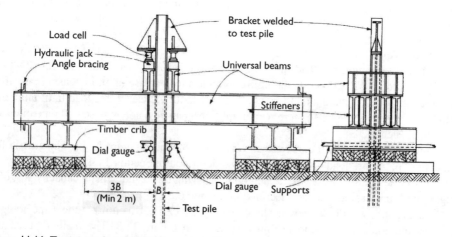

Figure 11.16 Testing rig for uplift on H-section pile using ground reaction.

ample distance from the test pile. If they are too close the lateral pressure on the pile induced by the load on the ground surface will increase the shaft friction on the pile shaft.

11.4.4 Lateral loading tests

Lateral loading tests are made by pulling a pair of piles together, or jacking them apart. If the expected movements are large, for example, when obtaining the load/deflection characteristics of breasting dolphin piles, a 'Tirfor' or block and tackle can be employed to pull the piles together and a graduated staff used to measure the horizontal movement, as shown in Figure 11.17. Where the lateral loads on piles are of a repetitive character, as in wave loading or traffic loads on a bridge, it is desirable to make cyclic loading tests. This involves alternately pushing and pulling of a pair of piles, using a rig of the type shown in Figure

Figure 11.17 Lateral loading test on two steel tubular piles forming part of a breasting dolphin.

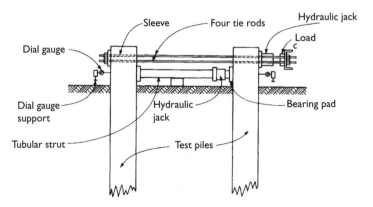

Figure 11.18 Testing rig for push and pull lateral loading test on a pair of piles.

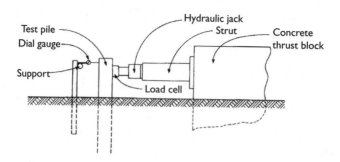

Figure 11.19 Testing rig for lateral loading test on single pile.

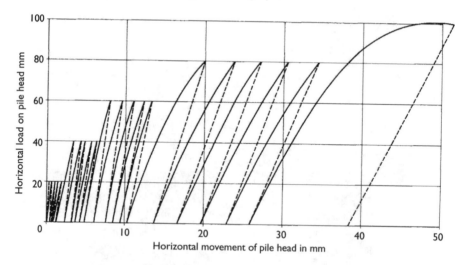

Figure 11.20 Load–deflection curve for cyclic horizontal loading test on pile (some load cycles omitted for clarity).

11.18. Instead of a pair of piles a single pile can be pushed or pulled against a thrust block (Figure 11.19). Where pushing methods are used restraining devices should be provided to ensure that the jack and strut assembly does not buckle during the application of load.

Where possible the lateral movement of the pile heads should be measured by dial gauges mounted on a frame supported independently of the test piles, as shown in Figure 11.19. This may not be feasible in marine piles since the oscillation of the piles and the structure supporting the frame in waves and currents may make it impossible to obtain readings with sufficient accuracy. Measurements made of the curvature of a pile by lowering an incli-nometer down a tube fixed to the wall of a hollow pile or cast centrally in a solid pile are helpful in checking the assumptions made on the point of fixity as described in Chapter 6. Highly accurate electro-levels can be mounted in a probe and lowered down a sleeve cast into the pile. The slope of the pile at the head can be measured by extending the pile above ground level by a stiff indicator rod. Dial gauge readings are made at the top and bottom of this rod. It is also helpful when testing piles in marine structures to make two sets of tests, applying the load at two different levels, say at the head and just above low water of spring tides. This provides two sets of curves relating deflections to bending moments.

Typical load–deflection curves for cyclic tests are shown in Figure 11.20.

EC7, Clause 7.7, does not prescribe a method of deriving the ultimate limit state design transverse load, F_{trd}, as is the case for compression and uplift loading. As for all ultimate limit state pile design EC7 does require the transverse ground resistance to be equal to or greater than the transverse load.

11.5   Tests for the structural integrity of piles

From time to time doubts are thrown on the soundness of pile shafts. Excavations for pile caps may show defective conditions of the type illustrated in Figures 3.40 and 3.41, and questions are immediately asked about the likelihood of similar defects at greater depths and in other piles on the site. Where preformed piles such as precast concrete or steel tubular sections are used, defects can readily be explored by lowering inclinometers down guides fixed to the interior (see Section 2.2.4) or by inserting a light or TV camera down the interior of a hollow pile. In the Netherlands it is the practice to check the soundness of precast prestressed concrete piles by embedding a thin electric cable down the shaft. After driving, a test for electrical continuity is made quite simply by incorporating a light bulb in the circuit. If the pile shaft is broken it shows as a break in the circuit. Such explorations cannot readily be made with cast-in-place concrete piles.

Turner[11.27] has classified and described integrity testing techniques under two main heads:

Direct examination
(1) Visual, during and after installation, including excavation and extraction of the pile
(2) Load testing, static, dynamic and internal compression
(3) Drilling, coring or probing, alongside the pile or into the pile.

Indirect examination
(4) Internal, using the drill holes or preformed ducts for sonic logging and nuclear backscatter and gamma-ray techniques, CCTV inspection, water or air pressure testing and calliper dimensional logging
(5) External from the top or side of an exposed pile using low strain acoustic integrity tests, dynamic load measurements for high strain integrity tests, ultrasonic pulses and electrical resistivity tests
(6) Remote, in a borehole alongside the pile and installation of sonic probe for 'parallel seismic' techniques.

Satisfactory evidence should be provided by the specialist that a particular method of non-destructive testing or integrity testing will be appropriate to the site and type of pile.

A complete pile can rarely be examined economically by excavation or extraction. Piles are frequently installed in soft or loose ground, making excavation difficult and costly particularly below the water table. It would be unthinkable to examine all piles on a site by excavation down the shaft.

Test loading is a positive method and its value in detecting defects in a shaft is illustrated in Figure 11.13e and f. However, test loading all the piles on a site is a costly operation, particularly if heavy kentledge loads are required (Figure 11.10). Figure 11.21 shows the much simpler arrangements required for the SIMBAT dynamic load test on small-diameter piles. The method of shaft compression has been described by Moon[11.28]. A rod or cable is

Figure 11.21 SIMBAT test using a mini rig for small diameter dynamic tests (courtesy Test Consult).

anchored at the base of the pile in a sleeve, and by jacking the tension member against the pile head the ability of the pile shaft to carry the design compressive load can be checked. It is possible to recover the tension members after each test.

Drilling, either by open hole methods using a percussion drill or rotary rock roller bit, or by rotary coring, can be used in piles of medium to large diameter, but it is difficult to keep the drill hole within the confines of the shaft of a small-diameter pile. If it is possible to flush an open hole clear of dirty water an inspection can be made by TV camera to look for cavities or honeycombed concrete. Heavy water losses when the drill hole is filled with water also indicate defective concrete. A cored hole provides a better indication of sound-ness, and compression tests can be made on the cores, but the method is more costly than open hole drilling. It should be noted that cores are only likely to be obtained from sound concrete and any defective zones may not be recovered for testing. Calliper logging down a drill hole gives an indication of overbreak caused by weak concrete or cavities.

Sonic pulse equipment can also be lowered down drill holes, when irregularities in the trace of the sonic log indicate a defective shaft. Pairs of ducts can be formed down a pile shaft at the time of placing the concrete and various logging devices used to scan the concrete between the ducts. These include sonic pulse measurements, gamma-ray logging and neutron emissions. The latter methods are believed to be reliable indicators of density changes and water content respectively, but are costly since they involve the use of skilled technicians and the transportation to site and operation of nuclear testing devices with their elaborate safety precautions. The seismic method consists of dropping a weight on the pile head and observing the time of return of the seismic wave reflected from the toe or anomaly in the pile. This method is quite widely used and has been shown by experience to give reasonably reliable results when operated and interpreted by specialists. The method does not give reliable results in jointed precast concrete piles, however. The dynamic response method consists of mounting a vibrating unit on the pile head and interpreting the oscillograph of the response from the pile. This method is again quite widely used.

Ground-probing radar techniques are being developed to assist in locating existing foundations and piles for potential reuse[2.21].

The main advantage of specifying integrity testing of all or randomly selected piles while pile installation is underway is that it encourages the piling contractor to keep a careful check on all the site operations. However, the methods do not replace the need for full-time supervision of the piling work by an experienced engineer or inspector.

Integrity testing will indicate if a pile is badly broken but not hair cracks; the anomalies shown up may need to be checked by another method. The limitations of integrity testing were demonstrated by experiences of a field trial competition in The Netherlands[7.5]. Ten different precast concrete pile shapes with different forms of defect were installed in drilled holes. The average score from 12 specialist firms competing in the trials was four correct identifications out of the 10 shapes, but the suitability of the test conditions has been criticized. Somewhat better results from a comparative blind testing are reported by Iskander et al.[11.29] for pulse echo and impulse response methods. Defects as small as 6% of the cross-sectional area of bored piles in varved clay were correctly identified. Cross-hole tomography was not as effective but was able to identify the pile lengths and lateral locations of the defects.

11.6   References

11.1   BINNS, A. Rotary coring in soils and soft rock for geotechnical engineering, *Geotechnical Engineering*, Vol. 131, No. 2, Institution of Civil Engineers, London, 1998, pp. 63–74.

11.2   TERZAGHI, K. and PECK, R. B. *Soil Mechanics in Engineering Practice*, 2nd edn, John Wiley, New York, 1974, p. 114.

11.3   JOUSTRA, K. Comparative measurements on the influence of the cone shape on results of soundings, *Proceedings of the European Symposium on Penetration Testing*, Stockholm, 1974.

11.4   CEARNS, P. J. and MCKENZIE, A. Application of dynamic cone penetrometer testing in East Anglia, *Symposium or Penetration Testing in U.K.*, Thomas Telford, London, 1988, pp. 123–7.

11.5   WROTH, C. P. and HUGHES, J. M. O. An instrument for the in-situ measurement of the properties of soft clays, *Proceedings of the 8th International Conference, ISSMFE*, Moscow, 1973, 1.2, pp. 487–94.

11.6   HUGHES J. M. O. and ERVIN, M. C. Development of a high pressure pressuremeter for determining the engineering properties of soft to medium strength rocks, *Proceedings of 3rd Australian – New Zealand Conference on Geomechanics, Wellington*, Vol. 1, 1980, pp. 243–7.

11.7   MADDISON, J. D., CHAMBERS, S., THOMAS, A., and JONES, D. B. The determination of deformation and shear strength characteristics of Trias and Carboniferous strata from in situ and laboratory testing for the Second Severn Crossing, *Advances in Site Investigation Practice*, C. Craig (ed.), Institution of Civil Engineers, Thomas Telford, London, 1996, pp. 598–609.

11.8   MARCHETTI, S. In situ tests by Flat Dilatometer, *Proceedings of the American Society of Civil Engineers*, Vol. 106, No. GT3, 1980, pp. 299–321.

11.9   WHITTLE, R. W. Recent developments in the cone pressuremeter, *Proceedings of the International Conference on Advances in Site Investigation Practice*, Thomas Telford, London, 1996, pp. 533–54.

11.10   MARSLAND, A. Large in-situ tests to measure the properties of stiff fissured clays, *Building Research Establishment*, Current Paper CP1/73, Department of the Environment, January 1973.

11.11 MARSLAND, A. and EASTON, B. J. Measurements of the displacement in the ground below loaded plates in deep boreholes, *Proceedings of the Symposium on Field Instrumentation*, British Geotechnical Society, London, 1973, pp. 304–17.

11.12 MCKINLAY, D. G., TOMLINSON, M. J., and ANDERSON, W. F. Observations on the undrained strength of a glacial till, *Geotechnique*, Vol. 24, No. 4, December 1974, pp. 503–16.

11.13 BUTLER, F. G. and LORD, J. A. Discussion, *Proceedings of the Conference on In-Situ Investigations in Soils and Rocks*, British Geotechnical Society, London, 1970, pp. 51–3.

11.14 Point load test – suggested method for determining point load strengths, *International Journal of Rock Mechanics and Mining Sciences*, Vol. 22, No. 2, 1985, pp. 53–60.

11.15 *Guidance Notes for Site Investigations for Offshore Renewable Energy Projects*, Society for Underwater Technology, 2005.

11.16 New Engineering Contract (NEC3). *The Engineering and Construction Contract*, 3rd edn, Institution of Civil Engineers, Thomas Telford, London, 2005.

11.17 *Civil Engineering Standard Method of Measurement*, 3rd edn, Institution of Civil Engineers, Thomas Telford, London, 1991.

11.18 WHITAKER, T. The constant rate of penetration test for the determination of the ultimate bearing capacity of a pile, *Proceedings of the Institution of Civil Engineers*, Vol. 26, September 1963, pp. 119–23.

11.19 DE COCK F., LEGRAND C., and HUYBRECHTS N. Axial static pile test in compression or in tension – Recommendations from ERTC – Piles, ISSMGE Subcommittee, *Proceedings of the 13th European Conference on Soil Mechanics and Geotechnical Engineering*, Prague, Vol. 3, 2003, pp. 717–41.

11.20 FLEMING, W. G. K. Static load tests of piles and their application, *Institution of Civil Engineers, Ground Engineering Board*, Notes for meeting on 22 May 1991.

11.21 HANNA, T. H. *Foundation Instrumentation*, Transtech Publications, Aedermansdorff, Switzerland, 1973, pp. 35–66.

11.22 HANNA, T. H. Personal communication to the authors.

11.23 WELTMAN, A. J. Pile load testing procedures. *Construction Industry Research and Information Association (CIRIA)*, Report PG7, 1980.

11.24 LONG, M. Assessment of SIMBAT dynamic pile tests, *ASCE Special Publication*, Vol. 113, 2002, pp. 539–53.

11.25 BERMINGHAM, P. and JANES, M. An innovative approach to load testing of high capacity piles, *Proceedings of the International Conference on Piling and Deep Foundations*, London, Balkema, Rotterdam, 1989.

11.26 CHIN, FUNG KEE. Diagnosis of pile condition, *Geotechnical Engineering*, Vol. 9, 1978, pp. 85–104.

11.27 TURNER, M. J. Integrity testing in piling practice, *Construction Industry Research and Information Association (CIRIA)*, Report 144, 1997.

11.28 MOON, R. A. Test method for the structural integrity of bored piles, *Civil Engineering and Public Works Review*, Vol. 67, No. 790, May 1972, pp. 476–80.

11.29 ISKANDER, M., ROY, D., SHELLY, S., and EALY, C. Drilled shaft defects: detection, and effects on capacity in varved clay, *ASCE Journal of Geotechnical and Geoenvironmental Engineering*, December 2003, pp. 1128–37.

Appendix

Properties of materials

A.1 Coarse-grained soils

	Density when drained above groundwater level γ (Mg/m³)	Density when submerged below groundwater level γ_{sat} (Mg/m³)	Angle of shearing resistance ϕ (degrees)
Loose gravel with low sand content	1.6–1.9	0.9	28–30
Medium dense gravel with low sand content	1.8–2.0	1.0	30–36
Dense to very dense gravel with low sand content	1.9–2.1	1.1	36–45
Loose well-graded sandy gravel	1.8–2.0	1.0	28–30
Medium-dense well-graded sandy gravel	1.9–2.1	1.1	30–36
Dense well-graded sandy gravel	2.0–2.2	1.2	36–45
Loose clayey sandy gravel	1.8–2.0	1.0	28–30
Medium-dense clayey sandy gravel	1.9–2.1	1.1	30–35
Dense to very dense clayey sandy gravel	2.1–2.2	1.2	35–40
Loose coarse to fine sand	1.7–2.0	1.0	28–30
Medium-dense coarse to fine sand	2.0–2.1	1.1	30–35
Dense to very dense coarse to fine sand	2.1–2.2	1,2	35–40
Loose fine and silty sand	1.5–1.7	0.7	28–30
Medium-dense fine and silty sand	1.7–1.9	0.9	30–35
Dense to very dense fine and silty sand	1.9–2.1	1.1	35–40

A.2 Fine-grained and organic soils

	Density when drained above groundwater level γ (Mg/m³)	Density when submerged below groundwater level γ_{sat} (Mg/m³)	Undrained shear strength (kN/m²)
Soft plastic clay	1.6–1.9	0.6–0.9	20–40
Firm plastic clay	1.75–2.0	0.75–1.1	40–75
Stiff plastic clay	1.8–2.1	0.8–1.1	75–150
Soft slightly plastic clay	1.7–2.0	0.7–1.0	20–40
Firm slightly plastic clay	1.8–2.1	0.8–1.1	40–75
Stiff slightly plastic clay	2.1–2.2	1.1–1.2	75–150
Stiff to very stiff clay	2.0–2.3	1.0–1.3	150–300
Organic clay	1.4–1.7	0.4–0.7	—
Peat	1.05–1.40	0.05–0.40	—

A.3 Rocks and other materials

Material	Density (mg/m³)
Granite	2.50
Sandstone	2.20
Basalts and dolerites	1.75–2.85
Shale	2.15–2.30
Stiff to hard mudstone	1.90–2.30
Limestone	2.0–2.70
Chalk	0.95–2.00
Broken brick	1.10–1.75
Solid brickwork	1.60–2.10
Ash and Clinker	0.65–1.00
Pulverized fuel ash	1.20–1.50
Loose coal	0.80
Compact stacked coal	1.20
Mass concrete	2.20
Reinforced concrete	2.40
Iron and steel	7.20–7.85

A.4   Engineering classification of chalk (Lord et al.[4.43])

Intact dry density scales of chalk

Density scale	Intact dry density $\gamma(Mg/m^3)$	Porosity n[a]	Saturation moisture content[a] (%)
Low density	<1.55	>0.43	>27.5
Medium density	1.55–1.70	0.43–0.37	27.5–21.8
High density	1.70–1.95	0.37–0.28	21.8–14.3
Very high density	>1.95	<0.28	<14.3

Note
a Based on the specific gravity of calcite of 2.70.

Classification of chalk by discontinuity aperture

Grade A	Discontinuities closed
Grade B	Typical discontinuity aperture <3 mm
Grade C	Typical discontinuity aperture >3 mm
Grade D	Structureless or remoulded mélange

Subdivisions of Grades A to C chalk by discontinuity spacing

Suffix	Typical discontinuity spacing (mm)
1	$t > 600$
2	$200 < t < 600$
3	$60 < t < 200$
4	$20 < t < 60$
5	$t < 20$

Subdivisions of Grade D Chalk by engineering behaviour

Suffix	Engineering behaviour	Dominant element	Comminuted chalk matrix (%)	Coarser fragments (%)
m	fine soil	matrix	approx >35	approx <65
c	coarse soil	clasts	approx <35	approx >65

Name index

Ackers, J. C. 476
Adams, J. L. 278, 308, 313, 356
Andersland, O. B. 449–51
Anderson, D. M. 449
Anderson, W. F. 538
Aydinoglu, N. 222

Baguelin, F. 346–9
Baldi, G. 268
Ballisager, C. G. 164
Barkan, D. D. 435
Baudin, C. 9
Beake, R. H. 202
Beattie, D. 392
Beebe, K. E. 424
Beetstra, G. W. 476
Bekaroglu, 222
Belloti, R. 289
Berezantsev, V. G. 165, 167, 168, 230
Bermingham, P. 538
Bessey, G. E. 489
Biddle, A. R. 70
Biggar, K. W. 453, 476
Binns, A. 502
Bittner, R. B. 476
Bjerrum, L. 42, 164, 218, 223, 259, 278
Black, D. R. 58
Bolognesi, A. J. L. 121–2
Bolton, M. D. 187, 458–61
Bond, A. J. 155
Bonnett, D. 418
Bowley, M. J. 489
Brandl, H. 474
Brassinga, H. E. 180
Bretschneider, C. L. 409
Britt, G. B. 70
Broadhead, A. 406
Broms, B. B. 289, 462
Broug, N. W. A. 180
Brown, P. R. 290
Brown, P. T. 256
Brown, T. G. 415

Browne, R. D. 491
Bruce, D. W. 494, 495
Bryden-Smith, D. W. 290
Brzezinski, L. S. 279–81
Buchanan, N. W. 462
Burbidge, M. C. 266
Burland, J. B. 243–4, 263, 265–7, 287, 289, 193–5
Bustamante, M. 58, 63
Butler. F. G. 222, 242, 256, 262, 274, 538
Byle, M. J. 476

Campanella, R. G. 267
Carder, D. R. 476
Carrier, W. D. 258, 268
Cearns, P. J. 504–5
Chambers, S. 537
Chandler, F. W. 274
Chandler, R. J. 222
Chao, W. T. 69, 493
Charue, N. 70
Chellis, R. D. 483, 486
Chin, Fung, Kee, 528–30
Chow, F. C. 181, 185, 221
Chow, Y. K. 278, 280
Christian, J. T. 258, 268
Christoffersen, H. P. 268
Churcher, D. W. 137
Claessen, A. J. M. 220
Clayton, C. R. I. 70, 222, 223
Coday, A. E. 497
Cole, K. W. 280, 281
Cooke, R. W. 161, 287
Cox, W. R. 357
Croasdale, K. R. 414–15
Curtis, R. L. 135, 281

Dailey, J. W. 409
Dalmatov, B. L. 450
Darley, P. 476
Davies, W. W. 396
Davis, A. G. 274

Subject index

eBooks – at www.eBookstore.tandf.co.uk

A library at your fingertips!

eBooks are electronic versions of printed books. You can store them on your PC/laptop or browse them online.

They have advantages for anyone needing rapid access to a wide variety of published, copyright information.

eBooks can help your research by enabling you to bookmark chapters, annotate text and use instant searches to find specific words or phrases. Several eBook files would fit on even a small laptop or PDA.

NEW: Save money by eSubscribing: cheap, online access to any eBook for as long as you need it.

Annual subscription packages

We now offer special low-cost bulk subscriptions to packages of eBooks in certain subject areas. These are available to libraries or to individuals.

For more information please contact webmaster.ebooks@tandf.co.uk

We're continually developing the eBook concept, so keep up to date by visiting the website.

www.eBookstore.tandf.co.uk